Preface

Sensor systems have been around for quite some time, and their obvious use in thermometer has gone unnoticed until recently. Their unique capability of monitoring a given area has attracted attention, and sensors have been useful in numerous areas. I envisioned the need for this book several years ago when started offering such a course. During the last four years, I collected papers on wireless sensor networks (WSNs) and prepared ppt files from them to deliver lectures. I was fortunate to be invited as a GIAN speaker at the IIT Kharagpur in summer 2015, a premier institute in India. This encouraged me to write two chapters every week when I offered this course again at the University of Cincinnati again during spring 2016. I was invited again to NIT Hamirpur, Himachal Pradesh, India, to deliver lectures under GIAN program and that encouraged me to redraw all pdf-based diagrams in black and white ppts. After that, I also visited Wuxi, China, to get first-hand information about GreenOrbs project and was able to add the newest info on large-scale WSN deployment.

Within engineering and computer science disciplines, wireless sensor network has recently attracted unparalleled interest. In particular, combinations of wireless communication and tiny sensor devices have revolutionized the world of telecommunications. To fully explore and utilize this new technology, universities need to offer new courses and train students in the field so that they could continue their graduate work in this area. However, the students in computer science and engineering (CSE) and electrical engineering (EE) are at best exposed to data communication aspects, while sensor-embedded systems remain untouched, as it is relatively difficult to learn about WSNs without having substantial background in wireless communication technology. On the other hand, EE students learn about the radio frequency (RF) communication aspect only, and the topic of data communication and computing system issues and their correlation in nomadic WSN remains untouched. I put in more efforts in streamlining all the chapters after my return to the USA in July 2016.

Many institutions offer course in WSN area, primarily for graduate students, and then only as special topics. Although there are many edited books related to WSNs, these books can be roughly classified into two groups. The first group focuses on readers in specific application areas, and the other covers only the general knowledge of environmental monitoring. The books in the first group require a detailed background in RF communication and signal processing and therefore are not suitable for students in CSE. Many recent texts emphasize microwave radar and sensor systems. However, the books in the second group do not provide any depth in the data communication aspects of the technology.

Creating such a unique instructional curriculum requires a great deal of efforts. Planning such a text is a relatively difficult task because of the diverse background requirements. The limitations of most existing books and courses affect the sensor industries in the USA. Companies must train newly hired college graduates for a long time before they can get into WSNs. To the best of our knowledge, such an organized course is not been taught anywhere in the world. Teaching WSN course strictly from research papers is difficult for the professor, which in turn causes students to learn the material inefficiently. Preparing systematic notes in this emerging area will enhance training, increase the availability of well-educated personnel, shorten the new employee training period within industries, and allow nations to continue to advance the research in this technological field.

This book explains how a WSN works in monitoring a given environment. I have selected chapter topics that focus on qualitative descriptions and realistic explanations of relationships between WSNs and performance parameters. Mathematical formulations are needed in engineering and computer science work, and we include some of the important concepts so that students can appreciate their usefulness in numerous WSNs. In all these applications, both security and privacy issues are important. The chapters are organized to provide a great deal of flexibility; emphasis can be given on different chapters, depending on the scope of the course and the instructor's own interests or emphasis.

In this textbook, I have tried to provide an overview of the basic principles behind WSNs and associated support infrastructure. A list of possible group simulation projects is included. The author has tried such projects for several years and has found them highly effective in training students. This book is written both for academic institutions and for working professionals. It can be used as a textbook for a one-semester or a one-quarter course. This book also can be used for training current or new employees of companies and could be adopted for short-term training courses. I hope I have been able to achieve our goal of helping students and others working in this area to have a detailed knowledge about this exciting technology.

Cincinnati, OH, USA Dharma Prakash Agrawal

Acknowledgements

This project would have not been possible without help from numerous individuals. Therefore, the authors would like to acknowledge the time and effort put in by all past and present members of our Research Center for Distributed and Mobile Computing at the University of Cincinnati. Special sincere thanks are due to Pallavi Meharia, Hailong Li, Amitabh Mishra, Suryadip Chakraborty, Aparna Venkataraman, Abhinav Prakash, and Peter Jun for letting me use their research results in some chapters. Thanks are also due to my wife for her patience and dedication. I am very grateful to Swati Meherishi, my editor at Springer, for persuading and convincing me to communicate with Springer for the possible publication of this book so quickly. The author welcomes any comments and suggestions for improvements or changes that could be incorporated in the forthcoming editions of this book. Please contact me at dpa@cs.uc.edu.

Dharma Prakash Agrawal

Contents

About the Author

Dr. Dharma Prakash Agrawal is the Ohio Board of Regents Distinguished Professor and founding director of the Center for Distributed and Mobile Computing at the School of Computing Sciences and Informatics. He has been a faculty member at the ECE Dept., Carnegie Mellon University (on sabbatical leave), NC State University, Raleigh and Wayne State University. His current research interests include resource allocation in wireless mesh networks, query processing and secured communication in sensor networks, environmental monitoring using sensor networks, and effective traffic handling in integrated wireless networks.

His most recent contribution in the form of a co-authored introductory textbook on Wireless and Mobile Computing has been widely accepted throughout the world and a fourth edition has just been published. The book has been reprinted both in China and India and translated into Korean and Chinese. His co-authored book on Ad hoc and Sensor Networks, 2nd edition, was published in the spring of 2011.

Professor Agrawal is a founding Editorial Board Member, International Journal on Distributed Sensor Networks, International Journal of Ad Hoc and Ubiquitous Computing (IJAHUC), International Journal of Ad Hoc and Sensor Wireless Networks, and the Journal of Information Assurance and Security (JIAS). He has served as an editor of the journal IEEE Computer, the Journal of Parallel and Distributed Systems, IEEE Transactions on Computers, and the International Journal of High Speed Computing. He has been the Program Chair and General Chair for numerous international conferences and meetings, and has received numerous certificates from the IEEE Computer Society, including a Third Millennium Medal for his outstanding contributions. He has published over 610 papers, given 38 different tutorials and extensive training courses in various conferences in USA, and numerous institutions in Taiwan, Korea, Jordan, UAE, Malaysia, and India in the areas of Ad hoc and Sensor Networks and Mesh Networks (including Globecom 2011 and ICC 2012). He has been nominated as an ISI Highly Cited Researcher, is a

Fellow of the IEEE, the ACM, the AAAS, and the World Innovation Foundation, and a recent recipient of the 2008 IEEE CS Harry Goode Award. In June 2011, he was selected as the best Mentor for Doctoral Students at the University of Cincinnati. In 2012, he was selected as a Founding fellow of the National Academy of Inventors.

Abbreviations

6LoWPANs	IPv6 low-power personal area networks
A/D	Analog to digital
ACK	Acknowledgement
ADL	Activities of daily living
ADV	Advertise
AES	*Advanced Encryption Standard*
AN SN	Anchor SN
ANOVA	Analysis of variance
AoA	Angle of arrival
AP	Access point
APIT	Approximate point-in-triangulation test
APR-REP	Anonymous path routing reply
APR-REQ	Anonymous path routing request
APTEEN	Adaptive periodic threshold sensitive energy-efficient sensor network protocol
BAN	Body area network
BC	Broadcast communication
BLAST	Base-station Location AnonymiTy
B-MAC	Berkeley MAC
BS	Base station
CA	Collision avoidance
CCA	Clear channel assessment
CCD	Charge-coupled device
CDMA	Code division multiple access
CH	Cluster head
CMOS	Complementary metal–oxide semiconductor
CO	Carbon monoxide
CO_2	Carbon dioxide
CoG	Center of gravity
CSMA	Carrier sense multiple access

C-SN	Camera sensor node
CSS	Coordinating sink station
CTS	Clear To Send
DCF	*Distributed coordination function*
DES	Data Encryption Standard
DIFS	*Distributed interframe space*
DoS	Denial of service
ECC	Elliptic curve cryptography
ECG	Electrocardiogram
EcoBot	Ecological robot
EEHF	Environmental energy harvesting framework
EM	Electromagnetic
EMG	Electromyography
EPKEM	Efficient pairwise key establishment and management
E-WME	Energy-opportunistic weighted minimum energy
FAR	False accept rate
FDMA	Frequency division multiple access
FoG	Freezing of gates
FRR	False reject rate
FS	Floor sensor
GAF	Geographic adaptive fidelity
GIS	Geographic information system
GMM	Gaussian mixture models
GPS	Global Positioning System
GSM	Global System for Mobile Communication
HD	High-power detection
HEDIS	HEterogeneous DIScovery
HEED	Hybrid, energy-efficient, distributed
HI	Hide identity
HLR	Home location register
HTTP	Hypertext Transfer Protocol
IBC	Identity-based cryptography
IoT	Internet-of-Things
IP	Internet Protocol
IPv6	Internet Protocol version 6
IR	Infrared
ISM	Industrial, scientific, and medical
ISO	International Organization for Standardization
LEACH	*Low-energy adaptive clustering hierarchy*
LL	Link layer
LPL	Low-power listening
LQI	Link quality indicator
LRWPAN	Low-rate wireless personal area network
LVDT	Linear variable differential transformer
MAC	Medium access control

ME	Minimum energy
MON	Mobile opportunistic network
MPR	Modified phantom routing
MS	Mobile station
MSC	Mobile switching center
MV	Machine vision
NACK	Negative ACKnowledgement
ND	Normal power detection
NDVI	Normalized difference vegetation index
OFDMA	Orthogonal frequency division multiple access
OOR	Optimal operating region
PAN	Personal area network
PDF	Probability distribution function
PEDAP	Power efficient data gathering and aggregation protocol
PIT	Point-in-triangulation
P-MAC	Pattern MAC
PP	Point to point
PSO	Particle swarm organization
QoS	Quality of service
RBS	Receiver-based synchronization
RE	Residual energy
RFID	Radio frequency identification
RMST	Reliable multi-segment transport
RN	Relay node
RS	Generic remote sensing
RSA	*Rivest–Shamir–Adleman*
RSN	Remote sensor node
RSSI	Received signal strength indication
RT	Round-trip
RTS	Request To Send
SDMA	Space division multiple access
SF	Selective forwarding
SIFS	*Short interframe space*
SMAC	Sensor MAC
S-MAC	Sensor-MAC
SN	Sensor node
SNS	Social network sites
SPIN	Sensor Protocol for Information via Negotiation
SWoT	Social Web of Things
SYNC	SYNChronization
TCP	Transmission control protocol
TDMA	Time division multiple access
TDoA	Time difference of arrival
TEEN	Threshold sensitive energy-efficient sensor network
ToA	Time of arrival

TODIS	Triple-Odd based DIScovery
TTL	Time-To-Live
UDP	User datagram protocol
UPD	Universal printer driver
UWB	Ultra-wide band
VLR	Visitor location register
WBAN	Wireless body area network
WCET	Worst-case execution time
WFP	Wait-for-First-Packet
WME	Weighted minimum energy
WoT	Web-of-Things
WPAN	Wireless personal area network
WS	Wearable sensor
WSN	Wireless sensor network

Notations

r_c	Communication transmission range
r_s	Sensing range

Equivalent Terms

Energy Harvesting	Energy scavenging
ECG	EKG
Forward path	Downlink from BS to MS
Handoff	Handover
Hello message	Beacon signals
Microwave tower	Base station
Reverse direction	Uplink from MS to BS

List of Figures

List of Tables

Part I
General Sensor Characteristics

Chapter 1
Introduction to Cell Phones and Wireless Technologies

1.1 Introduction

Cell phones or mobile phones have unprecedented impact on human lifestyle, and people are busy communicating with each other using such devices. It is unrealistic to say that all educated people carry at least one cell (mobile) phone and use it at least 110 times every day [1] to have a dialogue with someone known or unknown. This number increases to every 6 or 7 s during peak hours of 5 pm–8 pm, and some people unlock their phones up to 900 times per day. During off-peak period of 3 am–5 am, the intensity drops to 24% and still remains fairly high. It has been further quantified [1] that on an average, a person daily checks 23 times for messages, 22 times for voice calls, and 18 times to check for time. These are in the form of placing a call or sending a text message, and people have become so engaged that they even forget about watching their favorite television show or the game they are crazy about. People carry their cell phones not only while walking, but also running, jogging, and performing daily chores.

It is interesting to note that even an increased number of auto-accidents are attributed to diverted attention to texting rather than looking straight on the road even though texting while driving has been banned by law in most parts of the world. This impact is even seen equally among all age-group, not just children and young generation, and cell phones have become an integral part of daily routine. There are many devices that also monitor activities and health physiological conditions of human body in a 24×7 mode, and those devices commonly known as sensors are going to play a crucial role in enhancing the civilization. So, before we look into understanding and illustrating sensors, it is important to understand some important concepts in wireless technologies. This chapter is a short summary of these aspects of wireless systems and is not a comprehensive description of underlying characteristics.

In a wireless system, electromagnetic (EM) waves propagate from a source in all 360° directions with equal intensity using some medium (usually air) and spreads

D.P. Agrawal, *Embedded Sensor Systems*,
DOI 10.1007/978-981-10-3038-3_1

(travel) through air, water, walls, some objects, and the vacuum of outer space [2]. As the signal moves forward through the medium, the quality reduces and such attenuation is due to many factors. In order to communicate successfully, information should be received at adequate power level so that it could be correctly decoded by a receiving unit. One approach is to improve the medium so that EM signal attenuation could be minimized. Like people living in a given area, there is not much one can do to physically change or modify the medium that could minimize negative impact on signal propagation. So, the only thing that is easily feasible is to adjust the transmitting power level at the transmitter as per the distance between the transmitter–receiver pair. The area covered by a transmitter for a given power level is called the communication range $r_{c,}$ and all wireless devices lying within 360° area can receive and correctly decode the information. This is illustrated in Fig. 1.1 where once an airplane lands on the surface of earth, a cell phone user can connect to nearby serving station usually called as a base station (BS). The BS constantly transmits beacon signals, advertising its presence, and once heard and received, the cell phone can select the BS by sending a hello message back to the BS. Once this handshaking is completed, the cell phone can request channels from the BS, and once received, it can start communicating with the world.

We use a generic term MS (mobile station) as a receiving device. The BS typically insures an area of 5–20 kms radius, and the coverage area is usually called a cell. The shape of a cell is ideally considered circular in nature, while from analytical and simulation point of view, shape of a cell is considered hexagonal in nature. Sometimes, a rectangular shape is used to represent a coverage area of a cell and is illustrated in Fig. 1.2. But, from a realistic point of view, unless specified, the cell area is considered as hexagonal in shape and has been extensively used in both analytical and simulation results.

The EM signal strength reduces as one moves away from BS or center of a cell (Fig. 1.3). Actual shape of a cell could be in fact distorted as shown in Fig. 1.3. Signal strength of EM waves varies from the center of BS and is illustrated in

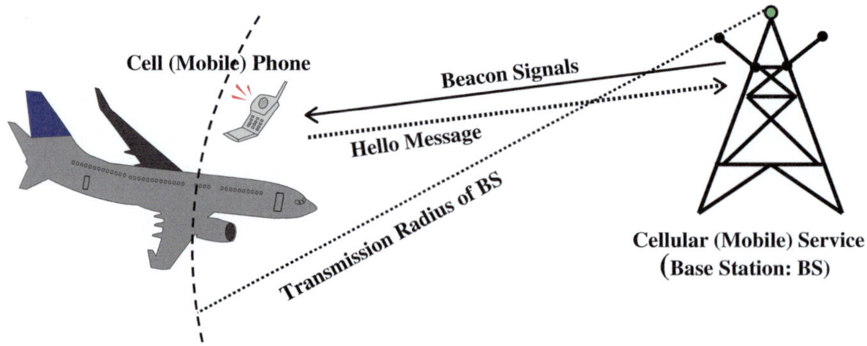

Fig. 1.1 Illustrating communication range of a wireless transmitter

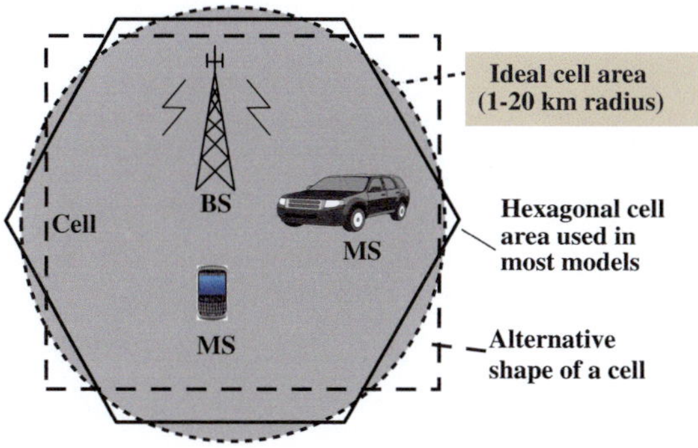

Fig. 1.2 Coverage area of a BS (Cell) and prevailing models

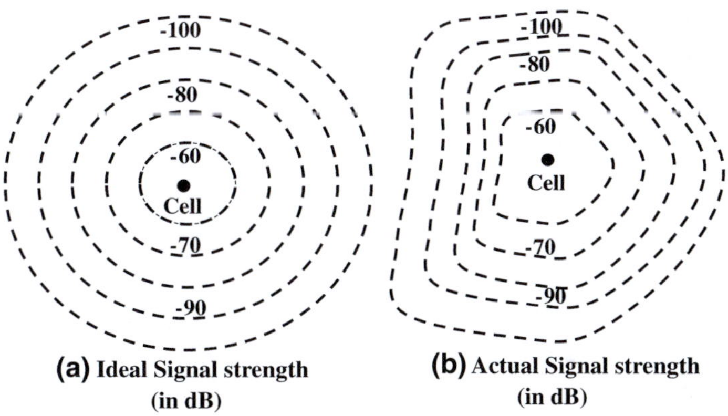

Fig. 1.3 Signal strength and coverage area of a cell

Fig. 1.3. The factors that contribute to variation of EM signal strength include terrain shape, atmospheric conditions, water vapor, and oxygen, while the medium permittivity, permeability, and conductivity do affect the signal strength [2]. In addition, signal attenuation also occurs due to multi-path reflections, scattering, diffraction, etc., and is illustrated in Fig. 1.4. On the one hand, such transmission phenomenon reduces contained energy, and EM signal goes through multiple reflections before reaching a MS. This causes EM signals to be received at any point in the space. Such transmission introduces a signal to follow different paths with slightly increased delay, and multiple copies of a signal are received as an envelope as illustrated in Fig. 1.5. This causes interference among each transmitted bit, and accordingly, the transmission rate has to be selected appropriately so as to avoid interference between successive bit broadcasts. It is interesting to note that the cell phone contains smart signal processing unit that only accepts strongest signal while rejecting all other delayed copies.

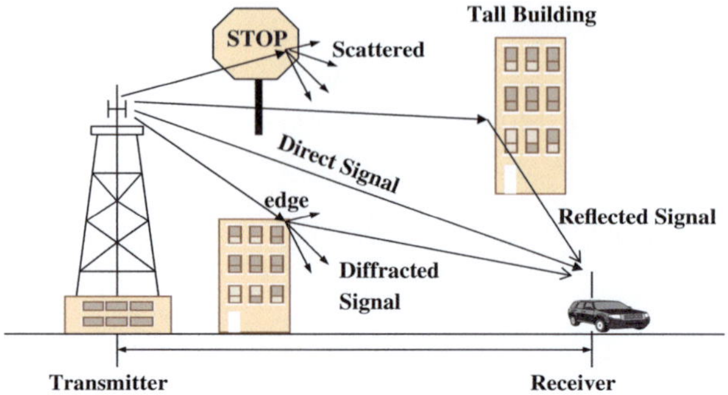

Fig. 1.4 Propagation of EM signal through open air space

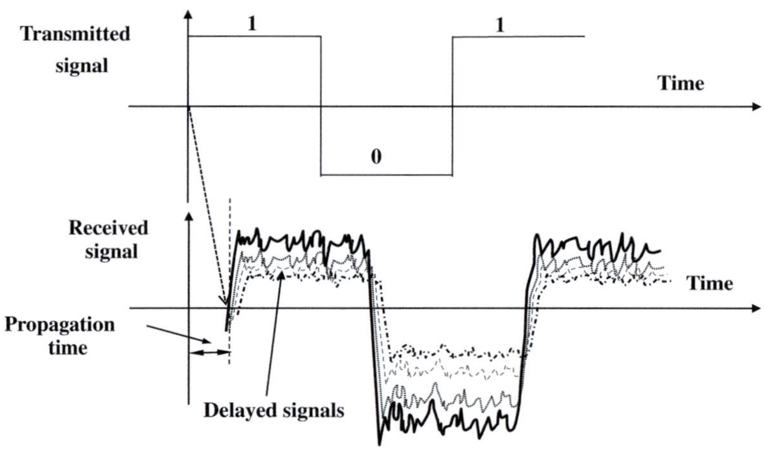

Fig. 1.5 Effect of EM signal propagation along multiple paths

The cell phones are supported by the infrastructure such that any incoming call can be forwarded automatically where the user is currently located. This is done by a technique similar to post office forwarding mail from old address to a new address via new post office registered at the old post office. This process is automated by a set of indirection using two registers home location register (HLR) and visitor location register (VLR) located at each Mobile Switching Center (MSC). HLR contains information about all MS registered at the BS and MSC, while VLR maintains IDs of all MSs currently visiting a BS and MSC. This pair of registers is present everywhere and allows redirection of all incoming call from registered MSC to visiting MSC as HLR of registered MSC points to VLR of current MSC. This has been illustrated in Fig. 1.6.

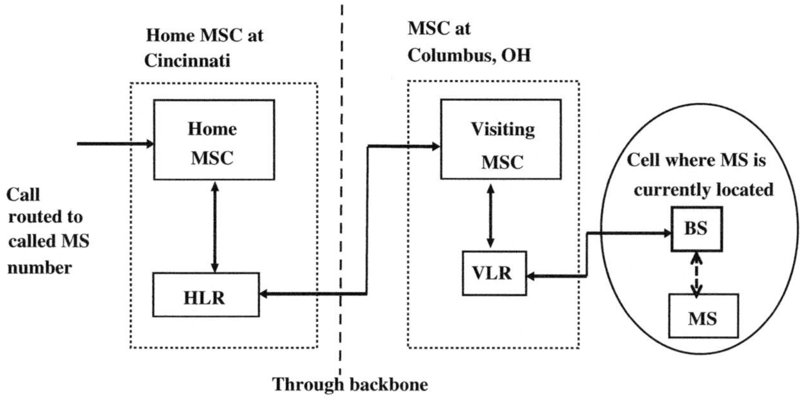

Fig. 1.6 Incoming call redirected from Cincinnati while visiting Columbus, OH by MS

The pointer from HLR of Cincinnati to VLR of Washington, DC is created when the MS answers beacon signal in Washington area. Such redirection pointer keeps on changing with current location of user, while the other end always points to HLR of Cincinnati where the MS (cell phone) is registered and billing information is maintained. On the other hand, all outgoing call from current location can be made directly from Washington, DC area and no need to route through Cincinnati. The only thing that needs to be sent back from Washington, DC is the billing information back to Cincinnati and that is how bidirectional pointer is very helpful in maintaining and managing a centralized account.

Establishing pointers between HLR and VLR for different cities of Cincinnati, Columbus, Atlanta, and Washington, DC is illustrated in Fig. 1.7. It is assumed that MSs ABCDE are registered in Cincinnati, while FGH are listed in Columbus, IJK in Atlanta, and LMNO in Washington, DC. Users BJN are visiting Columbus, while user M in Cincinnati, users DGO in Atlanta, and user C in Washington area. So, bidirectional pointers are created between HLRs and corresponding VLRs. This is summarized in Table 1.1. It may be noticed that the pointer is established from visiting MSC to home MSC as soon as the MS acknowledges beacon signals at the visiting location. Based on received beacon from visiting MS, the MSC at visitor area determines home MSC and sends request for necessary authentication information. After that, a bidirectional pointer is created. The forward direction pointer from home MSC HLR to visiting MSC VLR is needed to forward all incoming calls, and backward pointer is essential to send billing information back to home MSC.

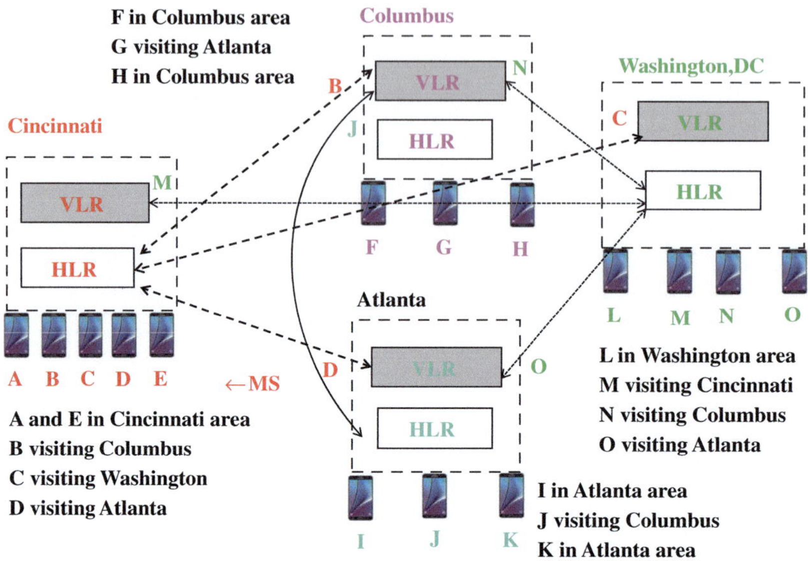

Fig. 1.7 Illustration of HLR–VLR pointers between different MSCs

Table 1.1 Establishing bidirectional pointer between home HLR and visiting VLR

MS	Bidirectional pointer established between 2 MSCs	
	Home MSC	Visiting MSC
A	Cincinnati	Cincinnati
B	Cincinnati	Columbus
C	Cincinnati	Washington
D	Cincinnati	Atlanta
E	Cincinnati	Cincinnati
F	Columbus	Columbus
G	Columbus	Atlanta
H	Columbus	Columbus
I	Atlanta	Atlanta
J	Atlanta	Columbus
K	Atlanta	Atlanta
L	Washington	Washington
M	Washington	Cincinnati
N	Washington	Columbus
O	Washington	Atlanta

1.2 MS Obtaining Traffic Channels from BS

Both BS and MS keep on sending beacon signals using a control channel (CC), indicating their presence and trying to determine whether they can communicate with each other. Usually, only one or two channels are used for control purpose. Once a MS selects a BS, it acknowledges that by sending a response beacon signal using a CC, containing its own phone number and ID of selected BS. As the cell phone is registered with the BS, it goes through authentication process at an associated MSC, and once approved, the BS employs CC to inform the use of two traffic channels (TCs) among a pool of available channels for transferring information from/to BS to/from MS. Thereafter, actual transfer of information between the BS and the MS can be started using these channels and is clearly illustrated in Fig. 1.8a. Steps for setting up a call from a BS to MS are shown in Fig. 1.8b as TCs can be used by MS only after the channels have been allocated by the BS to the MS via a CC message. A channel basically represents a frequency band, and three different schemes include frequency division multiple access (FDMA), TDMA (time division multiple access), and and Code Division Multiple Access (CDMA).

1.3 Multiplexing Schemes Used by a BS for Traffic Channels

In order to utilize channels in an efficient and effective way, some sort of multiplexing scheme is used. Initially, FDMA has been used where available bandwidth is divided into a set of equally spaced channels such that bandwidth of each channel can process voice traffic as given in Fig. 1.9a. Figure 1.9b shows how a pair of TCs are assigned to a given user, one for reverse direction (uplink from MS to BS) and another for forward path (downlink from BS to MS). The forward and reverse channel frequencies are usually indicated by the central frequencies and are separated by a fixed amount and are given in Fig. 1.10. In order to minimize

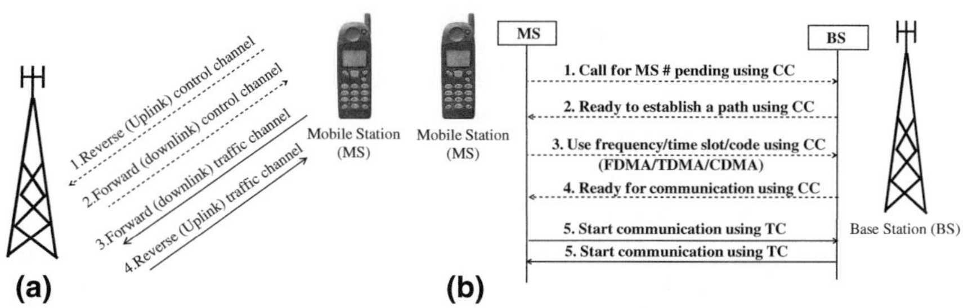

Fig. 1.8 **a** Establishing traffic channels for a MS. **b** Establishing traffic channels for a call from BS to MS

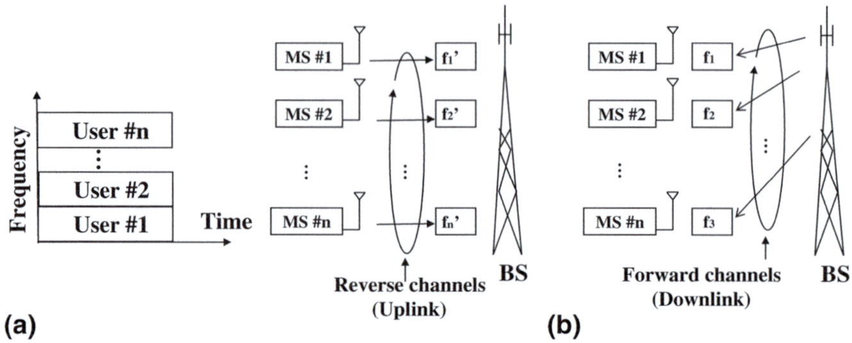

Fig. 1.9 **a** FDMA in a traffic channel. **b** Traffic channel allocation using FDMA

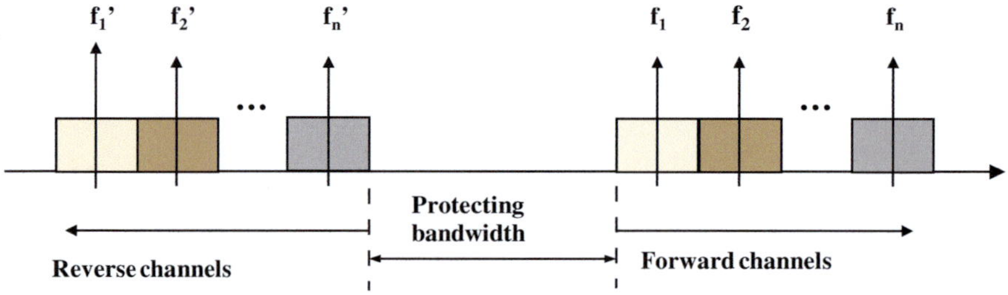

Fig. 1.10 Traffic channel allocation using FDMA

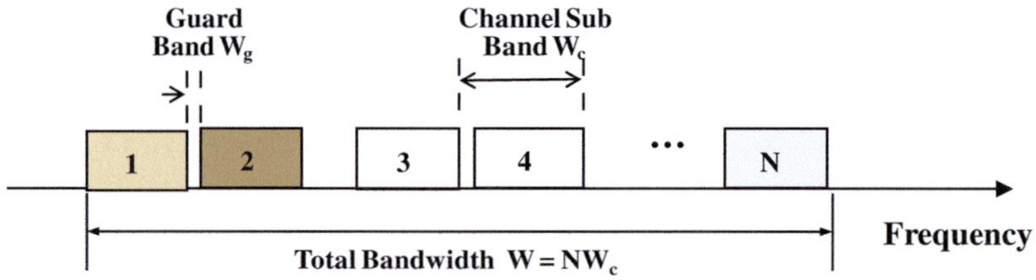

Fig. 1.11 Detailed traffic channel allocation using FDMA

interference between two successive channel frequencies, a gap is maintained and is known as a guard band. This helps in making sure that adjacent frequency components do not overlap (Fig. 1.11).

In real-life applications, such use of FDMA is not very efficient as the channel is kept reserved for a user whether it is used for actual transmission or not. Most of the time, people using cell phone, either talk or listen, and efficient data compression techniques further limit the period of channel use. This led to second-generation system of TDMA where frequency is subdivided in time domain and a slot is

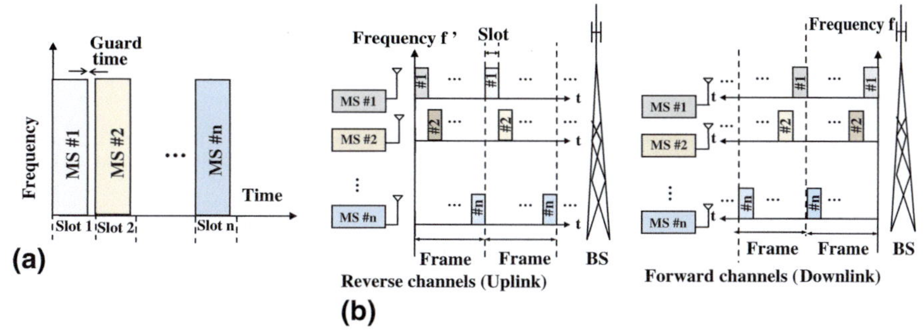

Fig. 1.12 **a** TDMA slot allocation. **b** Traffic channel allocation using TDMA

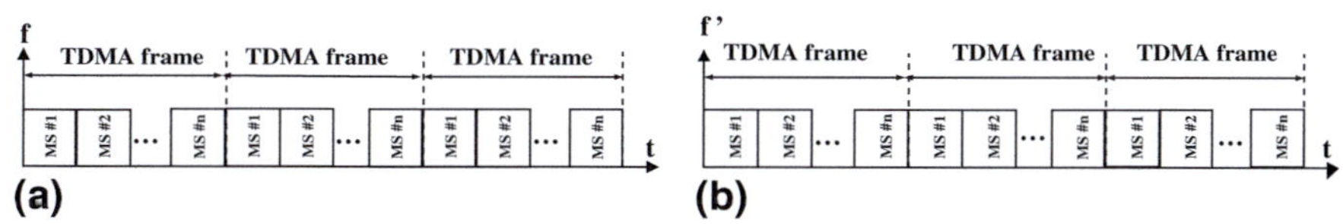

Fig. 1.13 **a** Forward channel (downlink) in TDMA. **b** Reverse channel (uplink) in TDMA

allocated to a given user only during prespecified time frame (Fig. 1.12a). Guard time between two successive slots is used to avoid overlapping and interference between signals. Two separate time slots are assigned to each user for forward direction and reverse direction information transfer (Fig. 1.12b). TDMA-based 2G system proved to be very useful in handling both voice calls and data transfer, and the forward and reverse channels are illustrated in Fig. 1.13. But, TDMA-inherent limitations are weaker digital signals at higher frequencies and increased call drop rates. A new scheme of CDMA (carrier division multiple access) has been introduced wherein each bit is Ex-ORed with a code word of size of 64 or 128 bits and is illustrated in Fig. 1.14.

Thus, the whole assigned bandwidth is used by each user, and unique orthogonal codes are assigned to different users using CDMA. Modified signal from users is mixed together in the air, and accumulated signal reaches the receiving station where associated code is used to extract original data corresponding to a given user

Fig. 1.14 Users assigned different codes in CDMA

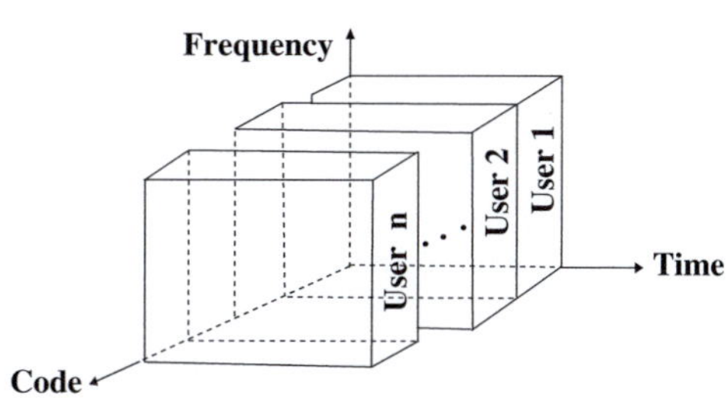

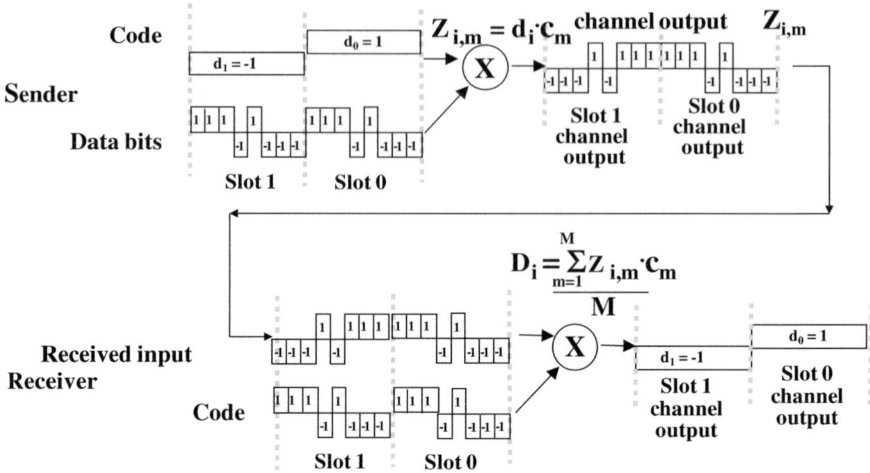

Fig. 1.15 **Modification of user's data in a CDMA system

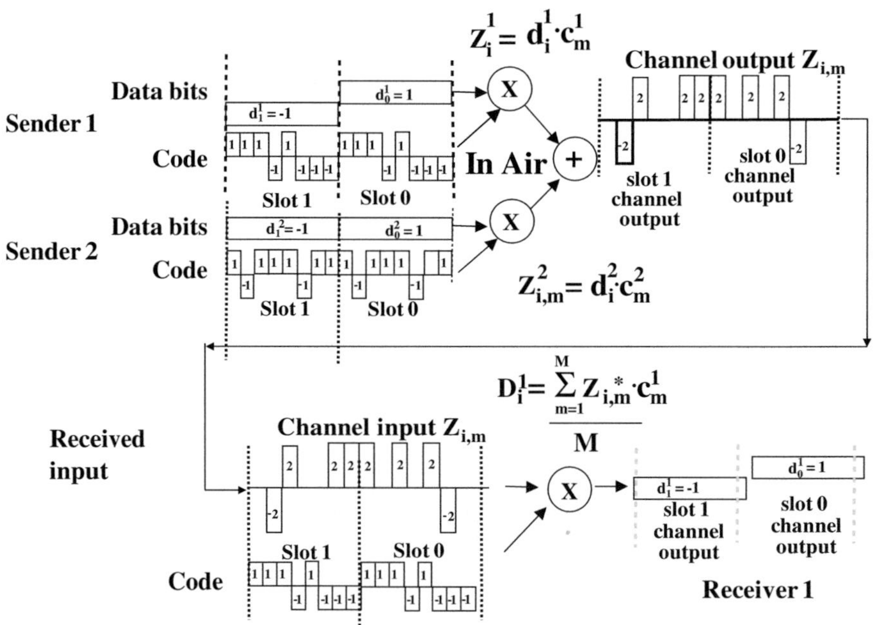

Fig. 1.16 Extracting user data from two users' mixed signal

as shown in Fig. 1.15. This way, the whole bandwidth is shared by all the users, and even though all signals are logically added in the medium, it is possible to extract given user's data using its unique CDMA code as shown in Fig. 1.16.

1.4 Orthogonal Frequency Division Multiplexing Access

The conventional FDMA requires partitioning of given bandwidth into multiple channels of adequate bandwidth to process voice calls effectively. A guard band is used between two adjacent channels to have separation in order to minimize interference and is an overhead. Similarly, guard time is used in TDMA which simply wastes bandwidth, and frequent switching is needed between users. These issues have been addressed by OFDMA by partitioning available bandwidth into many narrow orthogonal channels (100–8000) which are used in simultaneous transfer of information. The narrow bandwidth of each subchannel (less than 1 kHz) allows tolerance of multi-path delay spread and minimizes channel fading. Interchannel interference is avoided by the presence of orthogonal subchannels, while bandwidth is maintained by parallel use of all subchannels (Fig. 1.17).

1.5 Directional Antenna and SDMA

The antenna that is commonly designed and use radiate electromagnetic waves in all 360° with the same intensity and only one user can be served by such radiation. Further, throughput enhancement can be made feasible if directional antenna is employed that focuses EM beams in a small sector. Such deployment of many such omnidirectional antennas focusing in different sectors can serve multiple users using the same frequency band if users are located in different sector areas. Such space division multiple access (SDMA) allows the use of a single frequency band by antennas and can serve many users located in different areas and hence could provide improved bandwidth and is shown in Fig. 1.18. In a cellular system, the base station knows the distance of each user based on (received signal strengthRSSI indicator) and accordingly adjusts the transmitting power level. But, the direction of their location remains unknown. So, usefulness and effective bandwidth depend on where different users are spaced in the surrounding area. Frequency and wavelength of different power sources are given in Table 1.2.

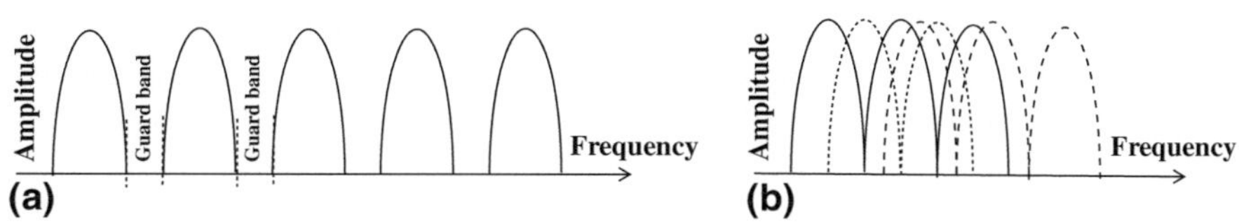

Fig. 1.17 **a** Conventional multi-carrier modulation in FDMA. **b** Orthogonal multi-carrier modulation used in OFDMA

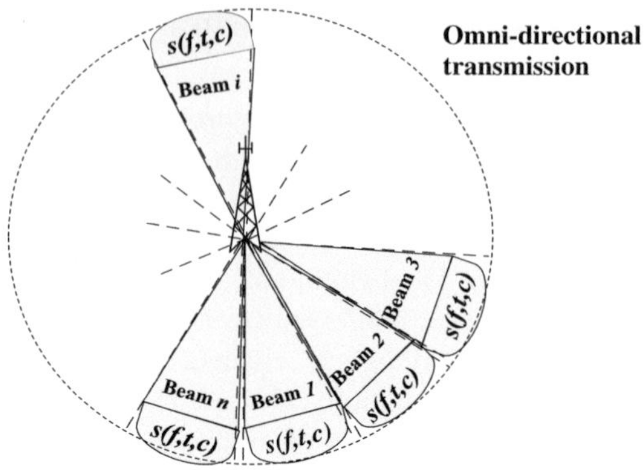

Fig. 1.18 SDMA concept of n-sectors using directional antennas [3]

Table 1.2 Frequency and wavelength of different sources

System	Frequency	Wavelength
AC current	60 Hz	5000 km
FM radio	100 MHz	3 m
Cellular	800 MHz	37.5 cm
Ka band satellite	20 GHz	15 mm
Ultraviolet light	10^{15} Hz	10^{-7} m

1.6 Cellular Transmission

Transmission of digital signal in a cellular system goes through many steps, and each is needed for integrity and correct transmission/reception of signals. These steps are summarized in Fig. 1.19. Voice and data information are converted into digital form and are coded appropriately for error detection/correction and possible fault tolerance. Then, the signal is appropriately modulated to transmit data at desired frequency band by the transmitter. EM waves are propagated through air or other medium where signal gets reflected, refracted, and scattered from obstructions, water drops in the air, and sharp objects as indicated in Fig. 1.4. In each of this phenomenon, signal strength gets attenuated before reaching the destination antenna. A reverse steps are followed on the received signal to interpret sent data correctly. Additional steps could be included to incorporate authentication, encryption, and secured communication . There has been concern whether EM radiation is hazardous to human health, even though it has not been proved. A general recommendation is to limit the amount of voice communication, or wherever possible, use SMS service to send messages. It is interesting to note that radiation level is higher in areas far away from the base station [4]. Cell phones are also allowed in airplanes once in cruising altitude of 30,000–40,000 feet as interference is possible at lower elevations. In many countries, cell phone use is banned while driving as the risk of an accident is more than drunk drivers [5]. Similar to debit and credit cards, there are two models of paying for cell phone usage. In the first prepaid mode, clients can prepay ahead of time using Internet or dedicated

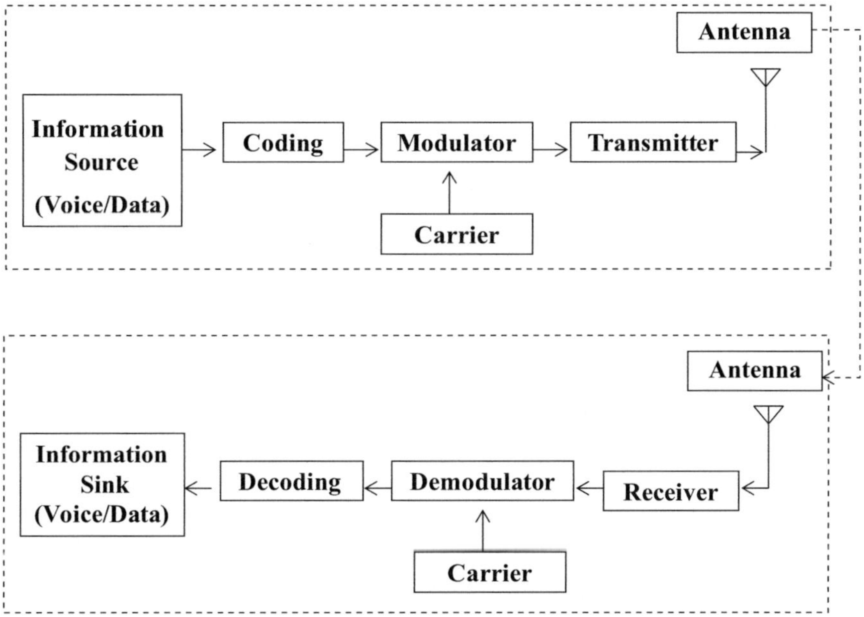

Fig. 1.19 Processing steps in a cellular system

shops and the balance is deducted based on its use. In the contract model, bills are paid on regular intervals after the facility has been used up. In some parts of the world, incoming calls are free of charge as calling party is billed.

1.6.1 Cellular Coverage Area for Traffic Channels

A typical cell phone has a coverage area of 0.8 km (cities) to 40 km (open areas) radius (Fig. 1.20a) from the center of a microwave tower (base station) [6], and to cover a larger area, many such towers have to be erected. These towers are static on the surface of earth and are connected hardwired on surface of earth or dedicated microwave links. So, when you move from one place to other, you have handoff (handover) from one tower to another and a new connection is established to the closest tower. One such example for travel from Cincinnati to Columbus, Ohio is illustrated in Fig. 1.20b.

Ideally, the cell shape is expected as circular in nature (Fig. 1.21a). Due to terrain and other factors, actual cell could be somewhat distorted as shown in Fig. 1.21b. Other shapes that are used in modeling coverage area of a cell are given in Fig. 1.21c. In most simulation and analytical work, hexagonal shape is used to represent a cell coverage area as replication of cells allows treatment of larger area without leaving any uncovered space. This is exemplified in Fig. 1.22a that shows a

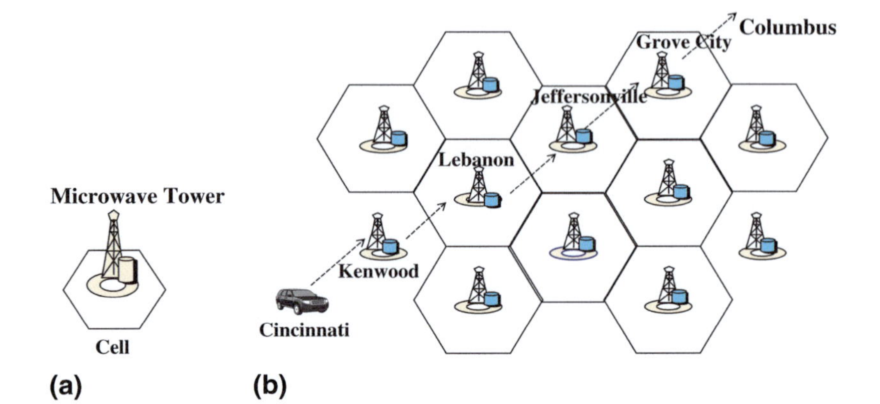

Fig. 1.20 **a** Cell Coverage area. **b** Handoff (handover) in a cellular system

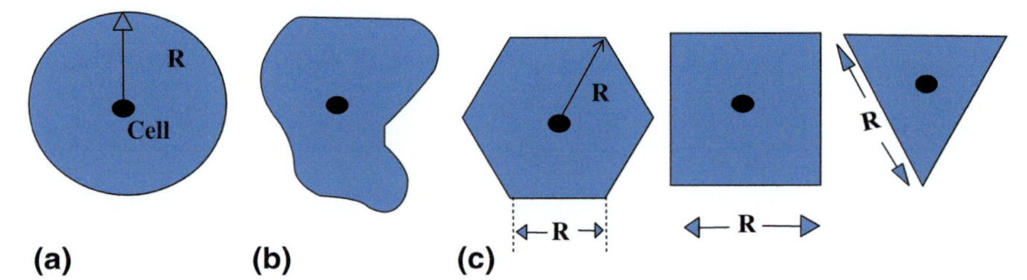

Fig. 1.21 **a** Ideal cell coverage area. **b** Realistic cell area. **c** Different models of cell coverage area

linear arrangement of 3 cells. Other possible arrangements of 3, 4, and 7 cells are given in Fig. 1.22b. From the design point of view, all allocated channels are divided equally among selected group commonly known as a cluster and fixed set of channels is assigned to each cell as shown in Fig. 1.22. For example, if 784 channels are assigned for use, each cell will be allocated 784/7 = 112 channels.

To keep the design of a cellular scheme simpler, selected pattern of a cluster is repeated to cover larger area and is shown in Fig. 1.23. This makes the design process simpler, and even though many cells use the same frequency band, they are kept far apart so that they do not interfere with each other. The distance between two close-by cells using the same frequency bands is known as reuse distance D and is clearly marked in Fig. 1.23. In order to maintain the same reuse distance between all pairs of channels, the shape of cluster is chosen as close to a circle as possible. Interference between cells using the same frequency bands commonly known as cochannels is illustrated in Fig. 1.23. It is interesting to note that higher

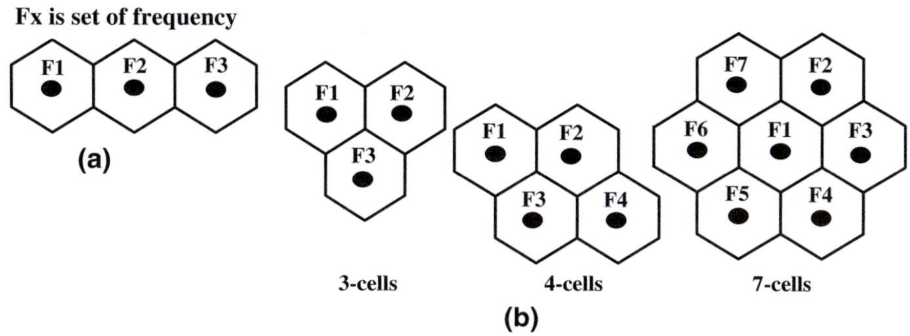

Fig. 1.22 **a** Linear 3 hexagonal cells. **b** Other hexagonal cell structures

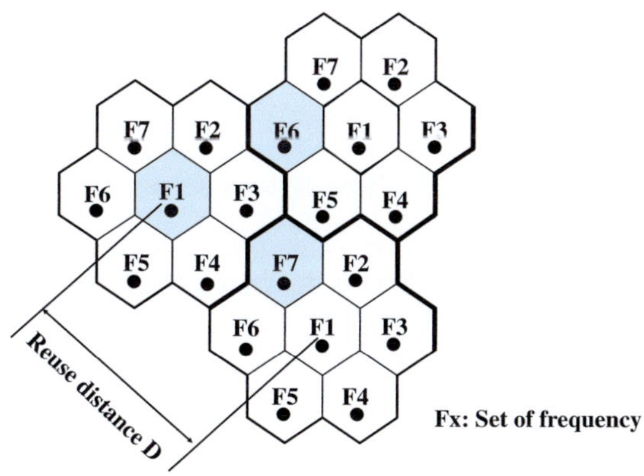

Fig. 1.23 Use of 7 hexagonal cells clusters and reuse distance

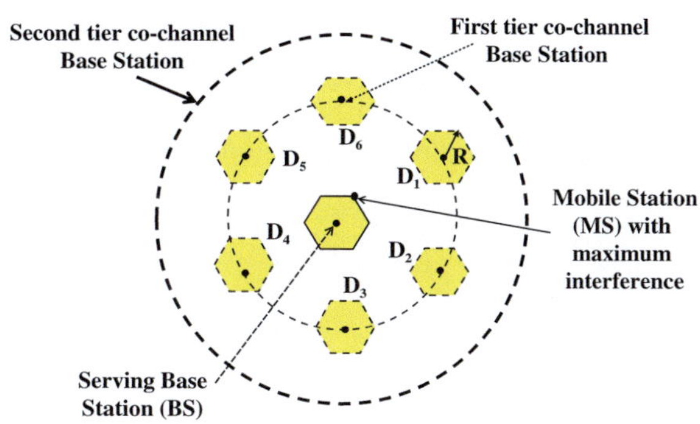

Fig. 1.24 First-tier and second-tier base stations and cochannel interference

order polygons illustrated in Fig. 1.25 are closer to circular nature of EM propagation. But, they do not provide non-overlapping and full coverage for larger coverage space (Fig. 1.24).

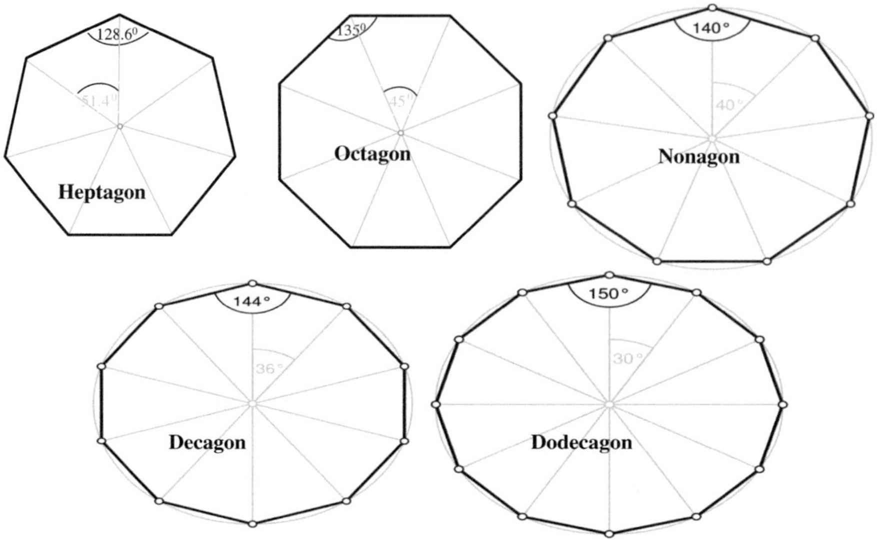

Fig. 1.25 Few other possible higher order polygons to represent cell area

1.6.2 Signal Strength in Cellular Area

The signal strength of EM waves reduces as you move away from the center of a BS (microwave tower) and is shown in Fig. 1.26. In fact, if you move away from BS_i, a handoff (handover) occurs to another close-by BS_j and is illustrated in Fig. 1.27. There are many locations of the MS where handoff is not desirable even though the signal strength due to two BSs may be equal. This is due to frequent handoff between two BSs when the path of MS crosses the boundary between two BSs and is known as a ping-pong effect. Therefore, MS is allowed to continue connection with existing BS, even though signal strength from adjacent BS could be slightly higher. Ping-pong region is also identified in Fig. 1.27.

Fig. 1.26 Variation of signal strength from location of BS in a cell area

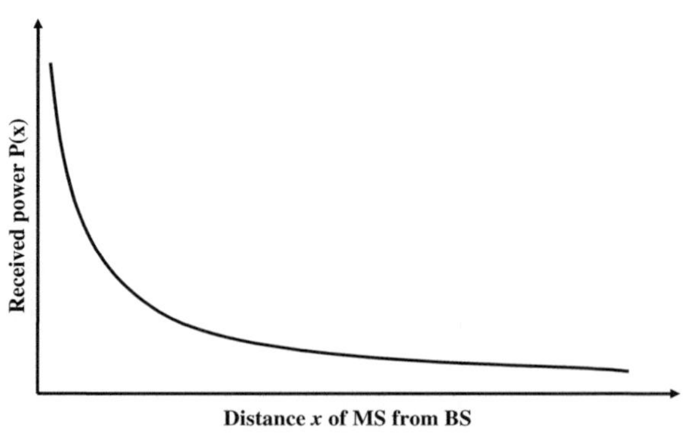

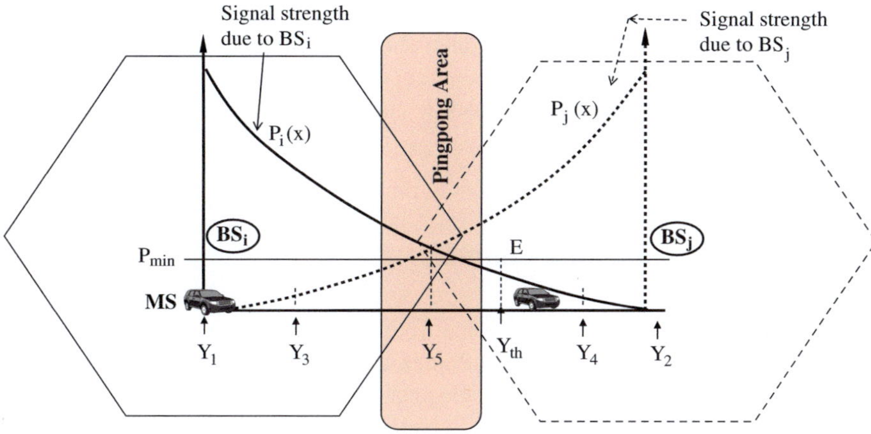

Fig. 1.27 Handoff between two adjacent BSs and ping-pong effect

1.6.3 Roaming Support

Forwarding of incoming call can be supported by HLR–VLR pair as indicated in Fig. 1.1. But, in some regions, if a service provider does not provide any support in a given area, then it may have a service agreement with another service provider that provides service in that area. The second company charges additional charge commonly known as roaming charges and is exemplified in Fig. 1.28. Then, software support of home agent–foreign agent pair provides redirection of incoming call, and received packet is encapsulated by the original service provider before

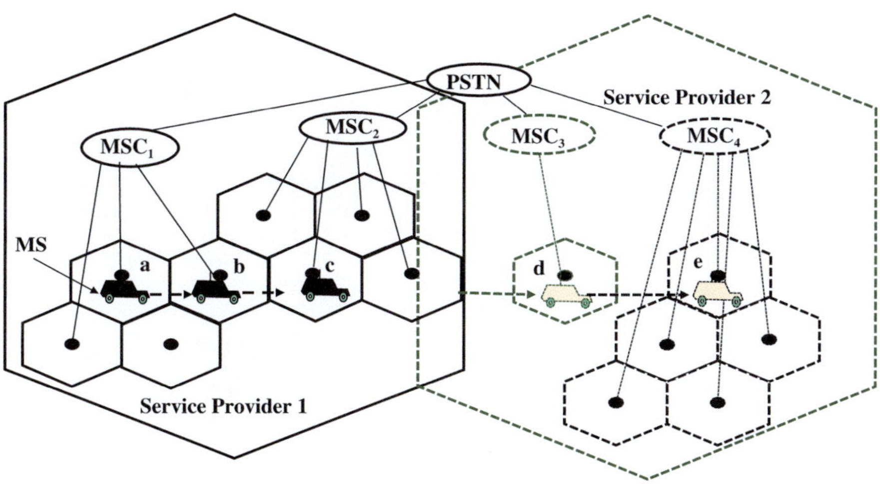

Fig. 1.28 Roaming support by different service providers

forwarded to serving company for forwarding to an appropriate BS. Outgoing call can be initiated directly by the serving area although roaming charges are added for all calls. Many service providers add roaming charges by dividing service area into zones and charge for roaming if the user is currently located outside its registered zone.

1.7 Access to Control Channels

Control channel is used to obtain service from a BS, including allocation of traffic channel to a given user. Usually, 1–3 duplex channels are there in each cell [7] to exchange useful information and are adequate to support a large number of users ($\sim$90 channels for active MSs). As it is hard to say when a user will be ready to communicate with a BS (or receive message from a BS), carrier sense multiple access (CSMA) scheme is used as shown in Fig. 1.29. An early version of CSMA has been used in slotted ALOHA and pure ALOHA where sender sends their packets when they have packets ready to transmit. In slotted ALOHA , user clocks are synchronized and packets of the same size are transmitted by each user during fixed-sized slot. In pure ALOHA, clocks are not synchronized and packets are of unequal size. Collision scenarios for slotted and pure ALOHA are summarized in Fig. 1.30. The throughput achievable as a function of load is illustrated in Fig. 1.31, and the maximum remains very low. Therefore, CSMA (carrier sense multiple access) scheme has been used by control channel of cell phones. The idea behind CSMA/CA is to sense the medium if some other user is using the control channel as wireless radio can either sense or transmit but not both functions at the same time. If someone is accessing the medium, defer wait till no one is using and then generate a random delay before accessing the control channel. Such a random delay reduces the probability of two users accessing the control channel at the same time. But, the collision cannot be totally eliminated as shown in Fig. 1.32.

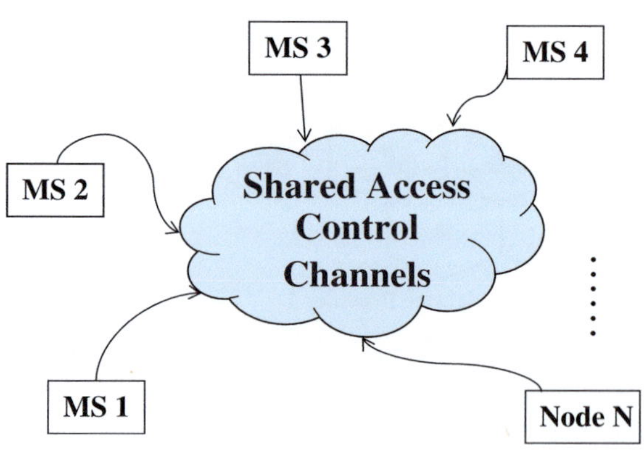

Fig. 1.29 Carrier sense multiple access for control channels

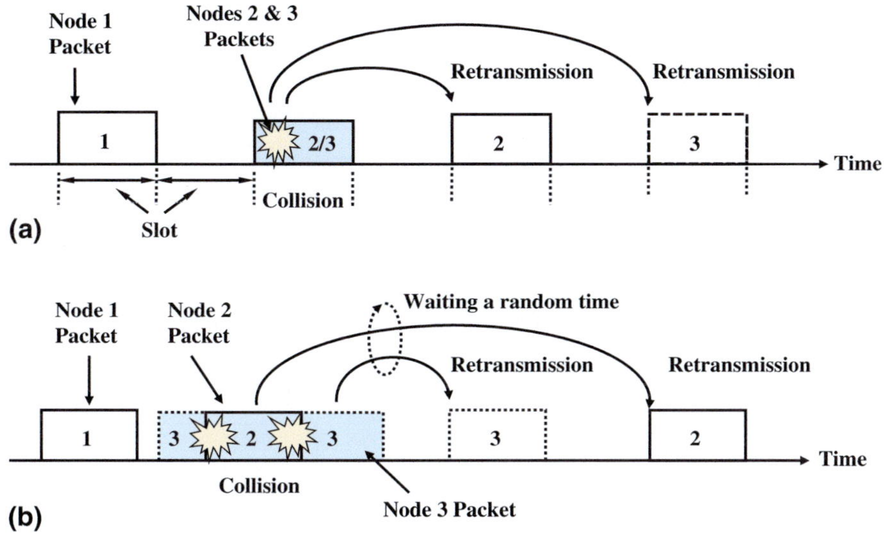

Fig. 1.30 **a** Collision mechanism in slotted ALOHA system. **b** Collision mechanism in pure ALOHA system

Fig. 1.31 Throughput in slotted and pure ALOHA systems

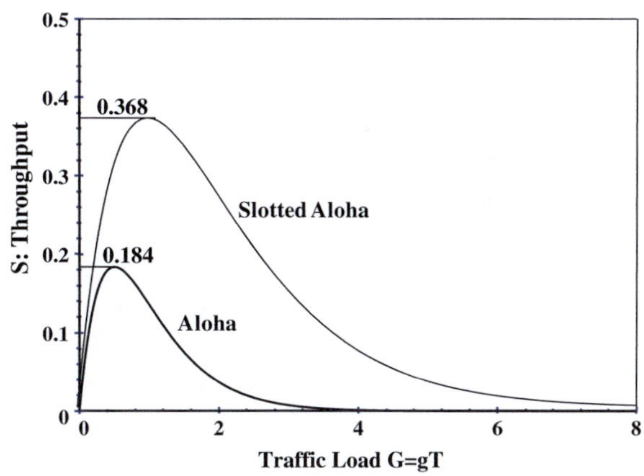

User can immediately seize the medium if no one is using the same. Another alternative is to wait for one time slot as all users' cell phones' clock is synchronized with that of the BS. A probability p is defined as the persistence value and represents the probability that a user will not wait for a time slot time T before using the control channel. Variation of throughput with varying traffic load for different p values is given in Fig. 1.33. In most cases, nonpersistent CSMA ($p = 0$) with wait time of one slot always provides good throughput. A more effective scheme of Fig. 1.34 is to define a contention window (CW) of 2^n and generate random delay between 0 and $(2^n - 1)$ slots to generate delay. If two users collide, the value of

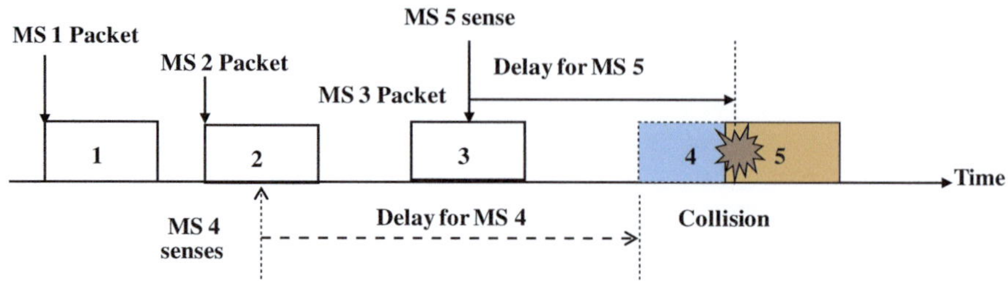

Fig. 1.32 Collision mechanism in CSMA

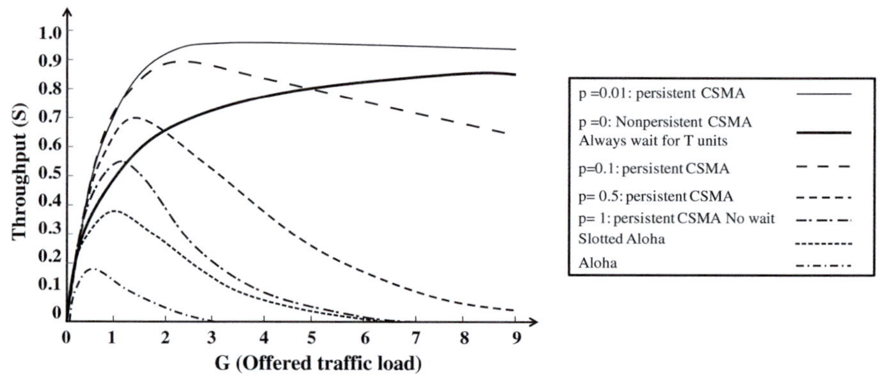

Fig. 1.33 Throughput using CSMA

CW is doubled with the hope that the two users will generate different values of random delays with doubled range. As soon as the counter of one user becomes zero, it starts using the control channel and then the counter for the second user is frozen. It will start decrementing only after current communication is finished and that way quickly gets access to the control channel. Thus, the CW-based random delay helps in minimizing collisions and hence CSMA/CA scheme, and this process is shown in Fig. 1.35. MS_A and MS_B are the back-off intervals of MSs A and B. We assume CW = 32. MS A and MS B have chosen a back-off interval of 25 and 20, respectively. MS B will reach zero before five units of time earlier than MS A

> When this happens, MS A will notice that the medium became busy and freezes its back-off interval currently at 5. As soon as the medium becomes idle again, MS A resumes its back-off countdown and transmits its data once the back-off interval reaches zero. In case if there happens to be further collision, the size of CW is doubled till some predefined upper limit, and after that, it is not increased any further.

Another approach is to utilize small packers of RTS–CTS (Request To Send–Clear To Send) in creating a handshaking mechanism between a sender and a receiver, and only after this step, actual transfer takes place. Two small interframes

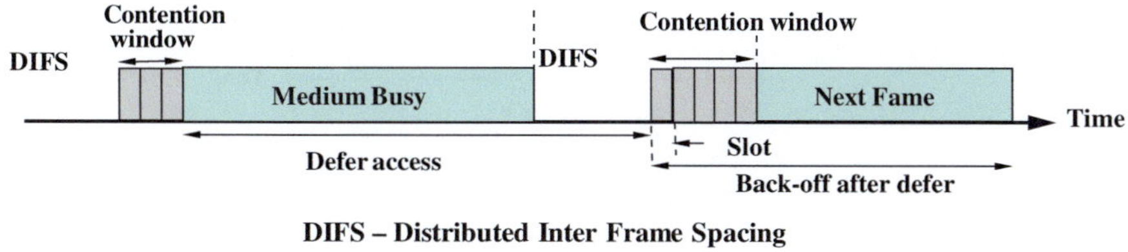

Fig. 1.34 Use of contention window with CSMA

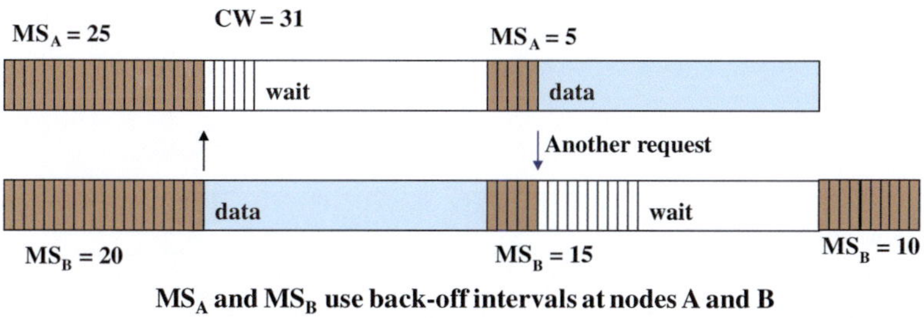

Fig. 1.35 Random delay helps in CSMA/CA

of DIFS (*distributed interframe space)* and SIFS (*short interframe space)* are added as additional delays, with DIFS = SIFS + 2T. This is done to ensure that no other user can access the medium when the sender sends RTS message, and then, the receiver could successfully send CTS message before any user can access the medium of control channel. Moreover, the receiver after receiving the packet from sender can send ACK signal as an acknowledgement before any other user can access the medium as explicit acknowledgement has to be sent by the receiver back to the sender as given in Fig. 1.36.

In ethernet, there is implicit ACK signal sent by the receiver to the sender. But, in wireless, no such signal is sent by the receiver and an explicit ACK packet has to

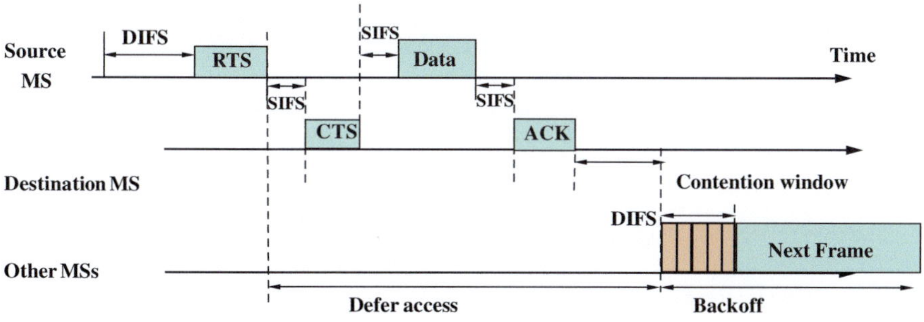

Fig. 1.36 CSMA/CA with RTS/CTS

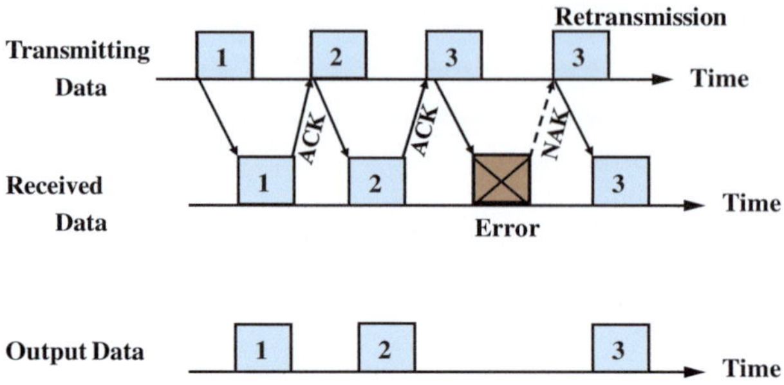

Fig. 1.37 STOP-AND-WAIT ACK signal in wireless system

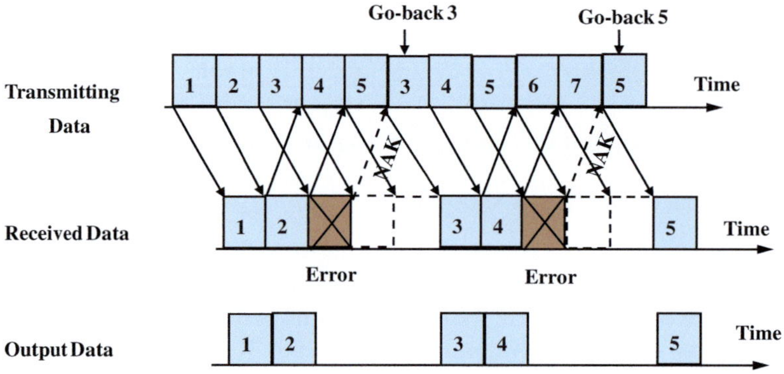

Fig. 1.38 GO BACK n-STEPS ACK signal in wireless system

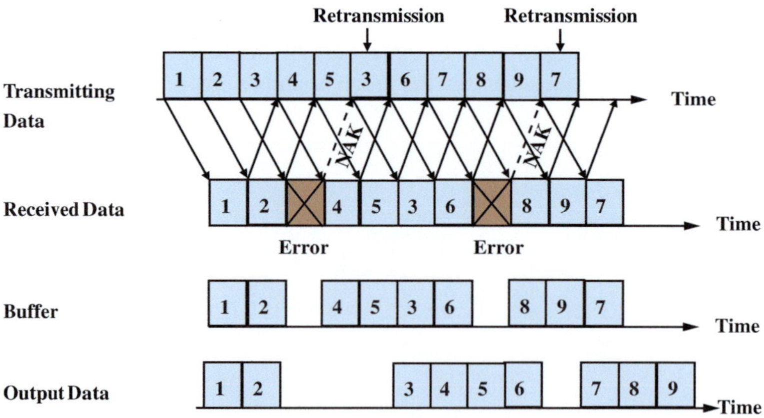

Fig. 1.39 Selective Repeat ACK signal in wireless system

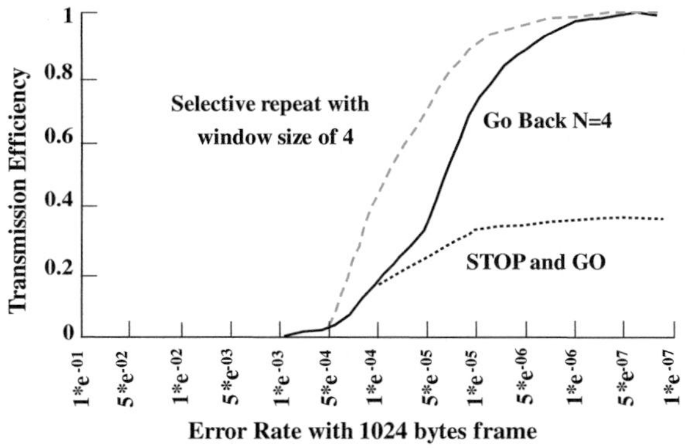

Fig. 1.40 Performance of three ACK protocols in wireless system

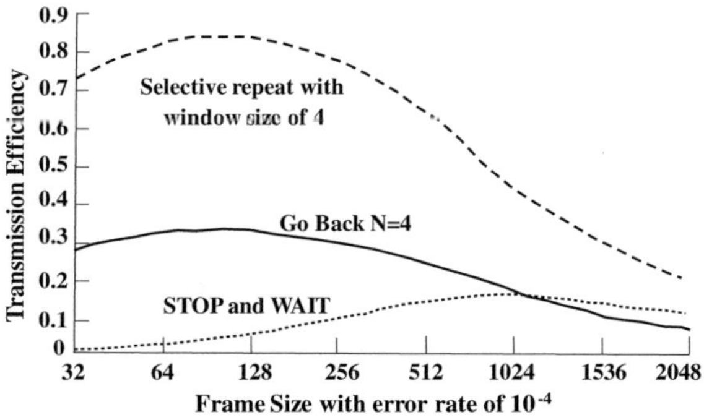

Fig. 1.41 Optimal packet size in 3 ACK protocols for wireless system

be sent by the receiver. There are three protocol schemes of STOP-AND-WAIT, GO BACK n-STEPS, and Selective Repeat that are followed in transmitting successive message packets and are shown in Figs. 1.37, 1.38 and 1.39. No NAK packet is even sent in wireless technology. Instead, if no ACK is received within prespecified time-out period, the sender assumes the presence of collision and initiates the process of resending the packet. This is the way forward error control in transmission medium is achieved. Relative throughput of these schemes under different error rates is given in Fig. 1.40. STOP-AND-WAIT scheme works well under noisy conditions while Selective Repeat provides best performance under lower error rates. So, there is a trade-off between different schemes and the user needs to select which scheme is better suited. Throughput efficiency for optimum frame size with an error rate of 10^{-4} in the three schemes is shown in Fig. 1.41.

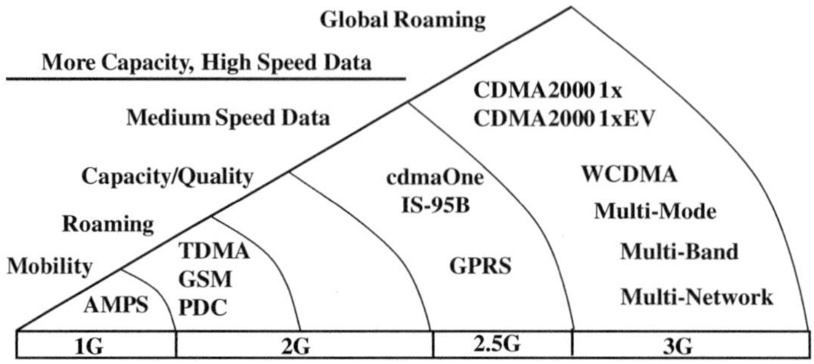

Fig. 1.42 Evolution of wireless technology and associated characteristics

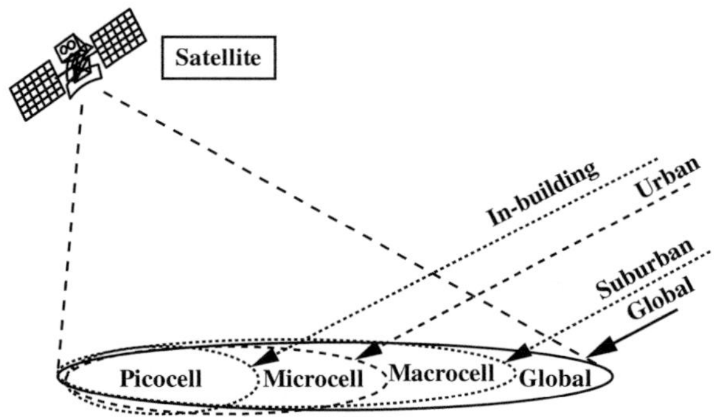

Fig. 1.43 Coverage by different wireless technologies

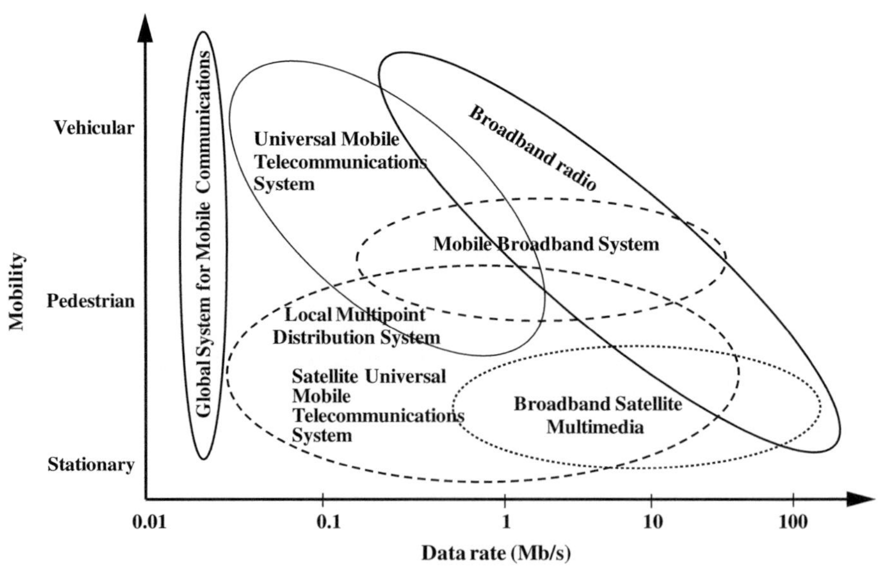

Fig. 1.44 Transmission capacity as a function of mobility of wireless technologies

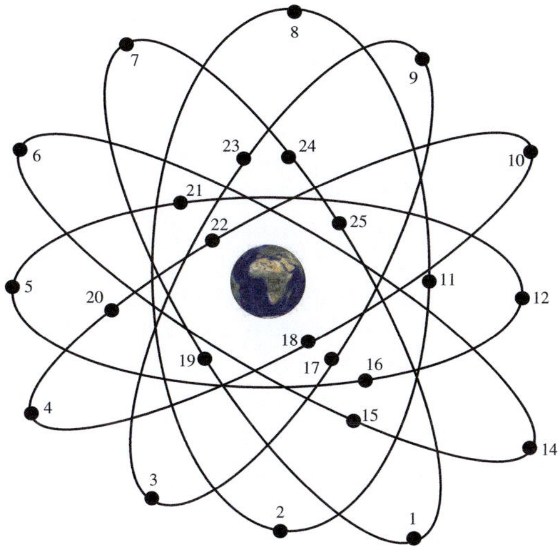

Fig. 1.45 24 geosynchronous satellites rotating around Earth

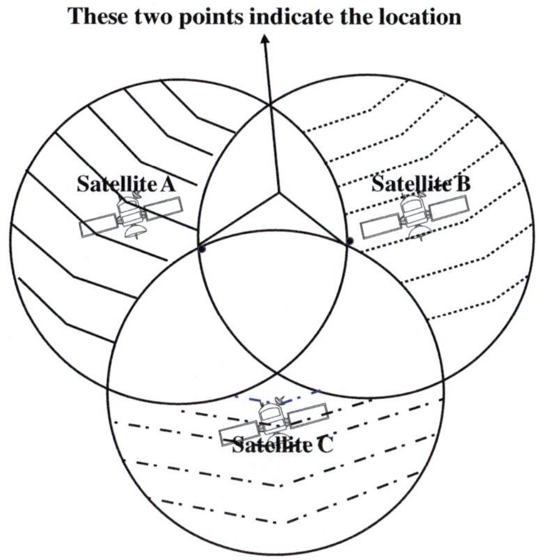

Fig. 1.46 Location determination by triangulation using distance from three geosatellites

1.8 Different Wireless Technologies

Wireless technology has evolved over the period of time, and capacity and mobility support have enhanced at a faster rate. This is shown in Fig. 1.42. Even the coverage area has changed, and depending on user's particular needs, appropriate wireless technology among many different options illustrated in Fig. 1.43 can be selected. Transmission capacity also changes with mobility support and is summarized in Fig. 1.44. Satellite systems have changed the way wireless communication is utilized. It provides wider coverage and transmits power at much lower level, requiring 80–100 db power gain. It is useful for connecting satellite phones around the world.

One special class of satellites specialized application in creating directions from one point to another using Global Positioning System (GPS) services. It is based on geosynchronous satellites which rotate around earth using six different orbits and by having four satellites on each orbit (Fig. 1.45). The earth rotates around itself and goes around sun as well. Movement of geosatellites is controlled in such a way that they appear stationary with respect to the surface of earth. The GPS devices on surface get signals from at least satellites and calculate distance of satellites from the device and determine its location using triangulation scheme (Fig. 1.46). As locations of satellites are fixed with respect to earth, triangulation gives exact coordinates of the GPS device. Once the current position of GEO device is known, rest of the direction to the destination is performed by the GPS device using local map stored in the GPS device. A similar triangulation scheme is done in cellular system in determining its location.

1.9 Access Points

Another wireless technology that is impacting human life is the AP (access points) used at homes, various institutions, and research organizations. The APs provide excellent bandwidth (4–48 Mbps) for a limited space area of 30–100 m and are very effective in providing needed Internet access both at homes and offices. As the coverage area for an AP is limited, when a user moves from one room to another, the signal strength may become much lower than that can be decoded and correctly interpreted. Then, handoff from previous AP to another close-by AP is necessary. ApAs work differently than BS of cell phones. Downloading of files from AP to users' laptops is done using programmed TDMA schedule, with each transfer

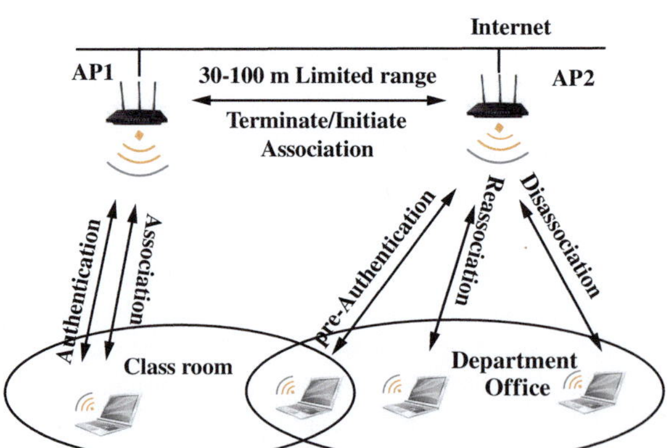

Fig. 1.47 Access point providing wireless Internet access to laptops

period based on bandwidth chucks requested by all active laptops to the AP. Each AP polls for all wireless devices every 20 ms [8] to update the active list of users and their bandwidth requirements. In this way, time slot for each user keeps on changing dynamically based on the needs of active users. Uploading from laptops to AP is done using CSMA/CA technique as it is hard to predict when a user is ready to send files to the AP. There are many different protocols introduced for APs and belong to 802.11 series such as 802.11a, 802.11b, 802.11c, 802.11g, 802.11n, and forthcoming 802.11ac. The bandwidth, modulation scheme, and number of antennas vary based on the type (Fig. 1.47).

1.10 Sensor Networks

Sensors are tiny devices deployed to measure physical parameters of the surrounding area and can be used 24x7 (24 h a day 7 days a week) to perform unattended monitoring of close-by space. This is achieved by having necessary functional units inside sensors that can be embedded for a given application. A generic sensor unit with different useful blocks is shown in Fig. 1.48a. The transducer type depends on the application under consideration and needs to be carefully selected. All these units get power from batteries (usually 2 AA), and energy consumption is of prime concern. Energy consumed in different parts in MW is given in Fig. 1.48b. It is interesting to note that most energy is consumed by the wireless radio. Even if the transceiver is in receive or idle mode, it consumes significant amount of power and is usually many order of magnitude than the CPU or sleep mode of the transceiver. Therefore, attempts must be made to keep the transceiver in sleep mode as long as possible. For example, if the sensor needs to send data every 2 min, the transceiver can be turned on for 20 ms and transmit the sense value and go back to sleep mode. It is observed that the energy consumed is proportional to the square of the distance of transmitted signal to be received correctly [3]. There are many transducers commercially available and include

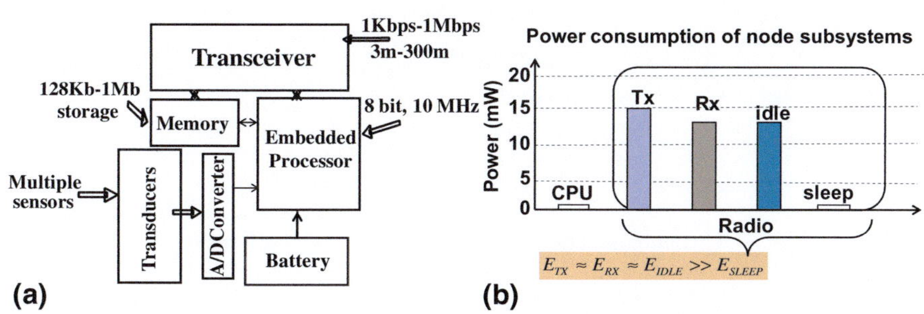

Fig. 1.48 **a** Functional blocks of a Sensor unit. **b** Energy consumption in various functional blocks [3]

pressure, temperature, light, biological and chemical agents, strain , fatigue, tilt, and many others. These are designed in such a way so as to withstand harsh environments such as excessive heat, overwhelming humidity, too much corrosion, constant pollution, and continuous radiation. All these environmental conditions must not affect the accuracy of sensed parameters.

Thermometer is one of the earlier transducers that have been used by our parents and grandparents. A single device indicates the temperature of the whole body. Thus, a unit sensing a physical parameter can be said to represent that value for a small circular region commonly known as a sensing range. So, if sensor needs to monitor a large region, many such units have to be deployed and are illustrated in Fig. 1.49. The question is where to place the second or third sensor with respect to the first one. The idea is to place sensors such that the sensing area covered by sensors does not overlap or leave uncovered space. With circular sensing range, it is hard to have either overlap or uncovered space. From these considerations, sensing ranges indicated by solid circles in Fig. 1.49 reflect the right positions of three adjacent sensors. Even though small gaps are left over between sensors, they are acceptable as a significant section of the area is covered by the three sensors combined together. Moreover, sensor data need to be collected at a central location known as a BS (base station) or sink node wherein collected data are analyzed and appropriate decision is made. Thus, in order to conserve consumed sensor energy, data are sent from sensors to BS in a multi-hop fashion.

Figure 1.50a shows unplanned deployment of 9 sensors to cover an area ABCD. As sensors are deployed randomly, overlap in sensing areas cannot be avoided, and

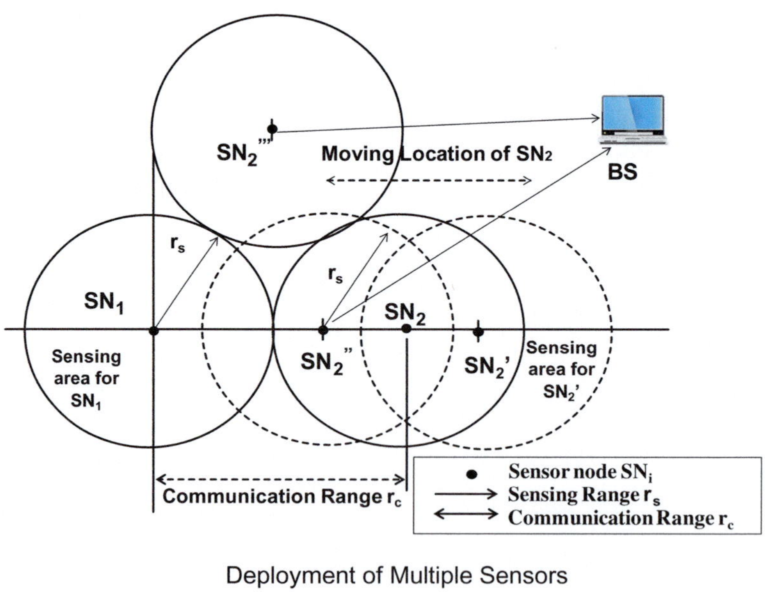

Fig. 1.49 Deployment of multiple sensors

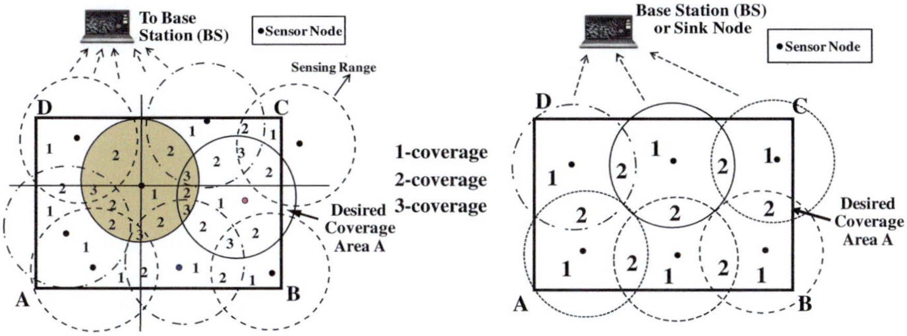

(a) Random Sensors for larger area and coverage degree

(b) Controlled Sensors for larger area and coverage degree

Fig. 1.50 **a** Random sensors for larger area and coverage degree. **b** Controlled sensors for larger area and coverage degree

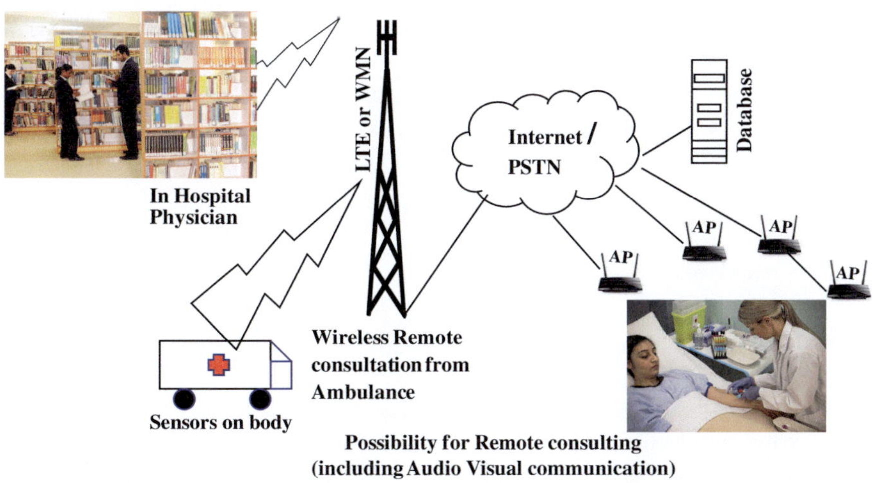

Fig. 1.51 Sensors for emergency assessment and necessary aid [taken from 3]

various sectors are marked 1, 2, or 3, indicating that the area parameter can be represented by reading from any of 1, 2, or 3 sensors as physical variables change very slowly and readings from adjacent sensors may be fairly close to each other. There, segments are called 1-, 2-, and 3-covered and could be used to provide reliability test among sensors or in having fault tolerance in the system. In this way, sensor ID is not very critical and what is important is the location of the sensors and the value indicated by them. Therefore, sensors are termed as "data centric" and not "address centric." If we have freedom to place sensors at any selected location such that the overlap areas could be minimized, then only six sensors can cover the same area as shown in Figs. 1.50b and 1.51)

Sensors were initially introduced for defense applications [9], and its use has been ever growing in numerous critical strategic areas; there are too many to list them. One obvious civilian use is to in ambulance by transmitting EKG and other vital signals by medical assistant to nearby hospitals and medical experts when an injured patient is being carried from the scene of an accident, in addition to only blood pressure and temperature. This enhances the possibility of saving of human life and can be said to be an excellent application of wireless technologies. Even a pill has been developed to measure rise in human temperature of firefighters to the captain when in action [10]. Another noteworthy pill contains a camera that could wirelessly transmit pictures inside an intestine to detect any possible tumor in stomach [11, 12]. So, there is no limit on potential use of sensors, and one needs to explore all possible wild imaginations.

1.11 Conclusions

Started with slow acceptance of wireless technology by people has now become an integral part of daily life. Everyone seems to carry at least one cell phone and the number of wireless devices, and human dependency on them is growing exponentially. Even though sensors seem to be a new member, it usefulness seems to be unlimited and potential impact seems to be unprecedented.

1.12 Questions

Q.1.1. How do you determine an antenna gain?
Q.1.2. What is refraction?
Q.1.3. What is the difference between diffraction and scattering?
Q.1.4. Why are electromagnetic waves not transmitted at very low frequency in wireless systems?
Q.1.5. What is the role of digital signal processing in a cellular system?
Q.1.6. Why is hexagonal cell shape preferred over square or triangular cell shapes to represent a cellular system?
Q.1.7. Show that it is not possible to cover larger area without any overlap or any gaps by cutting paper model of higher order polygons.
Q.1.8. What are the challenges in designing a cellular system?
Q.1.9. Why do you need beacon signals in cellular system?
Q.1.10. What is meant by propagation model of EM waves?
Q.1.11. Why do you have attenuation of EM signals?
Q.1.12. What happens to EM waves inside a building?
Q.1.13. What is the significance of RSSI in cell phones?
Q.1.14. Prove that doubling the transmission frequency power received is attenuated by 6 dB.

Q.1.15. Prove that doubling the distance between transmitting antenna and receiving antenna reduces the power received by 6 dB.
Q.1.16. What is meant by multiple access in a cellular system?
Q.1.17. Why is the information modulated before transmitted by a cellular system?
Q.1.18. What is meant by forward and reverse traffic channels?
Q.1.19. What is meant by cochannel interference?
Q.1.20. What is meant by handoff in a cellular system?
Q.1.21. How is the roaming supported by a cellular system?
Q.1.22. Why small-sized cells not used in a cellular systems?
Q.1.23. What is meant by frequency reuse?
Q.1.24. What is the difference between CSMA/CD and CSMA/CA?
Q.1.25. Why do you need CDMA codes to be orthogonal?
Q.1.26. How is the power control useful in a CDMA system?
Q.1.27. What is the role of guard band and guard time in a cellular system?
Q.1.28. What is the difference between CDMA and FDMA?
Q.1.29. What are the difference between ALOHA and slotted ALOHA protocols?
Q.1.30. What are the main functions of a base station in a cellular system?
Q.1.31. Why is CSMA/CA not used for traffic channels?
Q.1.32. What is meant by "ping-pong" effect?
Q.1.33. What are the advantages of using SDMA technique?
Q.1.34. Which multiple access technique is used by IEEE 802.11 standard for wireless AP?
Q.1.35. How do you provide NAK signal in a cell phone?
Q.1.36. What is the use of ARQ in cell phones?
Q.1.37. How do you secure your cell phones?
Q.1.38. What kind of Wi-fi routers used in your institution and how are they placed in a building?
Q.1.39. What are the design goals of a Wi-fi system?
Q.1.40. What are the differences between LANs and WLANs?
Q.1.41. What is meant by data encapsulation?
Q.1.42. What is the usefulness of home agent and foreign agent in a cellular system?
Q.1.43. What is piggybacking in context to sensor networks?
Q.1.44. What are the main functions of a base station in a sensor network?
Q.1.45. What is the use of analog-to-digital conversion in a sensor?
Q.1.46. How many sensors can connect to a BS, and what are the physical considerations?
Q.1.47. What are the roles of sensing range and communication range?
Q.1.48. Can a sensor network operate indoor, and what are the underlying limitations?
Q.1.49. Can there be interference in a sensor network?
Q.1.50. What mechanism can be used to program a sensor node?
Q.1.51. What are the differences between GPS, wireless cellular-assisted GPS, and indoor location systems?
Q.1.52. Why do you get errors in determining location by GPS?

References

1. http://www.dailymail.co.uk/sciencetech/article-2449632/How-check-phone-The-average-person-does-110-times-DAY-6-seconds-evening.html.
2. https://books.google.co.in/books?id=maUQAgAAQBAJ&pg=PA6&lpg=PA6&dq=EM+signal+attenuation+in+outer+space?&source=bl&ots=c2nBF7gLfE&sig=MHZGkOGj7NEXw0rQ9ITspHbHZE&hl=en&sa=X&ved=0ahUKEwj6iK7Mv8zJAhWNHI4KHa-nAToQ6AEIQjAH#v=onepage&q=EM%20signal%20attenuation%20in%20outer%20space%3F&f=false.
3. D. P. Agrawal and Q. A. Zeng, Introduction to Wireless and Mobile Systems, 4th Edition, Cengage Learning, 650 pages, 2016.
4. Rachel Lieberman, Brandel France de Bravo, and Diana Zuckerman, "Can Cell Phones Harm Our Health?," National Research Center for Women and Families, August 2008, Retrieved August 13, 2013.
5. David Strayer, Frank Drews, and Dennis Crouch, "Fatal Distraction? A Comparison Of The Cell-Phone Driver And The Drunk Driver," University of Utah Department of Psychology, 2003, Retrieved 2009-06-27.
6. https://en.wikipedia.org/wiki/Cellular_network.
7. http://www.pearsonhighered.com/samplechapter/0130422320.pdf.
8. www.google.co.in/patents/US7706399.
9. http://ieeexplore.ieee.org/xpl/articleDetails.jsp?reload=true&arnumber=6268958.
10. http://www.nasa.gov/topics/nasalife/thermometer_pill_prt.htm.
11. https://en.wikipedia.org/wiki/Capsule_endoscopy.
12. http://www.cdd.com.au/pages/procedures/capsule_endoscopy.html.

Chapter 2
Applications of Sensor Networks

2.1 Introduction

A wireless sensor network (WSN) consisting of a large number of randomly deployed sensor nodes (SNs) was initially introduced for intrusion detection for tactical applications [1], and a scheme determining type and number of talks in a battlefield is shown in Fig. 2.1. Since then, the use of WSNs has increased dramatically in many diverse disciplines and the list is growing every moment. SNs have become a central focus of daily routine, and as indicated in Fig. 2.1, sensing bridges the gap between an embedded system and networked collaborative activities. Thus, sensors are finding applications in areas people never thought about and multitude of benefits offered by sensors in different disciplines are summarized in Table 2.1 [2]. As is clear from the table, all aspects of daily life are being directly or indirectly affected by sensor devices and the influence is anticipated to grow in the near future.

Each SN consists of many different functional units, and main constituent components are illustrated in Fig. 2.2 (same as Fig. 1.48a). Transducer basically converts energy from one form to electrical voltage and current and is primarily responsible for measuring and representing physical parameter in the surrounding area. Thus, the type of transducer needed depends on the application as input signal type to a transducer could vary. The output of a transducer is in analog voltage or current form and needs to be converted by A/D converter in digital format. This signal is fed and processed by an embedded processor that could temporarily store the result in a local memory. The size of processor is kept small in order to keep the cost and power consumption to acceptable level. The heart of a SN is a wireless transmitter/receiver which can either transmit or receive data wireless at any desired time. It is interesting to note that all these units get power from batteries. A commercial version of a SN is called mote [3] and it has 4 different built-in transducers to measure sound, light, humidity, and temperature level in the surrounding area. Different amount of energy is consumed in different functional units and is shown in Table 2.2, and a lot more energy is consumed in transmission/

D.P. Agrawal, *Embedded Sensor Systems*,
DOI 10.1007/978-981-10-3038-3_2

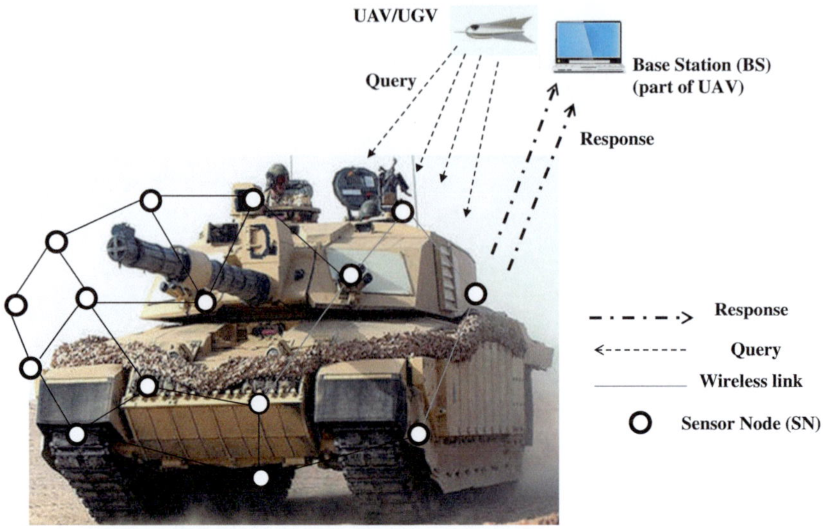

Fig. 2.1 A WSN using a large number of SNs

Table 2.1 WSNs bring multitude of benefits to our daily lives [2, 4]

Application	Benefit
Measure microclimates on farms	Increase crop yield per square km
Monitor traffic on road systems	Steer traffic away from jams, accidents, and construction zones; alert emergency services
Detect human presence in homes and offices	Reduce wasted power in HVAC and lighting
Electrical/gas/water metering	Optimize utility distribution systems and reduce inefficiencies

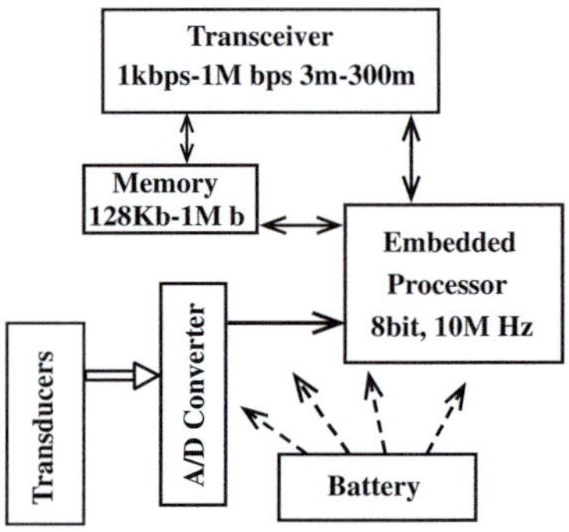

Fig. 2.2 Functional components of a SN device (Fig. 1.48a)

Table 2.2 Energy consumption in MICA2 motes' SN

Component	Current	Power
Nothing test	7.5 μA	0.02325 μW
Radio off	4 μA	0.0124 μW
Radio idle	16 mA	0.0496 mW
Radio receiver	16 mA	0.0496 mW
Radio transmit	21 mA	0.0651 mW
Computation only	10 μA	0.0310 μW
Transmit cost	–	1 μJ/bit

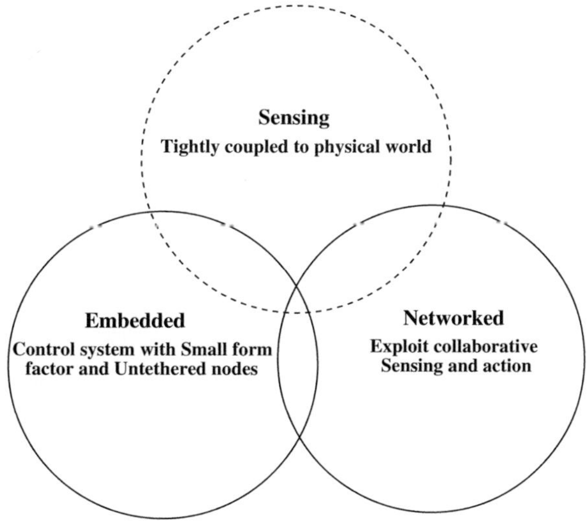

Fig. 2.3 Intersection of sensing, network, and embedded system

reception as compared to processing within a SN. It is also important to keep SNs in sleep mode in order to conserve energy consumption. As massive amount of SNs are to be deployed for a given application, many different aspects have become critical to usefulness and correct functioning. In some areas, digital cameras are used as sensing devices, utilizing them as SNs (Fig. 2.3).

2.2 Applications of WSNs

Numerous applications of WSNs have emerged, and a broad classification has been given in Table 2.1. It is rather hard to organize them in a systematic way, and some overlap is unavoidable. Figure 2.4 shows many areas of applications. However, they do not represent any chronological progression of development nor a complete list but a comprehensive classification of different areas. From functional point of view, WSNs can be divided into two complimentary steps: one is for collection of data from SNs (Fig. 2.5) and another is for dissemination of data to selected

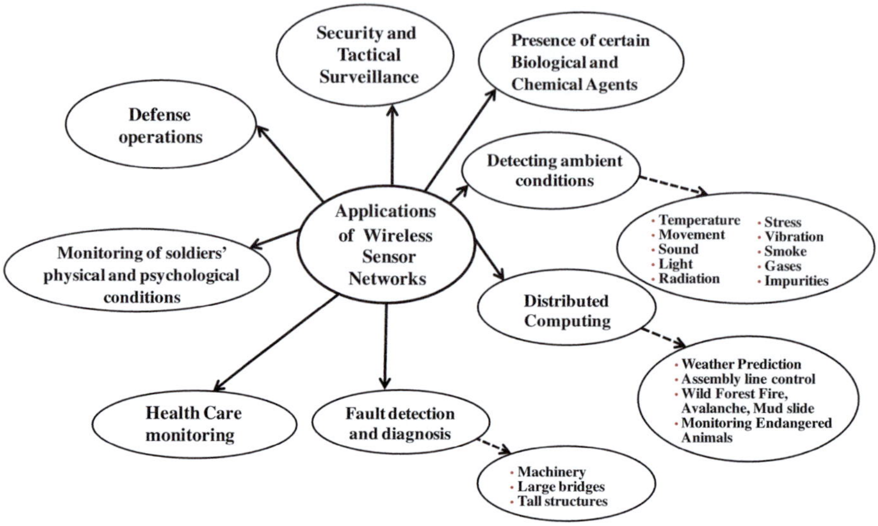

Fig. 2.4 Applications of WSNs

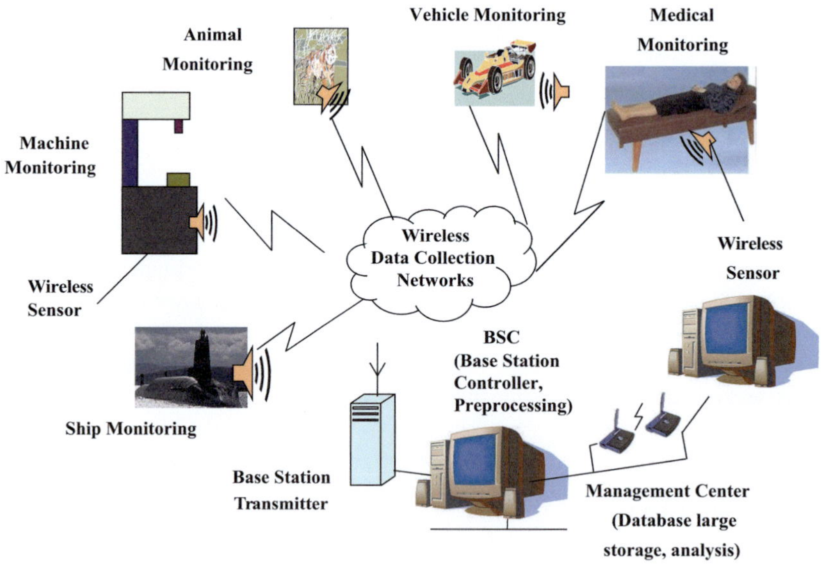

Fig. 2.5 Data acquisition using a WSN [5]

systems for appropriate action, primarily useful as an actuator (Fig. 2.6). The first step requires appropriate location of SNs, selection of appropriate and adequate data rate, harmonization between them for data transfer including SN clock synchronization, coordinated sleep–awake cycle sequence, aggregation of voluminous

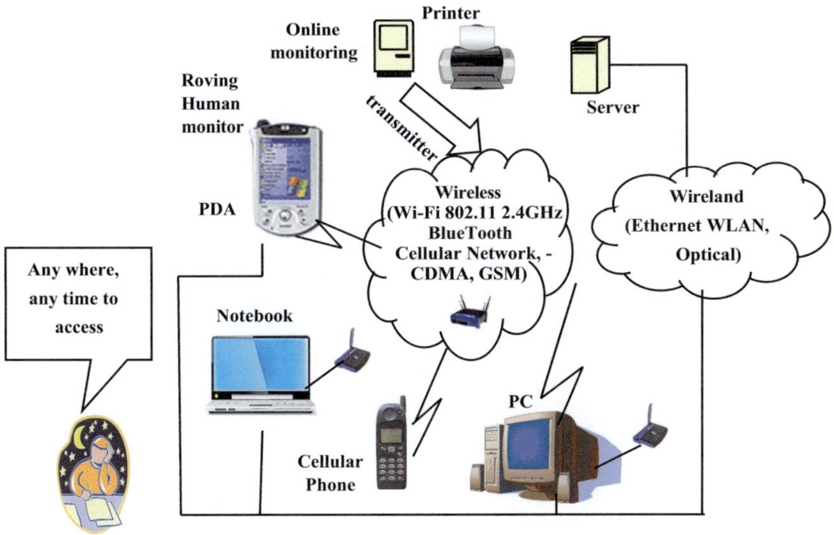

Fig. 2.6 Data distribution for a WSN

collected data, and delivery of data through storage for BS. The second step involves delivering desired collected data to appropriate devices and systems, including associated actuators.

2.2.1 Defense Applications of WSNs

WSNs were introduced for defense purpose, and we start with those types of applications. The idea here is to deploy SNs from low-flying airplanes or drones, and when SNs land on surface of land, they collect information from the surrounding area of war zone and send data to a powerful base station (BS) or sink node located inside the plane. Data are gathered and analyzed by the BS and determine strategic information such as type and number of tanks in the battlefield, number of soldiers, elevation of terrain, and types of hiding places such as bunkers. Such information is useful purely from defense perspectives and is shown in Fig. 2.7.

A more complex tactical system involves wireless communication between tanks and fighter planes and is shown in Fig. 2.8. WSN is also useful in detecting potential intrusion in a given area and is given in Fig. 2.9. A soldier in a battlefield is equipped with many different types of sensors and some of them are illustrated in Fig. 2.10. A closer look at night vision sensor is shown in Fig. 2.11. SNs are needed to protect soldier's face and thermal imaging, and night vision schemes are used to detect activities in dark (Figs. 2.12 and 2.13). Soldiers also have to face biological and chemical fighting, and SNs to detect and protect from such warfares are shown in

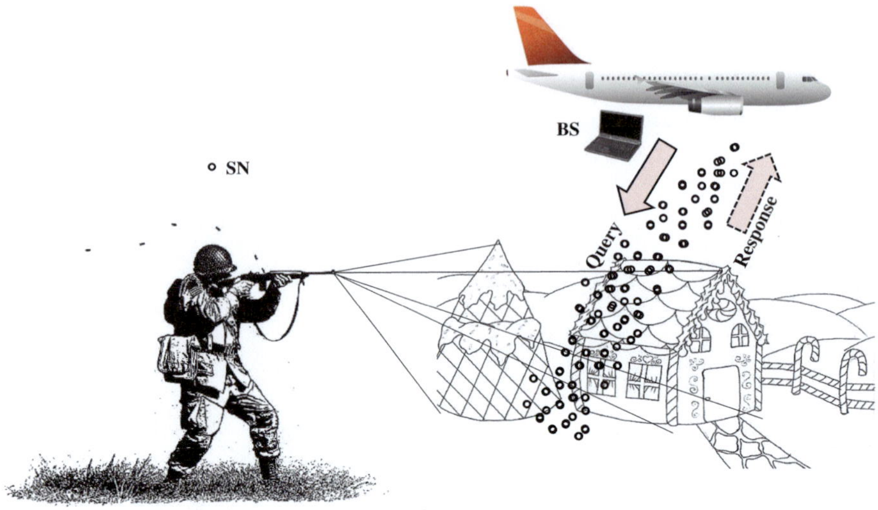

Fig. 2.7 Use of a WSN for defense application

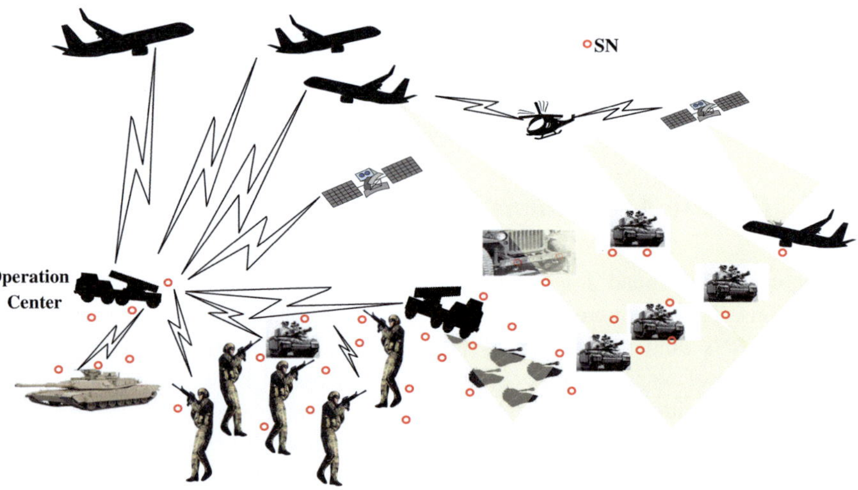

Fig. 2.8 Another defense application of WSN

Fig. 2.13. Another defense-oriented application that attracted a lot of attention is unmanned securing of a building by placing SNs on robots that move around the area and monitoring associated activities as shown in Figs. 2.14 and 2.15.

Another defense-based application involves detecting land mines (Fig. 2.16) which is now performed in a total manual way. Land mines are responsible for 73,576 casualties in 119 countries/areas in the past 10 years from 1999 to 2009 [11, 12]. Another peripheral area of interest to defense is to monitor national boarder for

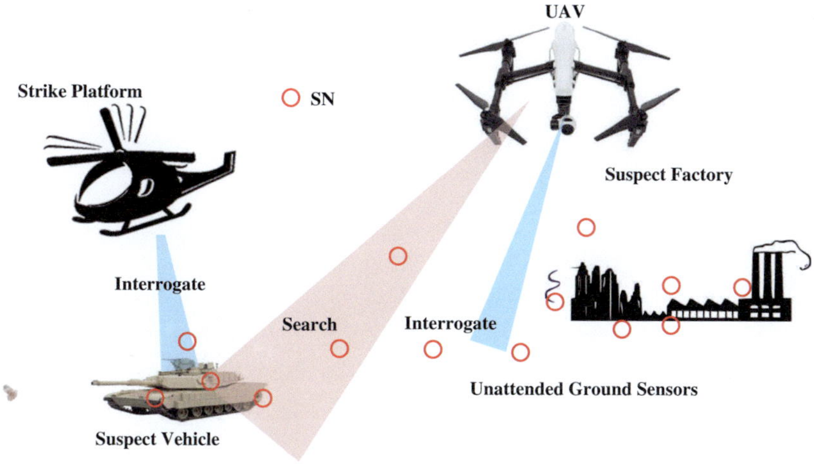

Fig. 2.9 Application of WSNs for defense against invasion

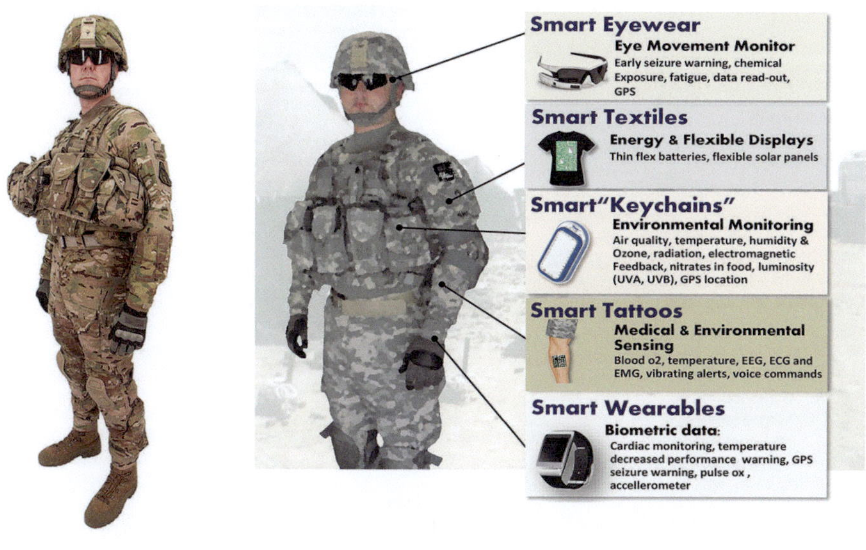

Fig. 2.10 **a** A soldier equipped with different SNs [6] and **b** closer look at a soldier with different SNs [7]

illegal trespassers that cross the countrywide border illegally [13] as indicated in Fig. 2.17. The idea is to deploy SNs around US–Mexico border that could detect illegal crossing of people by using heat and motion SNs that can send these parameters. Based on collected data, the BS can determine if there is substantial change in the measured temperature or increased mobilityin a short span of area.

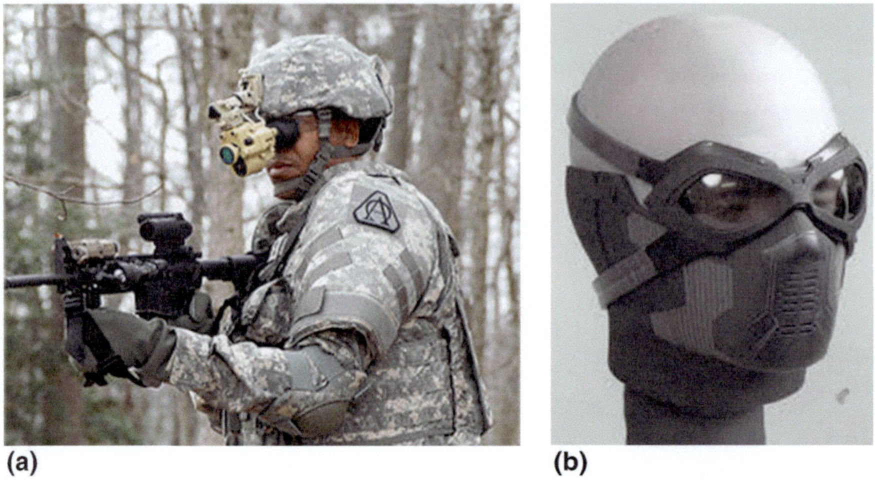

Fig. 2.11 **a** SNs used to protect soldier's face and **b** a soldier equipped with night vision SNs

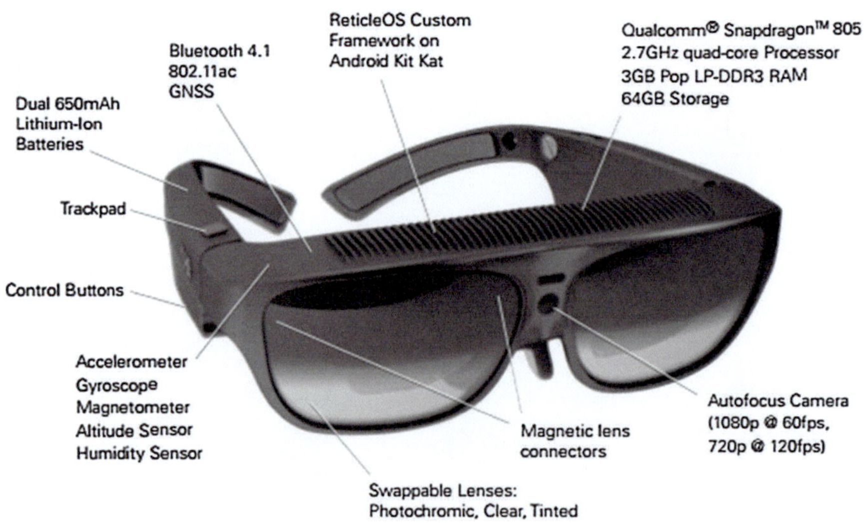

Fig. 2.12 **a** Night vision goggle [8] and **b** SNs used for night vision

That is how SNs help in determining unwanted activities as SNs do not have eyes and can only measure physical parameters from the surrounding area. But, human beings are much smarter, and as long as people do not come within sensing range of SNs, they can easily cross the border without being noticed by SNs. To keep undetected by SNs, many possible alternatives are possible such as digging tunnel under earth surface and following a path that cannot be detected by SNs.

(a)

(b)

Fig. 2.13 **a** SNs for thermal image and **b** SNs for another thermal image [9]

Another area of peripheral interest to defense department is the identification of accurate area in wild forest fires that occur frequently in USA and other parts of the world. Images taken from a satellite and an airplane provide an estimate aerial photograph of the territory as the space is commonly masked and occluded by smoke and fire waste. It is important to determine the location of fire initiation by measuring temperature distribution of the fire area. This is feasible by deploying SNs using low-flying airplanes or drones in the fire area, and once SNs land in the fire area, they send the temperature value to BS located at the plane or drone. Once data are collected at the BS from SNs, temperature distribution can indicate values and hence the origin of fire area can initiate the rescue operation from that area. This is illustrated in Fig. 2.18, with adequate number of SNs deployed as each SN has coverage area limited by its sensing range.

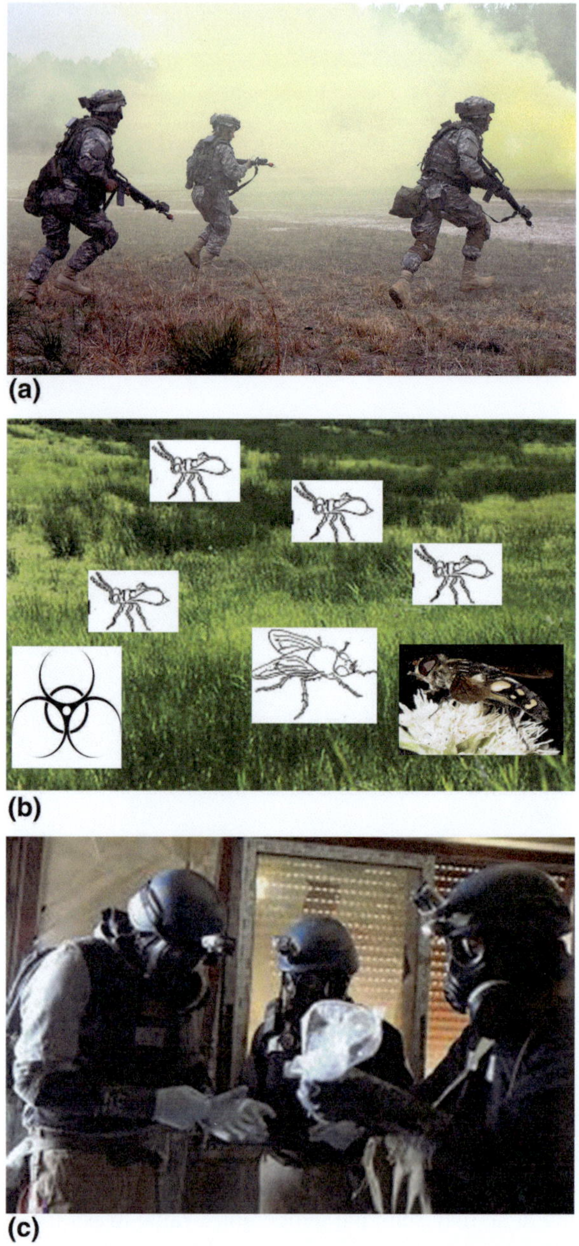

Fig. 2.14 **a** SNs for biological agents, **b** SNs for detecting biological agents and **c** SNs for chemical agents [10]

Fig. 2.15 Securing a building with SNs on robots

Fig. 2.16 SNs for detecting land mines

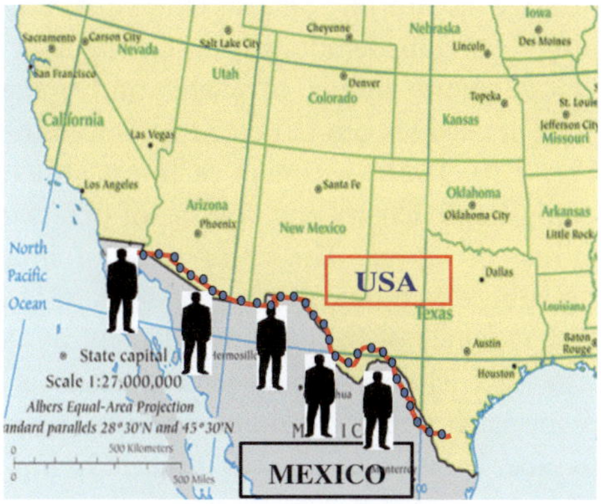

Fig. 2.17 SNs to detect illegal crossing

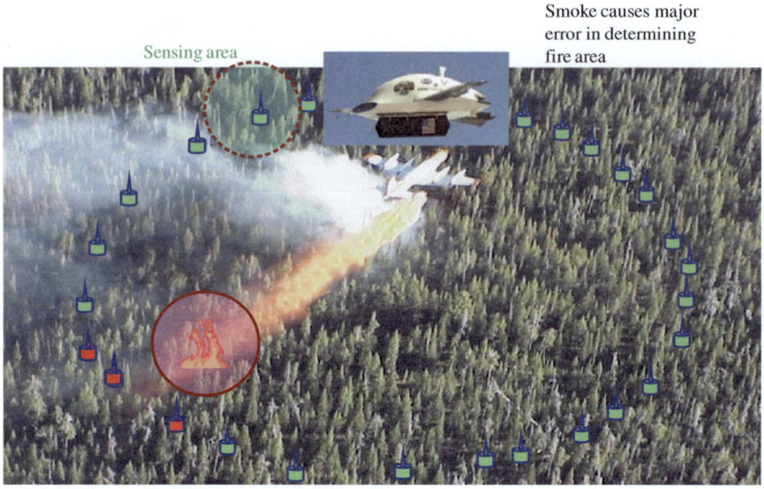

Fig. 2.18 Wild forest fire accurate area mapping

2.3 Civilian Applications

Many civilian applications have been suggested for WSNs. These can be divided into four main categories as shown in Fig. 2.19. These are discussed in the following paragraphs.

2.3.1 Weather Monitoring Applications

Weather monitoring is an important area of interest for our daily life and can be predicted by detecting numerous atmospheric weather-related parameters such as temperature, amount of rainwater, wind velocity, air pressure, wind velocity, natural disaster monitoring, and on land snow coverage and urban heat effects. One of the important considerations is collecting data related to the forests and their health can predict future falling of leaves and update harvesting information. High-density pixel intensities are collected to determine this phenomenon and are illustrated in Fig. 2.20. The assessment of forest structure is used to determine the forest conditions, and density and volume of pixels indicate past and future health in terms of falling of leaves. The color code of images could very well indicate such surrounding situations. Further analysis of forests is needed to monitor living animals, and activities of habitats are checked by determining landscape characteristics and structure. The landscape pattern such as structural components, patch geometry (e.g., patch size and amount of edge versus core), and spatial context of adjacency to other habitats are used to indicate living area for a given animal [15]. A patch is a closed area that differs from adjacent areas in terms of at least one attribute. Any

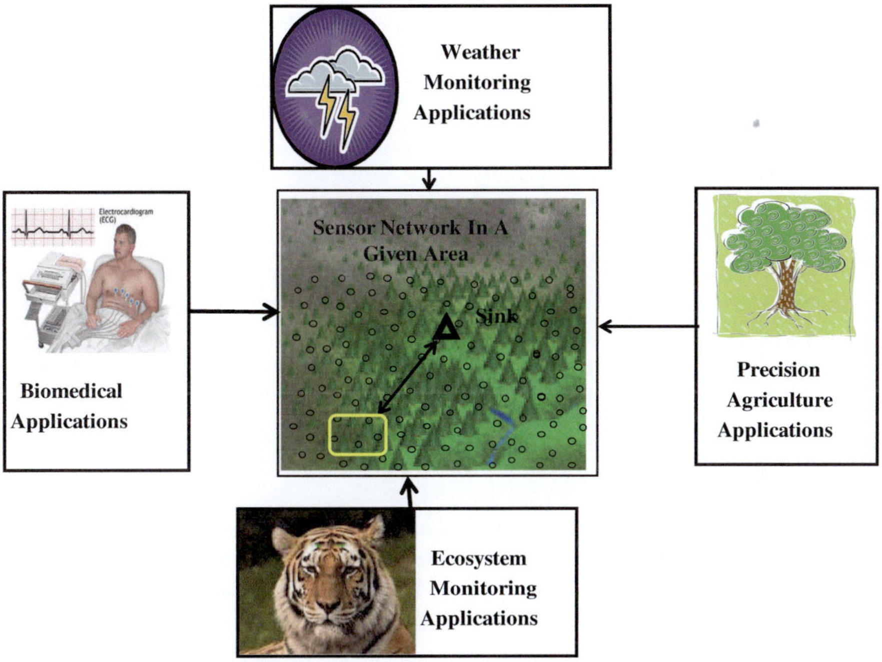

Fig. 2.19 Categorization of WSNs' civilian applications

Fig. 2.20 River in a forest area [14]

heterogeneity within a patch is usually neglected. The main attributes of a landscape are the presence of each class, the number of different patch types, membership concentration of different patch types, patch size distribution density, patch shape complexity, relative difference among patch types, spatial aggregation of patch types, etc. In this way, scaling is important in defining a landscape as it is directly related to how it is used by a species.

A qualified assessment can be made about water in the rivers and drainage basin in the surrounding areas. That also indicates the quality of life that indirectly controls the atmospheric conditions. In addition, microsensors can analyze data corresponding to air, water, and soil mixture on real-time basis and determine pollution level from collected data.

The quality of groundwater depends on how it moves through the unsaturated soil and finally to water area. Water is drawn by roots of plants, and many dissolved ingredients enter the plants in this way. SNs are designed to withstand harsh environment for a long period of time and are capable of performing real-time analysis of signals. Environmental hydrodynamics is illustrated in Fig. 2.21 and largely depends on sediments from the rivers and does affect the quality of rainwater [15]. Phosphorus content in rainwater also comes from contamination, and the environment is accordingly affected. Pesticides are also transported by the plants through their roots. Accurate boundary between forests and habitat living area can be easily determined using quality images. The ecological process is influenced by fertilization and draught. Archeological findings (Fig. 2.22) are also performed by satellite images, and exposed structures are analyzed for fault lines, identifying rocks, texture, pattern, and tone determination, fortitude of obscured structures, and the detection of hydrocarbons and oils and gases.

Monitoring of volcanos [16, 17] is also an important function that can be effectively achieved using multiple SNs and is illustrated in Fig. 2.23. The application is such that continuous transfer of data is not needed while it has to be triggered based on occurring events. This is also critical as a large area needs to be covered for monitoring the areas. Because of large area, a large number of sensors are needed and they need to be tightly synchronized.

National Science Foundation has been supporting highways research [18] to minimize delays in highways. The main idea is to explore the use of highway sensors networks for safety so that a warning signal can be generated to alert the driver of a possible danger in the forward direction. This is done by placing sensors at fixed locations on the highway for data collection from the cars. By measuring the distance between adjacent cars, the event information is forwarded about any accident or seriousness of an event such as traffic jam, occurrence of fog intensity, and duration. Another set of sensors are placed on each auto for receiving signals from the highway sensors and providing location information through GPS capability. A similar arrangement can be used to detect congested parts of a city, especially downtown or main shopping areas.

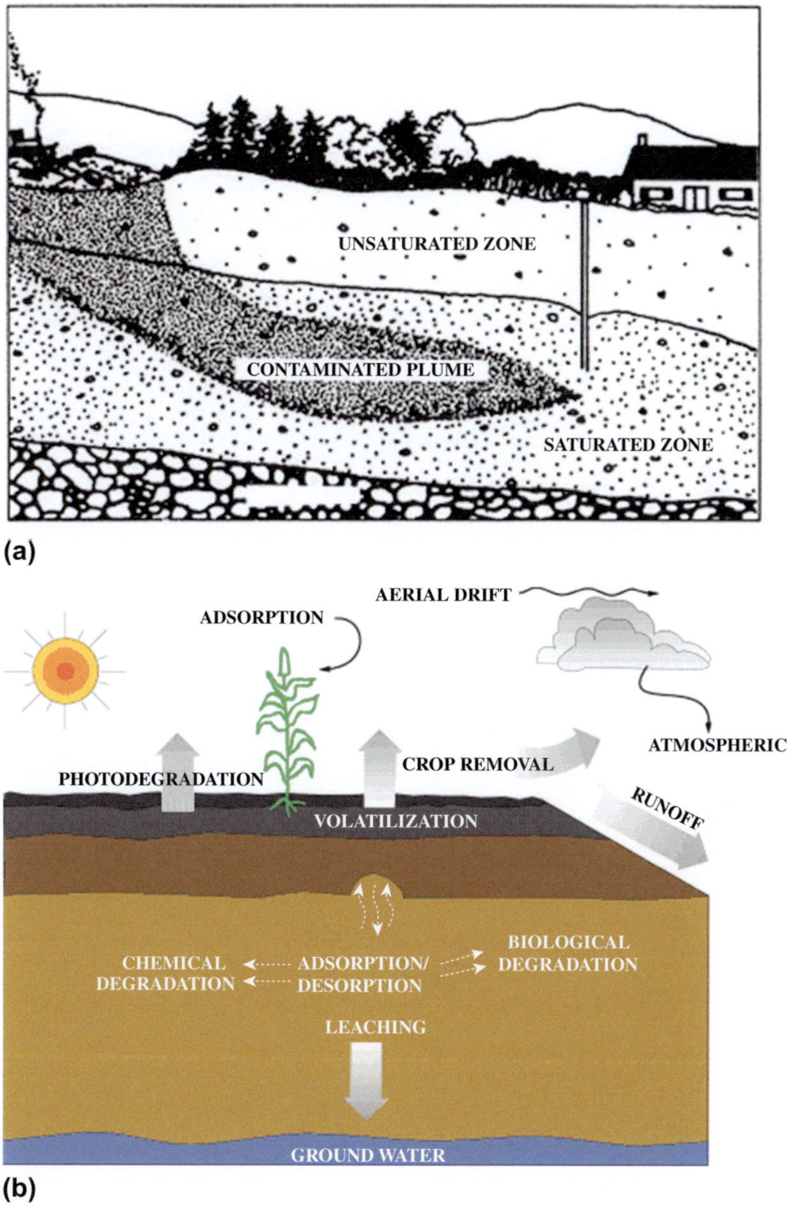

Fig. 2.21 **a** River in a forest area and **b** river in a forest area: another view

Fig. 2.22 **a** Archaeological images and **b** archaeological images: another view

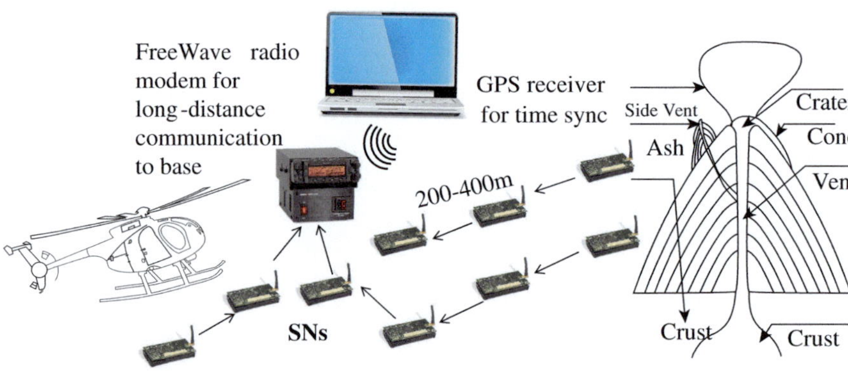

Fig. 2.23 Monitoring of volcanic activity using SNs

2.3.2 Precision Agriculture Applications

Precision agriculture has been a very hot topic and has been used for growing good-quality grains. A crop type is identified using spectral characteristics, image texture, and the knowledge of crop development over time. The crop condition is indicated by the health and vigor of the crop, detection of drought, pests, flooding, and disease. The most common remote sensing tool is a normalized difference vegetation index (NDVI). DARPA was the first to support assisting recovery of rare and endangered plant species by a comprehensive environmental measurement using a WSN. Each SN unit contains a computer and a wireless transceiver, with environmental sensors such as thermostat for temperature and photoresistor for light sensing while flexible piezoelectric strips are also employed for sensing the wind, relative humidity sensors, and some units with a high-resolution digital camera. Rechargeable batteries are employed in conjunction with thermoelectric unit for auto-charging. The weather data are collected and stored every ten minutes while high-resolution images are taken once an hour. Some of the SNs are also linked to the Internet to make it very useful in space exploration.

One noted example is that precision agriculture is growing grapes by measuring humidity and temperature at the root of grape trees (Fig. 2.24) and accordingly controlling amount and frequency of water irrigation. This is illustrated in Fig. 2.24b, c by placing SNs at the root of one tree selected every 4 blocks. Each SN senses three parameters such as light illumination, soil temperature, and soil moisture at different depths from 6″ and 14′. The soil moisture is obtained by capacitance sensors and is not affected in extreme dry conditions. The data collected also include the exact location of the SNs, and critical vineyard and wine grape information. The vineyard operations manager can view the data on any

(a)

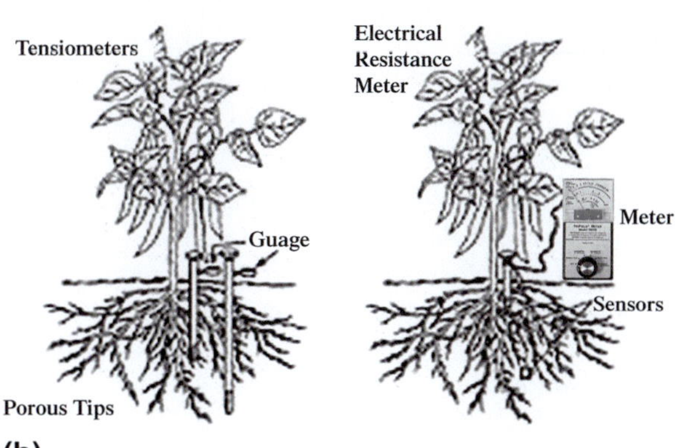

(b)

Fig. 2.24 **a** Grape trees field, **b** monitoring of grape trees using SNs and **c** placement of SN to monitor grape tree

Fig. 2.25 Monitoring of impurities in excessive drainage using SNs

Web-enabled cellular phone or PC and can also set the threshold values over Internet or via e-mail or SMS. These parameters are used to control pressure SNs at irrigation manifold, prefilter, and postfilter so as to control of water to be supplied to grape trees to grow better quality grapes. The grape project has been adopted not only in Napa Valley, CA, but also in Australia and many parts of the world.

It is interesting to note that when it rains heavily, the drainage system is commonly inadequate to handle a large volume of water, and overflow is diverted to lakes used for drinking water. In this way, unprocessed and dirty water is mixed in the creek from where drinking water is supplied after appropriate treatment. This may be hazardous to human health, and impurities can be checked if appropriate SNs are deployed to check impurity level as shown in Fig. 2.25. This could be very helpful in maintaining quality of drinking water supply.

2.3.3 *Echo System Monitoring Applications*

Echo system has become an ideal application for monitoring endangered animals and birds. The SN size can vary from small to large size and can be placed close to each other (1 cm) or far away from each other (up to 100 m). Data from SNs can be collected every ms or once in few days. If you deploy too many SNs, the amount of data collected at BS also grows with SNs and it may be desirable to process and compress data from SNs using signal processing algorithms. Data from SNs to BS are propagated through open airspace, and signal attenuation depends on the terrain and vegetation also affects the signal quality. Therefore, it is desirable to achieve reliable transfer of information in a dynamic WSN unattended for a long time in a harsh environment. To achieve these, high-resolution SNs could be deployed as scare resources.

As illustrated in Fig. 2.26, the idea is to place SNs in the neighborhood of species or embed inside their body to monitor their activities. Here, SNs are placed

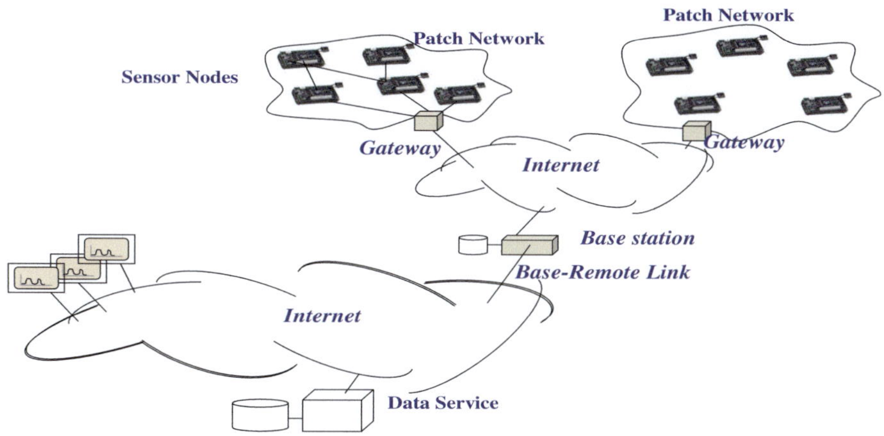

Fig. 2.26 Monitoring the behavior of storm petrel in Grand Duck Island

in the area to be sensed grouped into sensor patches to transmit sensed data to a gateway which is responsible for forwarding the information from the SNs patch to a remote BS through a local transit network. The BS logs the data and replicates the data every 15 min to a database located in Berkeley using a satellite link. Remote users can access the replicated database server in Berkeley, while local users make use of a small PDA-sized device to perform local interactions such as adjusting the sampling rates and power management parameters. Preprocessing is done to minimize data for habitat monitoring applications. A 2-tier network architecture is suggested that consists of micro-nodes and macro-nodes, wherein the micro-nodes perform local filtering and data to significantly reduce the amount of data transmitted to macro-nodes. In August 2002, researchers from the University of California at Berkeley (UCB) and Intel Research Laboratory deployed a mote-based tiered WSN in Great Duck Island (GDI), Maine, to monitor the behavior of storm petrel and similar schemes can be adopted for monitoring other animals in a given area of interest.

A similar approach has been utilized in UMass for monitoring activities of turtles by placing SNs on top of turtles and monitoring temperature and wetness over period of days as given in Fig. 2.27. Another project using SNs in monitoring behavior of animals under different environments by placing sensors at the collar of the zebra. This ecology project [19] investigates the migration of animals (Fig. 2.28) on the long-range basis, noting their interspecies interactions and determining their nightly behavior. There are many projects dealing with other animals.

A project at University of Hawaii at Manoa [20] looked at the issue of why endangered species of plants will grow in one area but not in neighboring areas. Inconspicuous SNs, each consisting of a computer, radio transceiver, and environmental sensors, sometimes including a high-resolution digital camera, were deployed in the Hawaii Volcanoes National Park. The sensed data were relayed back to the Internet with Bluetooth and 802.11b for delivering data packets through

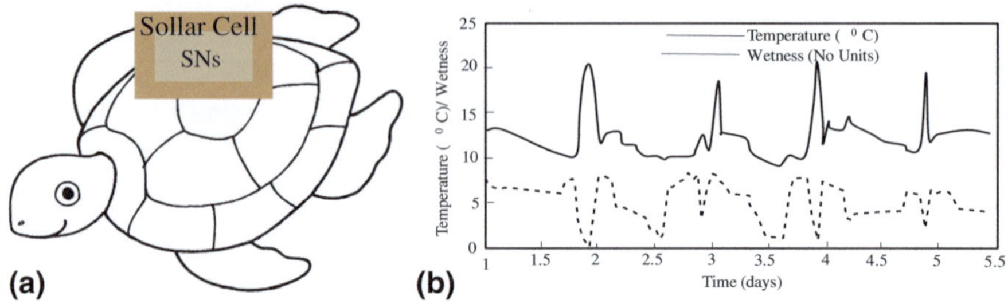

Fig. 2.27 **a** TurtleNet with SNs and **b** parameters of TurtleNet with SNs

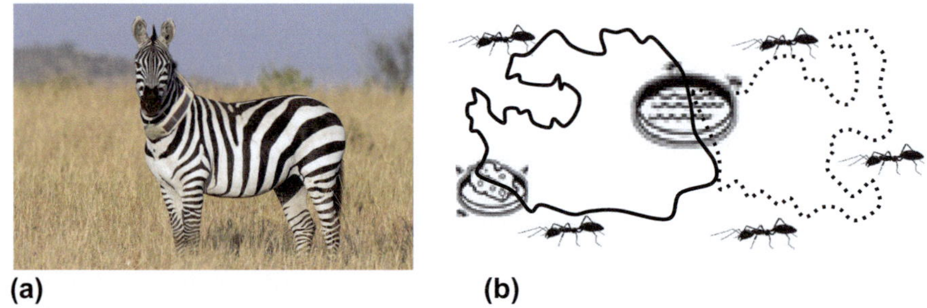

Fig. 2.28 **a** ZebraNet with SNs at Princeton University and **b** simulation of ants' food search process forming two trails

the IP. The placement strategy for the sensor nodes is then investigated, and various topologies of 1-dimensional and 2-dimensional regions such as triangle tile, square tile, hexagon tile, ring, star, and linear are explored. The sensor placement strategy evaluation is based on three goals: resilience to single point of failure, the area of interest to be covered by at least one sensor, and minimum number of nodes. Finally, the choice of placement of SNs depends on sensing and communication ranges and data are being interpreted in the near future.

Detecting structural faults in high-rise buildings in downtown area is a challenging question as interaction between ground motions and structure/foundation response is not well understood [20]. Current seismic network is not spatially dense enough to monitor structure deformation in response to ground motion. In understanding response of buildings and underlying soil to ground shaking, models are developed to predict structure response for earthquake scenarios. It is critical to identify seismic events that cause significant structure shaking, and processing of waveforms is to be done locally that requires a dense structure monitoring system. This work provides field data at sufficient densities to develop predictive models of structure, foundation, soil response, etc. A 17-story steel frame building has been instrumented with 100-node grid of seismometers at 100 m spacing across the UCLA campus and surrounding the building. Projects have explored the use of sensors in monitoring the health of buildings, bridges, and highways [21].

A Bluetooth-based scatternet has been proposed to monitor stress, vibration, temperature, humidity etc., in civil infrastructures. Simulation results are given to justify the effectiveness of their solution by having a set of rectangular Bluetooth-equipped sensor grids to model a portion of bridge span. Fiber optic-based sensors have been proposed for monitoring crack openings in concrete bridge decks of strain and corrosion of the reinforcement in concrete structures. Corrosion of steel bars is measured by using special super glue and angular strain sensors.

Smart small size SNs, equipped with a wireless communication interface, some sensors and an autonomous power supply constitute an embedded WSN has been proposed [22] that can be envisioned in the context of a smart house. The devices interact with other devices inside the house. It is observed that temporal and spatial validity intervals of sensed data are dependent on the validity interval of the data used in the application. The data from various SNs need to be conditioned to get metadata to be used for an event understanding. Another important project dealing with the use of WSNs is smart transportation [23] where poor signal-to-noise ratio is influenced by the traffic, construction, and explosions. Usually, insufficient data are available for large earthquakes, and structure response must be extrapolated from small and moderate size earthquakes.

2.3.4 Biomedical Applications

These are the areas that directly affects human life and for new applications are growing at an unprecedented rate. An initial application appeared in the first edition of the book back in 2000 [24] that suggested use of wireless technology when a patient is being transported to a close-by hospital. The idea is to support the medical personnel with ambulance with needed advice on the treatment of the patient based on ECG and other vital signals besides body temperature and blood pressure by wireless communication. The SNs play a vital role in generating ECG and other signals, and since then, many other biomedical applications have appeared.

Wireless technology facilitates the mobility of doctors, practitioners, and caregivers, and WSN is an important constituent of a wireless system. WSN allows access to patient information at any moment, everywhere, and on real-time basis. Wireless technology improves automatic data gathering through barcode or RFID reading and allows an immediate sharing of patient information and thereby improves the internal communication within the caregiver team and the support staff. This also helps in reducing paper work. A generic health monitoring involves checking of glucose level in blood, heart rate, and the detection of potential cancer, checking for chronic diseases, and a need for artificial retina, and cochlear implants used to provides a sense of sound to a person who is profoundly deaf or severely hard of hearing. Sensors in a hospital involve monitoring of vital signs and record anomalies.

Tele-monitoring involves monitoring of human physiological data, tracking and monitoring of doctors and patients inside a hospital, drug administrator in hospitals.

Sensors for physiological conditions are designed for personal health and general environmental monitoring, and temperature, pressure, humidity, and vibration/position are measured. This can be packaged in a wrist strap to make it a wearable system. Many versions of micro-cluster have been used in variety of military, navy, and marine corporation applications. There are many transducers (sensors) available off-the-shelf such as inertial and gyro systems by companies such as HoneyWell Sensing and Control, Analog Devices, MicroStrain, Memsense, Crossbow, and Empfasis. The structural monitoring systems include vibration sensors supplied by Columbia Research labs, Colibrys, Electro-sensors Inc., and Instrumented Sensor Technology. The accelerometers can be obtained from Tronics Microsystems and STMicroelectronics.

The first application of WSN is a project at USC [25] that utilizes a group of SNs to collect information about skin pigment that is used to determine oxygen intake and white cell count in the blood. This is useful during golden hour of a newborn baby when no one is allowed to insert any needle in baby's body (Fig. 2.29).

Another important invention is the creation of camera pill shown in Fig. 2.30a that could travel through body taking pictures for 6 h, helping diagnose a problem which doctor previously would have found only through surgery. The pill takes images as it is propelled forward by peristalsis. A wireless recorder, worn on a belt, receives the images transmitted by the pill. Movement of the capsule through digestive system is shown in Fig. 2.30b. Through the picture obtained, it is possible to detect Crohn's disease, malabsorption disorders, tumors of the small intestine, vascular disorders, and ulcerative colitis. The main advantage [26] of the camera pill is that it does not require trained staff and the process is relatively quick as no sedation is needed. On a similar line, a capsule has been developed by Philips [27] that could measure body temperature accurately and transmit live data. The pill is small enough to be swallowed with a glass of water, takes 6 h to pass through the body, and transmits temperature to a close by smart phone attached to the belt. This has extensively been used by astronauts and fire fighters.

Fig. 2.29 White cell count and oxygen level determination using SNs

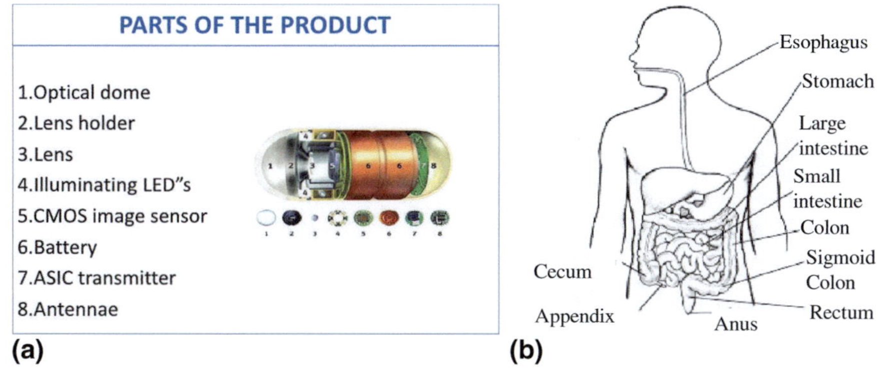

Fig. 2.30 **a** Camera pill or capsule and **b** movement of capsule through the digestive system

In this respect, body area networks (BAN) or wireless body area sensor networks (WBASN) [28] are enabling use of WSNs in many areas of monitoring human health. As a wearable computing, a WBASN has also been suggested to remotely monitor the progress of a physical therapy done at home and an initial prototype has been developed using electroluminescent strips indicating the range of human body's motion [29]. An indoor/outdoor wearable navigation system has been suggested for blind and visually impaired people through vocal interfaces about surrounding environment and changing the mode from indoor to outdoor and vice versa using simple vocal command [30]. A differential GPS receiver has been used to provide accurate location information outdoor while ultrasound position devices are used for indoor coverage. A wearable sensor network that finds environmental information and controls home electric appliances is by a Bluetooth-based network or a large scatternet network. The data throughput and communication delay have also been measured, and battery life is also observed. A thin multi-resolution flat SNs that adopt resolution based on the regions of interest or the information contents have been proposed which placed at a soldier's helmet provides entire scene simultaneously. A recent work allows the development of new vests that could optimally distribute soldiers' load, thereby alleviating discomfort and reducing the fatigue [31]. A recent US patent describes a cortisol sensor to detect level of hormone, indicating stress and fatigue levels of soldiers.

A generic personal BAN server shown in Fig. 2.31 can be implemented on an Internet-enabled PDA or a 3G mobile phone, or a regular laptop or desktop computer. It can communicate with remote upper-level services in hierarchical type architecture. It performs initialization, configuration, and synchronization of WBAN nodes, controls and monitors operation of WBAN nodes, collects readings from physiological SNs, processes and integrates data from the SNs, and securely communicates with remote healthcare provider. Specialized transducers and SNs are being developed to measure human body characterizing parameters in a non-invasive way to predict efficiently and accurately. Numerous proposals have recently been introduced in biomedical area, and the use of a micro-SNs array has

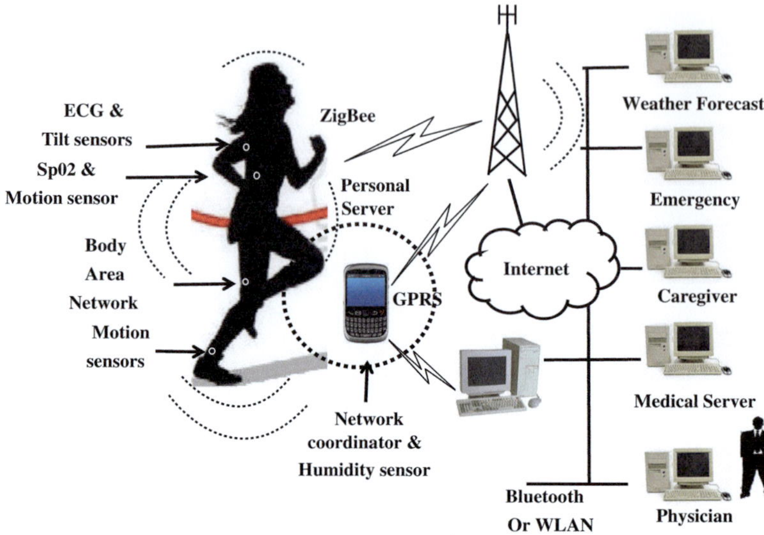

Fig. 2.31 A generic BAN or WBASN illustrated

Fig. 2.32 A generic remote monitoring

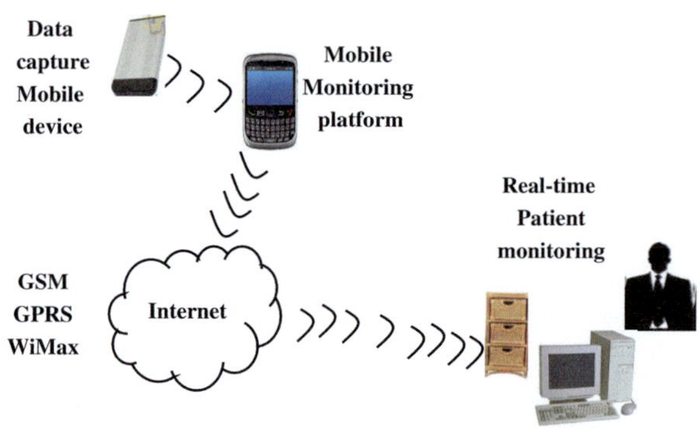

been suggested for artificial retina, glucose level monitoring, organ monitors, cancer detectors, and general health monitoring. Such remote monitoring of patients (Fig. 2.32) reduces the number of patients transferred to urban hospitals, allows tele-consultation and tele-diagnosis including the option of obtaining opinions of distant experts, facilitates the patient remote monitoring with instantaneous data transmission for analyses and follow-ups, allows remote handling of medical equipment (tele-surgery) and direct action of the expert on the patient, and improves coordination of first-responders workers during in the event of catastrophes or emergency cases. A detailed design of a wearable sensor vest has been introduced that measures, records, and transmits physical characteristics such as heart rate, temperature, and movement. This could also be very useful in assisted living of elderly and handicapped people or keeping track of endangered species.

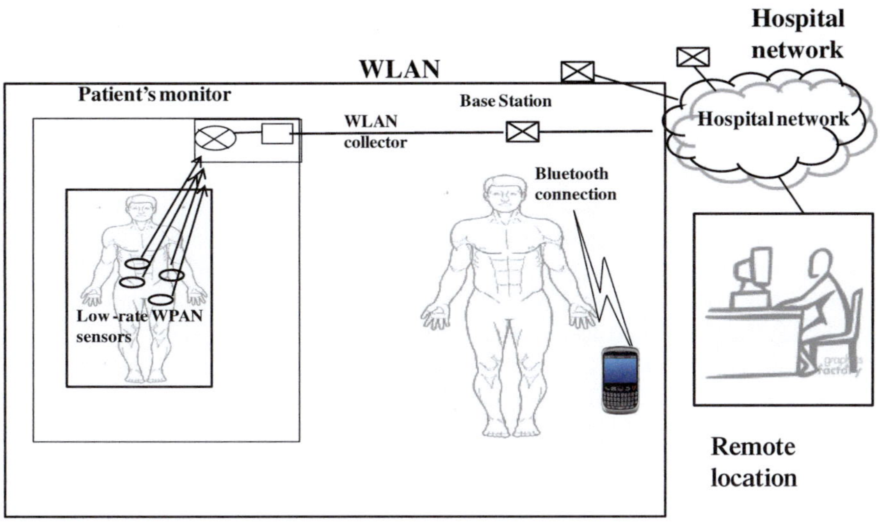

Fig. 2.33 Patient's room monitoring

The remote medical image repositories communicate through different types of network connections with the central computing site that coordinates the distributed analysis. Efforts have also been made to detect human daily life pattern by measuring physiological, behavioral, and environmental parameters using SNs such as accelerometers, audio sensors, and electrical signals and gathering data from SNs. Refining them by segmentation and integrating and finally interpreting the result as occurrence of an event detection could be useful in monitoring a given area without any human intervention.

Another example of use of WSN is to display medical information collected by SNs on the patient's body (WPAN) on a bedside monitor (Fig. 2.33) [32]. Such information is also transmitted to another location for remote monitoring, e.g., a nurses' station. In case of emergency, when the patient is moved from his/her room to the intensive care unit, these communications need to be maintained.

There is need for intercommunication among medical devices and clinical information systems. This has been accomplished with a number of medical products. Infusion pumps and ventilators commonly have RS-232 ports, and these devices can communicate with many physiological monitoring instruments. Products to link medical equipment and personal communication devices exist as well. However, virtually all of these are specialized applications—custom interfaces unique to the two devices being linked. To address the medical device plug-and-play interoperability problem, a single communication standard is needed.

Digital Imaging and Communications in Medicine (DICOM) standard [33] is created by the National Electrical Manufacturers Association (NEMA) to aid the distribution and viewing of medical images. DICOM is the most common standard for receiving scans from a hospital. A single DICOM file contains both a header (which stores information about the patient's name, the type of scan, image dimensions, etc.), and all of the image data. DICOM images can be compressed both by the common lossy JPEG compression scheme and by a lossless JPEG scheme. A single 500-slice MRI can produce a 68 MB image file and is useful in monitoring patients undergoing physical rehabilitation such as after a stroke.

The *Pluto* custom wearable designed at Harvard [34] incorporates the TI MSP430 microprocessor and ChipCon CC 2420 radio. Pluto can run continuously for almost 5 h on a rechargeable 120 mAh lithium battery. It has a Mini-B USB connector for programming and to recharge the battery. The software runs under TinyOS. Pulse oximeter [35] employs a noninvasive technology used to measure the heart rate (HR) and blood oxygen saturation (SpO_2). The technology projects infrared and near-infrared light through blood vessels near the skin. By detecting the amount of light absorbed by hemoglobin in the blood at two different wavelengths, the level of oxygen can be measured. The heart rate can also be measured since blood vessels contract and expand with the patient's pulse which affects the pattern of light absorbed over time. Computation of HR and SpO_2 from the light transmission waveforms can be performed using standard DSP algorithms.

The most common type of ECG involves the connection of several leads to a patient's chest, arms, and leg via adhesive foam pads (Fig. 2.34a). The device records a short sampling, e.g., 30s, of the heart's electric activity between different pairs of electrodes. When there is need to detect intermittent cardiac conditions, a continuous ECG measurement is used. This involves the use of a two- or three-electrode ECG to evaluate the patient's cardiac activity for an extended period. The ECG signal shown in Fig. 2.34b is small ($\sim$1 mV peak to peak). Before the signal is digitized, it has to be amplified (gain >1000) using low-noise amplifiers and filtered to remove noise. The P wave is associated with the contractions of the atria (the two chambers in the heart that receive blood from outside). The QRS is a series of waves associated with ventricular contractions (the ventricles are the two major pumping chambers in the heart). The T and U waves follow the ventricular contractions.

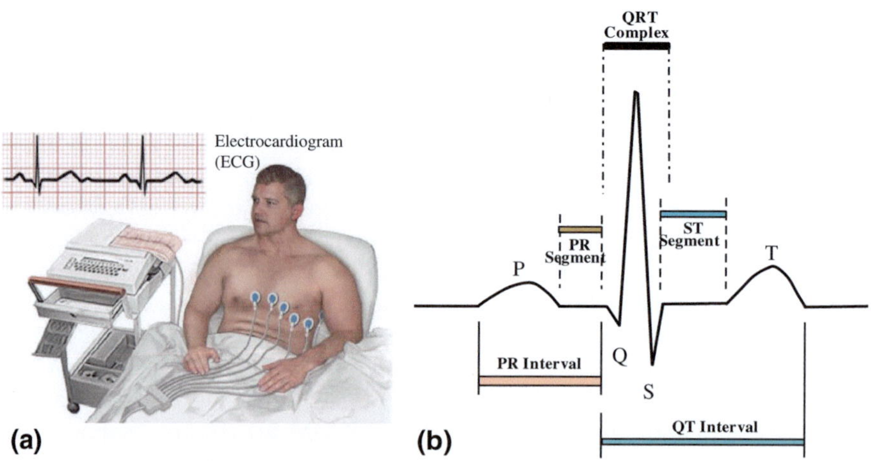

Fig. 2.34 **a** ECG test of a patient and **b** ECG pattern

IMEC [36] has recently developed a wireless, flexible, stretchable ECG patch for continuous cardiac monitoring. Placed on the arm or on the leg, the same system can be used to monitor muscle activity (EMG). The patch includes a microprocessor, a 2.4 GHz radio link, and a miniaturized rechargeable lithium-ion battery. The total size is 60×20 mm^2, and data are sampled between 250 and 1000 Hz and continuously transmitted. The battery has a capacity of 175 mAh which provides for continuous monitoring from one day to several days.

2.3.5 Other Applications

Societal-scale sensor networks can greatly improve the efficiency of energy provision chain, which consists of three components: the energy generation, distribution, and consumption infrastructure. It has been reported that 1% load reduction due to demand response can lead to a 10% reduction in wholesale prices, while a 5% load response can cut the wholesale price in half. Many other applications of WSNs are possible, and the list keeps growing. SNs are used extensively in various assembly line plants, and architecture depends on the application. Nowadays, we witness more and more electronic appliances in an average household. Therefore, great commercial opportunities exist for home automation and smart home/office environment cooling, heating, and humidity control. An example application is described where a "Smart Kindergarten" consisting of a sensor-based wireless network for early childhood education is discussed. It is envisioned that this interaction-based instruction method will soon replace the traditional stimulus-responses-based methods. In addition, emission of CO_2 from coal power plant can be minimized if efficient ways of distributing energy can be devised.

A project was undertaken at the University of Cincinnati to monitor emission of CO from vehicles around the campus [37]. The sensor boards were placed at 20′ height for protection purpose, and solar cells were used to charge the batteries. In another project, reflective markers are used to monitor human activities. Multiple cameras are employed that digitize different views of performance. An unusual application of SNs at University of Cincinnati has been to measure dancers' movements and accordingly generate music. A total of 23 SNs were used to measure movements of dancers having SNs attached to their hands [38]. Four different shows were conducted at the UC campus.

Another device that can be used as a SN is RFID and RF reader pair as they facilitate management of associated assets (wheel chairs, scanners, ambulatory equipment, etc.), improves patient localization and helps caregivers to provide services without delays, enhances the process of drug administration (identification, distribution, localization, returns and disposal), and facilitates the automatic data capture and the follow-up of blood and biological samples. Low-power mobile SNs are being developed that can move around to provide mobilityto SNs, and their future is yet to be established.

2.4 Conclusions

The WSN has been observed to be very useful in unattended monitoring of surrounding area 24 × 7 and is expected to grow at an unprecedented rate. There is no need to wait for any such confirmation as it has now become obvious. The challenge for the readers and researchers is to come up with unique applications that could prove to be overwhelming for the mankind.

2.5 Questions

Q.2.1. What are the differences and similarities between sensors and transducers?
Q.2.2. How do you select transducers for a given application?
Q.2.3. How do you determine the frequency of data to be collected by a sensor?
Q.2.4. How long you can keep the sensors in sleep mode?
Q.2.5. Can you use SNs for two different applications? Explain clearly.
Q.2.6. Can you think of two new applications of WSNs?
Q.2.7. Why are SNs appropriate for unusual applications?
Q.2.8. Why do you need synchronization among SNs?
Q.2.9. SNs can be hardwired together. Then, what is the need for having wireless connectivity among SNs?
Q.2.10. Multiple communication paths are available between SNs and BS. How do you select a path?
Q.2.11. A SN contains many functional components. Can you compare their cost?
Q.2.12. Do you include beacon signals with transceiver of each SN? Explain.

References

1. https://www.en.wikipedia.org/wiki/Wireless_sensor_network.
2. http://www.silabs.com/Support%20Documents/TechnicalDocs/evolution-of-wireless-sensor-networks.pdf.
3. http://www.xbow.com/Products.
4. https://en.wikipedia.org/wiki/Crossbow_Technology.
5. https://www.s-media-cache-ak0.pinimg.com/736x/f8/c6/9a/f8c69abf3018f5534997ec209a1c5b90.jpg.
6. https://www.pinterest.com/jmachaas/military-gear/.
7. https://www.google.co.in/imgres?imgurl=http://www.x20.org/wp-content/uploads/2014/04/thermal1-1.jpg&imgrefurl=http://www.x20.org/product/x27-clip-on-thermal-rifle-scope/&h=200&w=299&tbnid=IfZusWRzpHWOQM:&docid=1RKlWBCU7VSMiM&ei=f3CLVoiJOpCouwTv9Z_4Cw&tbm=isch&ved=0ahUKEwjIxPHqkpLKAhUQ1I4KHe_6B784ZBAzCAgoBTAF.

8. Pengda Cheng, Hongwei Zhu, Baochang Zhong, and Daozeng Wang ," Transport mechanisms of contaminants released from fine sediment in rivers," Acta Mechanica Sinica, vol. 31, no. 6, pp. 791–798, Dec. 2015.
9. http://www2.hawaii.edu/~esb/prof/pub/ijhpca02.html.
10. http://richardsonblog.dallasnews.com/2015/11/richardson-reports-two-major-wastewater-overflows.html/.
11. http://www.strategicstudiesinstitute.army.mil/pdffiles/ksil499.pdf.
12. http://www.ducks.org/hunting?id=2262&poe=RRGPPCAD&utm_source=google&utm_medium=cpc&utm_term=duck&utm_campaign=paid.
13. http://www.bobdillon33blog.com.
14. http://www.calit2.net/newsroom/article.php?id=214.
15. Mary M. Rowland Christina D. Vojta (editors), "A Technical Guide for Monitoring Wildlife Habitat," United States Department of Agriculture Forest Service Gen. Tech. Report WO-89 October 2013.
16. http://www.x20.org/product/l3-lwts-thermal-rifle-scope/.
17. http://www.slideshare.net/nidhin999/pill-camera-13267328.
18. http://www.x20.org/product/x27-clip-on-thermal-rifle-scope/.
19. http://www.princeton.edu/~equids/images/zebranet.pdf.
20. http://www2.hawaii.edu/~esb/pods/overview.html.
21. http://www.telfor.rs/telfor2002/radovi/4-19.pdf.
22. http://www.ee.ktu.lt/journal/2009/6/04 ISSN_1392-1215_Smart%20Dust%20Motes%20in%20Ubiquitous%20Computing%20Scenarios.pdf.
23. http://disalw3.epfl.ch/teaching/swarm_intelligence/ay_2006-07/exercises/SI_06-07_labhwk09_tutorial.pdf.
24. Dharma P. Agrawal and Q. A. Zeng, Introduction to Wireless and Mobile Systems, 4th edition, Cengage Learning, 650 pages, 2016.
25. https://www.marwadieducation.edu.in/UploadPdf/News/635836948227633617_TECHANZIA_ACADEMIC_YEAR_2014-2015.pdf.
26. http://www.nhs.uk/news/2013/01January/Pages/Pill-size-camera-may-make-cancer-diagnosis-easier.aspx.
27. http://www.bmedical.com.au/shop/activity-heat-research/core-body-temperature.htm.
28. https://www.en.wikipedia.org/wiki/Body_area_network.
29. http://www.instructables.com/id/PIR-Motion-Sensor-Tutorial/.
30. Nadia Kanwal, Erkan Bostanci, Keith Currie, and Adrian F. Clark, "A Navigation System for the Visually Impaired: A Fusion of Vision and Depth Sensor," Applied Bionics and Biomechanics, vol. 2015, pp. 1–16, 2015.
31. http://www.qualitydigest.com/inside/twitter-ed/pressure-mapping-body-armor-takes-load-soldiers.html.
32. http://www.ncbi.nlm.nih.gov/pmc/articles/PMC4208221/.
33. http://www.medicalconnections.co.uk/DicomObjects.
34. http://www.eecs.harvard.edu/~mdw/talks/ucsd-codebluc.pdf.
35. http://www.nxp.com/files/32bit/doc/app_note/AN4327.pdf?tid=AMdlDR.
36. http://www2.imec.be/be_en/research/wearable-health-monitoring.html.
37. Demin Wang, Dharma P. Agrawal, Wassana Toruksa, Chaichana Chaiwatpongsakorn, Mingming Lu and Tim C. Keener, "Monitoring Ambient Air Quality with Carbon Monoxide Sensor based Wireless Network," Communications of the ACM, vol. 53, no. 5, May 2010, pp. 138–141.
38. Mara Helmuth, Jung Hyun Jun, Talmai Oliveira, Vaibhav Pandit and Dharma Agrawal, "Wireless Sensor Networks and Percussion Tracking for Ocean Drum Processing," International Computer Music Conference (ICMC) 2012, Ljubljana Slovenia, September 9–15, 2012.

Chapter 3
Different Types of Transducers

3.1 Introduction

A transducer basically converts energy from one form to another, usually from many different forms such as energy, force, torque, light, motion, and position to electrical signals and is shown in Fig. 3.1. Active transducer such as a thermocouple converts temperature to voltage and do not require any external energy. However, a passive transducer is based on change in some passive electrical quantity, such as capacitance, resistance, or inductance, and requires external energy source. Transducer is an integral part of SNs and actuators in converting signals to appropriate form [1]. Many different transducers are commercially available and are summarized in references [2–4].

TransducersTransducers are expected to work linearly and consistently for a long range of values and are illustrated in Fig. 3.2. Figure 3.2a shows transformation from sound to EM waves, and Fig. 3.2b indicates translation from EM wave to sound. Conversion from moving intruder to optical signals and such conversion is useful in numerous application areas.

A SN will typically have many transducers (Fig. 3.3) and its type will depend on the application in hand. The data from any transducer can be selected under program control and value sent to the processor at desired time gaps. Once an appropriate transducer has been selected, data from many SNs need to be collected at the base station (BS). Data can be sent directly to the BS from a SN as shown in Fig. 3.4a. But, the energy consumed by a SN in transmitting k-bits message to BS at distance d can be given by:

$$E_{\text{Tx}}(k,d) = E_{\text{Tx}-\text{elec}}(k) + E_{\text{Tx}-\text{amp}}(k,d) = E_{\text{elec}} * k + * k * d^2, \qquad (3.1)$$

where $E_{\text{Tx}-\text{elec}}$ is the transmission electronics energy consumption, and $E_{\text{Tx}-\text{amp}}$ is the transmit amplifier energy consumption.

D.P. Agrawal, *Embedded Sensor Systems*,
DOI 10.1007/978-981-10-3038-3_3

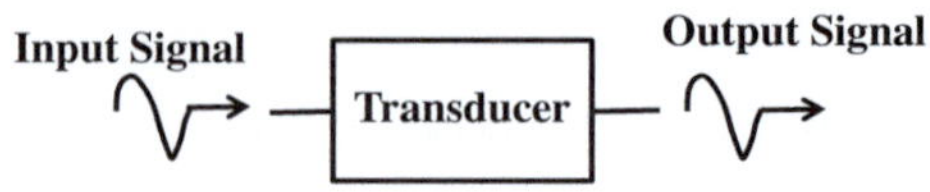

Fig. 3.1 Function of a transducer

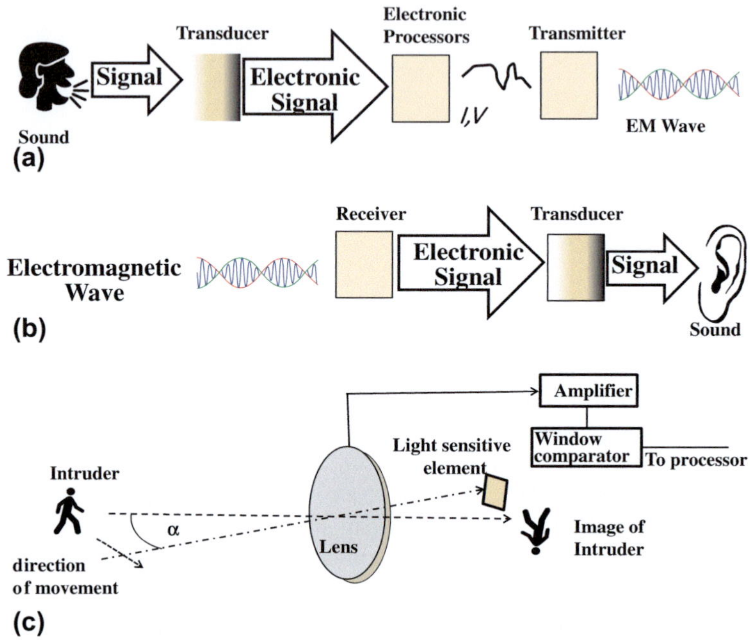

Fig. 3.2 Transducer illustrated **a** voice to EM waves. **b** EM waves to voice. **c** Optical signal conversion

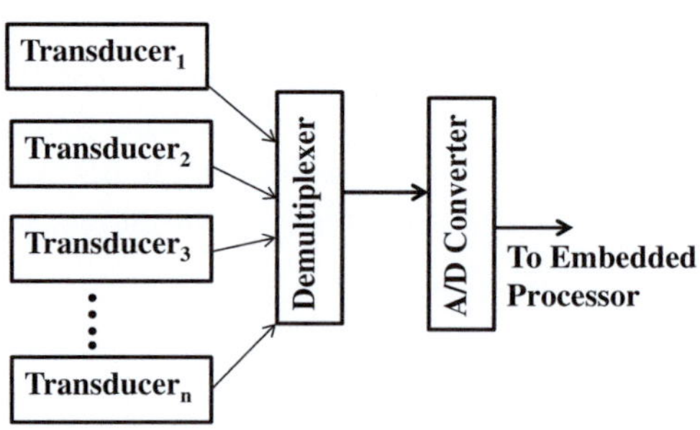

Fig. 3.3 SN with multiple transducers

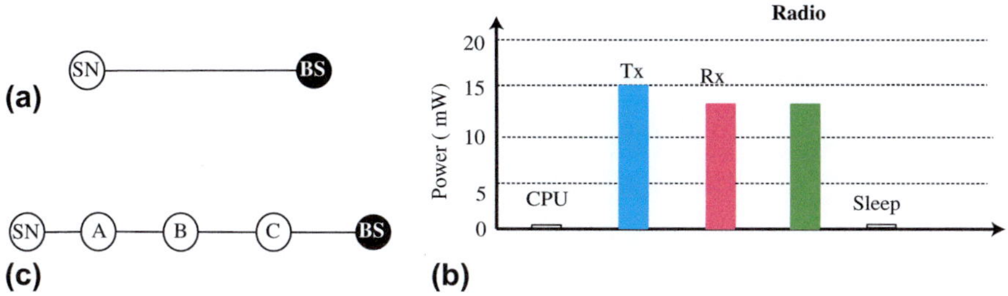

Fig. 3.4 **a** Direct transmission between SN and BS. **b** Power consumption in functional units of a SN. **c** Transmission using SNs A, B, C as intermediate between SN and BS

In a similar way,

$$\text{Receiver Energy, } E_{\text{Rx}}(k) = E_{\text{Rx-elec}}(k) = E_{\text{elec}} * k \tag{3.2}$$

and

$$\text{Computing/Processing Unit, } E_{\text{switch}} = C_{\text{total}} V_{\text{dd}}^2, \tag{3.3}$$

A good approximation can be taken as $E_{\text{Tx-elec}} = E_{\text{Rx-elec}} = E_{\text{elec}} = 50\,\text{nJ/bit} = 100\,\text{pJ/bit/m}^2$.

Energy consumption by transmitter is proportional to the square of its distance to BS (Fig. 3.4b), and energy is consumed even in receive and idle modes. Therefore, a better option is to transmit data from SN_1 to BS in a multi-hop fashion through other SNs A, B, and C as shown in Fig. 3.4c. Further attempts can be made to minimize energy consumption by keeping the transmitter connected only during the actual data transfer process. The corresponding delays are shown in Tables 3.1 and 3.2, and there is trade-off between time delay and energy depletion.

In fact a lot more energy is consumed in transmission than other functional units of a SN and is illustrated in Fig. 3.5 [5]. It is assumed that E_{fs} (free space fading

Table 3.1 Transmission delay in Scheme 3.4(a)

Sensor node	# Transmissions	# Receptions
SN_1	1	0
BS	0	1
Net time delay	Transmission of one packet	

Table 3.2 Transmission delay in Scheme 3.4(c)

Sensor node	# Transmissions	# Receptions
SN_1	1	0
A	1	1
B	1	1
C	1	1
BS	0	1
Net time delay	Transmission of four packets	

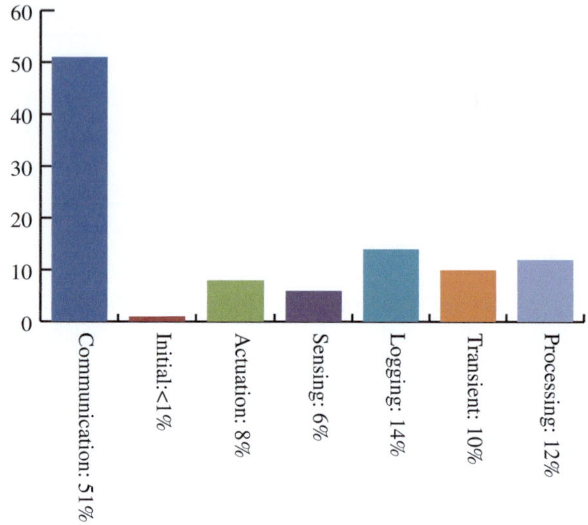

Fig. 3.5 Energy consumption in different operations of a SN

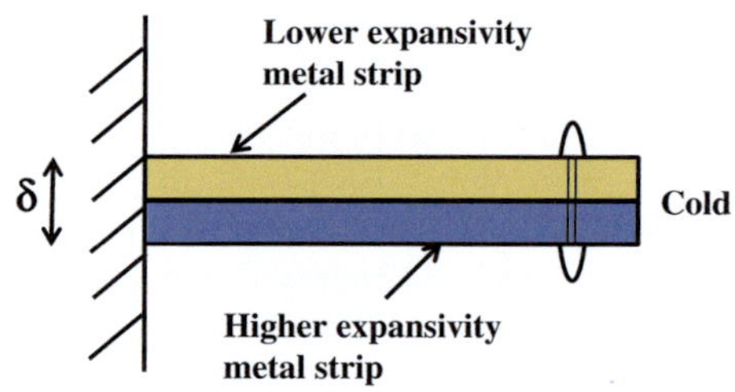

Fig. 3.6 Temperature transducer

energy) = 10 pJ/bit/m^2 and E_{mp} (multi-path fading energy) = 0.0013 pJ/bit/m^4. Energy consumption can be summarized as: communication = 51%, processing = 12%, transient = 10%, sensor loggings = 14% and transducer sensing = 6%. It is worth noticing that the technology improvements have impacted processor design to a large extent while not much enhancement is made in the communication field. So, it is anticipated that the relative impact of the communication subsystem on the system energy consumption will continue to grow (Fig. 3.6).

3.2 Types of Transducers

Human being has five sensors of sight, sound, smell, touch, and taste, and they all report to the brain as a central controller of a BS. Examples of commercially available transducers include thermometers, thermocouples, phototransistors, photo resistors, microphones, seismometers, and hydrophones and are adopted in numerous applications. Ambient conditions such as temperature, movement, sound, light, radiation, pressure/impact, stress, vibration, smoke, gases, and impurities are detected, and transducers such as temperature transducer, accelerometer, light transducer, pressure transducer, magnetic field transducer, ultrasonic transducer, photogate, CO_2 gas transducer, biometric transducer, and biological and chemical transducer can be utilized. There are factors [6] that affect the selection of transducers and are summarized in Table 3.3. Various detectable phenomenon and parameters that they could measure are given in Table 3.4. Carbon monoxide (CO) transducers can be useful in detecting the ozone level in the environment, the presence of CO in mines and underground structures, and wild forest fire. Other gases to be detected include sulphur dioxide, nitrogen oxides, carbon dioxide, ozone, and volatile organic compounds (VOCs).

Summary of different types of transducers is given in Tables 3.5, 3.6, 3.7, 3.8, 3.9, 3.10, 3.11 and 3.12 and are taken from [7]. It may be noted that we use the term transducer and not SN as the latter term is reserved for a system with many functionalities as shown in Fig. 3.2 (Table 3.13).

Table 3.3 Transducer characteristics and factors affecting their choice

Transducer characteristics	Affecting environmental factors	Economic factors
• Range • Linearity • Response time • Stability • Sensitivity • Error • Easiness in replication	• Size • Operating temperature range • Power consumption • Effect of humidity, corrosion • Electromagnetic interference • Ruggedness • Over range protection self-test capability	• Lifetime • Cost • Availability • Reliability

Table 3.4 Transducer types and associated characteristics

Stimulus	Quantity	Characteristics
Electric	Charge, voltage, current, electric field (amplitude, phase, polarization), conductivity, permittivity	Linear/rotational, proximity transducer
Magnetic	Magnetic field (amplitude, phase, polarization), flux, permeability	Rotational, proximity transducer
Optical	Refractive index, reflectivity, absorption	Flow, light, smart material, micro- and nano-transducer
Thermal	Temperature, flux, specific heat, thermal conductivity	Temperature, force, torque, and pressure transducer
Acoustic	Wave (amplitude, phase, polarization), spectrum, wave velocity	Flow transducer
Mechanical	Position, velocity, acceleration, force, strain, stress, pressure, torque	Acceleration transducer
Biological and chemical	Fluid concentrations (gas or liquid)	Smart material transducer

Table 3.5 Linear/rotational transducers and associated features [8]

Transducers	Features
• Linear/rotational variable differential transducer (LVDT/RVDT) • Optical encoder • Electrical tachometer • Hall effect sensor • Capacitive transducer • Strain gauge elements • Interferometer • Magnetic pickup • Gyroscope • Inductosyn • Acceleration sensors • Seismic accelerometer • Piezoelectric accelerometer	• High resolution with wide range capability; very stable in static and quasi-static applications • Simple, reliable, and low-cost solution; good for both absolute and incremental measurements • Resolution depends on type such as generator or magnetic pickups • High accuracy over a small-to-medium range • Very high resolution with high sensitivity, low-power requirements, good for high-frequency dynamic measurements • Very high accuracy in small ranges • Provides high resolution at low noise levels • Laser systems provide extremely high resolution in large ranges; very reliable and expensive • Output is sinusoidal • Very high resolution over small ranges • Good for measuring frequencies up to 40% of its natural frequency • High sensitivity, compact, and rugged • Very high natural frequency (100 kHz typical)

Table 3.6 Acceleration transducers and associated features [8]

Sensors	Features
Seismic accelerometer Piezoelectric accelerometer	Good for measuring frequencies up to 40% of its natural frequency High sensitivity, compact, and rugged; very high natural frequency (100 kHz typical)

Table 3.7 Force, torque, and pressure transducers and associated features [8]

Sensors	Features
• Strain gauge • Dynamometers/load cells • Piezoelectric load cells • Tactile sensor • Ultrasonic stress sensor	• Good for both static and dynamic measurements • They are also available as micro- and nano-sensors • Good for high-precision dynamic force measurements • Compact, has wide dynamic range, and • Good for small force measurements

Table 3.8 Flow transducers and associated features [8]

Sensors	Features
Pitot tube Orifice plate Flow nozzle, venture tubes Rotameter Ultrasonic type Turbine flow meter Electromagnetic flow meter	Widely used as a flow rate sensor to determine speed in aircrafts Least expensive with limited range Accurate on wide range of flow; more complex and expensive Good for upstream flow measurements; used in conjunction with variable inductance sensor Good for very high flow rates Can be used for both upstream and downstream flow measurements Not suited for fluids containing abrasive particles Relationship between flow rate and angular velocity is linear Least intrusive as it is noncontact type; can be used with fluids that are corrosive, contaminated. The fluid has to be electrically conductive

Table 3.9 Temperature transducers and associated features [8]

Sensors	Features
• Thermocouples • Thermistors • Thermodiodes, thermotransistors • RTD—resistance temperature detector • Infrared type • Infrared thermography	• This is the cheapest and the most versatile sensor; applicable over wide temperature range (−200∞C to 1200∞C typical) • Very high sensitivity in medium range (up to 100∞C typical); compact but nonlinear in nature • Ideally suited for chip temperature measurements; minimized self-heating • More stable over a long period of time compared to thermocouple • Linear over a wide range • Noncontact point sensor with resolution limited by wavelength • Measures whole-field temperature distribution

Table 3.10 Proximity transducers and associated features [8]

Sensors	Features
• Inductance, eddy current, hall effect • Photoelectric, capacitance, etc.	• Robust noncontact switching action • The digital outputs are often directly fed to the digital controller

Table 3.11 Light transducers and associated features [7]

Sensors	Features
Photoresistors, photodiodes, phototransistors, photoconductors, etc.	Measure light intensity with high sensitivity; inexpensive, reliable, and noncontact sensor
Charge-coupled diode	Captures digital image of a field of vision

Table 3.12 Smart material transducers and associated features [8]

Sensors	Features
• Optical fiber • As strain sensor • As level sensor • As force sensor • As temperature sensor • Piezoelectric • As strain sensor • As force sensor • As accelerometer • Magnetostrictive • As force sensors • As torque sensor	• Alternate to strain gages with very high accuracy and bandwidth • Sensitive to the reflecting surface's orientation and status • Reliable and accurate • High resolution in wide ranges • High resolution and range (up to 2000∞C) • Distributed sensing with high resolution and bandwidth • Most suitable for dynamic applications • Least hysteresis and good setpoint accuracy • Compact force sensor with high resolution and bandwidth, good for distributed and noncontact sensing applications • Accurate, high bandwidth, and noncontact sensor

Table 3.13 Smart micro- and nano-transducers and associated features [8]

Sensors	Features
• Micro-CCD image sensor	• Small size, full-field image sensor
• Fiberscope	• Small (0.2 mm diameter) field vision scope using SMA coil actuators
• Micro-ultrasonic sensor	• Detects flaws in small pipes
• Micro-tactile sensor	• Detects proximity between the end of catheter and blood vessels

3.3 Temperature Transducers

Temperature transducers are used in buildings, chemical process plants, engines, appliances, computers, and many other devices that involve monitoring of temperature and could avoid diverse effect such as fire, overheating, or failure of a freezer [9] value, and the temperature is measured indirectly by measuring pressure, volume, electrical resistance, and strain. A simple thermal sensor can be formed by combining two layers of metals having different linear expansively. A change in the length per degree of temperature is used to indicate temperature increase. The strip could bend due to the difference in their expansion. The length of a bimetallic strip L is changed depending on current temperature T as $L = L_0[1 + b(T - T_0)]$, where L_0 is the length at initial temperature T_0 and β is the coefficients of linear expansion. For example, a thermostat is used to make or break an electrical connection due to deflection.

Thermocouples, thermistors, and resistance temperature detectors are the most common transducers for measuring temperature, while the use of specialized fiber-optic transducer is also increasing [10]. Various associated characteristics are compared and shown in Table 3.14.

Gravity transducer indicates the influence of gravity, and the force due to gravity is mathematically expressed by the formula:

$$F = gm_1m_2/r^2 \tag{3.2}$$

where g is the gravitational constant representing acceleration as shown in Fig. 3.7.

Ultrasonic transducers are used for position measurements with sound waves emitted in the range of 2–13 MHz. Such scheme is also known as sound navigation and ranging (SONAR) and works based on radio detection and ranging (RADAR) of electromagnetic waves. Light transducers are used in cameras, infrared detectors, and ambient lighting applications. Here, a transducer is composed of photoconductor such as a photoresistor, photodiode, or phototransistor.

A photogate transducer employs infrared transmitter and receiver at opposite ends, and time is recorded when the light is broken to provide counting of a periodic motion. Night vision has been made possible by a combination of two approaches: sufficient spectral range and sufficient intensity range. Infrared illumination of spectral range 700–1000 nm is combined with CCD cameras which makes resulting scene appear as a monochrome image on a normal display device

Table 3.14 Various transducers and associated characteristics [10]

Sensing technique	Accuracy	Sensitivity	Required signal conditioning	Comparison
Thermocouple	Good	Good	• Amplification • Filtering • Cold junction • Compensation	• Self-powered • Inexpensive • Rugged • Large temperature range
Resistance temperature detector	Best	Better	• Amplification • Filtering • Current excitation	• Very accurate • Very stable
Thermistor	Better	Best	• Amplification • Filtering • Voltage excitation	• High resistance • Low thermal mass
Fiberoptics	Best	Best	• No amplification • Filtering	• Good for hazardous environments • Good for long distances • Immune to EM-induced noise • Small, lightweight

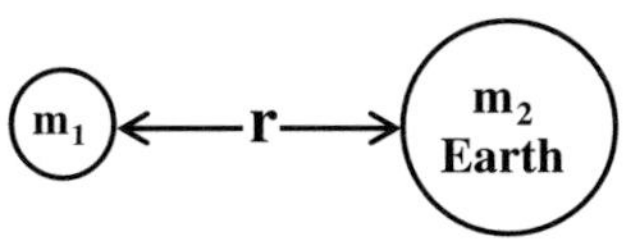

Fig. 3.7 Gravity transducer

[11]. An image intensifier is a vacuum-tube-based device that converts invisible light from an image to visible light to be viewed by naked eye [12]. A night vision device comprises of an image intensifier tube in a rigid casing and vehicle driver's perception and seeing distance is improved in darkness or poor weather.

Transducers play an important role in an automobile, and various types of such devices are shown in Fig. 3.8 [13].

3.4 Gas Transducers

Gas transducers are useful in numerous applications such as process control industries, environmental monitoring, boiler control, fire detection, alcohol breath test, detection of harmful gases in mines, home safety, and grading of agro-products such as coffee and spices. The parameters during use include operating temperature and humidity. Most of the gas transducers are bulky in nature and consumes enormous amount of power, usually at high voltage. Gas transducers can be classified as metal oxide-based gas transducers, capacitance-based gas transducers,

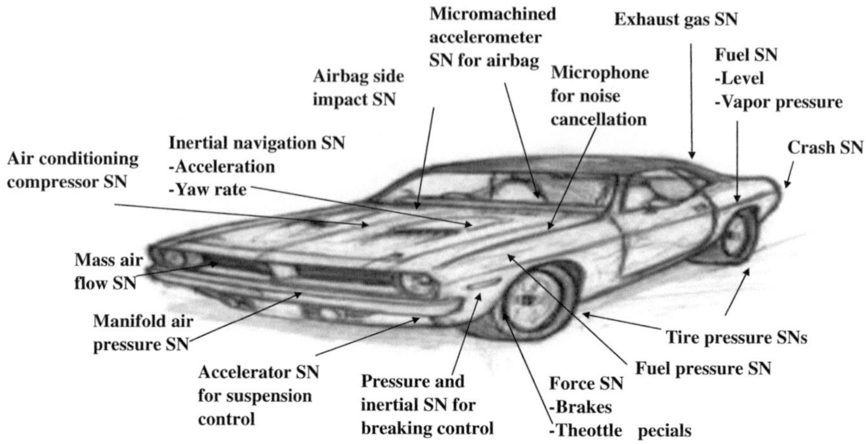

Fig. 3.8 Transducers in a car

acoustic wave-based gas transducers, calorimetric gas transducers, optical gas transducers, and electrochemical gas transducers.

Metal oxide transducers are also known as chemiresistors and are based on change in the resistance of a thin film when gas molecules are adsorbed at the surface of a semiconductor. Interaction between the gas and solid affects the resistance of the film. Capacitance-based gas transducer indicates the gas concentration value by measuring the change in dielectric constant of films between two electrodes and thus relies on interdigitized electrode structures.

The electrochemical transducer contains two terminals of an anode and a cathode, and anode does oxidization while cathode does the reduction process, causing current to flow. Reducible gases (such as oxygen, nitrogen oxides, and chlorine) are found at the cathode, while oxidizable gases (carbon monoxide, nitrogen dioxide, and hydrogen sulfide) are present at the anode. The output is proportional to the pressure of gases. Carbon monoxide gas sensor can be either battery-operated or AC-powered. No sound alarm is produced at lower density [e.g., 100 ppm (particles per million)], but sound is generated within few minutes for concentration at or above 400 ppm. Carbon monoxide CO transducer can be of different types such as semiconductor, electrochemical, digital, and biomimetic. CO_2 transducer measures gaseous CO_2 levels by absorbed infrared radiation in the environment in the range of 0–5000 ppm. CO_2 transducer contains a tube with an infrared source at one end and an infrared detector at the other end. Infrared radiation which is not being absorbed by CO_2 produces heat, and the infrared detector measures the temperature. Mostly, hydrogen gas transducer employs palladium to detect hydrogen as palladium selectively absorbs hydrogen gas to produce chemical palladium hydride. Various possible types are optical fiber hydrogen, nanoparticle-based hydrogen micro-transducers, diode-based and recent MEMS-based H_2 gas transducers that have unique advantages of speed, sensitivity, stability, and amenability.

Electric transducers are based on electric principle using capacitive transducers and may be static with charges that do not move or move at constant velocity or time-dependent if charges accelerate/decelerate. Magnetic transducers depend on electrostatic magnetic fields when currents are constant (dc), or time-dependent currents varying in time. Both fields are governed by Maxwell's equations.

3.5 Capacitive Transducers

Capacitance C between any two conducting bodies A and B, regardless of size and distance between them, is defined as the ratio between charge Q and potential V of a body and is measured as $C = Q/V$ in C/V or farad [F] and is illustrated in Fig. 3.9. Quantities influencing capacitance, charge, and stored energy may be sensed through changes in the electromagnetic field.

In a parallel plate capacitor, changes in the capacitance are due to stimuli like displacement (pressure, force), proximity, and permittivity (moisture), and are given by:

$$C = \frac{\varepsilon_0 \varepsilon_r S}{d}\ [\mathrm{F}] \tag{3.4}$$

where d is assumed small,

ε_0 the permittivity of vacuum,
ε_r the relative permittivity (dielectric constant) of the medium between plates,
S the area of the plates, and
D the distance between the plates

ε_0 is a constant equal to 8.854×10^{-12} F/m, and ε_r is the ratio between the permittivity of the medium to that of free space. Equation 3.4 is obtained by assuming that the electric field between the two plates does not leak outside the space between the plates. In more general case, when d is not small, plates can be arranged in a different configuration as a linear shown in Figs. 3.10 and 3.11, we cannot calculate the capacitance directly. But, we can still write the following:

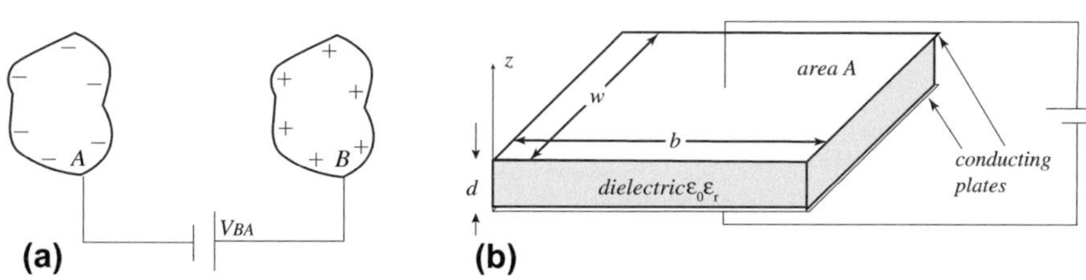

Fig. 3.9 **a** Illustration of a capacitance transducer. **b** Illustration of a parallel plate capacitor

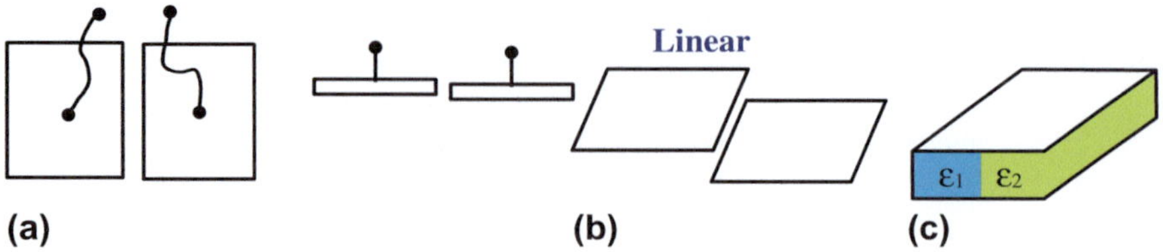

Fig. 3.10 Illustration of a linear capacitance transducer

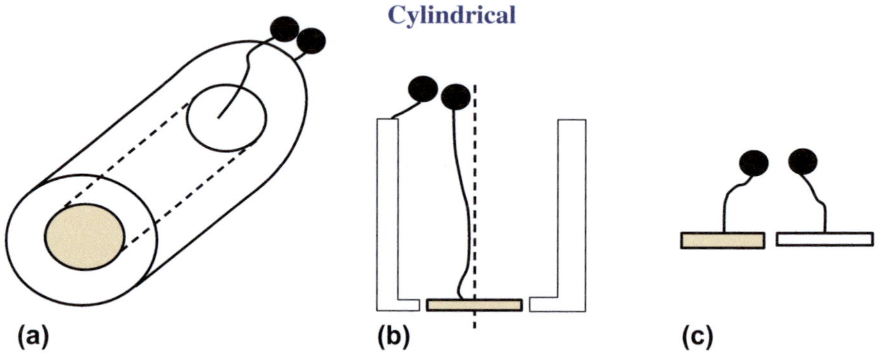

Fig. 3.11 Illustration of a cylindrical capacitance transducer

$$C = \alpha[\varepsilon_0, \varepsilon_r, S, 1/d]. \tag{3.5}$$

A transducer is made of a single plate while the second plate is a conductor that could move relative to the other (Fig. 3.12a), the distance is sensed by the capacitance that is inversely proportional to the motion and the output is linear for small distance. Another option is to keep two plates remain fixed, but the dielectric moves in or out by a device as shown in Fig. 3.12b. Another alternative is to keep the plates fixed and change the sense distance to a surface as given in Fig. 3.12c. Most capacitive transducers are a variation of this arrangement.

3.5.1 *Proximity Transducers*

When any material is present in a transducer, effective permittivity and the capacitance increase with conducting or non-conducting distance. The smaller the sensed distance d, the larger the sensitivity of the transducer as the dimensions of the transducer make a big difference in span and sensitivity. To overcome the nonlinearity, two fixed and one moving plate is used to constitute a proximity

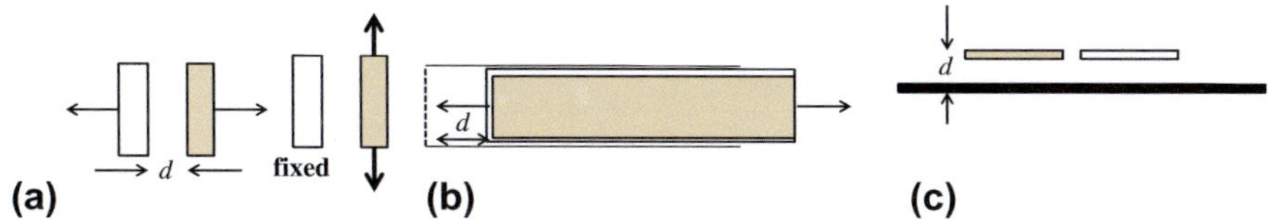

Fig. 3.12 Illustration of a displacement capacitance transducer **a** moving plates, **b** changing dielectric, **c** change sense distance *d*

capacitive transducer as shown in Fig. 3.13. As the portable plate moves up, its potential becomes positive and when it moves down, it is negative. Such a scheme is more linear than the previous transducers. A rotary position transducer is shown in Fig. 3.13c, and such linear displacement transducer could be an integrated comb-like, sideways sliding plates, or a plunger-type transducer. Other displacement transducer configurations are shown in Fig. 3.14.

3.6 Fluid-Level Transducers

By sensing the position of the fluid surface directly through a float, a fluid level may be sensed by any of the position or proximity transducers discussed earlier that changes the capacitance of a linear or a rotary capacitor. Another method is linear movement that can have a very large range of variation. As shown in Fig. 3.15a, a coaxial capacitor has two concentric cylinders, creating a capacitance C_0 as follows:

$$C_0 = \frac{2\pi\varepsilon_0 L}{\ln(b/a)} \, [\mathrm{F}] \tag{3.6}$$

where L is the length of a coaxial capacitor, a the inner radius, and b the outer radius, and the fluid fills the capacitor to a height h. Capacitance varies linearly with respect to height h. Even though capacitive fuel gages are of this type, the idea can be used for any nonconductive fluid such as oils as follows:

$$C_0 = \frac{2\pi\varepsilon_0}{\ln(b/a)} (h\varepsilon_r + L - h) \, [\mathrm{F}]. \tag{3.7}$$

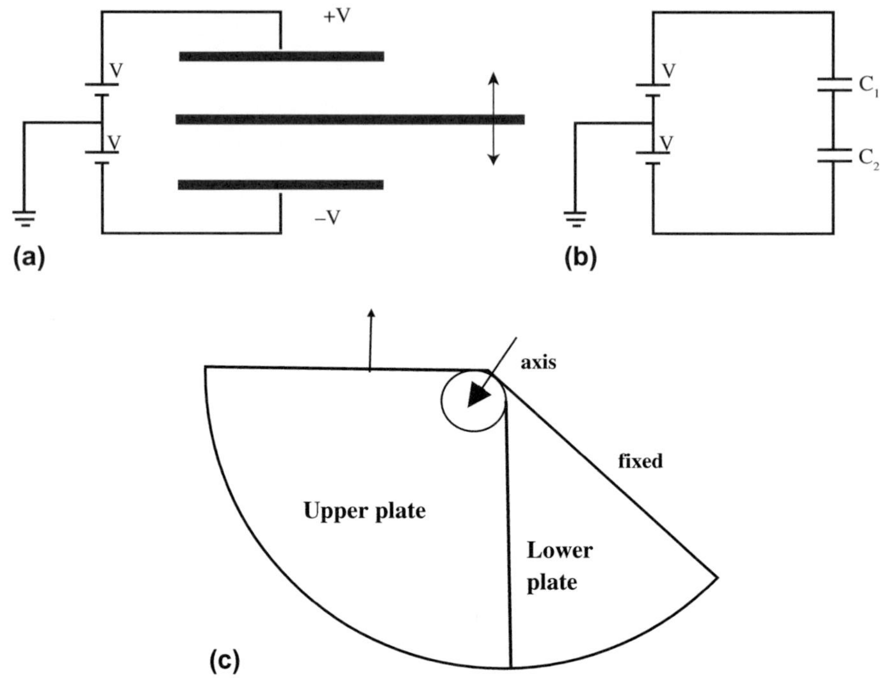

Fig. 3.13 Illustration of a proximity transducer **a** two fixed and one moving plate, **b** corresponding equivalent circuit, **c** example two rotary plates

3.7 Humidity Transducers

Humidity transducer shown in Fig. 3.16 [14] quantifies the presence of water in air as PPM which is a function of temperature and pressure and can be a mixture such as air, or a pure gas, such as nitrogen or argon. It consists of a hygroscopic dielectric material sandwiched between a pair of electrodes, forming a small capacitor. Lower electrode is formed on an alumina substrate, using gold or platinum, and a polymer layer is deposited on the electrode that senses humidity. On top of this polymer film, a gold layer is deposited which acts as top electrode that allows water vapor to pass through it, into the sensing layer, forming a capacitive-type transducer (Fig. 3.17).

A rain transducer [15] is a switching device activated by rainfall and has two main applications for rain transducers: water conservation device connected to an automatic irrigation system, and protecting the interior of an automobile from rain and support windscreen wipers of cars. Irrigation system rain transducer employs a hygroscopic disk that swells in the presence of rain and shrinks back as it dries out. Magnetic transducers and actuators are based on magnetic flux density called magnetic induction. These are also called inductive transducer as it relies on inductance and magnetic circuits, and magnetic forces can be qualitatively

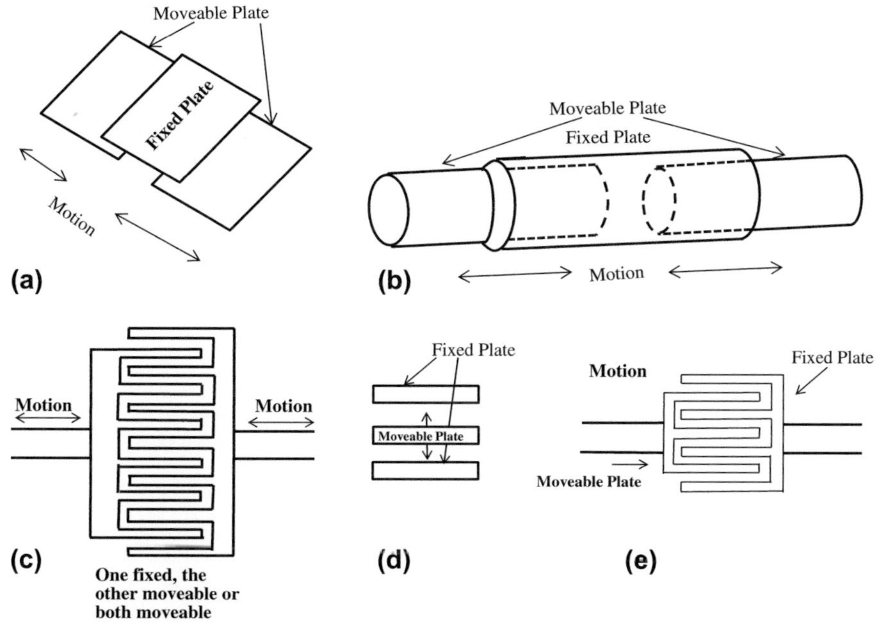

Fig. 3.14 Illustration of linear displacement transducers. **a** One fixed and two moving plates. **b** One fixed and two moving circular plates. **c** One fixed and other moving plates. **d** Two fixed and one moving plate. **e** Fixed and vertical moveable plates

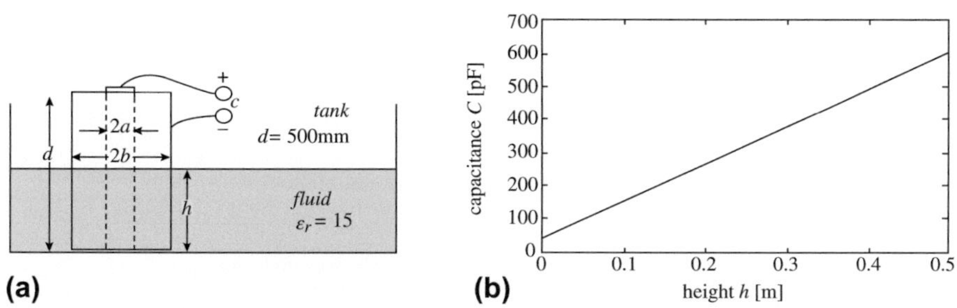

Fig. 3.15 Fluid-level transducers **a** example, **b** capacitance as a function of distance

explained, without depending on Maxwell's equations. A permanent magnet exerts a force on another magnet through space as a "field" exists around the magnet. The same effect is detected by driving a current through a coil as the two fields are identical, and any current generates a magnetic field.

A magnet attracts or repels another magnet. It attracts a piece of iron but not a piece of copper as governed by the material permeability μ as henry/meter. The strength of the magnetic field is given by flux density B [tesla] and the magnetic field intensity H as ampere/meter. The relation between the two is given as:

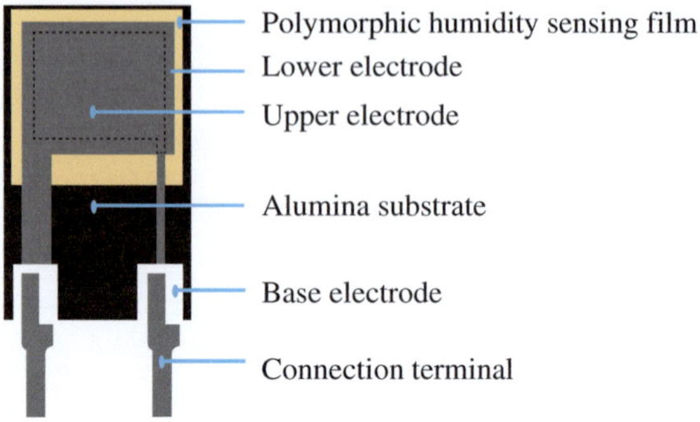

Fig. 3.16 Humidity transducer

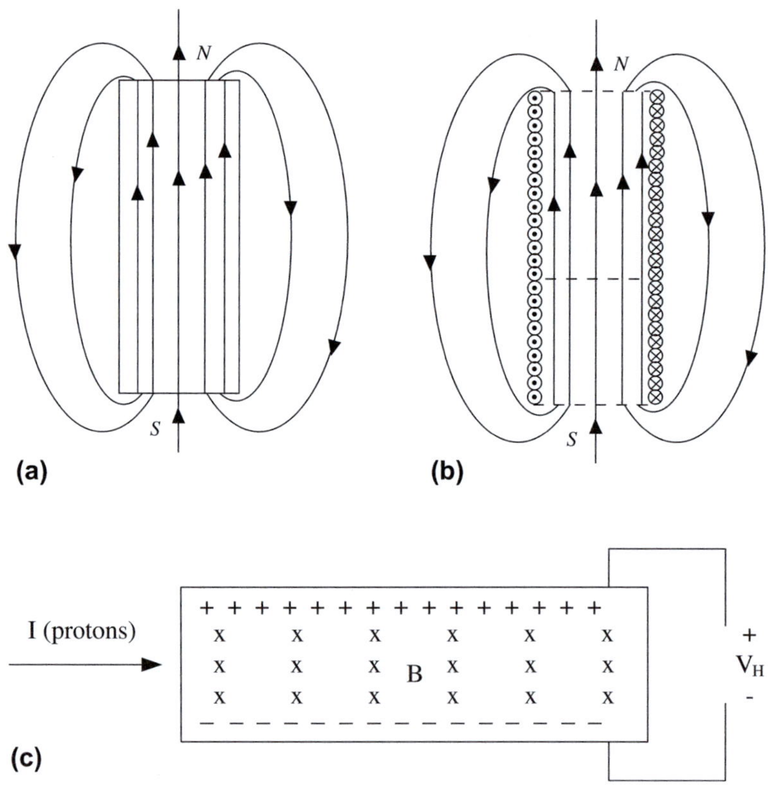

Fig. 3.17 Magnetic transducer **a** permanent magnet, **b** coil, **c** magnetic field transducer

$$B = \mu_0 \mu_r H \tag{3.7}$$

where μ_0 is the permeability of free space or magnetic constant.

The constant permeability of vacuum μ_0 is = $4\pi \times 10^{-7}$ [H/m]. Relative permeability of the medium is μ_r, and μ_r is the ratio between the permeability of the medium and that of vacuum. Permeability of some useful materials are diamagnetic $\mu_r < 1$, paramagnetic $\mu_r > 1$, and ferromagnetic $\mu_r \gg 1$(iron-like).

Permeability of diamagnetic, paramagnetic, ferromagnetic, and hard permanent materials can be obtained from corresponding references [16]. Soft magnetic materials are those for which magnetization is reversible. Hard magnetic materials are materials which retain magnetization and are therefore used for the production of permanent magnets. Hysteresis (Fig. 3.18a) is a property of ferromagnetic materials best explained through the magnetization curve while permeability is field-dependent.

For a long straight wire carrying a current I and placed in a medium of permeability $\mu = \mu_0 \mu_r$, the magnitude of the magnetic flux density is given as:

$$B = \mu_0 \mu_r \frac{I}{2\pi r} \tag{3.8}$$

where r is the distance from the current-carrying wire. The magnetic field is a vector perpendicular to I.

In a practical arrangement, the wire may not be very long or may be wound in a coil. The larger the current and/or the permeability, or the shorter the distance between current and the location where the magnetic field is needed, the larger is the magnetic field. Flux is the integral of flux density over an area S. If B is constant over an area S and at an angle θ to the surface, flux is $\Phi = BS \cos \theta$. The unit of flux is weber = [Tm^2]. While flux relates to the power and energy in the magnetic field, electrons move in conductors creating magnetic field B. Two wires carrying currents I in opposite (same) directions exert (attract) forces F on each other. For long parallel wires of length L, the force is given as:

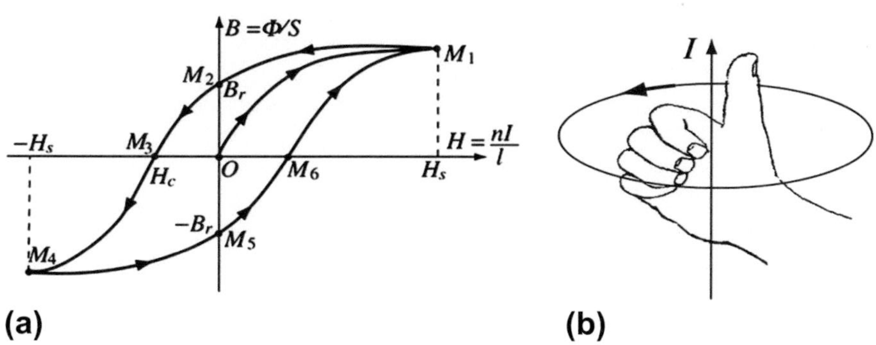

Fig. 3.18 **a** Hysteresis loop in ferromagnetic materials, **b** rules for currents, fields, and flux

$$F = BIL \tag{3.9}$$

For other configurations, the relation is much more complicated, but force is proportional to B, I, and L.

3.7.1 Inductive Transducers

Inductive transducer depends on two basic phenomena of inductance of a coil and changes in inductance. Inductance is a property of a magnetic device as capacitance is the property of an electric device usually associated with coils and conductors. Inductance L is defined as the ratio of flux Φ and the current I that produced it and can be changed either directly or indirectly as:

$$L = \frac{\Phi}{I}\left[\frac{\text{webber}}{\text{ampere}}\right] \text{ or } [\text{henry}] \tag{3.10}$$

All magnetic devices have an inductance, but inductance is most often associated with coils.

Self-inductance (Fig. 3.19a) and mutual inductance Fig. 3.19a are the two types of inductance, where self-inductance L_{ii} is the ratio of the flux produced by a conductor or a coil and the current that produces it while mutual inductance M_{ij} is the ratio of the flux produced by circuit i in circuit j and the current in circuit i that produced it. A mutual inductance Fig. 3.19c exists between any two circuits as long as there is a magnetic flux that couples the two. A coil is also called an inductor, and a transformer is made of two or more coils, coupled through mutual inductances. Transformer is an AC device of a primary (N_1 coils) and a secondary (N_2 coils) circuit with a transformer ratio of N_1/N_2.

Most inductive transducers rely on self-inductance, mutual inductance, or transformer concepts. The most common types of stimuli sensed by inductive transducers are position, displacement, and material composition. Inductive proximity transducer (Fig. 3.20a) consists of at least a coil to generate a magnetic field, and the coil's inductance depends on the diameter of the coil, number of turns, and materials around it. The current and the diameter of the coil define the extent to which the field projects away from the coil and therefore the span of the transducer. As the transducer gets closer to the sensed surface (Fig. 3.20b), the inductance of the coil increases if the sense surface is ferromagnetic.

A ferromagnetic core (powdered magnetic material such as Fe_2O_3, CoO_2) is added to increase the inductance of the transducer. A shield may be placed around to avoid indifference to nearby objects, thereby increasing the inductance and the span. In many transducers, two coils are used: one as a reference with fixed inductance, and the other as a transducer. When a surface is sensed, the sensing coil

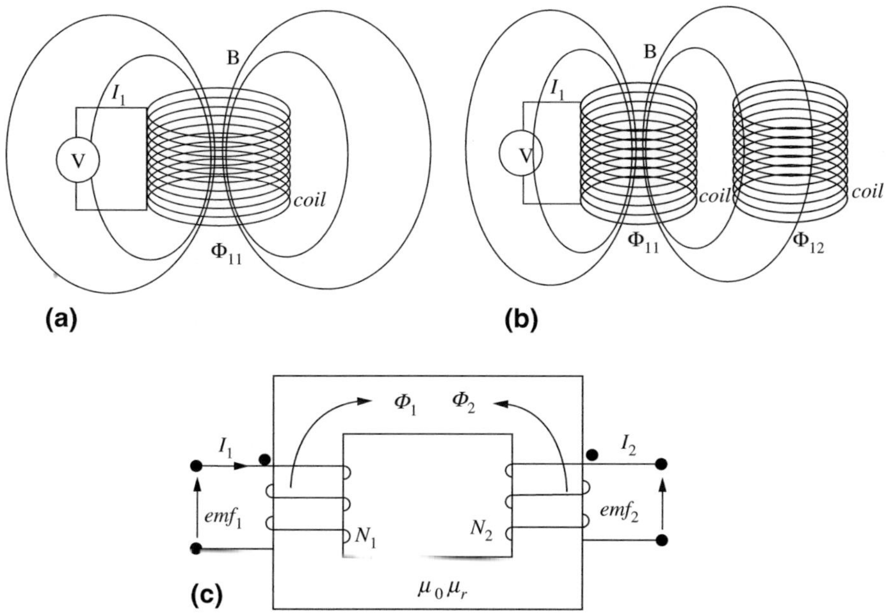

Fig. 3.19 **a** Self-inductance, **b** mutual inductance, **c** self-inductance

records a larger inductance and the imbalance represents measure of distance. Other transducers may employ a closed magnetic circuit with no shielding that concentrates on the magnetic field in the gaps.

Inductive proximity transducers are sensitive to proximity of non-conducting ferromagnetic materials or to any conducting media. Many inductive proximity transducers are of the eddy current type with two coils: one serving as reference and the other as a sensing coil with larger inductance. The imbalance between the coils measures the distance. The eddy currents (Fig. 3.21a) flow in closed loops, cause a field produced in the conductor and the net flux through the transducer coil is reduced, dissipating power.

In eddy current transducer, magnetic field penetrates into a conducting medium as:

$$B = B_0 e^{d/\delta}, \quad \text{or} \quad I = I_0 e^{d/\delta} \tag{3.11}$$

where B_0 and I_0 are the flux density and the current density at the surface, d is skin depth in the medium defined as the depth at which the field (or current) is attenuated to $1/e$ of its value at the surface. For planar surfaces, skin depth is given as:

$$\delta = \frac{1}{\sqrt{\pi f \mu \sigma}}\,[m] \tag{3.12}$$

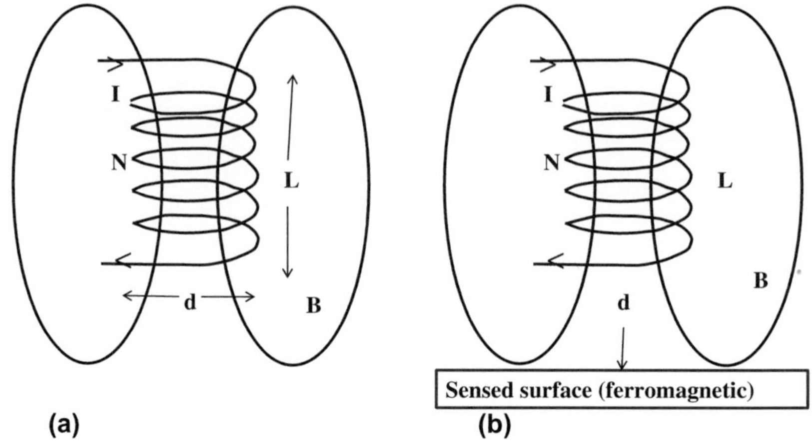

Fig. 3.20 **a** Inductive proximity transducer, **b** sensed surface by inductive proximity transducer

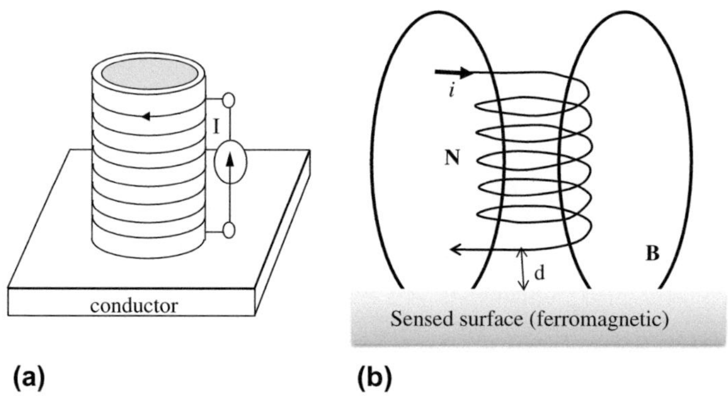

Fig. 3.21 **a** Eddy current transducer, **b** movable core transducer

where f is the frequency of the field.

The penetration depends on the frequency, conductivity, and permeability, and it is assumed that the sensed conductor is thick enough compared to skin depth. Either capacitive or inductive proximity transducer can be used to sense distance as a nonlinear function. Proximity transducers are usually used to indicate preset distance, while inductive transducers produce an electric output based on the change in their impedance.

3.8 Magnetometer Transducers

Magnetometer transducer [17] is widely used for measuring Earth's magnetic field and is used to measure magnetization of a magnetic material like a ferromagnet. It is being used as a compass in cell phones.

Magnetic reluctance transducer is an equivalent magnetic term to resistance as:

$$R = \frac{L}{\mu S}\left[\frac{1}{\mathrm{H}}\right] \tag{3.13}$$

Reluctance is smaller for a shorter magnetic path, larger-cross sectional area, and varies with its permeability. Typically, reluctance of a coil can be changed by adding a gap in the magnetic path. Movable core transducers (Fig. 3.21b) employ a movable core to change the inductance of a coil. Such a linear variable inductance transducer measures the position of the core that also measures force, pressure, etc.

A better displacement transducer is based on the idea of the transformer (Fig. 3.22a). As shown in Fig. 3.22b, distance between two transformer coils is varied, and the coupling coefficient between the two fixed coils is changed by moving the core. Output is zero if coupling is balanced, and motion to either side changes the output. This shielded variable reluctance transformer determines both distance and direction of change. These transducers are extremely rugged, and the phase is detected with a zero-crossing phase detector. Typical output is usually below 5 V and resolution is very high in the linear span of 10–20% length of the coil. Rotary version (Fig. 3.22c) is a variation and intended for angular displacement. The span is given in angles and can be up to ±30° to ±40°.

3.9 Optical and Underwater Transducers

Optical transducers can be used to measure numerous physical parameters usually by measuring some other quantity and that is translated by some equivalence. The light beam changes by the phenomena that being measured as the following five optical properties of intensity, phase, polarization, wavelength, and spectral distribution and is expressed as $E_P(t)\cos[\omega t + \theta(t)]$, where $E_P(t)$ represents intensity-based transducers, $\omega_P(t)$ as frequency-varying transducers, and $\theta(t)$ as phase modulation sensing.

Optical transducers can be classified as either extrinsic or intrinsic transducers (Fig. 3.23c, d). In extrinsic transducers, the light leaves the feed or transmitting fiber to be changed before it continues to the detector while in intrinsic transducers light beam does not leave the optical fiber but is changed when confined within the unit. Two types of transducers have been compared in Table 3.15.

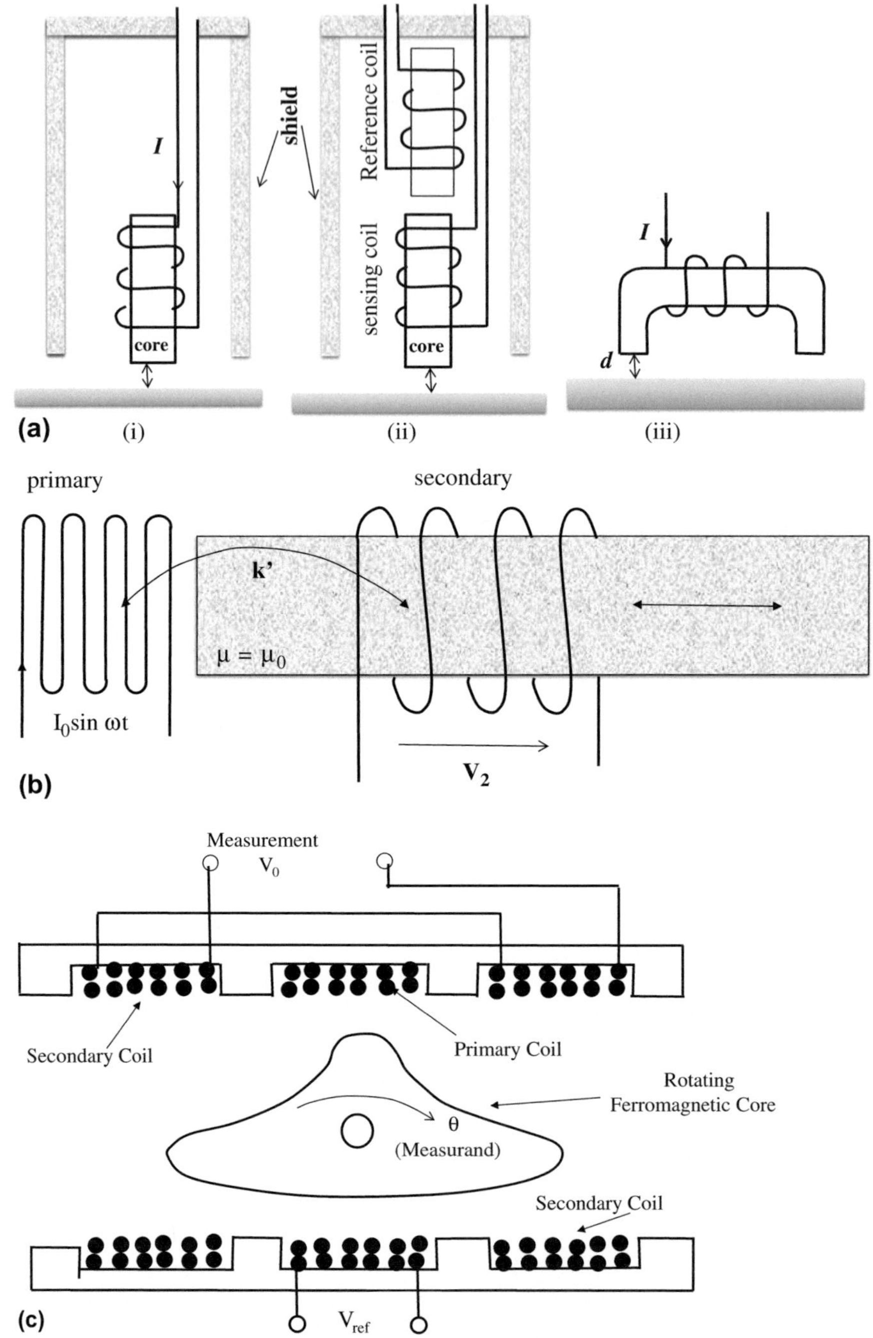

Fig. 3.22 **a**, **b** Linear variable differential transformer (LVDT) transducer, **c** rotary variable differential transformer transducer

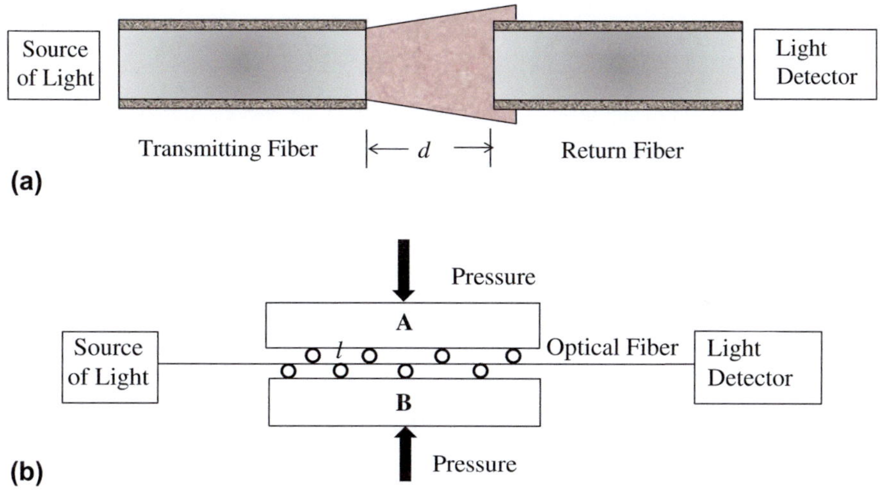

Fig. 3.23 **a** extrinsic optical transducer, **b** intrinsic optical transducer

Table 3.15 Opticaltransducers for measuring different parameters by measuring other quantities

Parameter to be measured	Based on quantity
Temperature	Chemical species
Pressure	Force
Flow	Radiation
Liquid level	pH
Displacement	Humidity
Vibration	Strain
Rotation	Velocity
Magnetic fields	Electric fields
Acceleration	Acoustic fields

3.9.1 Underwater Acoustic Transducers

Underwater acoustics [18] is a technique of sending and receiving message below water and the most common is using hydrophones. Under water communication is difficult due to factors such as multi-path propagation, time variations in the channel, small available bandwidth, and strong signal attenuation. Low data rates are used as compared to terrestrial communication, since underwater communication uses acoustic waves instead of electromagnetic waves (Table 3.16).

Table 3.16 Comparison between extrinsic or intrinsic transducers

Characteristics	Extrinsic	Intrinsic
Applications	Temperature, pressure, liquid level, and flow	Rotation, acceleration, strain, acoustic pressure and vibration
Sensitivity	Less sensitive	More sensitive
Multiplexing feasible?	Easily multiplexed	Tougher to multiplex
Ingress/egress	Ingress/egress connection problems	Reduces connection problems
Easier to use	Easier to use	More elaborate signal demodulation
Cost	Less expensive	More expensive

3.10 Strain and Biomedical Transducers

Strain transducers are based on Fiber Bragg Gratings (FBG) technology and senses as little as 9 micro-strain. Biomedical transducers use spectroscopic biomedical transducers, and CO_2, O_2, and pH can be measured simultaneously. Two types of gyroscopes are ring laser gyroscope (RLG) and fiber-optic gyroscope (FOG) (Table 3.17).

3.10.1 Pressure Transducers

Pressure transducers can be classified in terms of pressure ranges they measure (Fig. 3.24).

1. Absolute pressure transducer: This transducer measures the pressure relative to perfect vacuum.
2. Gauge pressure transducer: This transducer measures the pressure relative to atmospheric pressure, e.g., a tire pressure gauge is an example of gauge pressure measurement.
3. Vacuum pressure transducer: It may be used to describe a transducer that measures pressures below atmospheric pressure, showing the difference between that low pressure and atmospheric pressure, but it may also be used to describe a transducer that measures low pressure relative to perfect vacuum.

Table 3.17 *W* value for various gases for ionization chambers (eV/ion pair) [23]

Gas	Electrons (fast)	Alpha particles
Argon (A)	27.0	25.9
Helium (He)	32.5	31.7
Nitrogen (N_2)	35.8	36.0
Air	35.0	35.2
CH_4	30.2	29.0

4. Differential pressure transducer: This measures the difference between two pressures, one connected to each side of the transducer differential pressure. Such transducers are used to measure many properties, such as pressure drops across oil filters or air filters, fluid levels. Technically speaking, most pressure transducers are really differential pressure transducers, for example, a gauge pressure transducer.
5. Sealed pressure transducer: This transducer is similar to a gauge pressure transducer except that it measures pressure relative to some fixed pressure rather than the ambient atmospheric pressure.
6. Piezoresistive strain gauge: This uses the piezoresistive effect of bonded or formed strain gauges by a Wheatstone bridge circuit to detect strain due to applied pressure. Silicon, polysilicon thin film, bonded metal foil, thick film, and sputtered thin film are the most commonly used technologies to measure absolute, gauge, vacuum, and differential pressures.
7. Capacitive: Such a transducer uses a diaphragm and pressure cavity to create a variable capacitor that detects strain due to applied low pressure using metal, ceramic, and silicon diaphragms.
8. Electromagnetic: This measures the displacement of a diaphragm by means of changes in inductance using LVDT, Hall Effect or eddy current principle.
9. Piezoelectric: It uses the piezoelectric effect in certain materials such as quartz to measure the strain upon the sensing mechanism due to highly dynamic pressures.
10. Optical: This strain technology is used in applications where pressure measurement may be challenging such as highly remote area under high temperature by utilizing physical change of an optical fiber. Another similar technique utilizes a layered elastic film.
11. Potentiometric: This utilizes wiper motion to represent the strain.

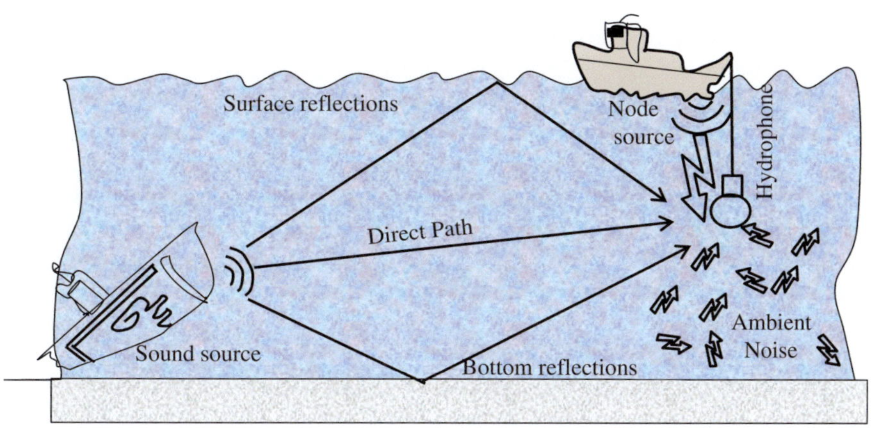

Fig. 3.24 Underwater sound transducer

12. Other types: Other techniques are used in the applications of remote locations at high temperature of gas or liquid and include the use of physical change in optical fiber to measure strain by applied pressure. A common example of this type is based on FBG. An equivalent scheme employs layered elastic film that redirects wavelengths as per applied pressure.

Force can be measured by designing a mechanical structure (Fig. 3.25) like strain gages or crystals or ceramics to reflect stiffness as force induces stress. Force can also be detected using a displacement transducer, such as an LVDT. In a strain gauge (Fig. 3.25b), electrical resistance of a wire varies with change in any strain applied and Wheatstone bridge is used to measure resistance accurately. The majority of strain gauges are resistive foil type mounted on a backing material and available in different shapes and sizes.

FlexiForce (Fig. 3.26) [19] is an ultra-thin (0.008 in.) piezoresistive force sensitive resistor (FSR) transducer with a round, 0.5″ diameter sensing area that is available in a variety of shapes and sizes with resistance reducing with force in the range of 100 g–10 kg. One side of the transducer (Fig. 3.26b) is printed with a polymer ink with embedded conductive particles [20]. When the transducer is straight, the particles give the ink a resistance of about 30 kΩ. The conductive particles move further apart (Fig. 3.26c) when it is bent away from the ink, the resistance increases to about 50 kΩ. By measuring the resistance, the bent in transducer can be determined. The resistor and the flex transducer form a voltage divider for VCC as shown in Fig. 3.27a. If 5 V is used for VCC, about 3.75 V is observed at the transducer if straight, and about 4.17 V when bent by 90°. When there is an impact on the device, it triggers a signal as shown in Fig. 3.27. A novel asynchronous interface [21] allows the transducer to be compatible with other systems. A multi-transducer control function uses a single-power-managed transducer node that detects various biomedical signals [22]. Both point-to-point and bus configurations with asynchronous and synchronous data transmission are possible.

In a surface acoustic wave (SAW) transducer unit, transmitter/receiver is held in close vicinity to have piezoelectric property, and measured data are sent to a physician using Bluetooth or GSM network. Passive SAW passive remote resonator transducer (Fig. 3.28b, c) is made from hermetically encapsulated inert silicon and employs ISM band measurable in 0–250 mm height with 1% accuracy.

Gyroscope [23] is a device for determining orientation, based on angular momentum. As the MEMS mass gets displaced, corresponding changes occur underneath capacitor plates that are converted into digital signals. Optical fiber transducers can be used in detecting partial discharges in transformers so as to prevent insulation breakdown and catastrophic failures.

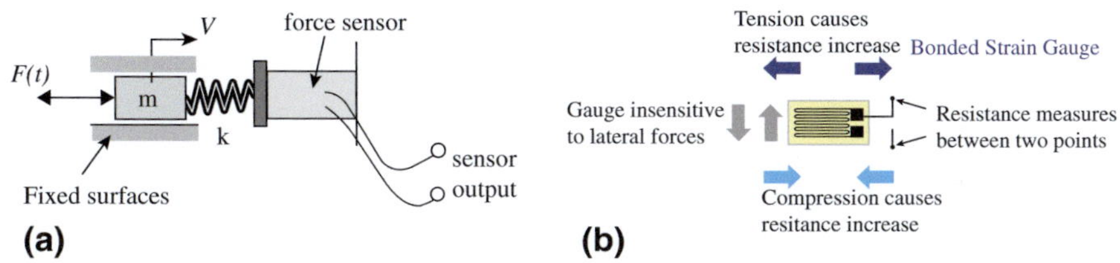

Fig. 3.25 **a** Transducer for force sensing, **b** strain gauge transducer

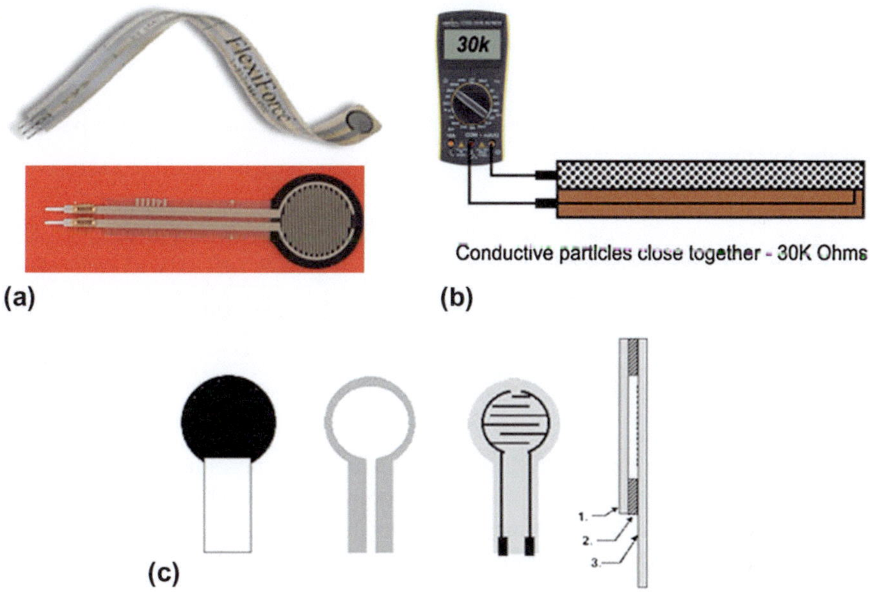

Fig. 3.26 **a** FlexiForce transducer [19], **b** conductive particle in a FlexiForce transducer [20], **c** changing resistance in a FlexiForce transducer [19]

3.11 Radiation Transducers

Radioactivity is characterized by units of activity, units of exposure, and units of absorbed dose; the basic unit of radioactivity is Becquerel [Bq]—defined as one transition per second—and it represents the rate of decay of a radionuclide. An older, non-SI (system of international) unit of activity was the curie (1 Ci = 3.7×10^{10} Bq. The basic unit of exposure is the coulomb per kilogram [C/kg] = [A s/kg]. There are three basic types of radiation transducers: ionization, scintillation, and semiconductor radiation transducers. These transducers are either detectors (detection without quantification) or transducer (both detection and quantification). *W*-value is defined as the average energy lost by the incident particle per ion pair formed.

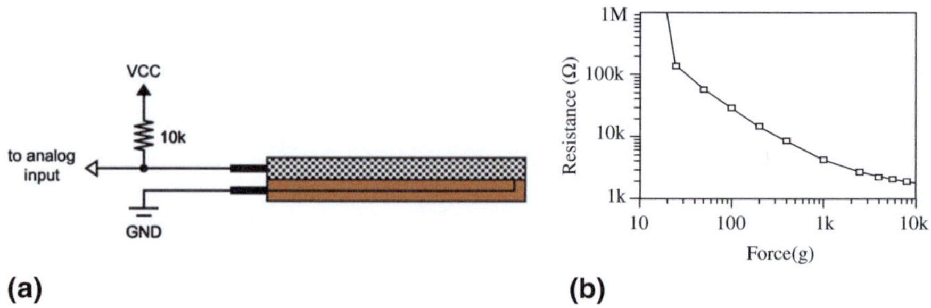

Fig. 3.27 **a** Voltage divider in a FlexiForce transducer [19], **b** variation in resistance with force in a FlexiForce transducer [19]

The strength of surfaces emitting radiation depends on the surface temperature, and the total emitted energy is:

$$M = \sigma \cdot T^4, \tag{3.14}$$

where Stefan's constant σ is 5.699×10^{-8} W m^{-2} K^{-4}, and the maximum wavelength of the emitted energy can be estimated from the Wien's displacement law as:

$$\Lambda_{max} \cdot T = C_3 \tag{3.15}$$

where $C_3 = 2897\,\mu\mathrm{m\,K}^{-1}$.

In infrared (IR) radiation, brightness temperature of the radiation is measured and is defined as the temperature of the black body which would emit the measured radiance which differs from the actual temperature of observed surface. For IR radiation, the emissivity of sea surface is between 0.98 and 0.99 and is characterized by ionization.

The frequencies are sufficiently high (above 750 THz) to ionize molecules based on Plank's equation, and many forms of radiation can penetrate through materials. On the other hand, below IR region, electromagnetic radiation can be generated and detected by simple antennas used as a transducer. At high frequencies, we can view them either as particles or as waves and the energy is given by the Planck equation as:

$$\lambda = \frac{h}{p} \tag{3.16}$$

where λ is the wavelength, h is the size of the cube, and p is the Planck constant.

Their wavelength is given by de Broglie's equation as momentum of the particle:

$$p = mv \tag{3.17}$$

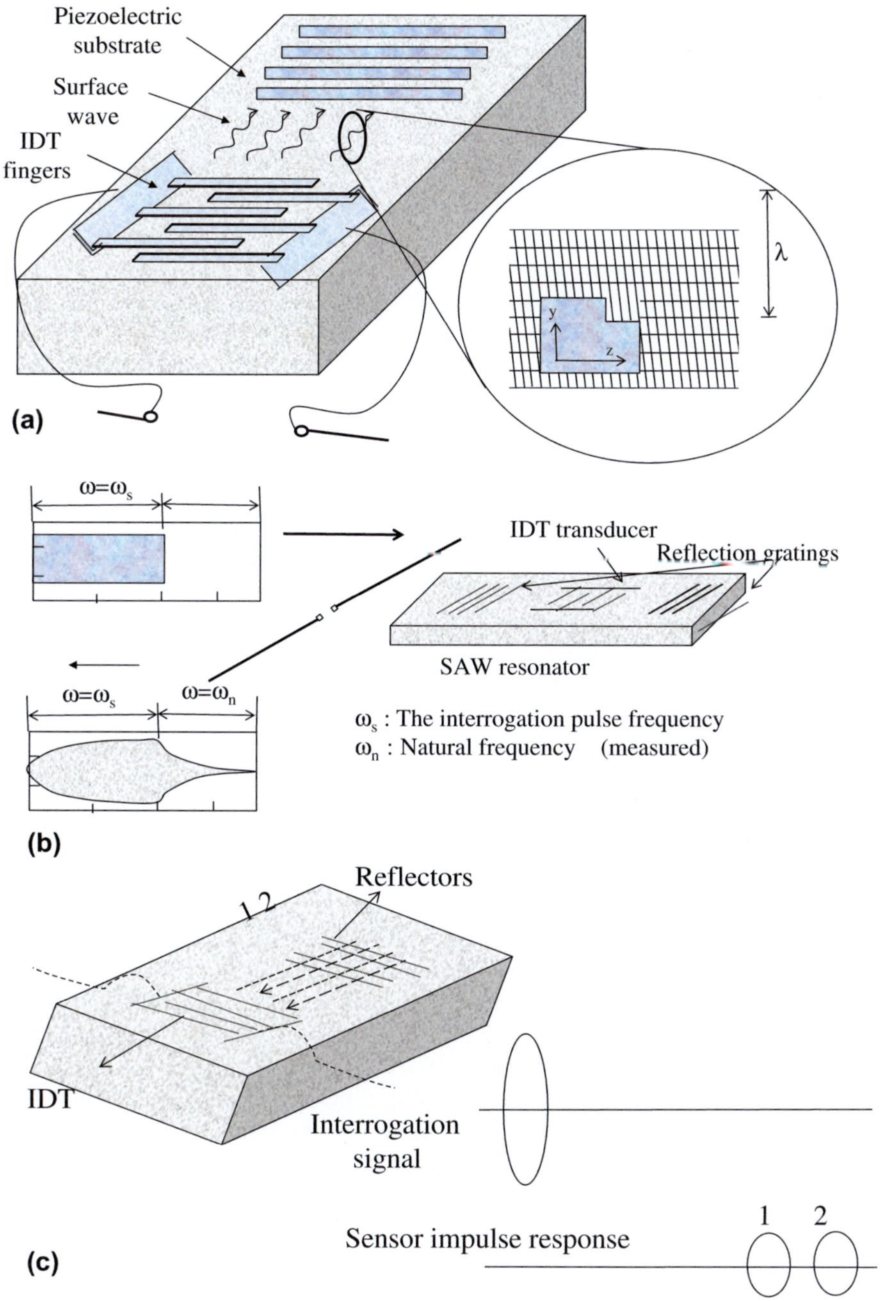

Fig. 3.28 **a** Surface acoustic wave (SAW) wireless passive remote transducer, **b** surface acoustic wave (SAW) resonator of transducer, **c** surface acoustic wave (SAW) reflective delay lines

The highest frequency in the microwave region is 300 GHz with the photon energy of 0.02 eV. The lowest frequency is approximately 3×10^{16}, and the photon energy is 2000 eV.

Thermographic cameras detect radiation in the IR range of the electromagnetic spectrum (roughly 9000–14,000 nm or 9–14 μm) and produce images called thermograms in the range −50 °C to over 2000 °C and used by firefighters. The amount of radiation increases with temperature. *Thermographic printing* [24, 25] is one example of IR imaging science. Firefighters use thermography to see through smoke, to find persons, and to localize the base of a fire. The maintenance technicians use thermography to locate overheating joints and sections of imminent failure of power lines. Building construction technicians use heat leaks in faulty thermal insulation. Night vision IR devices can produce image in the near-IR in complete darkness.

Spectroscopic sensing refers to the study of absorption, emission, or scattering of electromagnetic radiation by atoms, molecules, ions, solids, liquids, or gases [26]. Electromagnetic radiation from any matter can cause transitions between quantized energy levels, and spectroscopy enables chemical analysis and sensing. The list of such techniques is continuously increasing [27]. The chamber functions (Fig. 3.29b) in the saturation region of the *I–V* curve (Fig. 3.29c), with higher radiation frequency, lead to higher voltage across electrodes.

Two ionization chambers are used: The lower part contains a low-energy radioactive isotope (Krypton 85), while the upper part is an ionization chamber. The fabric passes between them, and the ionization current is calibrated in relation to density or thickness.

The potential across the electrodes of ionization chamber is high enough to produce an electric field beyond 10^6 V/m. Accelerated electrons collide with atoms, releasing additional electrons (and protons) called the Townsend avalanche which are collected by the anode to detect lower intensity radiation.

Geiger–Muller counter (Fig. 3.30a) is an ionization chamber with very high voltage. The output is dependent on the electric field. Semiconductor radiation transducer is essentially diode with reverse bias, and the reverse current formed by radiation is a measure of kinetic energy. The most common construction is similar to the PIN diode as shown in Fig. 3.30b. Properties of some common semiconductors are summarized in Table 3.18.

In a semiconductor radiation detector [29], a normal diode is built but with a much thicker intrinsic region doped with balanced impurities. This is achieved by an ion-drifting process by diffusing a compensating lithium throughout the layer. RADAR is the best known method of microwave sensing that is based on the reflection of waves by the target. The larger the target with intense source of waves, the larger is the signal received back from the target. Reception may be by the same or a separate antenna (Fig. 3.31a).

The scattering coefficient of electromagnetic waves called the scattering cross section or radar cross section σ is given by:

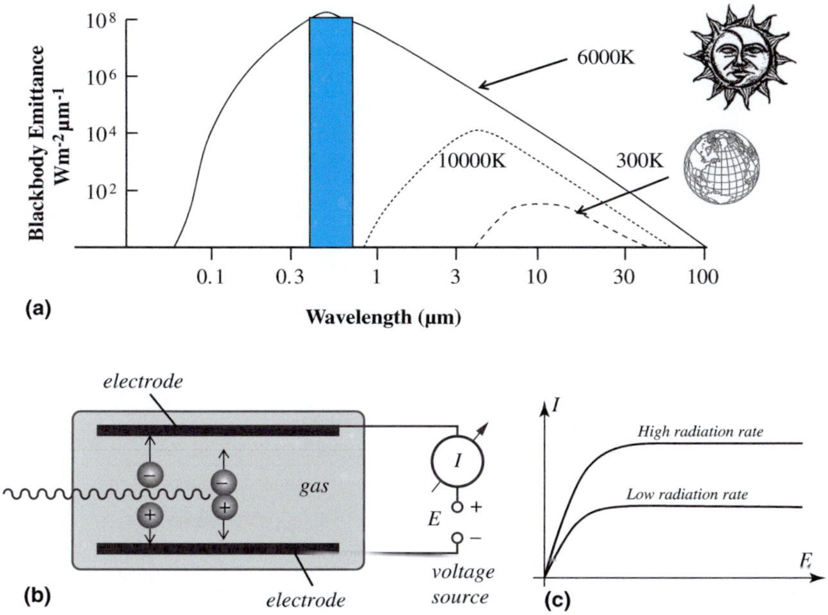

Fig. 3.29 **a** Infrared radiation as a function of wavelength, **b** ionization chamber, **c** performance of ionization chamber

$$\sigma = 4\pi R^2 \frac{P_\mathrm{s}}{P_\mathrm{i}} \tag{3.18}$$

where P_s is the scattered power density, P_i the incident power density, and R is the distance from source to target. For any object in the path of electromagnetic waves, the scattering coefficient σ is correlated as:

$$P_r = P_\mathrm{rad}\sigma \frac{\lambda^2 D^2}{(4\pi)^3 R^4} \tag{3.19}$$

where l is the wavelength, s the radar cross section, P_r the total received power, P_rad the total radiated power, and D the directivity of the antenna.

Directivity (Fig. 3.31b) is the property of antenna to sense the distance as well as size of object. A different approach (Fig. 3.31c) is based on the Doppler's effect where the amplitude and power involved are not important and rather measures a change in the frequency of the reflected waves.

Consider a moving target from a source at a velocity v as shown in Fig. 3.32a. The source transmits a signal at frequency f, and the reflected signal arrives back at the transmitter after a delay $2\Delta t$, where $\Delta t = \Delta S/v$. This delay causes a shift in the frequency of the received signal as follows:

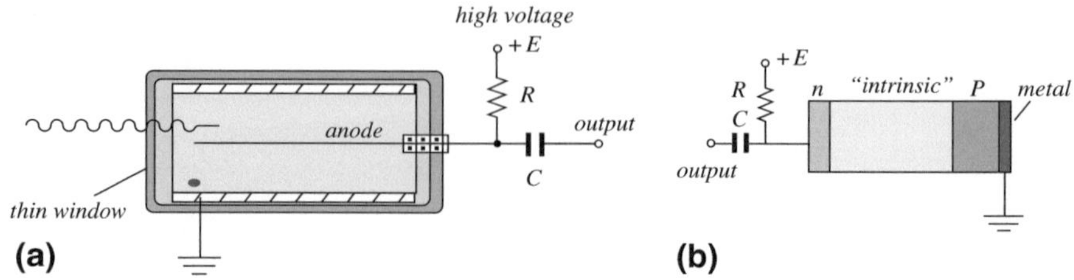

Fig. 3.30 **a** Geiger–Muller counter, **b** infrared radiation as a function of wavelength

Table 3.18 Properties of some common semiconductors [28]

Material	Operating temp [K]	Atomic number	Band gap [eV]	Energy per electron-hole pair [eV]
Silicon (Si)	300	14	1.12	3.61
Germanium (Ge)	77	32	0.74	2.98
Cadmium-teluride	300	48, 52	1.47	4.43
Mercury-Iodine (HgI_2)	300	80, 53	2.13	6.5
Gallium-Arsenide (GaAs)	300	31, 33	1.43	4.2

$$f' = \frac{f}{\left(1 + \frac{2v}{c}\right)} \tag{3.20}$$

Their names are related to the fields they produce which look like the fields of an electric dipole and a magnetic dipole, respectively, as shown in Fig. 3.32b–d. The two antennas are very similar, and magnetic field of the electric dipole is identical (in shape) to the magnetic field of the magnetic dipole and vice versa. The field radiated from a small dipole (electric or magnetic) of Fig. 3.32d is essentially the same as for an electrostatic dipole. When antennas are very close to a source (less than about one wavelength), they behave more or less like capacitors.

3.12 Transducers for Biomedical Applications

Transducers for biomedical applications need to follow some specific characteristics such as should be capable of real-time data acquisition and analysis, should be of lightweight, small size, consume ultra-low power, can do intelligent remote monitoring, provide security of patient information, should have reliability, interoperability characteristics, and be cost effective. The transducers could be classified into three areas (Fig. 3.33): health monitoring (glucose, heart rate (ECG), white blood cell count, blood pH, respiratory rate, etc.), chronic diseases (artificial retina,

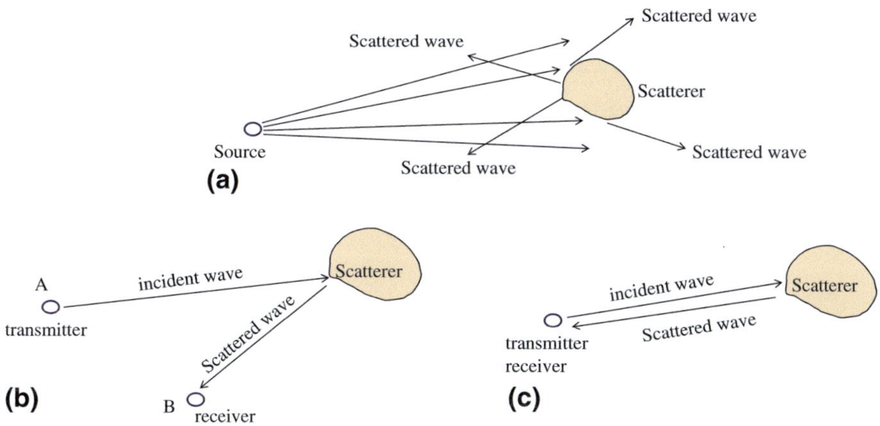

Fig. 3.31 **a** Another Geiger–Muller counter, **b** scattering coefficient, **c** Geiger–Muller counter

cochlear implants, etc.), and hospital transducers (monitor vital signs (body temperature, blood pressure, and record anomalies). Each day, more and more equipment is going "wireless" from pulse-oximeters to more complex patient vital signs monitors and ventilators. The supporting environments must scale from a few clients to 100s on a single subnet. External factors such as nearby TV and radio stations can interfere and can affect the overall performance. Therefore, interoperability profiles and standards are required to ensure plug-and-play operation in heterogeneous environments. Biomedical measurements and associated characteristics are summarized in Table 3.19.

Different biomedical data and typical data size are summarized in Table 3.20, while different wireless technologies are briefed in Table 3.21. The risk with symptoms varies with the type of health problem and real-time measurement criterion could be very subjective. So, type and number of transducers and frequency of collecting data could be selected appropriately. There are fabrics products and shirts commercially available that could connect electrodes to monitor ECG and respiratory activity of human body. Monitoring of elderly person and their daily chores is critical in taking care of them, and transducers with SNs can do unattended 24x7 monitoring without any human intervention. Data from transducers can be sent to a BS on way to physicians and medical personnel, nurses, and care givers. Elderly people suffering from Parkinson's disease need to be monitored day and night for freezing of gaits (FoG) as the patients trying to walk cannot move forward at all and frozen. Another area of interest is posture determination and monitoring. This was originally used for military tracking purposes for soldiers in finding physical location and psychological condition. This gives vital physical parameters indicating a person walking, running, or going upstairs, moreover, preventing fall for people suffering from Alzheimer and dementias. There is also a need for continuous monitoring of diabetic and cardiac patients. Monitoring health of older adults, who live at home, is also essential. Transducers and needed circuits

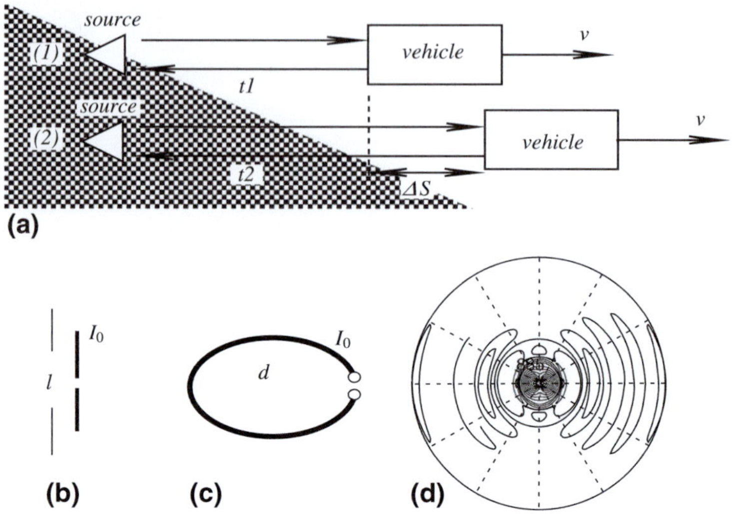

Fig. 3.32 **a** Doppler RADAR, **b**, **c**, **d** antenna as transducer

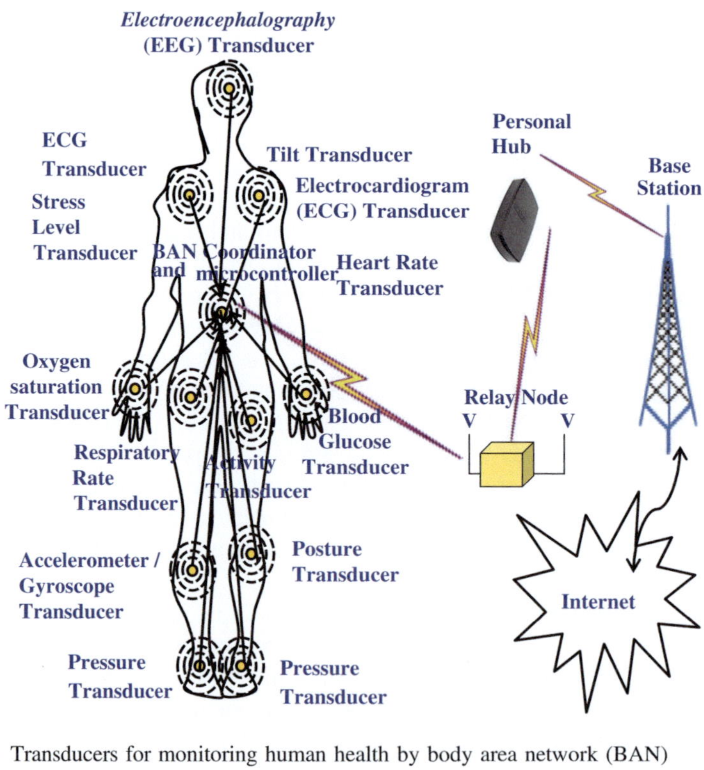

Fig. 3.33 Transducers for monitoring human health by body area network (BAN)

Table 3.19 Biomedical measurements and associated characteristics [30]

Biomedical measurements	Voltage range (V)	Number of users = K (SNs)	Bandwidth (Hz)	Sample rate (samples/s) = (Hz)	Resolution [b/sample]	Information rate [b/s]
ECG	0.5–4 m	5–9	0.01–250	1250	12	15,000
Heart sound	Extremely small	2–4	5–2000	10,000	12	120,000
Heart rate	0.5–4 m	2	0.4–5	25	24	600
EEG	2–200 μ	20	0.5–70	350	12	4200
EMG	0.1–5 m	2+	0–10,000	50,000	12	600,000
Respiratory rate	Small	1	0.1–10	50	16	800
Temperature of body	1–100 m	1+	0–1	5	16	80

Bandwidth = $f_{max} - f_{min}$; sample rate = 5_f_{max}; information rate = R_b = resolution sample rate

are installed in seniors' homes to detect changes in behavior and physical activity, including walking and sleeping patterns, which could be a advantage. Transducers can also be used to monitor the number of repeats of a given exercise for rehabilitation purpose. In sports, transducers can be used as a tool for mechanical functioning of the body, like the heart and muscles conditions. So, it can play a vital role in human health monitoring.

Another area of interest is the use of reflective markers in determining and capturing mobility of dancers. Even, multiple cameras that digitize different views of performance can be very useful as a camera is equivalent to have a large number of transducers at pixels constituting a frame. This is also useful in posture determination and steps in determining postures using camera transducer is given in Fig. 3.34.

Posture determination can be done for the human whole body, portions of the body, facial animation, animals, puppets, and other objects. Recently, a chair with 19 transducers has been designed [32] which is robust and provides *up to 87% accuracy in classifying 10 postures with new subjects. This l*ow-cost chair utilizes a *n*ovel scheme of *domain knowledge and near-optimal transducer placement* methodology and requires only *19 pressure transducers to provide n*ear-real-time performance on a *10-Hz standard desktop computer.*

Another critical issue in healthcare is concussion that is characterized by clinical syndrome with immediate and transient damage of neural function. Transducers can be used to determine the effectiveness of a helmet in protecting brain from any potential Fig. 3.35 injuries. A head impact telemetry system (HITS) has been developed [33, 34] for concussion determination and a combination of 6 transducers to detect the magnitude, duration, location, and direction of up to as many as 100 football hits and communicate to a working station within 150 yards. When the impact force is more than 10 g, the HITS transducers are activated and 6 transducers detect magnitude, duration, location, and direction.

A transducer can be permanently placed inside human body [35] as a single intervention can greatly reduce the risk of any infection while continuous monitoring

Table 3.20 Biomedical data and file size [31]

Biomedical data	Type	Typical file size
ECG recording	Electrical signal	100 kB
Electronic stethoscope	Audio	100 kB
X-ray	Still image	1 MB
30 s of ultrasound image	Moving image	10 MB

Table 3.21 Wireless technologies and data rates [31]

Technology	Data rate	Frequency spectrum (MHz)
GSM	9.6 kbps	900/1800/1900
GPRS	171.2 kbps	900/1800/1900
EDGE	384 kbps	900/1800/1900
3G/UMTS	2 Mbps	1885–2200

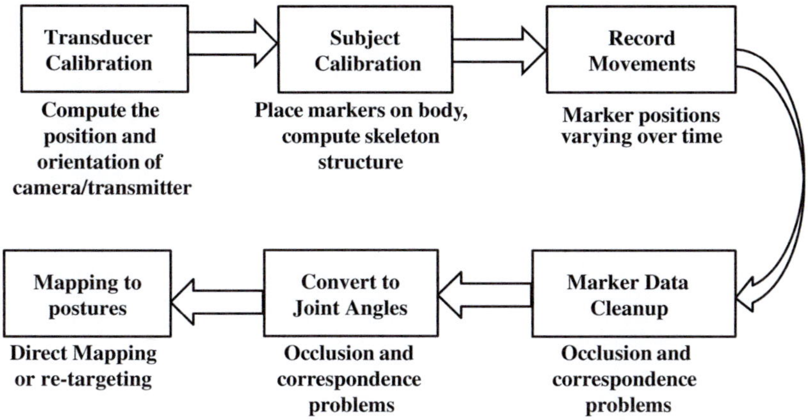

Fig. 3.34 Steps for posture determination using camera transducers

Fig. 3.35 Chair with transducers for posture determination

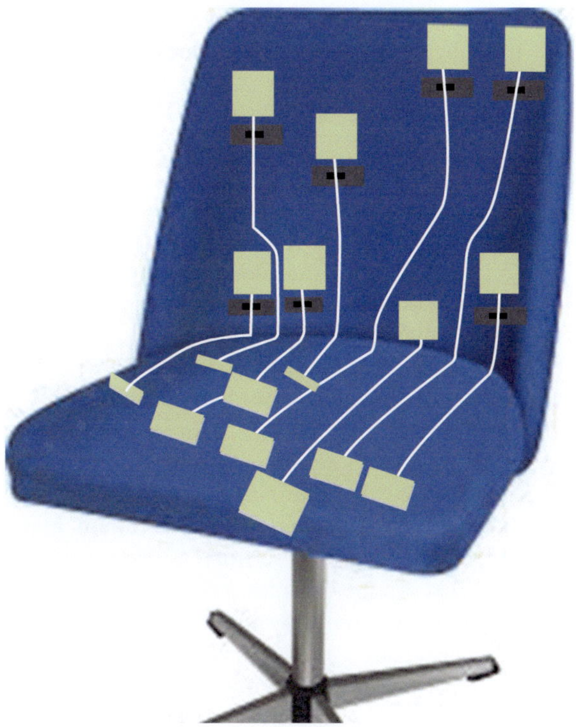

is possible over years without interfering with normal physiology. Use of existing SAW (standing acoustic wave) pressure transducer oscillator crystal was loaded in several orientations to determine the basic stress sensitivity [36]. In the dynamic mode test, the SAW oscillator crystal was exposed to propagation sound waves in the water tank of a hydrophone calibrator. It was concluded that cantilever-sensor configurations are more sensitive than diaphragm-sensor arrangements [36].

RFID tag also acts as a transducer as it automatically identifies and tracks electronically stored tag attached to an object. RFID is the wireless use of electromagnetic fields to transfer data [36] as tags are powered by and read at short range (a few meters) via electromagnetic induction reader. RFID tags are used in many industries such as attached to an automobile during production can be used to track its progress through the assembly line; pharmaceuticals can be tracked through warehouses; livestock and pets may have tags injected, allowing positive identification of the animals. It is also been added to a passport for quick identification.

3.13 Conclusion

Transducer plays a very important role in making SN effective for numerous applications. New transducers are being developed daily, enhancing the usefulness of SNs and are expected to grow at an unprecedented rate. The challenge is to define new transducers and it is up to researchers to make WSNs versatile for human life.

3.14 Questions

Q.3.1. Why do you need a transducer?
Q.3.2. What makes a transducer appropriate for a given application?
Q.3.3. How do you compare an analog and a digital transducer?
Q.3.4. Is an electric lamp a transducer?
Q.3.5. Is a Microphone a transducer?
Q.3.6. How can you calibrate a transducer?
Q.3.7. How can you calibrate a transducer?
Q.3.8. How do you define resolution of a transducer?
Q.3.9. How do you determine sensing range of a transducer?
Q.3.10. What term describes the maximum expected error associated with a measurement or a sensor?
Q.3.11. What are the qualities a transducer should possess?
Q.3.12. How do you determine the frequency of data to be collected by a transducer?
Q.3.13. What are the limitations of a transducer?
Q.3.14. Are sensors and actuators examples of transducers?
Q.3.15. Resolution of a transducer depends on?
Q.3.16. What are the parameters that dictate a transducer capability?
Q.3.17. How do you use the capacitance microphone for detection of heart sound?
Q.3.18. What is a filter and what is its usefulness?
Q.3.19. What is a strain gauge? What are its types and applications?

Q.3.20. What are active transducers?
Q.3.21. What are thermocouples? What is their principle?
Q.3.22. What are thermistors? What are their applications?
Q.3.23. What are merits and demerits of capacitive transducers?
Q.3.24. What are passive transducers? Discuss working of a passive transducer.
Q.3.25. What is meant by biometrics?
Q.3.26. What is meant by patient monitoring?
Q.3.27. Why do you use polymer probes as compared to conventional ceramic probes?
Q.3.28. How can you detect small intergranular stress corrosion in an austenitic allow in or near a weld?
Q.3.29. You want to do flaw detection in a plastic pipe and the material is very lossy with high attenuation even at 2 MHz. The pipe is 10 mm outer diameter and 5 mm inner diameter and you must inspect it at a high rate. What sort of transducers would you choose?
Q.3.30. Why is flow measurement necessary?
Q.3.31. What kind of method can you use in measuring level of molten glass?
Q.3.32. What is meant by signal conditioning in a transducer?
Q.3.33. What is the role of noise in a transducer?
Q.3.34. Under what conditions a transducer is said to have good frequency response?
Q.3.35. How can you classify transducers?
Q.3.36. On what principles do the variable inductance type of transducers work?
Q.3.37. What are potential uses of thermistors?
Q.3.38. Why does galvanometer give slower response than an oscilloscope?
Q.3.39. Why are capacitive transducers used for dynamic measurements?
Q.3.40. What is principle of photoelectric transducers?
Q.3.41. What is piezoelectric transducer?
Q.3.42. What is a piezoelectric transducer?
Q.3.43. What are advantages and disadvantages of a piezoelectric transducer?
Q.3.44. What are the advantages and limitations of foil strain gauges?
Q.3.45. What are principles on which inductive transducers work?

References

1. https://en.wikipedia.org/wiki/Transducer.
2. https://www.google.com/search?q=sensor+specifications&biw=1097&bih=555&tbm=isch&imgil=_c0LDIkt3Tuj1M%253A%253BkuY9nZDt-kvE3M%253Bhttp%25253A%25252F%25252Fwww1.keyence.com.sg%25252Fproducts%25252Fpressure%25252Fpressure%25252Fap40%25252Fap40_specifications_1.php&source=iu&pf=m&fir=_c0LDIkt3Tuj1M%253A%252CkuY9nZDt-kvE3M%252C_&usg=__zzTON31FHs2vZLPBUwmSZMie6IU%3D&ved=0ahUKEwjL7_3W2LPKAhXJ8j4KHdBEAvsQyjcIKw&ei=7AadVsvJA8 nl-wHQiYnYDw#tbm=isch&q=types+of+sensors+and+specifications&imgrc=_lTmDLQTJtGLmM%3A

3. https://www.watlow.com/reference/guides/0205.cfm.
4. https://www.google.com/search?q=comparison+of+temperature+sensors&sa=X&biw=1097&bih=555&tbm=isch&tbo=u&source=univ&ved=0ahUKEwjP5cu2k7bKAhVB5yYKHZBVAm0QsAQIOg&dpr=1.75.
5. M. N. Halgamuge, M. Zuckerman, and K. Ramamohanarao, "An estimation of sensor energy consumption," Progress in Electromagnetic Research B, 2009, vol. 12, pp. 259–295.
6. http://www.allsyllabus.com/aj/note/ECE/ELECTRONIC%20INSTRUMENTATION/Unit7/SELECTING%20A%20TRANSDUCER.php#.Vp1dM_krJD8.
7. http://nesl.ee.ucla.edu/tutorials/mobicom02/slides/Mobicom-Tutorial-4-DE.pdf.
8. M. Anjanappa, K. Datta, and T. Song, Introduction to Sensors and Actuators, Chapter 16, The Mechatronics Handbook – CRC Press 2002.
9. https://sensorsandtransducers.wordpress.com/2012/02/06/bimetallic-strip/.
10. http://www.ni.com/white-paper/13654/en/.
11. http://www.cctv-information.co.uk/i/CCTV?url=http://www.cctv-information.co.uk/constant/infrared.html.
12. http://en.wikipedia.org/wiki/Night_vision.
13. www.ann.ece.ufl.edu/courses/eel6935_09spr/…/Automotive-Sensors.ppt.
14. http://www.engineersgarage.com/articles/humidity-Transducer.
15. http://en.wikipedia.org/wiki/Rain_sensor .
16. https://en.wikipedia.org/wiki/Permeability_(electromagnetism).
17. http://en.wikipedia.org/wiki/Magnetometer.
18. I. F. Akyildiz, D. Pompili, and T. Melodia, "Underwater Acoustic Sensor Networks: Research Challenges," Ad Hoc Networks (Elsevier), vol. 3, no. 3, pp. 257-279, March 2005.
19. http://www.tekscan.com/flexible-force-sensors.
20. https://www.sparkfun.com/tutorials/270.
21. Chiung-An Chen, Shih-Lun Chen, Hong-Yi Huang, and Ching-Hsing Luo, "An Asynchronous Multi-Sensor Micro Control Unit for Wireless Body Sensor Networks (WBSNs)," Sensors 2011, vol. 11, pp. 7022–7036, 2011.
22. http://en.wikipedia.org/wiki/Peripheral_Sensor_Interface_5.
23. https://www.ifixit.com/Teardown/iPhone+4+Gyroscope+Teardown/3156.
24. https://www.science.mcmaster.ca/medphys/images/files/courses/4R06/4R6Notes3_GasFilled_Detectors.pdf.
25. http://www.epa.gov/radiation/docs/cleanup/nanotechnology/chapter-3-spectroscopic-sensors.pdf.
26. http://en.wikipedia.org/wiki/Thermography.
27. http://www.epa.gov/radiation/docs/cleanup/nanotechnology/chapter-3-spectroscopic-sensors.pdf.
28. https://books.google.com/books?isbn=0070583692.
29. https://en.wikipedia.org/wiki/Semiconductor_detector.
30. S. Arnon, et al., "A Comparative Study of Wireless Communication Network Configurations for Medical Applications," IEEE Wireless Communications, pp. 56–61, February 2003.
31. M. Fadlee, et al., "Bluetooth Telemedicine Processor for Multichannel Biomedical Signal Transmission via Mobile Cellular Networks," IEEE Transactions on Information Technology in Biomedicine, pp. 35–43, March 2005.
32. Bilge Mutlu, Andreas Krause, Jodi Forlizzi, Carlos Guestrin, Jessica Hodgins, "Robust, Low-cost, Non-intrusive Sensing and Recognition of Seated Postures," CMU, repository.cmu.edu/cgi/viewcontent.cgi?article=1040&context=hcii.
33. http://www.simbex.com/hit-system.html.
34. http://en.wikipedia.org/wiki/Radio-frequency_identification.
35. "Applications for implantable SAW pressure sensors," ubimon.doc.ic.ac.uk/bsn/public/mcleod-slides.ppt.
36. E. J. Staples, "A surface acoustic wave (SAW) pressure sensor evaluation," Final Report by Rockwell International, 1979, www.dtic.mil/dtic/tr/fulltext/u2/a325816.pdf.

Chapter 4
Transducers' Range Modeling

4.1 Introduction

Transducers have been proposed for many unattended surveillance applications, forest fire detection, and healthcare monitoring. As we use a thermometer to measure temperature, the reading represents for the whole body and there is no need to use a second thermometer. Thus, the body area represents sensing range of a thermometer transducer. So, in most common applications, each transducer represents that parameter for a given area and it is said to detect events within its *sensing range* . The event could include presence of fire, presence of an animal, or no. of soldiers, or tanks or flow of oxygen. Transducers collaborate other functional units of a SN to deliver data to processing centre commonly known as a base station (BS or sink node) via one or more intermediate SNs. Many previous works assume *disk* sensing model [1] wherein a transducer is assumed to accurately indicate value of sensed parameter in a circular area of radius r_s (Fig. 4.1a) and do not have any knowledge outside the area as shown in Fig. 4.1b. A laser transducer emits laser beam transducer in a linear form and is illustrated in Fig. 4.2a, and a digital camera coverage area is in the form of a fan as shown in Fig. 4.1b. The wiggle effect shown in Fig. 4.2c confirms the fact that nothing is perfect in the world and exact location of the transducer may not be accurate. So, still assuming circular sensing area, the circle wiggles at the exact location of the transducer. The area covered by a transducer is simply approximated by a wiggle from its central location.

Additional area covered by the wiggle effect of an *i*th transducer may have error $w_{i.}$ Let disk radius be represented by *X and* wiggle radius *y as an* (*X*, *y*) *disk.* Then, an (*R*, *w*) disk is functionally equivalent to an (*R*–*w*, 0) disk. For any point *P of distance greater than R–w from the center of the* disk *O, there exists a point Q whose distance from O is less than w and whose distance* from *P* is greater than *R*. Point *P* is not covered if the center of the disk is on *Q*, and it is better to take an intersection of the two sensing ranges: normal range and wiggle range and are

D.P. Agrawal, *Embedded Sensor Systems*,
DOI 10.1007/978-981-10-3038-3_4

illustrated in Fig. 4.4. 1-D wiggle basically takes error in one dimension, say x; while 2-D wiggle incorporates error in both x and y dimensions (Fig. 4.3).

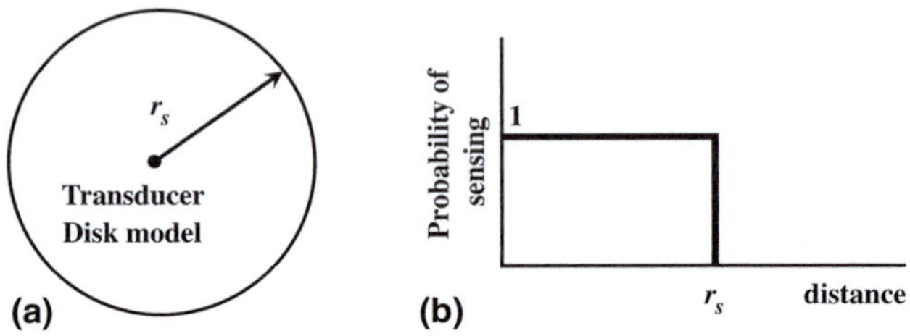

Fig. 4.1 **a** Disk model of a transducer and **b** probability of sensing by a transducer

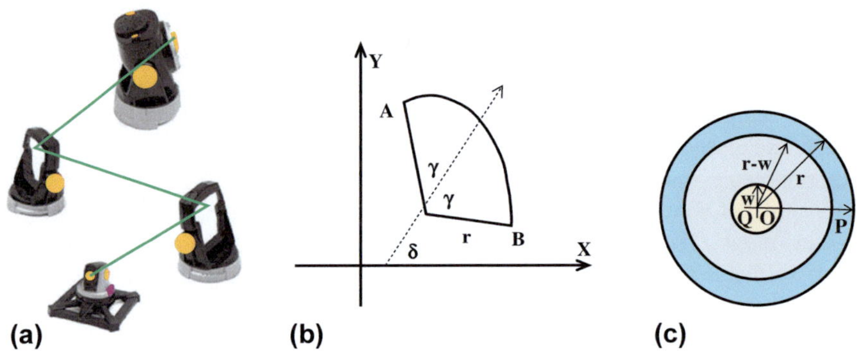

Fig. 4.2 **a** Linear segment of a laser transducer [2], **b** Fan model of a camera transducers, and **c** Wiggle effect of a disk model for a transducer

Fig. 4.3 **a** Sensing range of a transducer, **b** sensing range of a transducer with 1-D giggle, and **c** sensing range of a transducer with 2-D giggle [3]

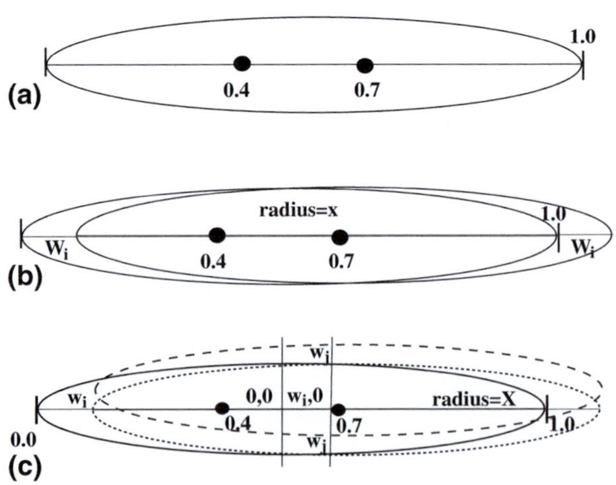

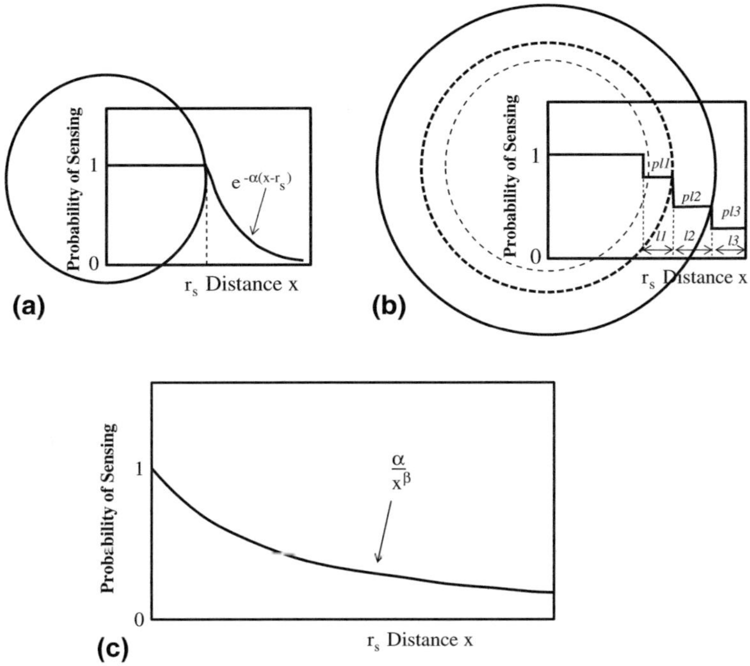

Fig. 4.4 Three sensing range models of a transducer **a** Zou et al. [4], **b** Ahmed et al. [5], and **c** Liu et al. [7]

4.2 Modeling of a Transducers' Sensing Range

Disk sensing model of a transducer is being used as it is easier to design and analyze coverage area while it is not realistic [4–6]. For example, you take a temperature of a child and an adult and tend to use body area as the sensing coverage area of the thermometer transducer that varies drastically. Moreover, nothing is perfect in the world and saying that the transducer can only represent of an area r_s and cannot say beyond r_s is rather incorrect as we are measuring physical parameter of an environment where value is expected to change slowly and rather smoothly. Moreover, signals do not fall abruptly and there are fair chances that the transducer will estimate the event parameter beyond r_s. So, many other models have been suggested in the literature and three most important ones are shown in Fig. 4.5.

A major assumption in disk model is that if an event occurs at a distance up to r_s, the transducer will correctly record and represent occurrence of the event and will be detected. The disk sensing model of a transducer is appealing as it is simple and less complicated to analyze and simulate. However, it is quite unlikely that physical signal will drop abruptly from full value to zero and there is some chance to detect an event at distance greater than r_s from the transducer. Thus, a disk model does not

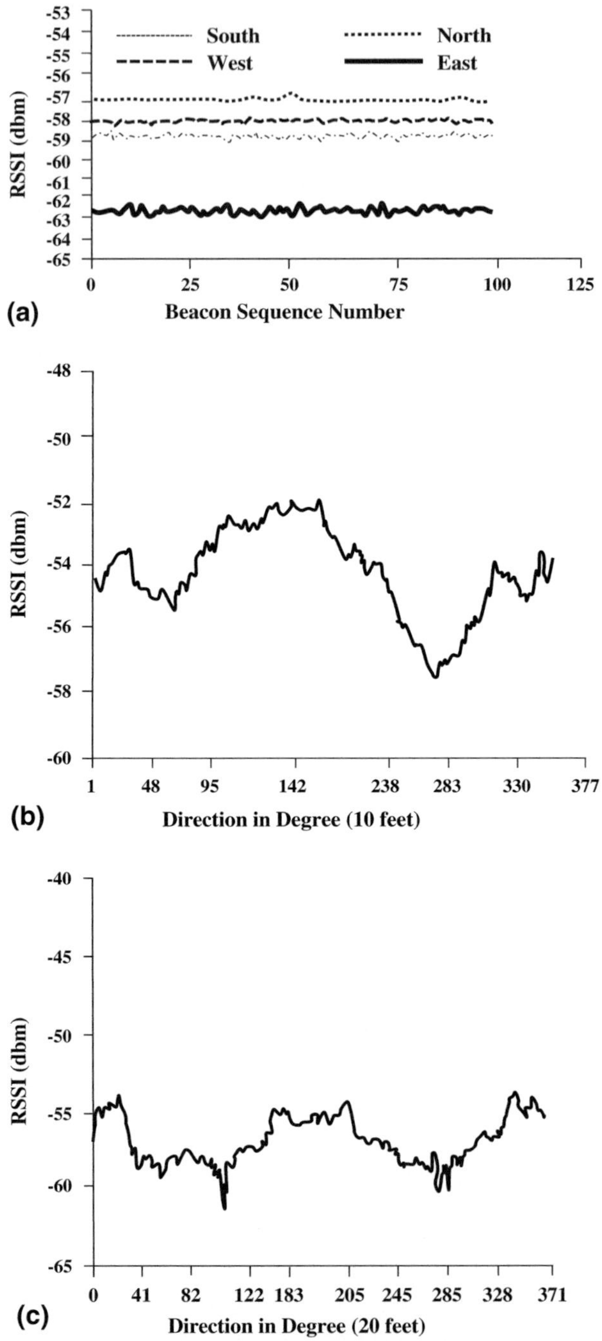

Fig. 4.5 **a** Signal strength in 4 directions as a function of time, **b** received signal strength (RSSI) measured at 10 feet, and **c** received signal strength (RSSI) measured at 20 feet [9]

fully utilize transducers' sensing ability and more transducers may be deployed than needed, requiring more money. The redundant transducers cause increased interference and waste energy. In the first model [4], the probability of sensing correctly up to r_s distance is assumed to be 1 and then decays exponentially. In the second model, a step function is used and the probability of correctly predicting sensed parameter is assumed to be fixed (<1 for distance larger than r_s) for different ranges. In the third model, the probability of correctly measuring the parameter varies exponentially with distance of point from the transducer.

A probabilistic sensing model of a transducer is assumed to realistic as the parameter being sensed the transducer design and the ecological settings are stochastic in nature. Identical SNs and transducers do not exhibit exactly identical behavior, and sensing range of transducers no longer follows nice regular disks. A probabilistic coverage (PCP) model has been proposed [8] wherein no single sensing model can accurately model different transducer types and the surrounding environment could impact the results. Simulation results indicate that PCP outperforms other schemes in terms of transducer (SN) requirements and energy consumption and robust against random failure and imprecise location determination and inaccurate synchronization. In PCP, an area is covered by all n-deployed transducers collectively. The least-covered probability θ of a point in a given subarea is the centre of three transducers forming a triangle, and the maximum separation between any two transducers dictate the number of SNs and transducers needed. PCP has been implemented PCP [8] using ns-2 for up to 1000 nodes. Later on up to 20,000 nodes deployed in a 1 km × 1 km area.

Characteristics of different types of transducers are summarized in Table 4.1. In most applications, disk model is adopted. But, radio propagation follows anisotropic diffusion (nonlinear) process and the value changes continuously, even with a small change in the direction. A pair of MICA motes is used [9] to observe such

Table 4.1 Characteristics of different types of transducers [9]

Transducer type	Data rate	Sensing range	Sensing power	Coverage shape
Temperature/humidity	Low	Short	Low	Disk
Vibration	High	Short	Low	Disk
Pressure	Low	Short	Low	Disk
Microphone	High	High	High	Disk
Smoke detector	Low	Long	Low	Disk
Accelerometer	High	Short	Low	Disk
Radiation	Low	Long	Low	Disk
Scanner	High	Short	High	Disk
Motion using accelerometer	Low	Long	High	Disk
Infrared laser tripwire	High	Short	High	Line segment
Camera	High	Long	High	Fan
Motion using camera	Low	Short	Low	Fan
Magnetic	Low	Short	Low	Fan
Electrical Field	Low	Short	Low	Fan

irregular variations in received signal strength, packet reception ratio, and communication range and is shown in Fig. 4.5a. Thus, the radio irregularity is a common phenomenon and it is hard to neglect. Variation in the signal strength in different directions also affects the packet loss rate.

Irregularity is present, indicating different communication ranges (Fig. 4.6. RSS threshold is selected between −55.5 and −59 dBm. dBm could be written more completely as dBmW (Fig. 4.7).

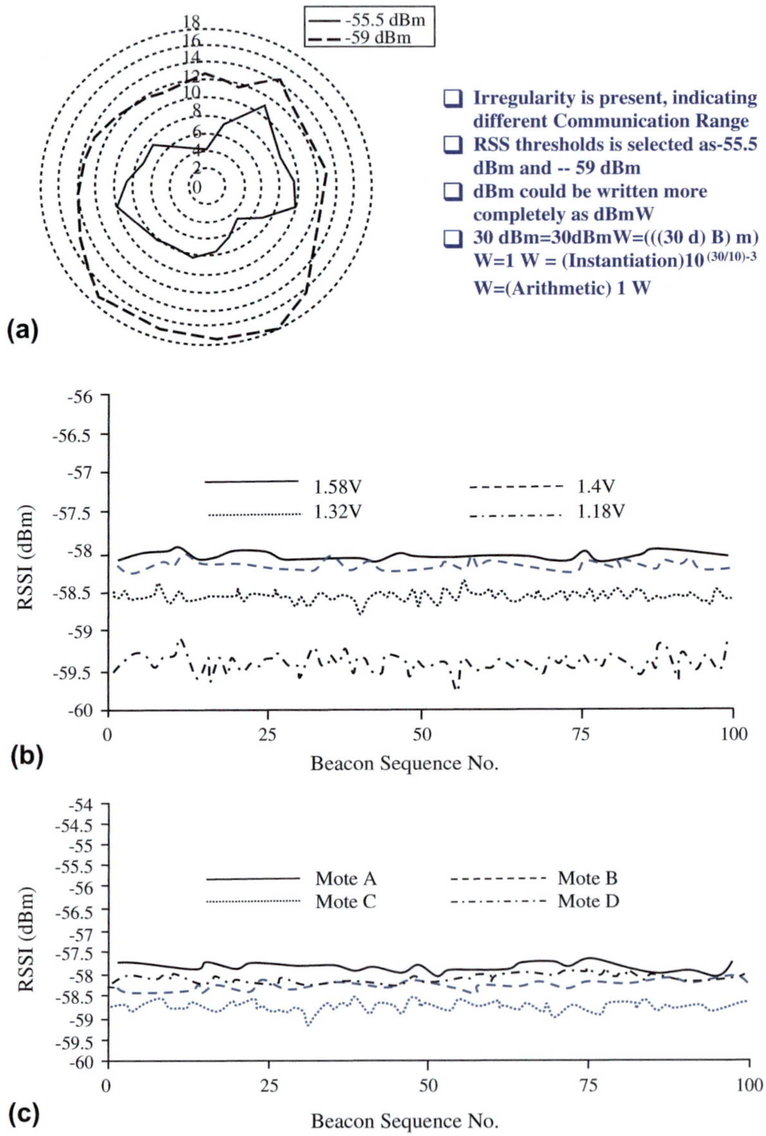

Fig. 4.6 **a** Anisotropic ratio range, **b** one mote with different battery status, and **c** many motes with the same battery status [9]

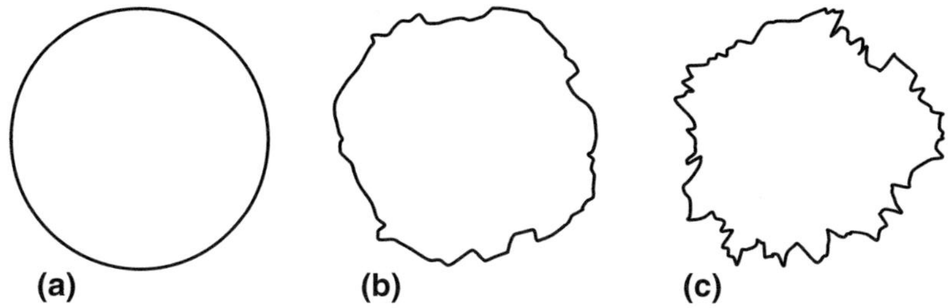

Fig. 4.7 **a** Degree of irregularity = 0, **b** degree of irregularity = 0.003, and **c** degree of irregularity = 0.01 [9]

Battery power level (Fig. 4.8a) does not remain fixed and does not fit a normal distribution. SNs are divided into two groups: sentry SNs and non-sentry SNs. Sentry SNs are supposed to work all the time. Non-sentry SNs are put to sleep to save power and are only awakened when an important event occurs. The degree of irregularity (DOI) assumes an upper and lower bounds on signal propagation. RIM (radio irregularity model) takes into account radio sending energy, energy loss, background noise, and interference among different communication signals. Figure 4.9b, c show

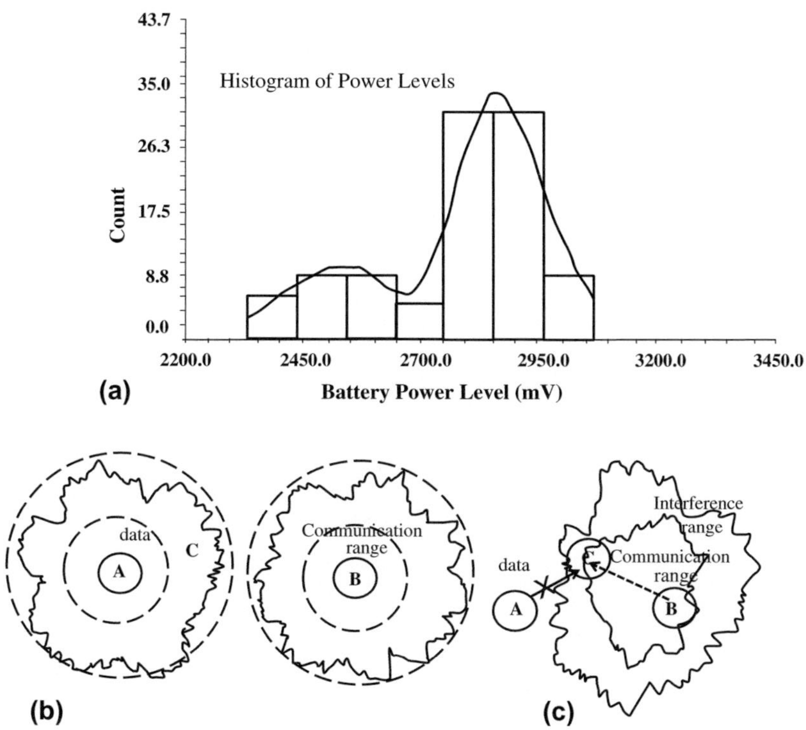

Fig. 4.8 **a** Battery power snapshot, **b** no interference in DOI, and **c** interference in radio irregularity model [9]

Fig. 4.9 **a** Impact of topology with symmetric connectivity, **b** impact of topology with asymmetric connectivity, and **c** impact of topology with no connectivity [9]

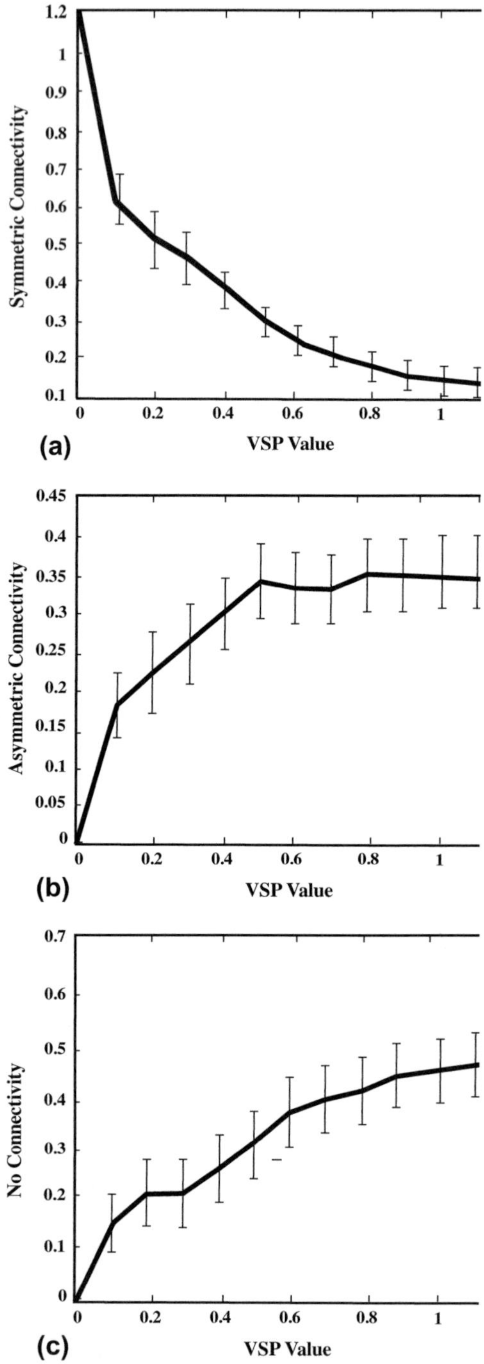

the impact of communication interference. Geographical adaptive fidelity (GAF) is also considered as a function of symmetric, asymmetric, and no connectivity (Fig. 4.9) between adjacent SNs in a terrain divided into virtual grids, and in each grid, one SN remains awake while others sleep to save power.

Many transducers exhibit directional properties such as measuring sound with voice, wind speed, direction mass airflow, wind direction, traffic counting, vibration, tilt, occupancy, directional LEDs, pressure, heat and temperature. The corresponding sensing area of a directional transducer SN under the disk and sector models is shown in Fig. 4.10. A directional sensing area approximated by the polygon model is shown in Fig. 4.11a. The sensing range is approximated by a set of nodes $[(R_1, q_1)\ldots(R_i, q_i)\ldots(R_x, q_x)]$, where R_i is the distance of i**th** node from the SN and q_i is counterclockwise angle. The calculation of $R_s(S_n, P_i)$ under the polygon model (Fig. 4.11b) can be given by the sensing range of S_n in the direction of P_i that can be given by:

$$R_s(S_n, P_i) = \frac{(R_{sp} \cdot R_{sq}) \times \sin(\theta_{sq} - \theta_{sp})}{R_{sp} \times \sin(\theta(s_n, P_i) - \theta_{sp}) - R_{sq} \times \sin(\theta(s_n, P_i) - \theta_{sq})}, \tag{4.1}$$

where $q_{sp} < q(S_n, P_{i)} < q_{sq}$, $R_s(S_n, P_{i)} = d(P_a, P_j)$, and P_j is the intersection point of ray $P_a{\cdot}P_j$ and line segment vex$_p$ and vex$_q$. The sensing coverage level (SCL) of a SN S_n that can be used by topology control mechanism (Fig. 4.11c) is given by:

$$\mathrm{SCL}(S_n, P_i) = [R_s(S_n, P_i)/d(S_n, P_i)]^2 \tag{4.2}$$

The sensing range S_n in the direction P_i can be given by:

$$R_s(S_n, P_i) = \frac{(R_{sp} \cdot R_{sq}) \times \sin(\theta_{sq} - \theta_{sp})}{R_{sp} \times \sin(\theta(s_n, P_i) - \theta_{sp}) - R_{sq} \times \sin(\theta(s_n, P_i) - \theta_{sq})} \tag{4.3}$$

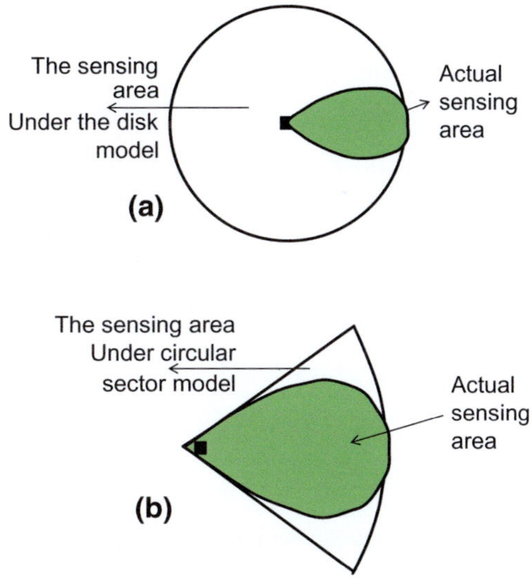

Fig. 4.10 **a** Sensing area of a directional transducer and **b** sensing area of a circular sector directional transducer [10]

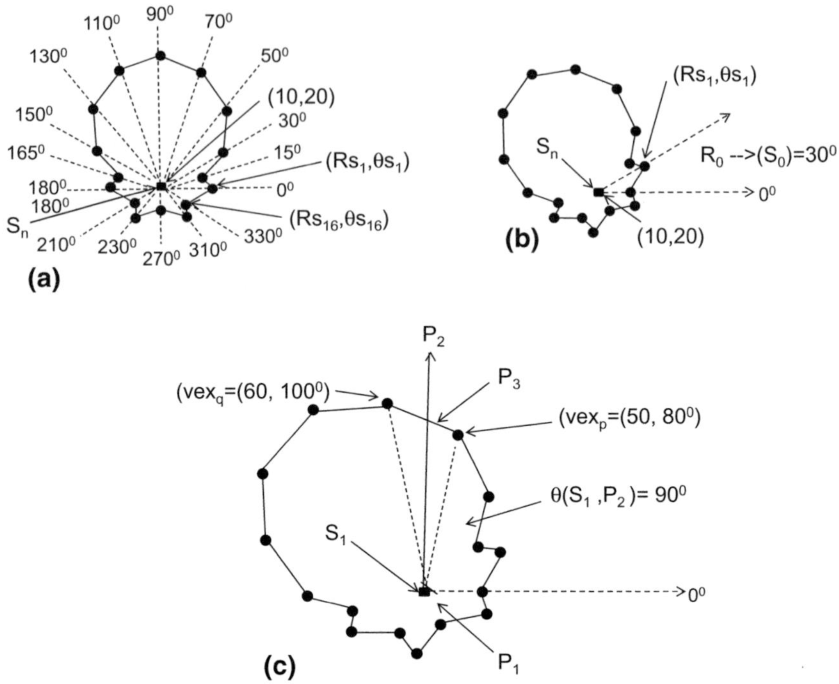

Fig. 4.11 **a** Directional sensing area by the polygon model, **b** directional sensing area approximated by the polygon model, and **c** calculation of sensing coverage range $R_s(S_n, P_i)$ [10]

This is plotted in Fig. 4.12. If $d(S_n, P_i) \leq R_s(S_n, P_i)$, P_i is within the sensing area of S_n. Otherwise, it cannot be covered by sensing area of S_n. In case of obstacles, the calculation of the CCL (communication coverage level) and SCL follows the "line-of-sight" property.

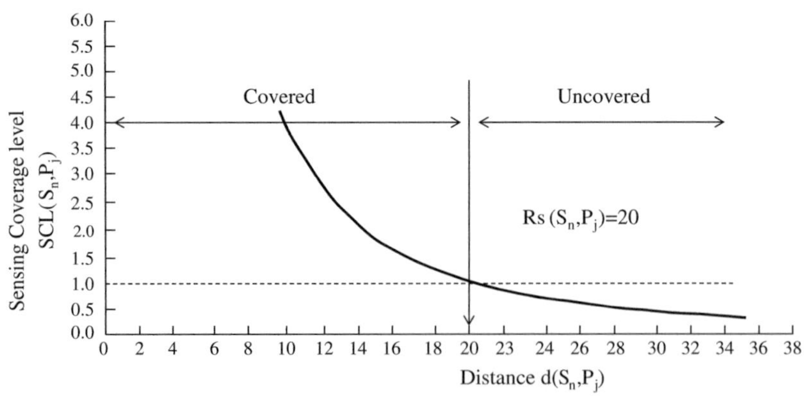

Fig. 4.12 Relationship between sensing coverage level and distance from S_n [10]

Table 4.2 Subdivisions of IR [12]

Division name	Abbreviation	Wavelength (μm)	Frequency (THz)	Photon energy (meV)
Near-infrared	NIR, IR-A*DIN*	0.75–1.4	214–400	886–1653
Short-wavelength infrared	SWIR, IR-B*DIN*	1.4–3	100–214	413–886
Mid-wavelength infrared	MWIR, IR-C*DIN*; MidIR. also called intermediate infrared (IIR)	3–8	37–100	155–413
Long-wavelength infrared	LWIR, IR-C*DIN*	8–15	20–37	83–155
Far-infrared	FIR	15–1000	0.3–20	1.2–83

In most applications, disk-type model is used except in few cases where either infrared beam is used to have point-to-point communication or camera with pixel-based transducers is used in a camera SN (C-SN) to provide a fan-type coverage. Infrared (IR) is invisible radiant electromagnetic energy that has longer wavelength of 700 nm to 1 mm (frequency 430 THz to 300 GHz) than visible light [11]. It is worth mentioning that most thermal radiation by objects in room temperature is IR. IR is commonly divided into the schemes shown in Table 4.2.

Near-IR is defined by water absorption in fiber optical communication. Water absorption increases to 1450 nm in short-wavelength IR while 3–5 μm band is used in mid-wavelength IR missiles. The thermal imaging provides complete image of objects in long-wavelength IR. The IR communication band is divided [13] as shown in Table 4.3.

C is the dominant band for long-distance communication networks while S and L bands are not widely deployed. IR is also employed in short-range communication between peripherals by LEDs where a beam is switched off and on to encode data. IR cannot penetrate walls but does not interfere with other devices. IR lasers operate in 4 gigabits/s in urban areas. Strong IR in high temperature could be harmful to eyes.

Table 4.3 Telecommunication bands in IR [13]

Band	Descriptor	Wavelength range (nm)
O band	Original	1260–1360
E band	Extended	1360–1460
S band	Short wavelength	1460–1530
C band	Conventional	1530–1565
L band	Long wavelength	1565–1625
U band	Ultra-long wavelength	1625–1675

4.3 Modeling of Camera Transducers' (C-SN) Sensing Range

A digital camera (C-SN) is a solid-state system that captures light and converts through a viewfinder or LCD monitor into an image and an equivalent of a film [14]. Received image can be manipulated or scaled. Characterizing parameters of a C-SN are image size, resolution, low-light performance, depth of field, dynamic range, lenses, and physical size. Smaller C-SN utilizes crop factor to lenses, capturing less of a scene. In that way, sensing range is limited while a large number of transducers can be considered to be present in a camera. So, the problem in such transducer is not the sensing model but the orientation of the camera coverage area. The most common C-SNs are CCD (charge-coupled device) and CMOS (complementary metal–oxide–semiconductor). The largest size C-SN is called a full frame (36 mm × 24 mm), and larger sensor has a bigger body. Active pixel sensor (28.7 mm × 19 mm) is the most popular type with fixed and interchangeable lens. 17.3 mm × 13 mm is a quarter-sized camera. 1 inch (13.2 mm × 8 mm) camera is a pocket-sized camera with a crop factor larger than 2.7. Smartphones such as iPhone 5 employ 1/3 inch (4.8 mm × 3.6 mm) lens. Thus, a camera can be said to be equivalent to have a large number of transducers, with each transducer sensing only one pixel area among millions of pixels.

The pixel size in digital C-SN is affected by the dynamic range defined as maximum signal divided by noise floor in a pixel which is a combination of noise, A/D conversion limitations, and amplifier noise. The number of pixels impacts the image quality than single pixel [15]. So, the problem with C-SN is different from a SN transducer and the sensing area/range for a given C-SN is pretty much fixed. So, the problem with C-SN is different from SN as a large number of C-SNs are deployed and their orientation given in Fig. 4.13b plays an important role in determining overall sensing area.

Figure 4.13a shows two randomly deployed C-SNs in a 10 × 10 grid area and can be calculated by diving the area into small grids and then the coverage c can be defined as:

$$c = \frac{\text{Number of covered grids}}{\text{Number of total grids}}, \tag{4.4}$$

$$c = \frac{20}{100} = 0.2. \tag{4.5}$$

An area C covered with N random C-SNs is shown in Fig. 4.13b and can be expressed as:

$$c = 1 - \left(1 - \frac{\alpha R^2}{S}\right)N, \tag{4.6}$$

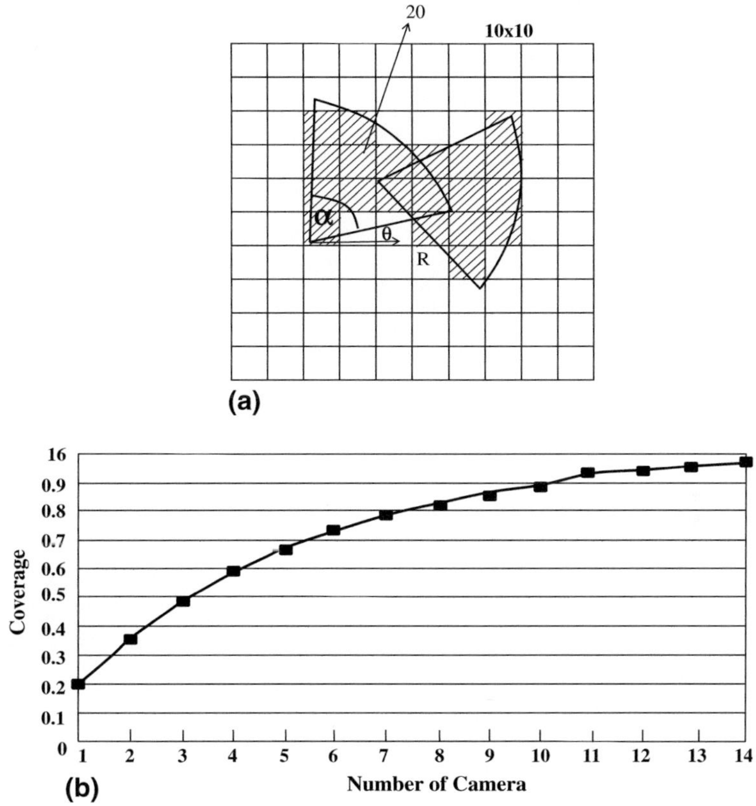

Fig. 4.13 **a** Two C-SNs and coverage area and **b** coverage area as a function of randomly deployed C-SNs

where R is the radius of area covered and α coverage angle.

In other words, given c, N, the number of desired C-SNs can be given by:

$$N = \frac{\ln(1-c)}{\ln(S-\alpha R^2) - \ln S}. \tag{4.7}$$

Coverage area as a function of number of C-SNs is given in Fig. 4.13b. The first 7 cameras can increase the coverage to 0.8 from zero, but the next 7 C-SNs can only improve the coverage by only 0.16. Each C-SN cannot change its position but can change its orientation, and reports that to the controlling station. A simple optimization problem is to determine orientations $(\theta_1, \theta_2, \theta_3, \ldots, \theta_n)$ such that the coverage is maximized. This is done by using a particle swarm organization (PSO) algorithm that uses M particles, with each particle having position vector x and velocity vector v; *remembering a private best position p for each particle, global best g is obtained by doing iterations by following two steps:*

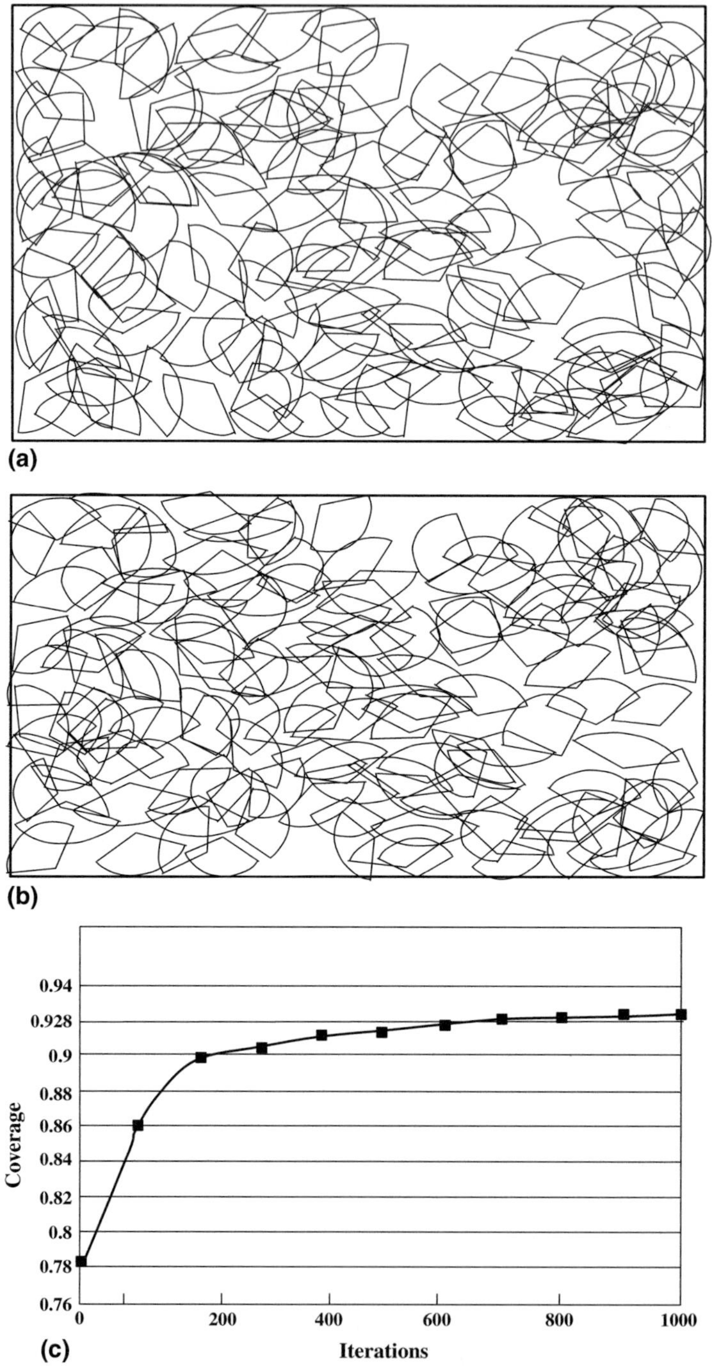

Fig. 4.14 **a** Initial layout of 152 C-SNs with *coverage of 0.782052*, **b** improved layout of 152 cameras to *0.92336*, and **c** variations of coverage with iterations

$$\begin{aligned} v &= C_1 \times v + C_2 \times \text{rnd}() \times (p - x) + C_3 \times \text{rnd}() + (g - x) \\ x &= x + v \end{aligned}, \tag{4.8}$$

where C_1, C_2, and C_3 are constants. Based on prior work [16], we set $C_1 = 0.729$, $C_2 = C_3 = 1.49445$. In an experiment, 500 × 500 rectangular area is monitored with 152 randomly distributed C-SNs, with each having $R = 50$, $\alpha = \pi/3$. 20 particles are used with maximum iteration = 1000. The global best coverage of 0.782052 is improved to 0.92336 after 1000 iterations (14% improvement) which will require 244 C-SNs if no orientation improvement is applied (Fig. 4.14).

4.4 Conclusions

As transducer plays a very important role in numerous applications, it is important to model the same appropriately. Precisely, modeling makes the simulation and analysis more accurate but also helps in minimizing required number of SNs. The challenge is to define a good model that is simple enough to be used while leading to reasonable results in a time effective manner.

4.5 Questions

Q.4.1. What is meant by modeling of a transducer sensing range?
Q.4.2. How do I test my sensor to make sure it is working properly?
Q.4.3. What is the role of noise in a transducer?
Q.4.4. Do transducers require calibration?
Q.4.5. How do you select the right transducer/sensor range for your application?
Q.4.6. Do you need a computer to use these transducers?
Q.4.7. How do thermal modeling by snake help infrared reception?
Q.4.8. How much rain does your sensor need to shut off the sprinklers?
Q.4.9. Which metrics do you use to compare transducer models for localization?
Q.4.10. Why isn't your camera sensing up to 50 feet as advertised?
Q.4.11. What are the effects of temperature on piezoceramic transducers?
Q.4.12. Why do you need to have a breath alcohol tester?
Q.4.13. What type of maintenance is required to maintain a personal house monitoring unit?
Q.4.14. What are the advantages of ultrasonic level transducers?
Q.4.15. What is the impact of coverage area if you have 10% wiggle distance?
Q.4.16. How much difference you have if you have 2-D wiggle in problem 3.14?
Q.4.17. What is the impact of irregularity on transducer coverage area?

Q.4.18. Can you envision usefulness of irregularity in helping reduce the number of transducers required to cover a given area?

Q.4.19. What will be the impact if wiggle is also present with irregularity in sensing area?

References

1. https://books.google.com/books?isbn=1849960593.
2. http://144.206.159.178/FT/CONF/16438452/16438496.pdf.
3. Amotz Bar-Noy, Theodore Brown, Matthew P. Johnson, and Ou Liu, "Cheap or Flexible Sensor Coverage," Eds.): DCOSS 2009, LNCS 5516, pp. 245–258, Springer-Verlag Berlin Heidelberg 2009.
4. Yi Zou and Krishnendu Chakrabarty, "Sensor Deployment and Target Localization Based on Virtual Forces," INFOCOM 2003, pp. 1293–1303.
5. N. Ahmed, S. S. Kanhere, and S. Jha, "Probabilistic Coverage in Wireless Sensor Networks," 30th Anniversary. The IEEE Conference on Local Computer Networks, 2005, pp. 681–688.
6. G. Wang, G. Cao, T. L. Porta, and W. Zhang, "Sensor relocation in Mobile Sensor Networks," IEEE INFOCOM'05, vol. 4, 2005, pp. 2302–2312.
7. Liping Liu, Guidan Li, Zhi Wang, and Yu-geng Sun, "Energy-Efficient Collaborative Target Coverage in Wireless Sensor Networks," Computer Engineering and Applications, Dec.2008, vol. 44, no. 34, pp.:31–34.
8. Mohamed Hefeeda and Hossein Ahmadi, "A Probabilistic Coverage Protocol for Wireless Sensor Networks," https://www.cs.sfu.ca/~mhefeeda/.../icnp07.pdf.
9. G. Zhou, T. He, S. Krishnamurthy, and J. A. Stankovic, "Models and Solutions for Radio Irregularity in Wireless Sensor Networks," ACM Transactions on Sensor Networks, vol. 2, no. 2, pp. 221–262, 2006.
10. Chun-Hsien Wu and Yeh-Ching Chung, "A Polygon Model for Wireless Sensor Network Deployment with Directional Sensing Areas," Sensors, vol. 9, pp. 9998–10022, 2009.
11. R Mulligan and A. Ammari, "Coverage in Wireless Sensor Networks: A Survey," Network Protocols and Algorithms, 2010, vol. 2, no. 2, pp. 27–53 and http://en.wikipedia.org/wiki/Radio-frequency_identification.
12. https://en.wikipedia.org/wiki/Infrared.
13. Rajiv Ramaswami, "Optical Fiber Communication: From Transmission to Networking" (PDF). IEEE Communications Magazine, vol. 40, no. 5, May 2002.
14. http://www.techhive.com/article/2052159/demystifying-digital-camera-sensors-once-and-for-all.html.
15. http://www.clarkvision.com/articles/digital.sensor.performance.summary/.
16. Yichun Xu and Bangjun Lei, "Particle swarm optimization to improve the coverage of a camera network," *Proc. SPIE* 7497, MIPPR 2009: Medical Imaging, Parallel Processing of Images, and Optimization Techniques, 749718 October 30, 2009.

Chapter 5
Clock Synchronization and Localization

5.1 Introduction

WSNs have numerous applications and are usually embedded in the environment. The basic objective in a WSN is to collect sensed data from a large number of SNs at one location commonly known as BS (or sink node), analyze the data to monitor the environment, and eventually control the actuator. This is to be done for a long period of time without any human intervention. The SNs are usually miniature, lightweight, inexpensive, and energy-efficient and need to keep transceiver in sleep mode as long as possible in order to conserve energy. On the other hand, when data are sent from one SN to another SN on way to BS, both SNs need to be awake. So, there is a need for very tight synchronization between SNs. In that respect, any SN's clock can be used as a reference. But from overall access and logistic point of view, it is rather better to consider clock of a BS as the reference node. BS sends a query to all SNs, indicating which physical parameters are desirable and at what frequency rate. BS usually transmits that in one transmission at higher power level such that all SNs receive that requested query in a single broadcast.

Each SN has a sensing range r_s and the communication transmission range r_c (Fig. 5.1a), which determines distance to SN's neighbors. Such a WSN can be represented by a network graph as shown in Fig. 5.1b. The communication is assumed to be symmetric. This means if SN A can communicate with SN B, then SN B can also communicate with SN A and undirected graph simply indicates that.

If two neighboring SNs transmit data at the same time, collision occurs and received data cannot be interpreted correctly (Fig. 5.2a). That means, all collided packets need to be retransmitted. In Fig. 5.2b, adjacent SNs a and b know the presence of each other and so do SNs b and c. But, a does not know the presence of c, and if both a and c transmit data to b at the same time, collision occurs. This is known as hidden terminal problem, and a is termed as hidden from c and vice versa. Figure 5.2c illustrates exposed terminal problem where SN c cannot communicate with SN d as SN b is having ongoing communication with SN a and detecting and

D.P. Agrawal, *Embedded Sensor Systems*,
DOI 10.1007/978-981-10-3038-3_5

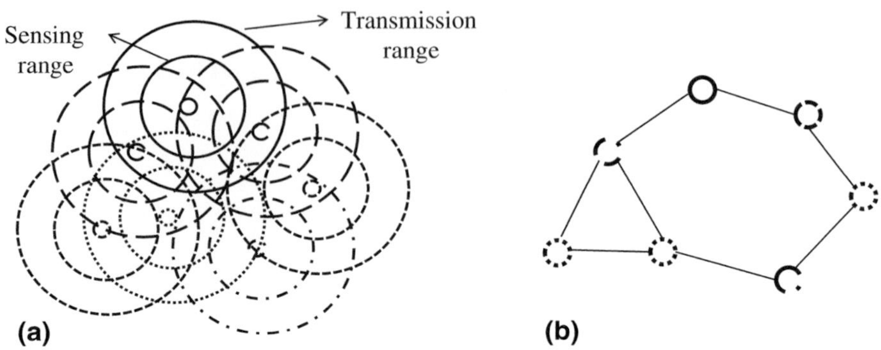

Fig. 5.1 **a** Disk model of a transducer/SN and **b** graph model of a WSN

sensing current transmission from SN b, and SN c will not transmit. This is unnecessary as transmission from SN c to d does not interfere with current communication from SN a to SN b.

Each SN has a transmission range which determines its neighbors SNs with whom the SN can communicate. The *k*-neighborhood of a node *v* is given by $\Delta_k(v)$, *and the* size of neighboring SNs is given by *v*:

$$\delta_k(v) = |\Delta_k(v)|. \tag{5.1}$$

Then, max *k*-neighborhood size in the network can be given by:

$$\delta_k = \max_v \delta_k(v). \tag{5.2}$$

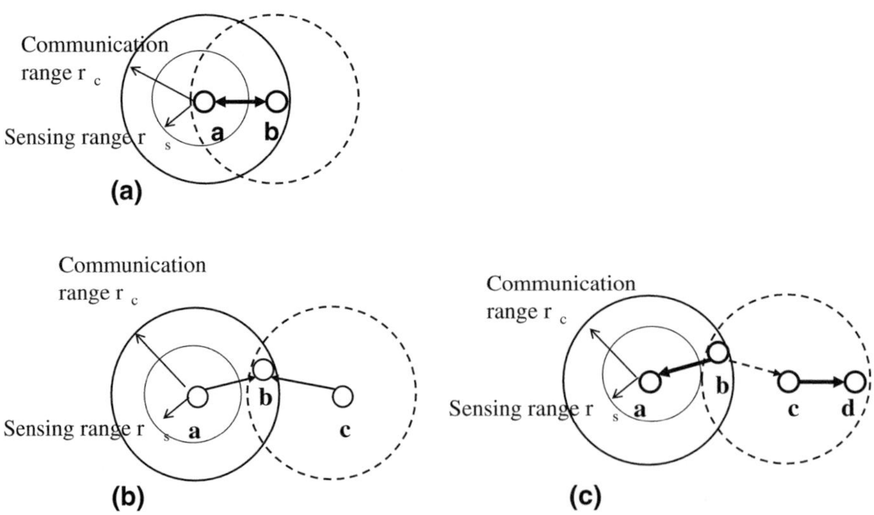

Fig. 5.2 **a** Collision between 2 SNs, **b** hidden terminal problem and **c** exposed terminal problem

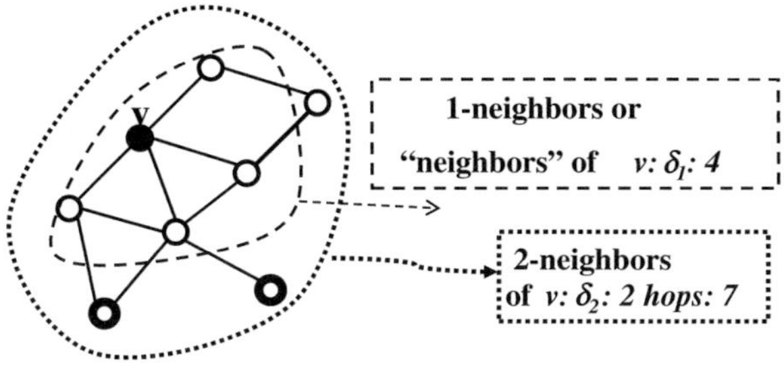

Fig. 5.3 1-hop and 2-hops neighbors of a reference SN v

Let n be the number of nodes in the network, then 1- and 2-neighbors of a SN v are shown in Fig. 5.3.

5.2 Clock and Signal Propagation in a WSN

Often, a hardware clock is present in a SN, with an oscillator generating pulses at a given frequency. A counter register is usually incremented after a fixed number of pulses, and only register content is available to software. A change rate of the register indicates resolution of the clock. $H_i(t)$ is defined as real time t of a SN i. Small letters (like t, t') denote real physical times, while capital letters represent time stamps or anything else visible to SNs. A SN software local clock is usually determined as follows:

$$L_i(t) = q_i H_i(t) + f_i, \tag{5.3}$$

where q_i is the drift rate and f_i is the phase shift.

Time synchronization algorithms do modify q_i and f_i, but not the counter register. If $C(t)$ is the value of a clock reported at time t, there exists $C'(t) = \mathrm{d}C(t)/\mathrm{d}t$ and $C'' = \mathrm{d}^2C(t)/\mathrm{d}t^2$ for $t \geq 0$. The difference between the time reported by two clocks is called offset. If time at two SNs clocks C_1 and C_2 are $C_1(t)$ and $C_2(t)$, respectively, the offset of the clock C_1 relative to clock C_2 at time $t \geq 0$ is $C_1(t) - C_2(t)$. The rate at which the clock progresses is called frequency. The frequency at time t of C_1 is $C_1'(t)$. The frequency ratio between two clocks is defined as clock ratio (a). The ratio of C_1 relative to C_2 at time t is $a = C_1'(t)/C_2'(t)$. The difference in the frequencies of two clocks is called as skew(d). The skew of C_1 relative to C_2 at time t is $d = C_1'(t) - C_2'(t) = aC_2'(t) - C_2'(t) = (a-1)C_2'(t)$. The drift of clock C_1 relative to the clock C_2 at time t is $C_1''(t) - C_2''(t)$ and basically represents the difference of change in the frequency rate. In a WSN, there are many sources of unknown, non-deterministic SNs, and the latency between time stamp of a message from a SN and its reception at another SN/BS ought to be considered. So, it is better if all SNs are synchronized together, and one SN can be taken as a reference for synchronization (Fig. 5.4).

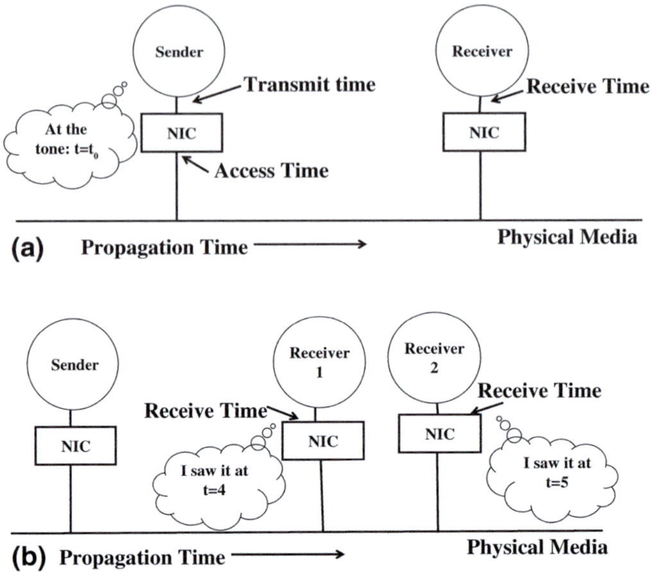

Fig. 5.4 **a** Propagation of signals between two SNs and **b** sync reference broadcast

So, receiver-based synchronization (RBS) considers the time the sender SN reads its clock to when the receiver SN reads its clock and is defined as traditional critical path. RBS is only sensitive to the differences in receive time and propagation delays as shown in Fig. 5.5. RBS removes *send* and *access time* errors. Broadcast is used as a *relative* time reference, and each receiver synchronizes to a *reference packet. A r*eference packet is injected into the channel at the same time *for all receivers, and* time stamp is not there in the message. Any broadcast packet can be used as a reference such as RTS/CTS packets and route discovery packets. It is natural to use BS clock as a reference as it usually broadcasts the query. The phase offset can be estimated as follows. For simplest case of a single pulse being sent to two receivers, the transmitter SN broadcasts reference packet and each receiver SN

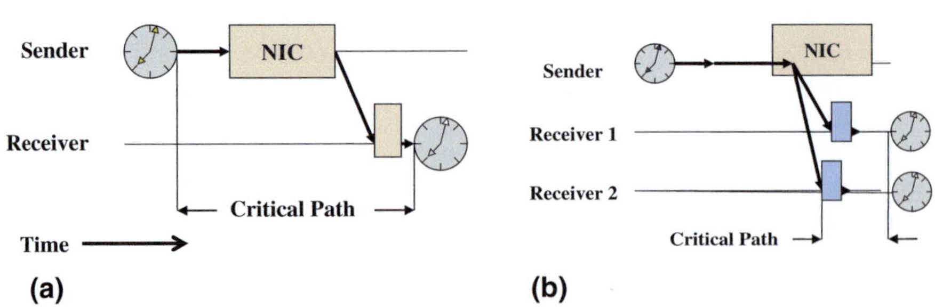

Fig. 5.5 **a** Synchronization between a sender SN and a receiver SN [1] and **b** synchronization between a sender SN and multiple receiver SNs

records the time that beacon was received according to its *local clock*. Receiver SNs exchange their observations to have sufficient information to form a *local* (relative) timescale. However, global timescales are also important; simple case can be extended to many receivers assuming propagation delay to be zero, and no clock skew is present. In this way, receiver error follows *non*-determinism and is Gaussian in nature. Sending multiple messages always increases the precision. In such a case, a transmitter SN broadcasts m packets; each receiver SN records time the beacon was observed and exchange observed time; and receiver SN i computes phase offset to another receiver SN j as the *average* of the offsets implied by each pulse received by both SNs. This can be mathematically expressed as [1]:

$$\forall i \in n,\ j \in n: \quad \text{Offset}[i,j] = \frac{1}{m}\sum_{k=1}^{m}(T_{j,k} - T_{i,k}) \tag{5.4}$$

The packet reception time is shown in Fig. 5.6. The metrics for indicating effectiveness of synchronization mechanism are precision, longevity of synchronization, time and power budget available for synchronization , geographical span of the area to be monitored, and size and WSN topology. As receiver cannot predict when the packet will arrive and there could be clock skew, guard band is used in between. Time sync is critical at *many* layers in WSNs such as beam-forming, localization, tracking, distributed DSP, data aggregation , and caching. A clock in a computer at time t is given by an expression:

$$C(t) = k\int_{0}^{t}\omega(\tau)\,\mathrm{d}\tau + C(tl_0), \tag{5.5}$$

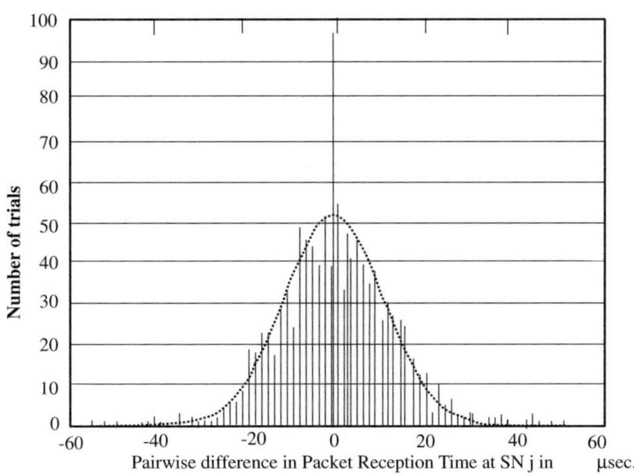

Fig. 5.6 Pairwise difference in packet reception time at SN j in µsec [1]

where ω is frequency of the oscillator and $C(t_0)$ is the time of the computer click implemented based on a hardware oscillator. The computer clock (Fig. 5.7a) is an approximation of a real time t as:

$$C(t) = q * t + b, \tag{5.6}$$

where q is a clock drift and b is an offset of the clock. In a perfect clock, rate $q = 1$ and offset $b = 0$.

The clock offset is defined as the difference between the local times of two SNs, and synchronization is required to adjust clock readings such that they match. The clock rate is the frequency at which a clock progresses and clock skew is the difference in frequencies of two clocks. The clock rate dC/dt depends on temperature, humidity, supply voltage, age of quartz, etc., resulting in drift rate (dC/dt − 1). The sources of inaccuracies are primarily that SNs are switched on at random times and phase difference θ_i can be random. Actual oscillators do have random deviations from nominal frequency (drift, skew), and deviations are specified in ppm (pulses per million). The ppm value counts the additional pulses or lost pulses over the time of one million pulses at a nominal rate. The cheaper is the oscillator, and larger is the average deviation (Fig. 5.7b). For SNs, values between 1 ppm (one second every 11 days) and 100 ppm (one second every 2.8 h) are commonly assumed. Berkeley motes have an average drift of 40 ppm. The oscillator frequency depends on the time (oscillator aging) and environment (temperature, pressure, supply voltage, etc.). One-time synchronization is not sufficient for the time dependent as drift rates need to be resynchronized frequently. However, assuming stability over tens of minutes is considered reasonable.

General properties considered for time synchronization algorithms include physical time versus logical time, external versus internal synchronization , global versus local algorithms, synchronizing all SNs of a WSN synchronized or only a local neighborhood, absolute versus relative time synchronization, and hardware-versus software-based mechanisms. A GPS receiver is a hardware solution, but is too costly and energy-consuming in WSN SNs, and at least four satellites need to be on line of sight. It is important to consider whether a priori or a posteriori synchronization is needed after an interesting event, or whether a deterministic v/s

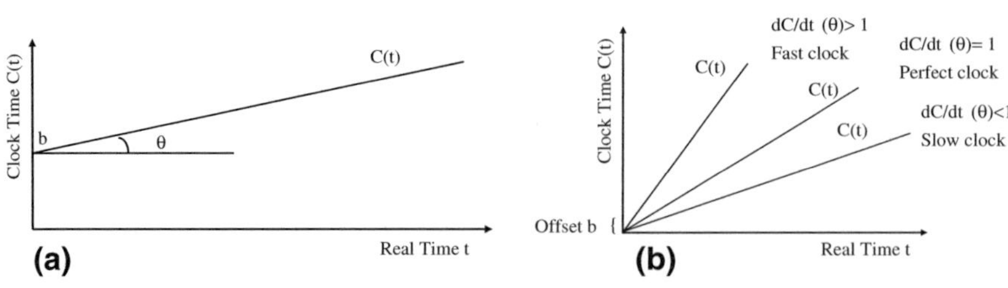

Fig. 5.7 **a** Approximation of a computer clock and **b** computer clock parameters

stochastic precision bound is desirable. It should be noted that a backward jumps of local clocks be avoided and sudden jumps need to be avoided.

The performance metrics that ought to be considered are precision, energy costs, and fault tolerance. The precision indicates maximum synchronization error for deterministic algorithms, error mean/standard deviation/quintiles for stochastic values. The energy costs include number of exchanged packets, computational costs, and memory requirements. The maximum drift rate ρ given by manufacturer (typical 1–100 ppm) guarantees that

$$1 - \rho \leq \frac{\mathrm{d}C}{\mathrm{d}t} \leq 1 + \rho. \tag{5.7}$$

The drift rate causes clocks to differ even after synchronization , and two synchronized identical clocks can drift from each other at rate of at most $2\rho_{\max}$. To limit relative offset to δ seconds, the resynchronization interval τ_{sync} must meet the requirement:

$$\tau_{\text{sync}} \leq \frac{\delta}{2\rho_{\max}}. \tag{5.8}$$

A fundamental building block of time synchronization is when to trigger resynchronization, what should be periodically, and should it be due to an external event? Estimating clocks of other SNs with exchanged packets and correcting and adjusting its own clock accordingly. Determining which SNs need to be synchronized with other SNs is an interesting challenge. What are the constraints for time synchronization and whether an algorithm can scale to larger WSNs of unreliable SNs? The precision requirements could vary from ms to tens of seconds. The use of extra hardware such as GPS receivers is mostly not an option. The SNs have no or very low mobility ; often, there are no fixed upper bounds on packet delivery times due to MAC delays, buffering, etc.; propagation delay between neighboring nodes is negligible; and manual node configuration is not an option. The basic idea is that substantial energy costs are associated with frequent resynchronization if all SNs are kept synchronized all the time. This is specially overkill if a WSN becomes active very rarely. When a SN observes an external event at time t, it stores its local time stamp $L_i(t)$, achieves synchronization with neighbor SN/sink node, and converts $L_i(t)$ accordingly.

Synchronization can be implemented in many different ways, and challenges for time synchronization are the environmental effects, energy constraints, wireless medium, mobility characteristics, etc. It could be either external or internal. External synchronization between a sender and a receiver and at least one node must have access to the external time scale. In internal receiver/receiver synchronization, no external timescale is used while nodes must agree on common time. For example, in Fig. 5.8a, when SN A senses an event, it starts counting time with its clock. SN A sends SN B a message regarding the event, and a time stamp includes how long the time has elapsed since the event. SN B estimates the

Fig. 5.8 **a** Synchronization after an event and **b** message forwarding for synchronization after an event

transmission delay to the time stamp and continues counting the time. Messages are forwarded in the same fashion as A → B to SN N (Fig. 5.8b). N is able to recover the time by looking at the time stamp. The assumption is that the maximum clock skew is known, and the link can survive long enough so that a synchronization message can be sent after the application message (for given maximum drift rate ρ) as:

$$1 - \rho \leq \frac{\Delta C}{\Delta t} \leq 1 + \rho, \tag{5.9}$$

where ΔC is the drift in time Δt.

In sender/receiver synchronization, a receiver SN synchronizes its clock with a sender SN and pairwise synchronization involves synchronization of a single receiver SN to a single sender SN. Sources of inaccuracies [2] include MAC delay, interrupt latencies upon receiving packets, delays between packet interrupts and time-stamping operation, and delay in operating system and protocol stack. Improvement is possible if its packet is time-stamped after the MAC delay, immediately before transmitting the first bit. This removes the delay between packet interrupts and time-stamping from the uncertainty, and leaves only interrupt latencies. WSN-wide synchronization involves who synchronizes with whom to keep the whole WSN synchronized. The classical Network Time Protocol [3] belongs to this class. Lightweight Time Synchronization synchronizes the clocks of all SNs to one reference clock (e.g., equipped with GPS receivers) and is based on sender/receiver technique. A WSN-wide synchronization requires minimum spanning tree construction with reference node as root and synchronizes them in step by step mechanism. Here, a reference node R triggers construction of a spanning tree, it first synchronizes its neighbors, the first-level neighbors synchronize second-level neighbors, and so on. Different distributed algorithms for construction of spanning tree can be used, e.g., DDFS (distributed depth-first search). The cost consists of costs for construction of spanning tree and synchronizing two nodes costs 3 packets, while synchronizing n nodes costs $\sim 3n$ packets. The other alternative is to use distributed multi-hop synchronization by not having an explicit spanning tree, but each SN knows identity of reference node(s) and routes to them. Minimum spanning tree is implicitly constructed as SNs need to know shortest route to the closest reference node. Other sender/receiver-based protocols are similar except the way spanning tree is constructed and how and when time stamps are incorporated.

5.3 Localization of a SN

Localization simply means determining where a given SN is physically located within a WSN. This process is needed to identify the location at which sensed reading originated from, and most applications use communication protocol that queries about geographic area, instead of identity of a SN. The purpose of a WSN is to monitor and report events take place in a particular area. Hence, the main parameters define how well the network observes a given area coverage. Localization techniques can be classified as shown in Fig. 5.9.

Coordinate-Free Localization is based on received signal strength indication (RSSI) measurement as strength of radio signal reduces with square of distance from the source that can indicate the distance of a SN from the source SN. But, as the signal travels through open space and radio propagation is highly non-uniform as wall, furniture, grass, asphalts, etc. absorb and reflect radio waves. The received signal usually contains noise in the order of several meters. The accuracy can be enhanced by either modifying measured distance or combining RSSI with some other parameters. RSSI is measured in dBm, which is 10 times the logarithm of the ratio of the received power (P) and the referenced transmitted power, and the equation leads to a straightforward indicator of distance estimation.

RSSI is known to be an inadequate estimator for determining the distance as the link quality between SNs behaves very inconsistently in estimating. Indoor is even worse (Fig. 5.10) as multi-path fading occurs due to reflection, refraction, and scattering of radio waves by structures inside a building even if multi-path medium is well characterized and modeled. Experimental results [4] have shown (Figs. 5.11 and 5.12) that RSSI estimates with imperfect links are difficult as RSSI patterns are observed to be complex. A slightest movement affects the received value significantly. RSSI serves at best as bounds for the distances. In fact, a single RSSI value can represent many distant locations in the room, where the same value is obtained.

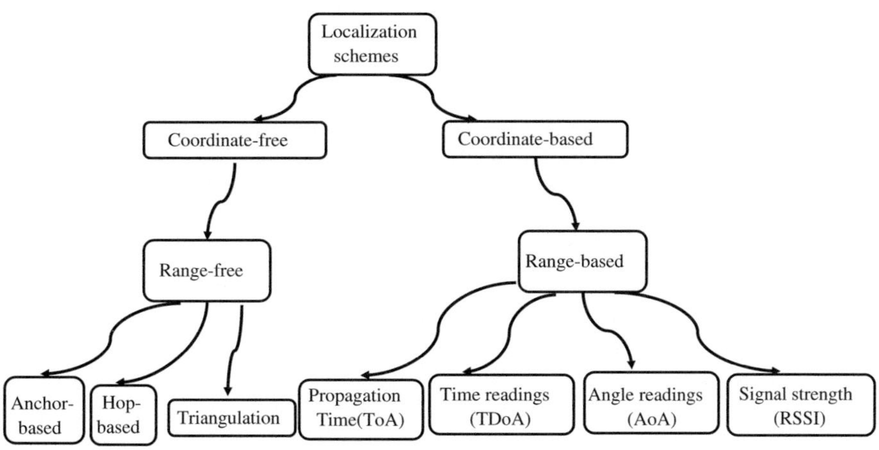

Fig. 5.9 Characterizing localization techniques for SNs

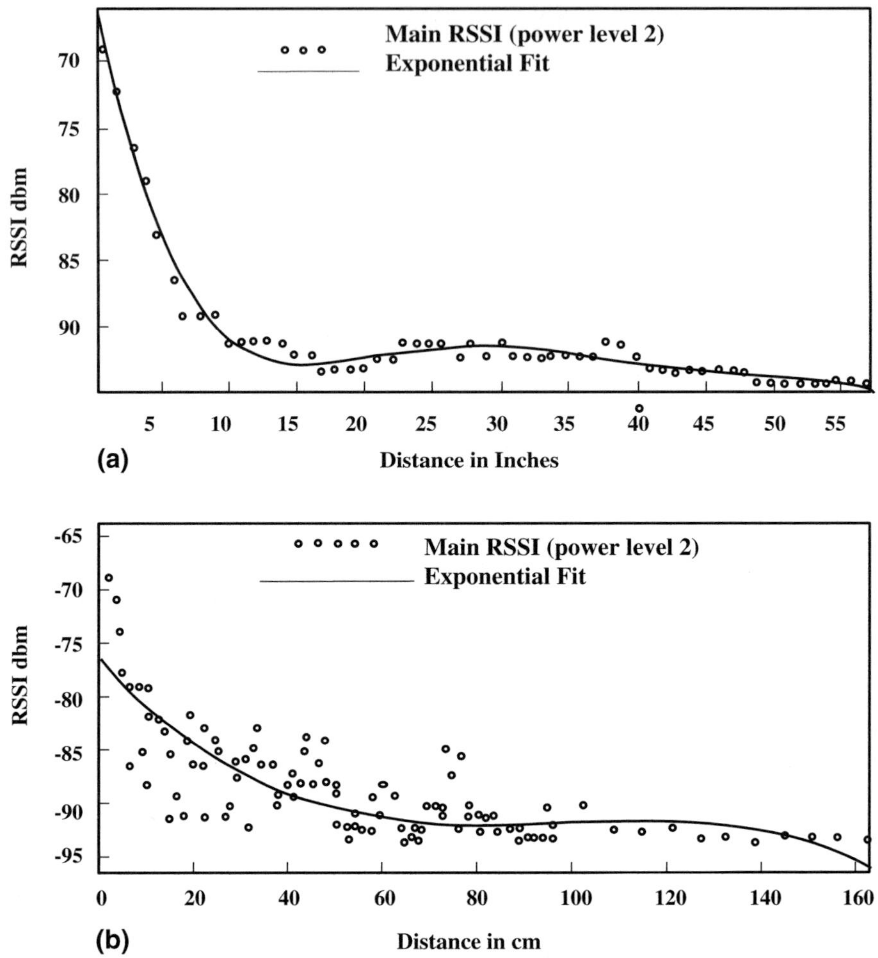

Fig. 5.10 **a** Outdoor variation of RSSI and **b** indoor variation of RSSI [4]

Algorithms may fail, since strongly nonlinear distribution of RSSI makes this process error-prone. RSSI and the sender SN-announced power level do not necessarily match. The environmental factors (temperature and humidity) have been shown to interfere with RSSI readings as well. The link quality indicator (LQI), which is effectively a measure of chip error rate commonly available in commercial SNs, is observed to be highly correlated with both RSSI and with packet reception rate. Therefore, a combination of RSSI + LQI seems better suited for indoor localization. Distance estimation system has been proposed [4] that uses only the RSSI and LQI values. The test hypothesis was verified through experimental setup for both indoor and outdoor experiments. RSSI and LQI values were measured using Crossbow's TelosB mote (TPR2420CA) that has an integrated IEEE 802.15.4/ZigBee-compliant RF transceiver (CC2420 chipset) by varying distances

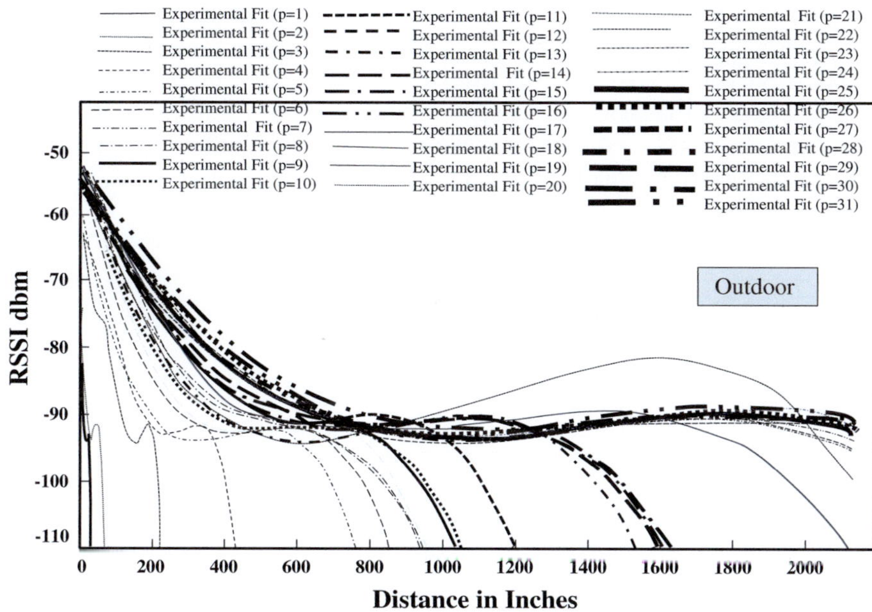

Fig. 5.11 Outdoor variation of RSSI at different power levels [4]

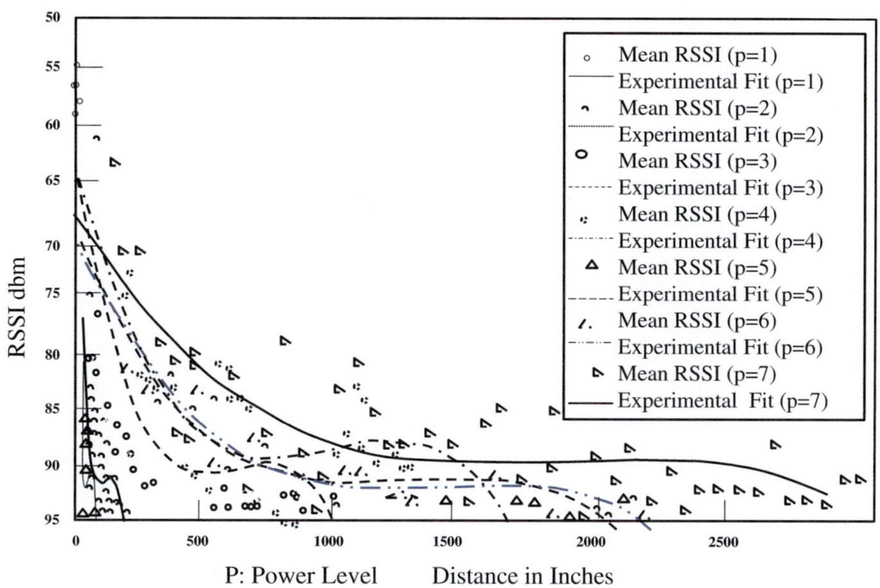

Fig. 5.12 Indoor variation of RSSI at different power levels [4]

and power levels, and the results are summarized in Fig. 5.12. For each power level and distance, the mean of the RSSI at the receiver SN is measured that is observed to fit an exponential curve data. −93 dBm RSSI is observed at the distances [10, 13, 36, 37]. Figure 5.13a shows the number of elements when experiment performed indoor just for RSSI = −100 dBm measurement. When RSSI is combined with LQI, the membership reduces drastically as shown in Fig. 5.13b. Thus, a combination of (RSSI + LQI) seems better suited for indoor localization.

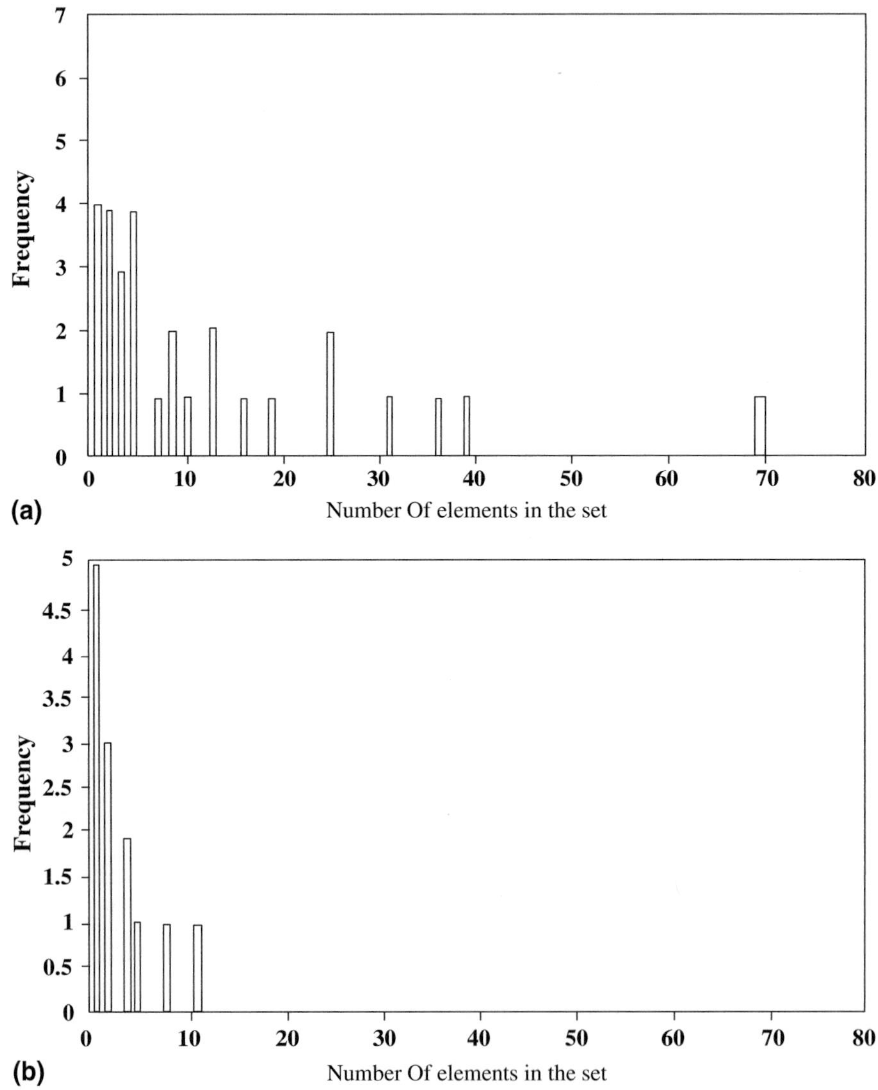

Fig. 5.13 **a** Number of elements in RSSI set for indoor and **b** number of elements in RSSI + LQI set for indoor [4]

In a realistic WSN, RSSI is still used as a measure of distance. Assuming two SNs A and B can communicate with each other with distance less than R, then the hop count h_{ij} can be used to compute internode distance between A and other SNs. The distance d_{ij} between two SNs at h_{ij} hops away will be less than $h_{ij} * d$ (Fig. 5.14), and the distance covered by one hop is:

$$d = R\left(1 + e^{-n} - \int_{-1}^{1} e^{-\frac{n}{\pi}(\arccos t - t\sqrt{1-t^2})}\,dt\right), \tag{5.10}$$

Triangulation is one scheme to determine the location by measuring angles from known points. Trilateration determines the location by measuring distance between reference points, while multilateration determines the location by measuring time difference of signal from reference points. Many range-free localization algorithms for the WSNs depend on trilateration algorithm and employs few SNs as reference anchors (Fig. 5.15a), assuming their position to be known. Other SNs use the information from the anchor nodes to estimate their own position (Fig. 5.15b) as:

$$\begin{bmatrix} (x-x_1)^2 + (y-y_1)^2 \\ (x-x_2)^2 + (y-y_2)^2 \\ (x-x_3)^2 + (y-y_3)^2 \end{bmatrix} = \begin{bmatrix} d_1^2 \\ d_2^2 \\ d_3^2 \end{bmatrix}, \tag{5.11}$$

leading to:

$$2x(x_i - x_1) + 2y(y_i - y_1) = d_1^2 - d_i^2 + x_i^2 - x_1^2 + y_i^2 - y_1^2 \quad (i = 2, 3) \tag{5.12}$$

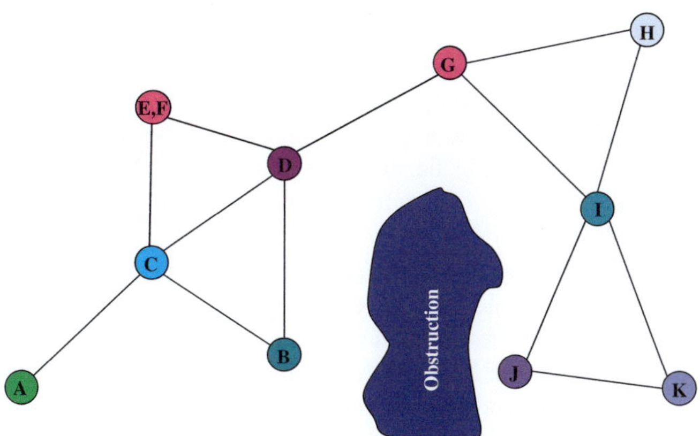

Fig. 5.14 Number of elements in RSSI set for indoor [5]

Fig. 5.15 **a** Trilateration in a WSN with selected anchors and **b** trilateration algorithm with 3 anchors

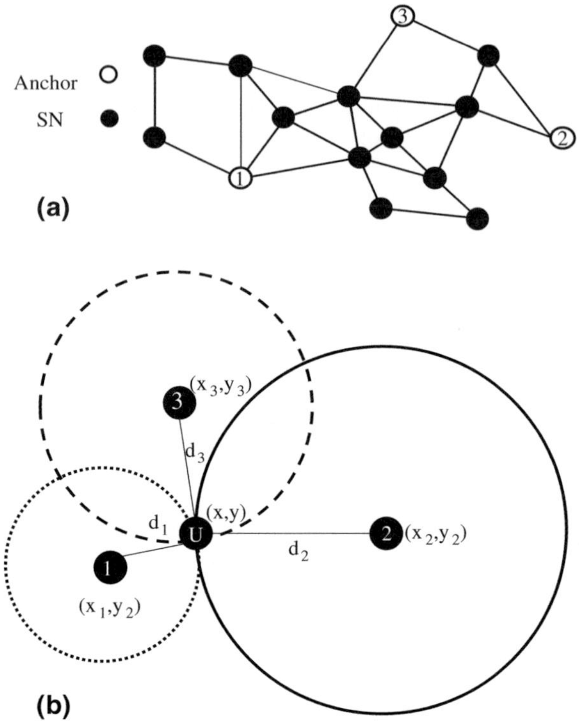

Besides RSSI that has multi-path fading, background interference, and irregular signal propagation , there are many ways to measure the distance between a SN and anchor SNs. These include time of arrival (ToA) (time of arrival), GPS with expensive in hardware and energy-consuming, TDoA (time difference of arrival), and AoA (angle of arrival). AoA is based on estimating relative angles between neighbors and requires additional hardware. It could be expensive to deploy in large WSNs. One way of measuring ToA with GPS is to use satellite constellation of at least 4 satellites with atomic clocks (Fig. 5.16a). As satellites broadcast precise time, the ToA is an estimated distance to satellite and trilateration determines location of a SN. This process is expensive, and it is rather impractical to use the absolute address. Locations can be determined by using AoA by using antenna array to measure the direction of neighbors. Special landmarks or anchors usually have compass and GPS capability. They broadcast location and flood beacons, and update bearing along the way. Once beacons from three landmarks are known, the position of the SN can be calculated as shown in Fig. 5.16b.

In a basic APIT (approximate point-in-triangulation test) Scheme [7], anchors are location aware SNs in the WSN. APIT employs area-based approach to isolate triangular regions between beaconing nodes. Once the area is known, the CoG (center of gravity) calculation is performed for the location of the SN as illustrated in Fig. 5.17a. In a perfect PIT (point-in-triangulation test) [8] scheme (Fig. 5.17b),

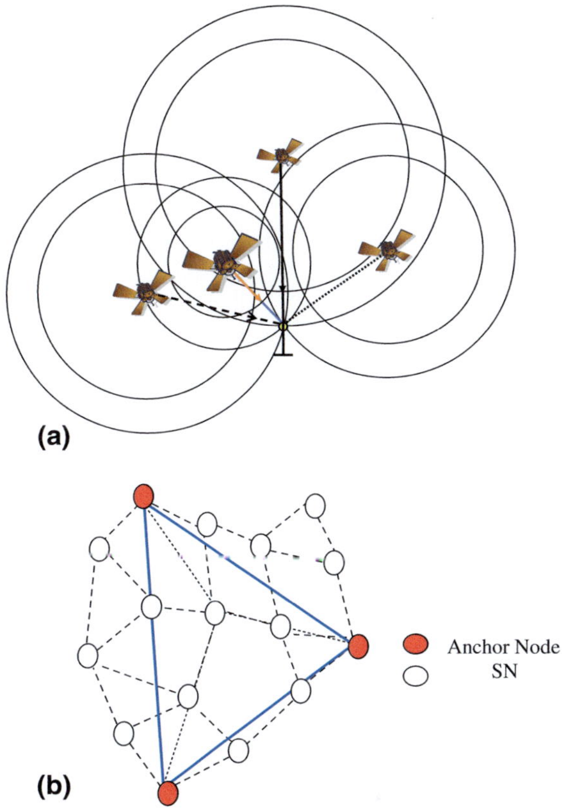

Fig. 5.16 **a** Angle of arrival in a WSN with selected satellites as anchors [6] and **b** time of arrival in a WSN with selected anchors

given anchor nodes A, B, and C, if a SN M is inside the triangle ABC, when M is shifted in any direction, the new position must be nearer to (further from) at least one anchor node A, B, or C. If the SN M is outside the triangle ABC; when M is shifted, there must exist a direction in which the position of M is further from or closer to all three anchors A, B, and C. If there exists a direction such that a point adjacent to M is further/closer to points A, B, and C simultaneously, then M is outside of ABC. Otherwise, M is inside ABC. A perfect PIT test is infeasible in practice. Experiments show that the received signal strength is decreasing in an environment without obstacles. Therefore, further away a node is from the anchor, and weaker will be the received signal strength.

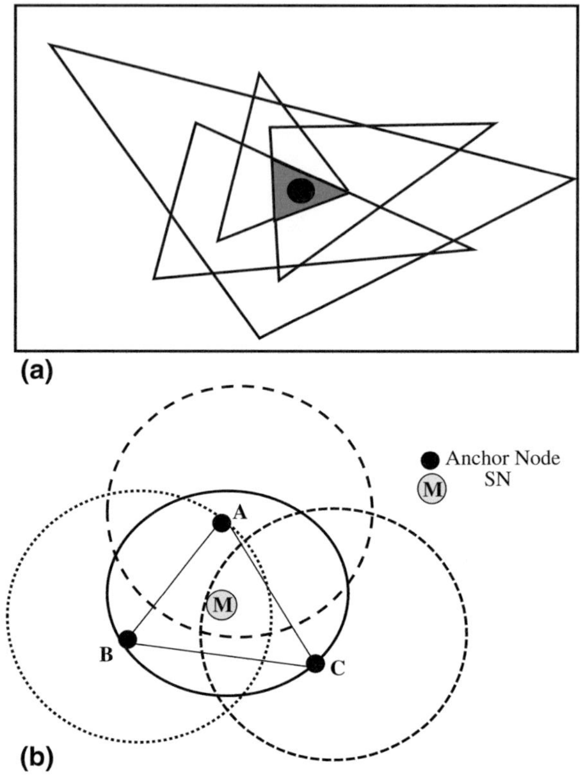

Fig. 5.17 **a** Approximate triangulation scheme and **b** point-in-triangulation scheme

5.4 Conclusions

As a large number of SNs ought to be deployed in a typical WSN application, synchronization among SNs has become an important aspect. This is especially critical if SNs are forced to go to sleep mode and are awakened when SN or neighboring SN wants to transmit data to BS in a multi-hop fashion. Both transmitting SN and receiving SN need to be awake and synchronized for any successful information transfer. In a WSN, 1-hop synchronization is easier to implement while multi-hop synchronization is relatively complex to achieve, even though that has to be eventually achieved directly or indirectly. As received data determine the presence of an event, the localization is needed to determine where the desired information is coming from, indicating the presence of an event.

It is important to determine how many events (moving objects) have been detected, and what is the direction and speed of the moving object? Thus, synchronization and localization are two issues are complimentary to each other and are essential for successful operation of a large WSN.

5.5 Questions

Q.5.1. How can time synchronization be propagated across the network by using a spanning tree favoring direct connections with reliable delays?

Q.5.2. Assuming that the clock of each CH is synchronized with the BS, what are the limits of maximum and minimum errors? Can you determine similar value for a sensor node?

Q.5.3. What is the key idea of the interval method in a WSN?

Q.5.4. Why does synchronization remain especially challenging problems in a WSN?

Q.5.5. Assume that two nodes have a maximum drift rate from the real time of 200 ppm each. Your goal is to synchronize their clocks such that their relative offset does not exceed 2 s. What is the necessary resynchronization interval?

Q.5.6. A network of five nodes is synchronized to an external reference time with maximum errors of 1, 3, 4, 1, and 2 time units, respectively. What is the maximum precision that can be obtained in this network?

Q.5.7. Node A sends a synchronization request to node B at 3150 (on node A's clock). At 3250, node A receives the reply from node B with a time stamp of 3120. (a) What is node A's clock offset with respect to the time at node B (you can ignore any processing delays at either node)? (b) Is node A's clock going too slow or too fast? (c) How should node A adjust the clock?

Q.5.8. Three beacons are located at $a = (1, 1)$, $b = (1, -1)$, and $c = (-1, 1)$. The received powers from nodes a, b, and c are 1.2, 1.5, and 1.7, respectively. Calculate the unknown position of the receiver through a weighted centroid computation.

Q.5.9. Assume that five reference nodes are known at (0, 3, 0), (6, 0, 0), (3, 4, 0), (−4, −3, 0), and (0, 0, −8), respectively; also, $t12 = 0$ s, $t13 = 1$ s, $t14 = 0.7$ s, $t15 = 0.7$ s, and $t16 = 1.7$ s. The velocity of propagation is v. (a) Find the unknown location (xt, yt, zt). (b) Now assume that the propagation speed is known to be 8.7 m/s. Find the unknown location (xt, yt, zt).

Q.5.10. How is localizing and tracking of moving objects an essential capability for a WSN?

Q.5.11. What is the goal of localization or tracking in obtaining a good estimate of the target state?

Q.5.12. Can you have joint time synchronization and localization of an unknown node in a WSN?

References

1. J. Elson, slides, cs.uccs.edu/~cs526/mote/mobilcome/Mobicom-Tutorial-3-MS.ppt.
2. Noh Kyoung-lae Noh, E. Serpedin, and K. Qaraqe, "A New Approach for Time Synchronization in Wireless Sensor Networks: Pairwise Broadcast Synchronization," IEEE Wireless Communications, vol. 7, no. 9, pp. 3318–3322, Sept. 2008.
3. Network Time Protocol (Version 3) Specification, Implementation and Analysis, https://tools.ietf.org/html/rfc1305.
4. Madhanmohan Raju, Talmai B Oliveira, and Dharma P Agrawal, *"A practical distance estimator through distributed RSSI/LQI processing - An experimental study," IEEE International Conference on Communications,* WS-SCPA (2nd IEEE International Workshop on Smart Communication Protocols and Algorithms), *Ottawa, Canada, June 15, 2012, pp. 8155–8159.*
5. Asma Mesmoudi, Mohammed Feham, and Nabila Labraoui, "Wireless Sensor Networks Localization Algorithms: A Comprehensive Survey," International Journal of Computer Networks & Communications (IJCNC), vol.5, no. 6, pp. 45–64, November 2013.
6. J. Bachrach and C. Taylor, "Localization in Sensor Networks," Handbook of Sensor Networks: Algorithms and Architectures, Wiley-Interscience, 2005.
7. Waltenegus Dargie and Christian Poellabauer, Fundamentals of Wireless Sensor Networks: Theory and Practice, John Wiley & Sons Ltd., 2010.
8. M. Venkatakrishnan and A. K. Yuvashree, "Survey on Range Free-Localization Schemes Based on Anchors Count in WSN Model," IJIRST–International Journal for Innovative Research in Science & Technology, vol. 1, no. 9, February 2015.

Chapter 6
Topology Discovery, Residual Energy, and Energy Harvesting

6.1 Introduction

A WSN commonly has a large number of randomly deployed SNs that send their sensed data to a BS. As each SN uses battery as the energy source, energy consumption is of prime concern. As the energy consumed is a function of square of distance, a transmitted signal can be received, and information is transmitted from a SN to BS in a multi-hop fashion. This is illustrated in Fig. 6.1. As the SNs are deployed randomly, the first thing that SNs do is to send beacon signals to determine neighboring SNs, if any. As SNs operate in a distributed manner, it is rather difficult to control when signals are sent by a SN, and beacons could possibly collide with each other. As shown in Fig. 6.2a, if beacon signals from SNs *x* and *y* overlap with each other, a collision occurs and no response is sent by none of the devices. If only SN *x* sends beacon signal, then both SNs *y* and *z* can hear and respond back independently to SN *x*. In this way, SN *x* determines the presence of SNs *y* and *z* in the neighborhood, and the corresponding RSSI value indicates distance between *y* and *z* from *x*. Once SN *x* successfully receives acknowledgment from *y* and *z*, it can piggyback this information in future beacon signals to determine other SNs in the close by area.

It may be noted that a large number of SNs are deployed which could lead to increased message transmission in the network. So, from this point of view, just minimum but adequate number of SNs ought to be used. It is interesting to note that the SN in the wait mode consumes almost the same amount of energy as the transmitting mode and the wait period should be minimized. Any collision in beacon signals and resultant acknowledge signals result in waste of time and energy and should be avoided as much as possible. Therefore, simple mechanism ought to be used for message transfer between SNs that could avoid collision in a distributed manner. Minimum energy is consumed in a SN only during the sleep mode, and efforts should be made to lengthen the sleep time of each SN. But, when SN sends data to another SN on way to BS, both the SNs need to be awake. Otherwise,

D.P. Agrawal, *Embedded Sensor Systems*,
DOI 10.1007/978-981-10-3038-3_6

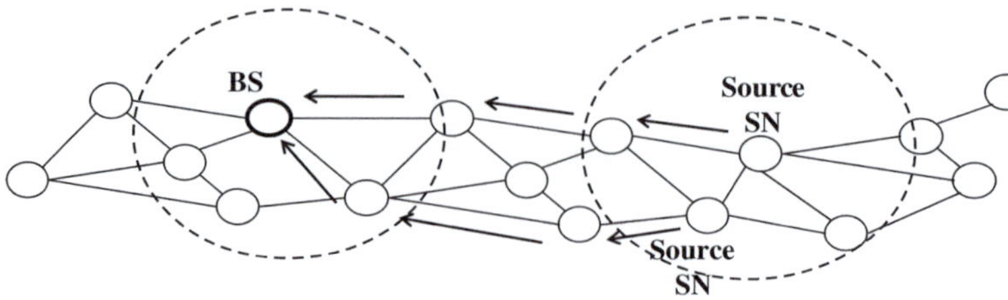

Fig. 6.1 Multi-hop communication from a SN to the BS

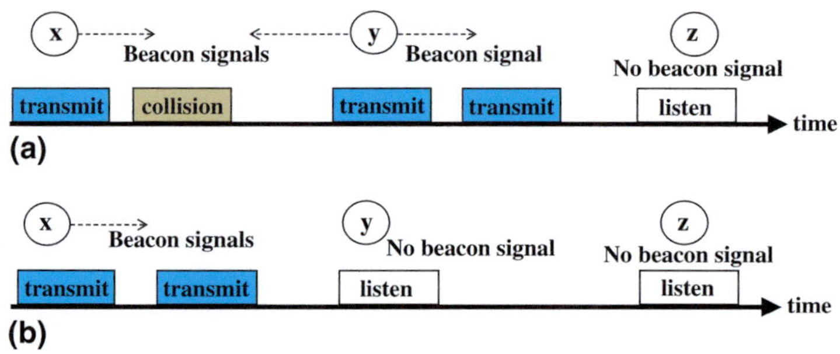

Fig. 6.2 Use of beacon signals to detect neighboring SNs

the message will be lost and cannot be propagated to BS. So, SNs remain in sleep state most of the time (98%) and are awake for a very short duration (2%) to conserve energy. So, there is always frequent transition from awake to sleep mode and vice versa and that causes increased power consumption.

Conventional 7-layer ISO architecture cannot be used directly and simplicity of ALU and accompanying small storage encourage the use of simple algorithms that require small amount of storage. As SNs are autonomous, any centralized scheme cannot work for them. Any overhead in the medium access protocol needs to be minimized and simple design for the encoder/decoder has to be preferred due to silicon space requirements and run-time complexity. Even simple error-correcting code is appropriate to keep the complexity to lowest. In addition, to keep the cost of the SNs little, the crystal that is used may be a cheaper one and may not be able to support accurate synchronization. Even TDMA-based scheduling for SN may not be feasible until clustering has been done, and CH for each cluster has been identified.

The SN responding to a query may depend on the event that could occur anywhere in the network and only SNs in the affected area would respond to such situation. Thus, the data are event based and both spatial and temporal correlation to the event ought to be explored for both relevance and data compression. So, the bottom line is that existing MAC protocols are not appropriate for direct use in a WSN. Although SNs are static, changes in WSN topology could occur due to many factors. One of the events is an object or person passing by or having dust storm or

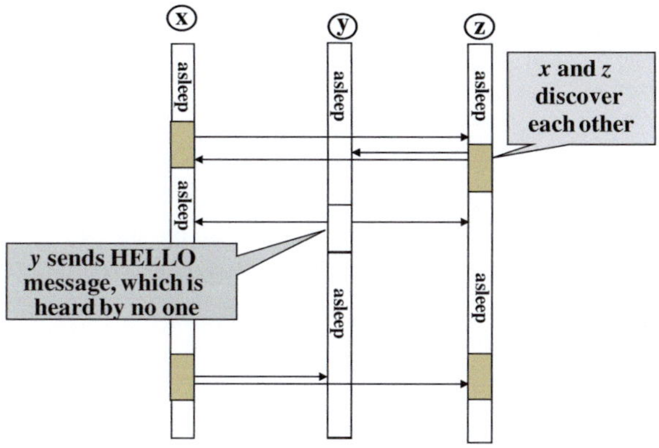

Fig. 6.3 Impact of sleep cycles on neighbor SNs discovery process

Fig. 6.4 Differentiating between neighbor discovery and topology maintenance

rain and fog and affecting a direct transmission of EM waves. Some adjacent SNs may run out of energy and could not receive/forward the data. Loss of local synchronization due to accumulated clock drifts is also attributed to affect the level of received signal. Figure 6.3 shows the impact of sleep cycle in discovering close by SNs. When SNs x and z are awake at the same time, they can send beacon signals to each other and recognize the presence of each other. Otherwise, if either is in sleep mode, the beacon signal from the other SN will go unnoticed. It is also important to distinguish between neighbor discovery in the initial phase (all SNs active) and the ongoing topology maintenance at the operation phase (SNs active only 2% of time, rest remain sleeping) and is shown in Fig. 6.4.

6.2 Neighbor Determination and Hop Distance

As discussed earlier, SN transfers data to BS in a WSN following a multi-hop path. Therefore, destination in a WSN is always the BS while the source SN could change based on the location of an event. Therefore, location of a SN is important in determining the path to follow to the BS. In a WSN, source SN x needs to decide an intended intermediate SN y within its transmission range. Using the unit disk graph model [1], SN x needs to find how many sensors are within communication range and which SN is farthest away from the SN toward the BS. A first-order radio energy consumption model of SN has been proposed by Dongjin et al. [2] that shows energy for transmitting one bit of data over distance u between them is given by:

$$E_{tx}(xy) = E_{\text{elec}} + E_{\text{amp}} u^k, \tag{6.1}$$

where E_{elec} is the energy consumed by the transmitter, E_{amp} is the transmitting amplifier, and k ($k > 2$) is the propagation loss exponent. The energy consumed by the receiver for each bit of the data is E_{rx}. The energy consumed to relay one bit data over distance u is given by:

$$E_{\text{relay}} = E_{\text{elec}} + E_{\text{amp}} u^k + E_{rx} = E + E_{\text{amp}} u^k, \tag{6.2}$$

where $E = E_{\text{elec}} + E_{rx}$.The consumed energy E can be minimized [3] if the number of h hops from SN u_0 and BS can be defined as:

$$h = \left\lfloor \frac{u}{u_0} \right\rfloor \quad \text{or} \quad \left\lceil \frac{u}{u_0} \right\rceil, \tag{6.3}$$

where

$$u_0 = \sqrt[k]{\frac{E}{E_{\text{amp}}(k - 2^{1-k})}}. \tag{6.4}$$

6.3 MAC Protocols

Three different types of MAC protocols have been introduced in the literature as contention (Random/CSMA)-based, reservation-based, and hybrid (CSMA/TDMA) protocols. As "Idle Listening" consumes significant energy, the solution that has been suggested in S-MAC protocol [4] is to Periodic listen and sleep as given in Fig. 6.5. The SN listens for about 120 ms and then sleeps till 2 s. Radio is turned off during sleeping time, thereby reducing duty cycle to 10%. Each SN goes into periodic sleep mode during which it switches the radio off and sets a timer to awake later (Fig. 6.5). When the timer expires, it wakes up and listens to see whether any other SN wants to talk to it. The duration of the sleep and listen cycles is application dependent and they are set the same for all SNs. For example, if temperature is to be measured, then once every 1 h or so may be adequate. But, if nuclear radiation level is to be determined, it has to be done even a fraction of a second. This scheme

Fig. 6.5 Sleep–listen cycle in S-MAC protocol

requires a periodic synchronization among SNs to take care of any type of clock drift. Therefore, SYNC packets are exchanged periodically to maintain schedule synchronization (Fig. 6.6a) and receivers adjust their timer counters immediately after they receive the SYNC packet . In order to calculate the HELLO message frequency, SN y should estimate the number of in-segment stations that participate in the discovery of a hidden SN x. This is equal to $\deg_S(x)$, namely the number of neighbors that the hidden SN x has inside the connected segment. There are 3 methods considered (Fig. 6.6b): Average in-segment degree of the segment's SNs (=4) is used as an estimate of the in-segment degree of x, the number of y's in-segment neighbors ($\deg_S(y) = 3$) is used as an estimate of $\deg_S(x)$, and *a* linear combination of the two.

Suppose that SN x is discovering neighbors as shown in Fig. 6.7. When SN x wakes up every T_I seconds for an average period of time equal to H, it announces HELLO messages. The SNs will discover x within a time period T with probability P. Each SN y in topology maintenance state wakes up every $T_N(y)$ seconds for H time and broadcasts HELLO message. Active state of SNs x and y should overlap at least δ, $0 < \delta < 1$ within active time H. Therefore, SN y awakened at t should remain up between $t - H(1 - \delta)$ and $t + H(1 - \delta)$ time. Then, the probability that x and y discover each other is $2H(1 - \delta)/T_N(v)$.

A detailed simulation has been done to determine the ratio of hidden neighbors as a function of time by placing 2000 SNs in an area of 10,000 × 10,000. Sensing range r_s for SN is assumed to be 300 units and half of the SNs are considered in hidden. A hidden SN is detected with a probability P between 0.3 and 0.7 within $T = 100$ time. The hidden SNs are assumed to be in neighbor discovery mode and wake up randomly every $T_I = 20$ average time units while non-hidden SNs to be in topology maintenance state and wake up randomly to discover hidden SNs, on an

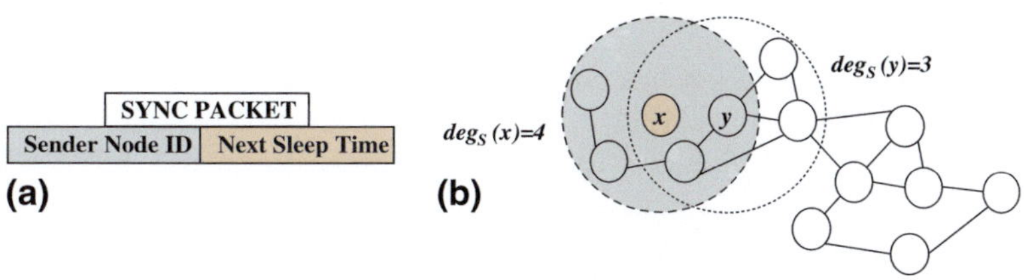

Fig. 6.6 **a** Use of SYNC packet, **b** SN degree estimation for message frequency

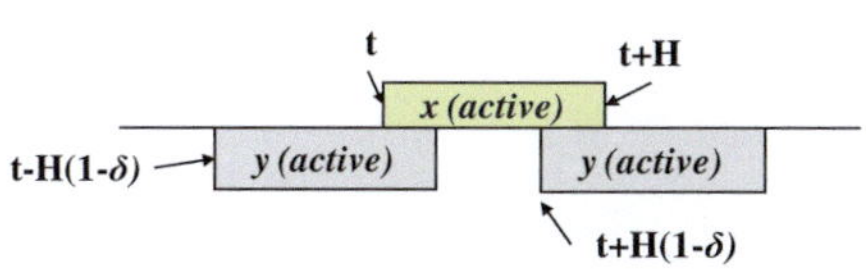

Fig. 6.7 Topology maintenance strategy

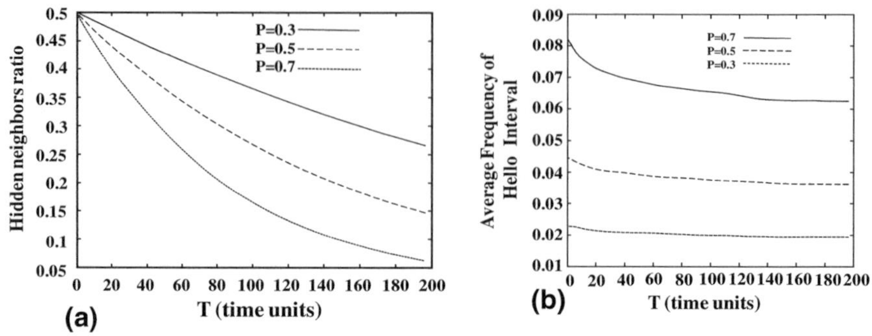

Fig. 6.8 **a** Ratio of hidden neighbors as a function of time [19]. **b** Average frequency of Hello interval as a function of time [19]

average every $T_N(y)$ time units. When a hidden SN is detected, it joins the segment while two hidden SNs that detect each other remain in the neighbor discovery state. The results are shown in Fig. 6.8.

6.4 Residual Energy Mapping

In a randomly deployed WSN, once the neighborhood SNs have been determined, the next step is to cluster the SNs. Before such clustering can be done, the next step is to determine energy remaining with SNs such that it could help in designating cluster heads (CHs) in the network as they have to do more work than other SNs. Such residual energy (RE) is a continuous core process that can be separated into three phases of determining RE graph; performing in-network aggregation to propagate to BS; and rebuilding topology tree from the root to have justifiable energy depletion. The SN sends message about RE in the form as shown in Fig. 6.9a and the details of polygon information indicating the presence of holes are shown in Fig. 6.9b [5]. Polygon for Part 2 is shown to have 2 holes with no connectivity and that area is indicated in Fig. 6.9b.

In a WSN, energy consumption of the SNs is random, and then the corresponding energy map will be quite irregular. In general, events tend to follow both

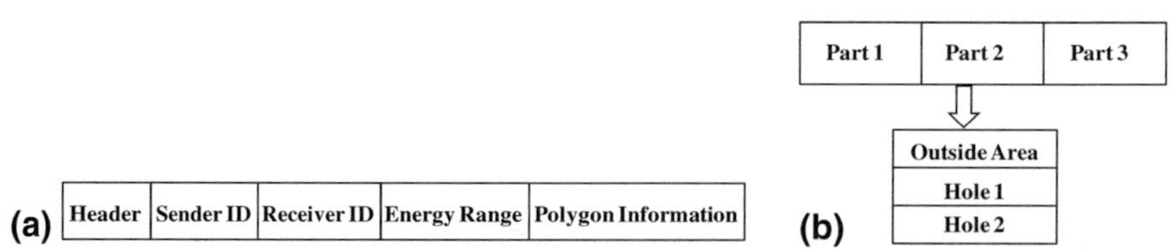

Fig. 6.9 **a** Message format for SNs' residual energy [5]. **b** Details of polygon holes

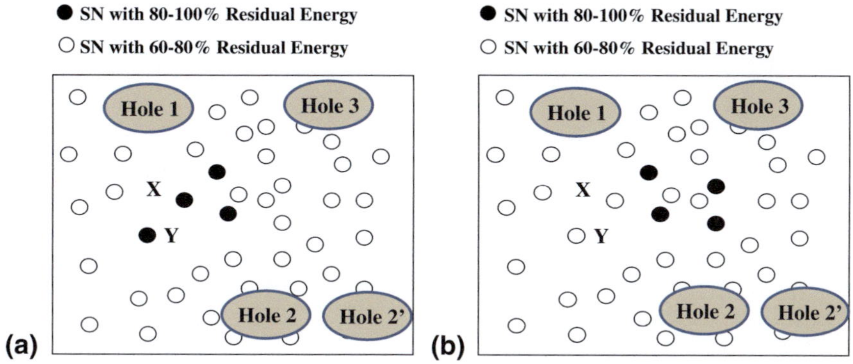

Fig. 6.10 New CH based on energy level in an area with holes. **a** Initial state and **b** state after energy level at SN *X* has dropped

temporal locality and spatial locality, and one such example is illustrated in Fig. 6.10. The RE among SNs could change animatedly and hence cluster formation and CH determination has to be performed at different times.

Conventional Sensor MAC (S-MAC) [6] adjusts its operations as changes take place in the surrounding area while Berkeley MAC (B-MAC) employs RTS/CTS and ACK as high-level task to change applications on and off [7]. B-MAC suffers from high value of latency as the receiver ought to be awake before any data transfer and the receiver could possibly wake up much earlier than the end of preamble. An aggressive RTS is introduced in [8] to replace multiple RTS packets to RTS burst divided by small gaps so as to receive a CTS packet to initiate transfer of data. This way, the message latency between two SNs could be condensed to half. MICA2 motes have been used to implement B-MAC and S-MAC using TinyOS in [7] and placed in an open area with no obstruction 15 cms above the ground. T-MAC [9] has been simulated using MATLAB.

Table 6.1 shows the memory requirements for S-MAC and B-MAC protocols and code storage needed by the two protocols is shown in Table 6.2. Performance of these two protocols in terms of measured throughput and power consumption is given in Fig. 6.11.

As there are no GPS devices attached to each SN (even if GPS does not work indoors), location of each SN needs to be determine its location. This is done by having some reference SNs [called Anchor SNs (ANs)] and determining relative

Table 6.1 Comparing memory requirements in bytes for S-MAC and T-MAC using TinyOS

Protocol	ROM	RAM
B-MAC	3046	166
B-MAC w/ACK	3340	168
B-MAC w/LPL	4092	170
B-MAC w/LPL and ACK	4386	172
B-MAC w/LPL and ACK + RTS-CTS	4616	277
S-MAC	6274	516

Table 6.2 B-MAC and S-MAC implementation in TinyOS

Length (bytes)	B-MAC	S-MAC
Preamble	8	18
Synchronization	2	2
Header	5	9
Footer (CRC)	2	2
Data length	29	29
Total	46	60

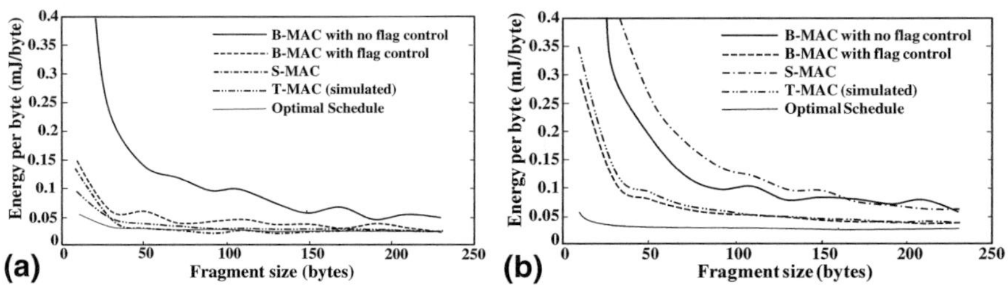

Fig. 6.11 **a** Measured throughput of S-MAC and B-MAC protocols with no duty cycle under a contended channel. **b** Measured power consumption in S-MAC and B-MAC for maintaining throughput in 10-SNs WSN [8]

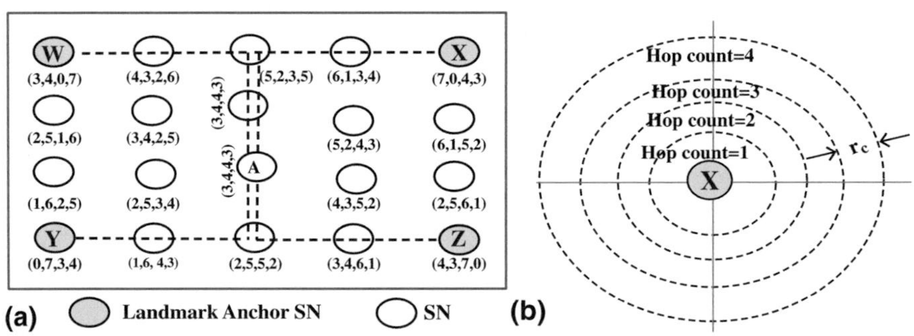

Fig. 6.12 **a** Logical coordinate system representing distance of each SN as # hops from selected ANs [10]. **b** Hop counts from the landmark AN *X* [10]

addresses of other SNs with respect to these selected ANs. One such example is illustrated in Fig. 6.12a, where SNs are arranged in 2D mesh and SNs at four corners WXYZ are considered as reference ANs and distance of each sensor in terms of number of hops is given by four tuples. SN A has a logical coordinate vector of (3,4,4,3), showing that SN A is 3-hops from ANs *W* and *Z*, 4 hops from ANs *X* and *Y*. The hop is also presented by circular disk as shown in Fig. 6.12b where distance between each successive circle represents the communication distance r_c.

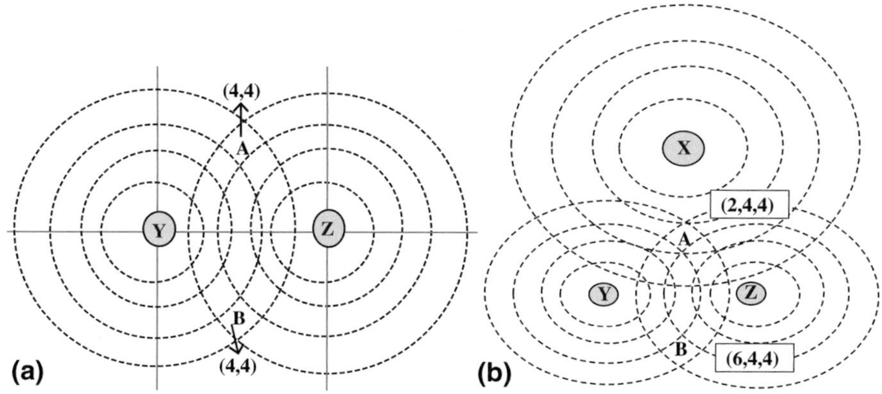

Fig. 6.13 **a** SNs in zones A and B share the same logical coordinate from ANs *Y* and *Z* [10]. **b** Zones A and B can be identified by using a third landmark AN *X* [10]

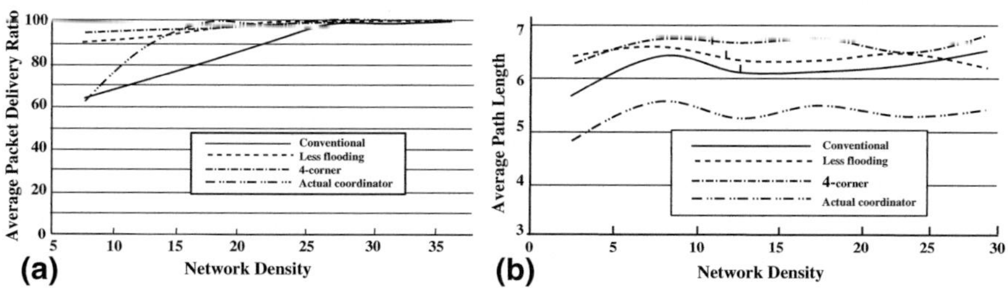

Fig. 6.14 **a** Average packet delivery ratio with network size 1000 m × 500 m [20]. **b** Average path length with network size 1000 m × 500 m [20]

If only two ANs are used in the WSN, two zones A and B in Fig. 6.13a will share the same coordinates of (4,4) [10]. This can be resolved by adding another AN *X* to get (2,4,4) and (6,4,4) as their addresses. A new protocol with reduced flooding has been proposed in Caruso et al. [10] using Glomosim and uniformly mesh-distributed in 100 m × 500 m area, with number of SNs varying from 150 to 900. Each SN has r_t = 100 m, and propagation delay is assumed to be 1 μs. Four ANs are considered at four corners. The average packet delivery ratio for randomly selected 100 pairs of sources and destinations and path length is shown in Fig. 6.14a, b.

6.5 Routing with Energy Harvesting

In a WSN, SNs forward their data to the BS in a multi-hop fashion, possibly along shortest paths. But, no attempts are paid to check whether the intermediate SNs have enough energy or if there is any provision for recharging them using

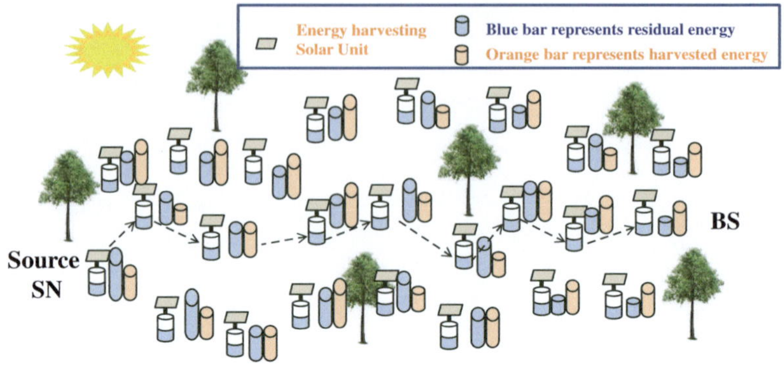

Fig. 6.15 WSN with residual energy harvested by solar units

energy-harvesting technique. Most of the existing works have been solely concerned with the design of different forwarding mechanisms for performance improvement or duty-cycled SNs for energy efficiency. However, little work had been done looking at geographic routing which taking harvested energy into account. Particularly, no solution has been designed precisely to solve the uncertainty of remaining energy in charging SNs of a WSN by energy harvesting. One such example of a real scenario is illustrated in Fig. 5.15, where residual energy of each SN is explicitly shown. Batteries of SNs are recharged by harvesting solar energy using lunar panels. As is clear from the diagram, available energy is not homogeneous at all SNs and tasks can be distributed to the SNs based on the availability of energy. Current energy-aware task allocation policy is based on SN residual energy and does not take replenishing capability of SNs into account. Therefore, in allocating task, spatio-temporal property supply should be browbeaten. A novel 2-hop SN-disjoint multi-path geographic routing algorithm TPGFPlus [11] has recently been introduced where a SN selects its next-hop forwarder SN that is closest to the BS among all 1-hop and 2-hop neighbor SNs. In addition, sleep–awake scheme is adopted to increase the effective lifetime (Fig. 6.15).

Energy-harvesting problem involves taking out maximum effort out of a given energy atmosphere and a distributed framework is called environmental energy-harvesting framework (EEHF) [12] where local residual energy is considered in recharging SNs and accordingly tasks are assigned and leaders elected and data transfer path selected. A WSN with multi-hop message forwarding can be described by a directed graph $G(V,E)$, *with* V as a set of vertices signifying SNs and E is a set of edges indicating the communication links between them. A path from source SN to the BS consists of one or multiple edges, and energy is consumed in both transmitting data and receiving any packet by a SN. Assuming all the SNs start with the same amount of energy, and the remaining energy at SNs consists of energy expanded when minimum energy (ME) routing is followed that opt for a route that costs least total energy cost, and max–min routing that picks the route with maximum residual energy. So, if SN is recharged with energy harvesting, then the objective is to optimize this current power depletion index of SNs.

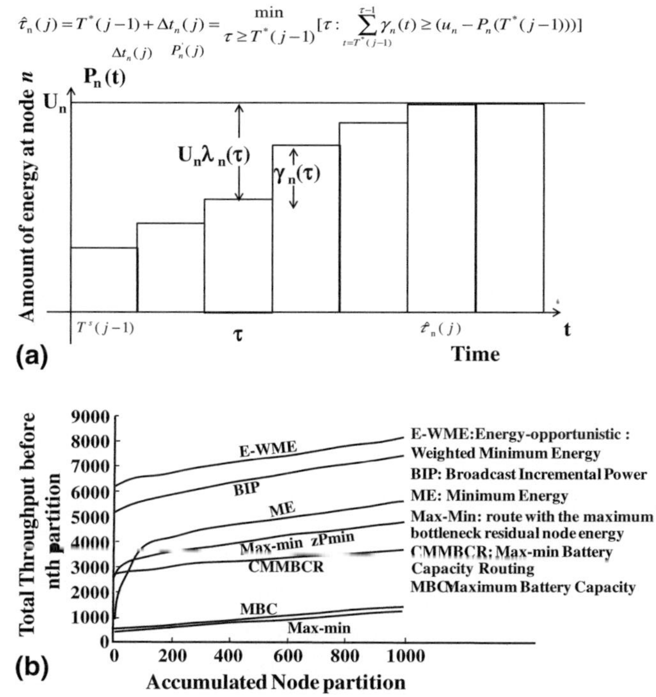

Fig. 6.16 **a** Energy at SN n assuming no request accepted after request $(j - 1)$ [12]. **b** End-to-end throughput against the number of SN pairs that have experienced partition [12]

A new energy-opportunistic weighted minimum energy (E-WME) algorithm introduced in [12] is based on only local information and without any detailed statistical information. An exponential function is assumed for SN residual energy, and transmit and receive energies vary as a linear function (Fig. 6.16a) while an inversely linear function is assumed for the replenishment rate. Due to energy drainage, the WSN with 200 randomly deploy SNs in a 10×10 field could tend to partition in multiple subnetworks and end-to-end throughput is given in Fig. 6.16b. SN energy distribution after sending 1000 packets is shown in Fig. 6.17a [12] while Fig. 6.17b shows energy spent per packet for ME and E-WME routing [12].

6.6 Energy Harvesting by Fuel Cells and Healthcare Applications

Many different techniques can be used to harvest energy and a comparison of energy sources is given in Table 6.3 [13].

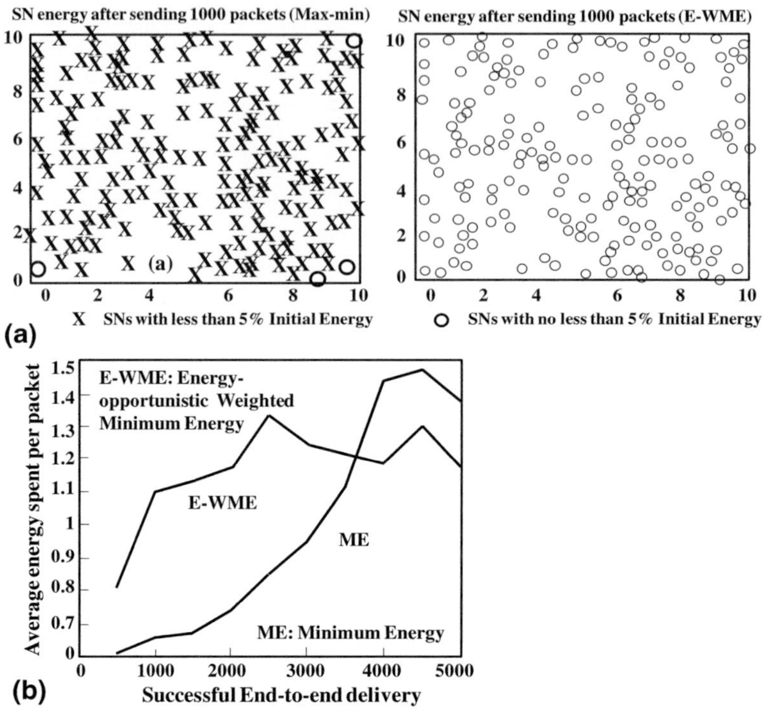

Fig. 6.17 **a** SN energy distribution after sending 1000 packets [12]. **b** Average energy spent per packet for ME and E-WME [12]

Table 6.3 A comparison of energy sources [13]

Energy source	Power density	Energy density
Batteries (zinc-air)		1050–1560 mWh/cm^3
Batteries (rechargeable lithium)		300 mWh/cm^3 (3–4 V)
Solar (outdoors)	15 mW/cm^2 (direct sun) 0.15 mW/cm^2 (cloudy day)	
Solar (indoor)	0.006 mW/cm^2 (standard office desk) 0.57 mW/cm^2 (<69 W desk lamp)	
Vibrations	0.01–0.1 mW/cm^3	
Acoustic noise	3E−6 mW/cm^2 at 75 db 9.6E−4 mW/cm^2 at 100 db	
Passive human-powered systems	1.8 mW (shoe inserts)	
Nuclear reaction	80 mW/cm^3 1E6 mWh/cm^3	

Table 6.4 Alkaline battery versus microbial fuel cell [16]

Type	Voltage open circuit (V)	Capacity	Energy	Weight (g)	Energy density (J/g)	Cost
Alkaline cell-AA	1.5	2.8 Ah	4.2 Wh	25	604	£0.30
Microbial fuel cell	0.8	1.05 Ah (~120 μA for 1 year)	1 Wh (100 μW for 1 year)	10	360	<£1

Table 6.5 Energy-harvesting sources today [25]

Energy source	Characteristics	Efficiency (%)	Harvested power (mW/cm^2)
Light	Outdoor Indoor	10–24	100 100
Thermal	Human Industrial	~0.1 ~3	60 ~1–10
Vibration	~Hz—human ~kHz—machine	25–50	4 800
RF	GSM 900 MHz Wi-fi	~50	0.1 0.001

Ecological Robot (EcoBot) [14] means a class of autonomous robots that can collect energy from environmental waste material, produce carbon dioxide, and continue to be self-sustainable. The anode is formed by the bacteria found in sludge that acts as catalysts to generate energy from the given substrate while O_2 from free air acts as the oxidizing agent to work as a cathode and combines electrons and protons to produce H_2O [15]. A microbial fuel cell cannot match with a typical alkaline standard battery or a solar panel and Table 6.4 summarizes a comparison after 1 year of continuous operation [16]. Energy harvesting is especially important for a body area network (BAN) which poses a real challenge in identifying energy source. Some of the specific sources of energy are [17] electromagnetic waves, heat, potential energy, and electric potential difference besides the solar power and are summarized in Table 6.5. But, most of these schemes are not appropriate for the BAN.

One sample example of BAN in healthcare application is shown in Fig. 6.18 where SNs are placed on different parts of the body that send data to a coordinator that acts as a gateway to collect data and send to a BS. Another example is the use of rise in body temperature of firefighters when in action. Firefighters experience a very high physical and physiological stress during firefighting and are at a high risk of cardiac arrest due to heat stress. It is important to constantly monitor their heart rate as blood pressure needs to be measured during active maneuvering in protecting their lives. The body temperature may rise to 104 °F during an active operation. Commercially available ECG sensors have low battery life and scavenging energy from body heat and surrounding heat could enhance the lifetime of

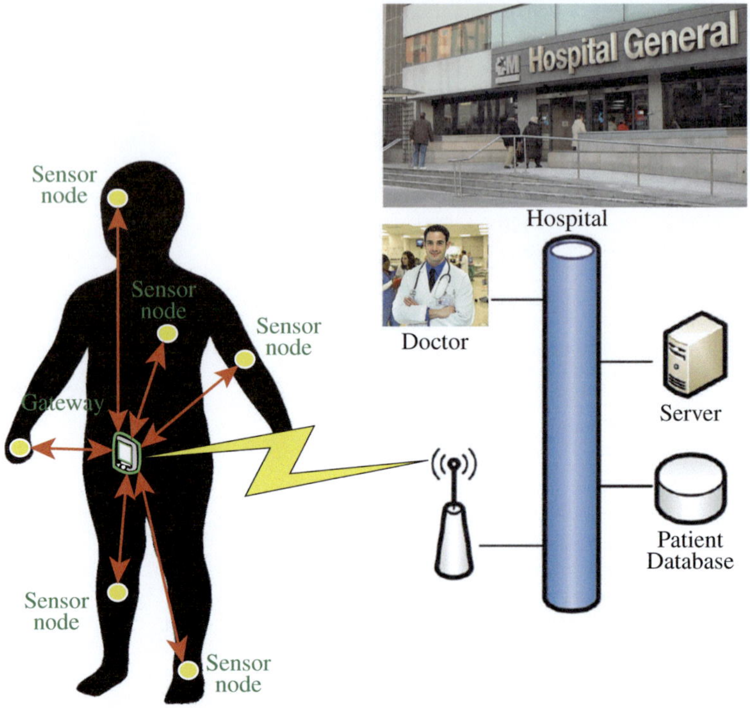

Fig. 6.18 Use of BAN in healthcare application

sensors for its perpetual operation. Data transmission to the gateway consumes most of the energy and frequent data transmission causes to drain of energy fast. Energy spent ETx(k,d) for transmitting a k-bit message to a distance d can be computed as ETx-elec(k) + ETx-amp(k,d).

Thus, body heat is observed to increase as a result of physical and psychological stress and body heat also increases with increase in atmospheric temperature beyond 130 °F.

The average specific heat of human body is 0.83 kcal, which means that 0.83 kcal of body heat is lost or gained when one degree change in temperature occurs. The famous Seebeck's Theory [18] states that a difference in temperature in two dissimilar materials generates potential difference as per first law of thermodynamics: $Q = mC\,\Delta T$. The generated voltage is given by $V = \alpha \times \Delta T$, where α is the Seebeck's coefficient. Seebeck's coefficient for good thermoelectric material can range from 100 to 300 μV/K. This leads to about 25 mV to be generated by every 1° rise in temperature, and by having them in series, adequate voltage can be generated to power SN which can be stored in a capacitor for constant operation of the SN, irrespective of the frequency of data transmission. Simulation experiments

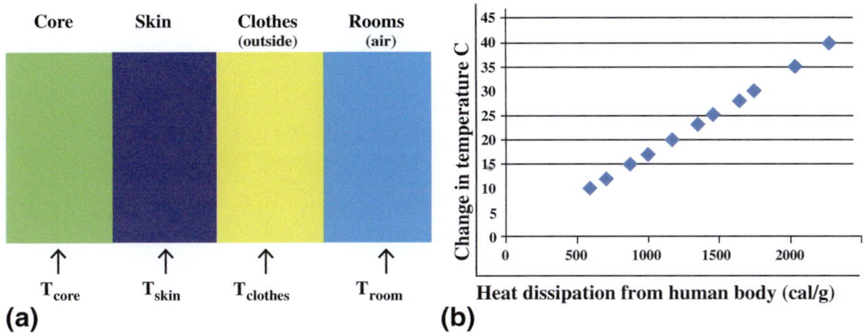

Fig. 6.19 **a** Coverage of two C-SNs in an area [24]. **b** Coverage as a function of the number of C-SNs

were performed and measured the output voltage generated across thermoelectric materials with varying Seebeck's coefficient randomly, ranging from 100 to 300 μV/K, for different differences in temperatures as shown in Fig. 6.19a. It is observed that greater the Seebeck's coefficient and difference in temperature, higher is the amount of voltage generated. The output power generated depends on the different values of load resistance and temperature difference across and Fig. 6.19b shows that maximum power is generated when the load varies and difference in temperature created across the two ends of thermoelectric generators is high. If communication has to be secured, then some security measures need to be adopted. This leads to increased packet size as illustrated in Figs. 6.20 and 6.21.

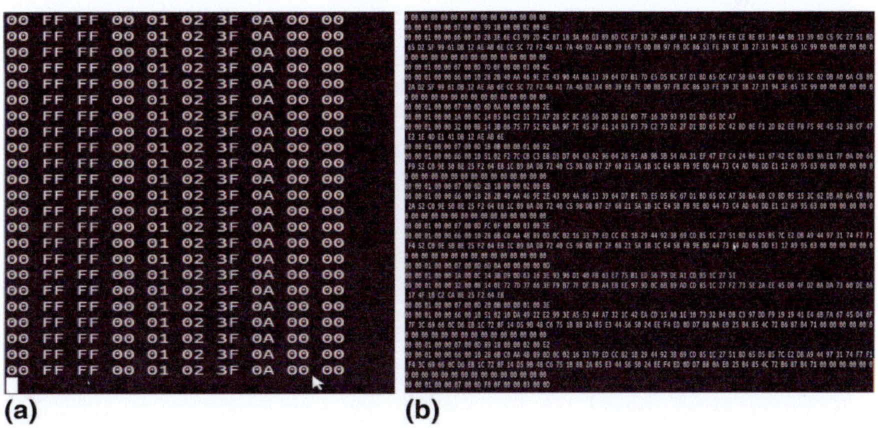

Fig. 6.20 **a** Normal dummy packet sent from a sensor SN [26]. **b** Packet size increases considerably on application of security measures [23]

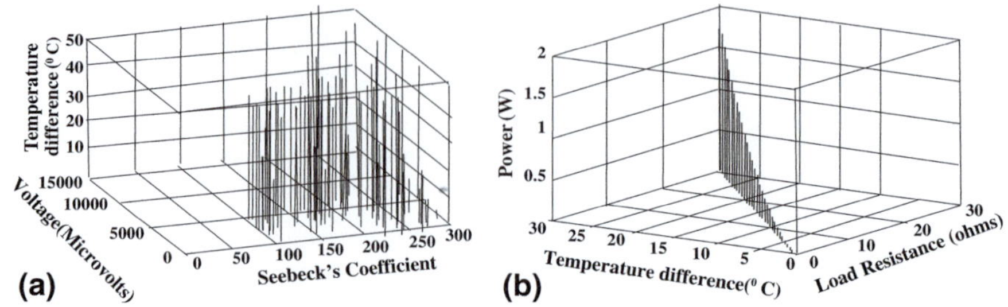

Fig. 6.21 **a** % Area enclosed with 0.782052 coverage by initial layout [27]. **b** % Area enclosed by improved layout with 0.92336 coverage [21]

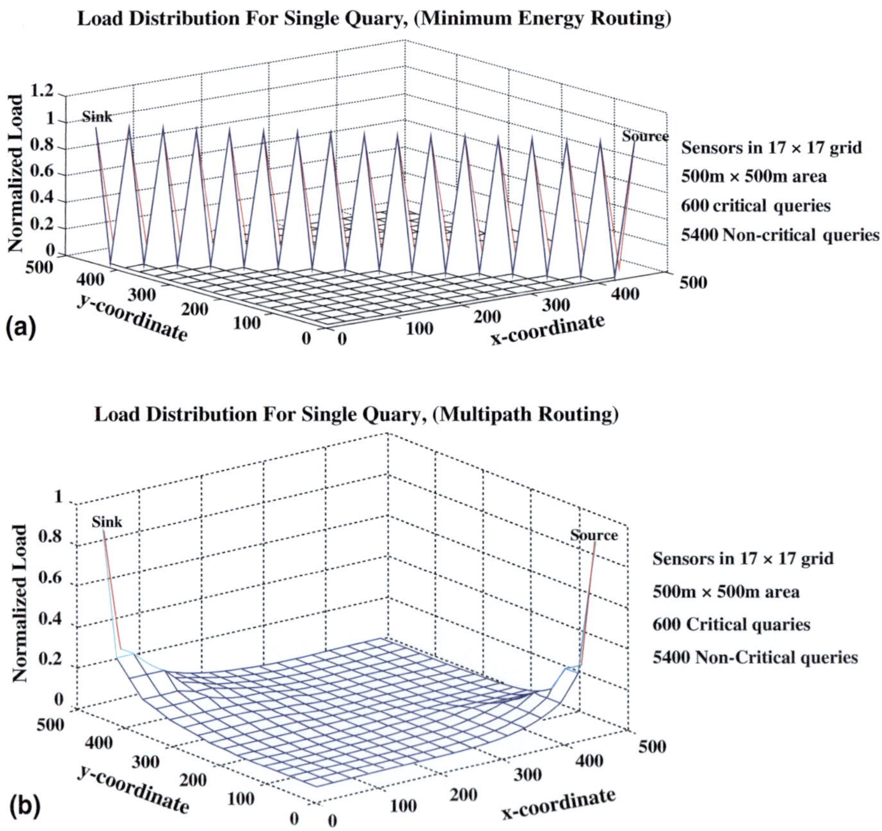

Fig. 6.22 **a** Unequal energy consumption due to continuous use of a path [24]. **b** Balancing energy consumption with multiple path [24]

6.7 Balanced Energy Consumption with Multiple Paths

In a WSN, if an event is detected, the SNs discovering the occurrence could continuously transmit data along a selected path to BS and could deplete all SNs along the path for transmitting and receiving data. This could potentially partition the WSN if some SNs run out of energy. Such an effect could be minimized if shortest path is used only for grave data and non-critical data can be sent along longer path. One such example is shown in Fig. 6.22 where 17 × 17 grid is simulated and BS is located in one corner while source SN is in an opposite corner and shortest path is along the diagonal.

6.8 Conclusions

Neighbor discovery is an initial and important step in a randomly deployed WSN and a lot of energy is consumed and time spent and cannot be avoided due to organization style. As most energy is consumed in transmission, reception, and idle listening, it is useful to keep the SNs sleeping as much as possible. Lifetime longevity necessitates that the SNs need to be kept active for a very short period. The path from SN to BS should not be selected based on hop length, but attempts should be made to keep residual energy in mind for keeping selected path connected for a longer period of time. In this respect, energy harvesting is useful in charging vigor-depleted SNs even though it is difficult to implement. Excessive energy consumption could lead to partitioning of WSNs and could be avoided by following multiple path routing if only critical data follow shortest path while non-essential data could be forwarded along longer paths, making other SNs to participate in data receiving and forwarding process.

6.9 Questions

Q.6.1. A WSN consists of 100 randomly deployed SNs. If each SN takes transmission of 5 beacon signals on the average in determining neighboring SNs, how much energy is consumed in determining the topology if each beacon signal is transmitted for 500 ms? Assume each beacon signal to consume 200 mW.

Q.6.2. If beacon signals in Q.6.1 can be received up to a distance of 100 m, how much energy will be needed if power level is increased to cover a distance of 500 m?

Q.6.3. The beacon frame of Q.6.1 consists of 60 bytes. How much energy is saved if the beacon frame size is reduced to 30 bytes?

Q.6.4. How much energy can be saved if the beacon frame duration of Q.6.1 is changed from 10 to 5 ms?

Q.6.5. A SN has an option to use 10 different frequencies as beacon signals. What is the probability that two adjacent SNs will use the same frequency for beacon signal to create collision?

Q.6.6. In Q. 6.5, what is the probability of beacon collision between any two SNs if there are 4 SNs in the vicinity?

Q.6.7. A WSN is given in the following figure, SN_1 needs to send data to the BS. In the second of routing, a path can be selected either passing through SN_4, SN_5, or SN_6. Under what condition, a particular path will be followed?

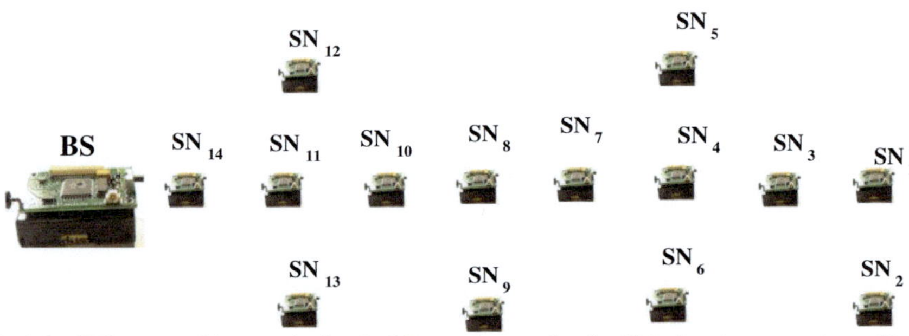

Q.6.8. What would you do in Q.6.7 so that path via SN_5 is always selected?

References

1. B. N. Clark, C. J. Colbourn, and D. S. Johnson, "Unit disk graphs," *Discrete Mathematics* 1990; no. 86, pp. 165–177.
2. S. Dongjin, B. Krishnamachari, and J. Heidemann, " Experimental study of the effects of transmission power control and blacklisting in wireless sensor networks," *1st IEEE Communications Society Conference on Sensor and Ad HocCommunications and Networks*, Santa Carla, CA, 2004; 289–298.
3. Oswald Jumira, Riaan Wolhuter, and Sherali Zeadally, "Energy-efficient beaconless geographic routing in energy harvested wireless sensor networks," Concurrency and Computation: Practice and Experience, 2012.
4. W. Ye, John Heidemann, and Deborah Estrin, "Medium Access Control with Coordinated Adaptive Sleeping for Wireless Sensor Networks," IEEE/ACM Trans. on Networking, June 2004.
5. Edward Chani and Song Han, "Energy Efficient Residual Energy Monitoring in Wireless Sensor Networks," International Journal of Distributed Sensor Networks, 2009, no. 5, pp. 1–23.
6. W. Ye, J. Heidemann, and D. Estrin, "Medium Access Control with Coordinated Adaptive Sleeping for Wireless Sensor Networks," *IEEE/ACM Trans. Networking*, vol. 12, no. 3, pp. 493–506, June 2004.
7. Joseph Polastre, Jason Hill, and David Culler, "Versatile Low Power Media Access for Wireless Sensor Networks," SENSYS, 2004.

8. Sha Liu, Kai-Wei Fan and Prasun Sinha, "CMAC: An Energy Efficient MAC Layer Protocol Using Convergent Packet Forwarding for Wireless Sensor Networks," in Proc. SECON, 2007.
9. T. van Dam and K. Langendoen. An adaptive energy-efficient mac protocol for wireless sensor networks. In *Proceedings of the First ACM Conference on Embedded Networked Sensor Systems*, Nov. 2003.
10. A. Caruso, S. Chessa, S. De, and A. Urpi, "GPS Free Coordinate Assignment and Routing in Wireless Sensor Networks," in Proceedings of IEEE International Conference on Computer Communication (INFOCOM), Miami, March, 2005.
11. Guangjie Han, Yuhui Dong, Lei Shu, Hui Guo, and Jianwei Niu, "Geographic Multipath Routing in Duty-Cycled Wireless Sensor Networks with Energy Harvesting," Green Computing and Communications (GreenCom), 2013 IEEE and Internet of Things (iThings/CPSCom), IEEE International Conference on and IEEE Cyber, Physical and Social Computing, 20–23 Aug. 2013, pp. 31–39.
12. Longbi Lin, Ness B. Shroff, and R. Srikant, "Asymptotically Optimal Power-Aware Routing for Multihop Wireless Networks with Renewable Energy Sources," Globecom 2012, pp. 3418–3424.
13. J. M. Rabaey, M. J. Ammer, J. L. da Silva, and D. Patel, and S. Roundy, "PicoRadio supports ad hoc ultra-low power wireless networking, IEEE Computer, vol. 33, no. 7, pp. 42–48.
14. http://en.wikipedia.org/wiki/EcoBot.
15. http://www.brl.ac.uk.
16. http://www.brl.ac.uk/brlresearchprojects/microbialfuelcells/ecobotii.aspx.
17. C. Cordeiro and D. P. Agrawal, "Ad hoc & Sensor Networks: Theory and Applications," *World Scientific Publishing,* Spring 2006, ISBN no. 81-256-681-3; 81-256-682-1, 2nd ed. 2011.
18. Cooler Peltier Cooler datasheet [Online]. Available http://www.hebeiltd.com.cn/peltier.datasheet/TEC1-12710.pdf.
19. Reuven Cohen and Boris Kapchits, "Continuous Neighbor Discovery in Asynchronous Sensor Networks," IEEE/ACM Transactions on Networking, Feb. 2011, vol. 19, no. 1, pp. 69–79.
20. Jang-Ping Sheu, Yu-Chia Chang, and Gang-Hua Song, "Logical Coordinate Assignment for Geographic Routing in Wireless Sensor Networks," *International Journal of Pervasive Computing and Communications, vol.* 3, no. 3, 2008, pp. 274–288.
21. Anagha Jamthe and Dharma P. Agrawal, "Approaches for Energy Harvesting and Power Management in Wireless Healthcare Sensor Networks," International Journal of Computer and Communication Engineering, vol. 2, no. 5, September 2013, pp. 596–600.
22. D. Smith, G. Horn, E. Goldstein, S. Petruzzello *et al.*, "Firefighter fatalities and injuries, the role of heat stress and PPE," University of Illinois Fire Service Institute, Fire Fighter Life Safety Research Center.
23. P. Herman, "Physics of Human Body: A Physical view of Physiology," Springer Berlin Heidelberg New York 2007, ISBN no. 978-3-540-29603-4, Third edition 2007.
24. Neha Jain, Dilip K. Madathil, and Dharma P. Agrawal, "MidHopRoute: A Multiple Path Routing Framework for Load Balancing with Service Differentiation in Wireless Sensor Networks," Special Issue on Wireless Sensor Networks of the International Journal of Ad Hoc and Ubiquitous Computing, vol. 1, no. 4, 2006, pp. 210–221.
25. R. O. Bonow, D. L. Mann, D. P. Zipes, and P. Libby, "Braunwald's Heart Disease: A Textbook of Cardiovascular Medicine," 9th Edition, 2012.
26. M. Malik and D. P. Agrawal, "Secure Web Framework For Mobile Devices," *GC'12 Workshop: The 4th IEEE International Workshop on Management of Emerging Networks and Services (GC'12 Workshop—MENS 2012)*, Anheim, Ca, Dec. 3–7, 2012.
27. G. J. Synder, "Small thermoelectric generators," The Electrochemical Society *Interface*, Fall 2008, pp. 54–56.

Chapter 7
TCP, Neighborhood Formation, Reliable Transport, and Simulators for WSNs

7.1 Introduction

TCP is the widely used transport protocol and most of the IP-based technologies already well known and proven to be working. The pervasive nature of IP networks allows use of an existing infrastructure. But, TCP needs to distinguish between the natures of errors. WSN are built exclusively with wireless links as SNs need to send data to a single station commonly known as BS or sink node. TCP has been developed primarily for wire-line networks, and the design is greatly influenced by end-to-end connection set up agreement between sender and receiver (Fig. 7.1a), and associated transmission and acknowledgment mechanisms. In wire line, adequate intelligence in physical and links that could take care of error control, encryption, or flow control as the TCP performance depends on flow and congestion control as most losses is due to congestion. Moreover, all TCP segments carry a checksum that is utilized to detect errors with either the TCP header or data. TCP also keeps track of bytes received in order to discard duplicate copies of data that has already been received. In TCP retransmission for lost or damaged data, ACK signal is included for positive acknowledgment, and retransmission is performed if ACK not received with a timeout period. As successive packets follow the same path in TCP, the byte sequence is maintained.

In a WSN, a large number of SNs are deployed either randomly, and transmission from each SN to BS is achieved in multi-hop fashion in order to conserve energy. It may be noted that in WSN, energy consumed is proportional to square of distance of intermediate receiver SN from the source SN. In a WSN, the error rate is relatively higher as signal passes through noisy air medium, and there is hop-by-hop transmission between SNs (Fig. 7.1b). If no explicit ACK is received from an adjacent SN within timeout period, it is interpreted as NACK, and error is controlled by retransmission of a packet. This causes high error rates and longer delays are present in a WSN and it becomes worst if mobility is present.

D.P. Agrawal, *Embedded Sensor Systems*,
DOI 10.1007/978-981-10-3038-3_7

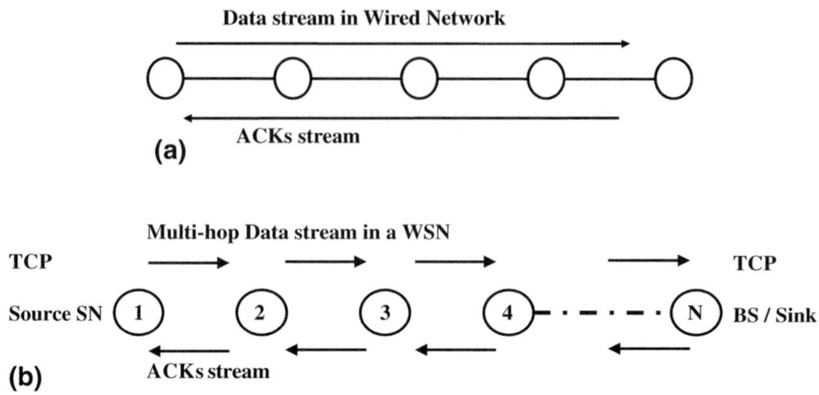

Fig. 7.1 **a** TCP in a wired network. **b** Adopting TCP for a WSN

7.2 Identifying Neighboring SNs

Information transfer could be in the form of a single packet, or multiple packets, or stream of packets between a source and destination, and as signal goes through open space, errors are present and up to 50% packets are received correctly (Fig. 7.2a). The packet transfer could be between SNs and BSs or between different set of SNs (Fig. 7.2b). As SNs are deployed randomly in a WSN, the first thing SNs do is to send beacon signals to determine neighboring SNs, if any. This is illustrated in Fig. 7.3. As SNs operate in a distributed manner, it is rather difficult to control when signals are sent by a SN and beacons could possibly collide with each other. Beacons could possibly collide with each other. As shown in Fig. 7.2a, if beacon signals from SNs x and y overlap with each other, a collision occurs, and no response is sent by none of the devices. If only SN x sends beacon signal, then both SNs y and z can hear that and respond back independently to SN x. In this way, SN x determines the presence of SNs y and z in the neighborhood , and the corresponding RSSI value indicates distance between y and z from x. Once SN x successfully receives acknowledgment from y and z, it can piggyback this information in future beacon signals to determine other SNs in the close by area.

It may be noted that a large number of SNs are deployed which could lead to increased message transmission in the network. So, from this point of view, just minimum but adequate number of SNs ought to be used. It is interesting to note that the SN in the wait mode consumes almost the same amount of energy as the transmitting mode, and the wait period should be minimized. Any collision in beacon signals and resultant acknowledge signals results in waste of time and energy and should be avoided as much as possible. SNs could be asynchronous when deployed with respect to sleep-awake cycle that plays an important role in receiving beacon signals. Figure 7.4 shows 4 SNs and sleep-awake cycles for them. Therefore, it is important to synchronize sleep-awake cycles of SNs. The SNs use time slots for sleep-awake cycles . Discovery of two adjacent SNs occurs when

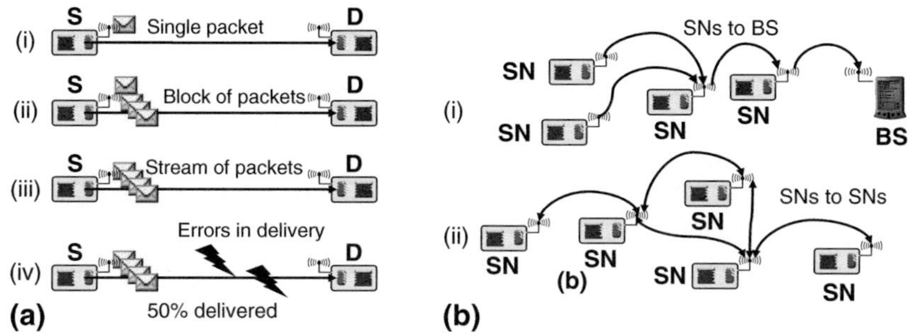

Fig. 7.2 Communication in a WSN **a** packets between S-D pair, **b** multi-hop communication between SNs and BS

Octets: 2	1	4 or 10	2	variable	variable	variable	2
Frame Control	Beacon sequence number	Source address information	Superframe specification	GTS field	Pending address field	Beacon Payload	Frame check sequence
MAC header			MAC Payload				MAC footer

Bits: 0-3	4-7	8-11	12	12	14	15
Beacon Order	Superframe order	Final CAP slot	Battery life extension	Reserved	PAN coordinator	Association permit

(a)

Octets: 2	41	4 to 20	variable	2
Frame control	Data sequence number	Address information	Data payload	Frame check sequence
MAC header			MAC payload	MAC footer

(b)

Octets: 2	1	2
Frame control	Data sequence number	Frame check sequence
MAC header		MAC footer

(c)

Fig. 7.3 **a** Beacon frame format, **b** data format, and **c** ACK format

two active slots overlap (Fig. 7.4) with sleep-awake duty cycles synchronized or remain asynchronous. A simple mechanism to be used for message transfer between SNs is that it could avoid collision in a distributed manner. Minimum energy is consumed in a SN only during the sleep mode, and efforts should be made to lengthen the sleep time of each SN. But, when SN sends data to another SN on way to BS, both the SNs need to be awake; otherwise, the message will be lost and cannot be propagated to BS. So, SNs remain in sleep state most of the time (98%) and are awake for a very short duration (2%) to conserve energy. So, there is always frequent transition from awake to sleep mode and vice versa and that causes increased power consumption.

Assuming the clocks for SNs are synchronized and the same time slot is used by SNs, SNs selectively remain awake for full slot duration and beacon starts at the beginning and end of an awake slot [1] as shown in Fig. 7.5. The fraction of neighboring SNs discovery in Birthday algorithm varies with time and is illustrated in Fig. 7.6 (Fig. 7.7).

Many algorithms have been introduced to discover neighboring SN. It may be noted that clocks of SNs are synchronized while sleep-awake start–end time may be synchronized or may be out of sync. In Disco system, [2] employs two primes p_1 and p_2 by two SNs p_i and p_j and both wake up every p_1th and p_2th slot (5th and 7th) and discovery is guaranteed in $p_{1i} \times p_{1j}$ slots. In U-Connect [3], each SN selects one prime p, every SN wakes up every pth slot and $(p - 1)/2$ slots every $p * p$ slots and overlap is guaranteed within $p \times p$ slots. Both Disco and U-Connect work with symmetric and asymmetric duty cycles. These three algorithms do satisfy worst-case limits, and average performance is observed to be poor as shown in Fig. 7.8.

Existing deterministic discovery schedule with a pseudo-random component considers two SNs with the same (symmetric) duty cycles, and offset between slots

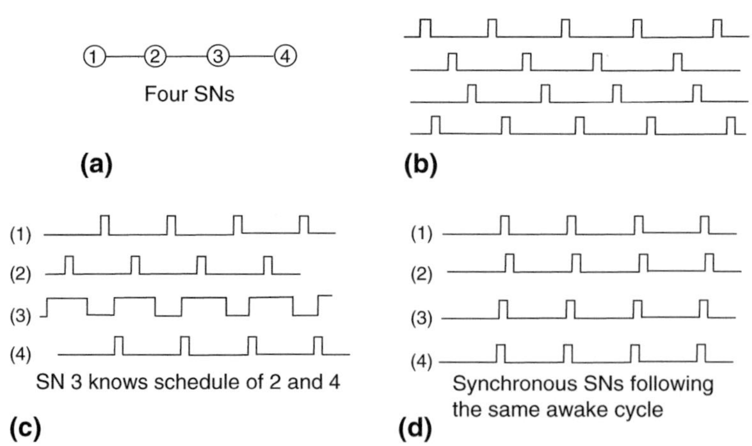

Fig. 7.4 **a** Four SNs, **b** awake cycle for 4 SNs, **c** SN3 knows schedule for SNs 2, and 4, **d** synchronous SNs

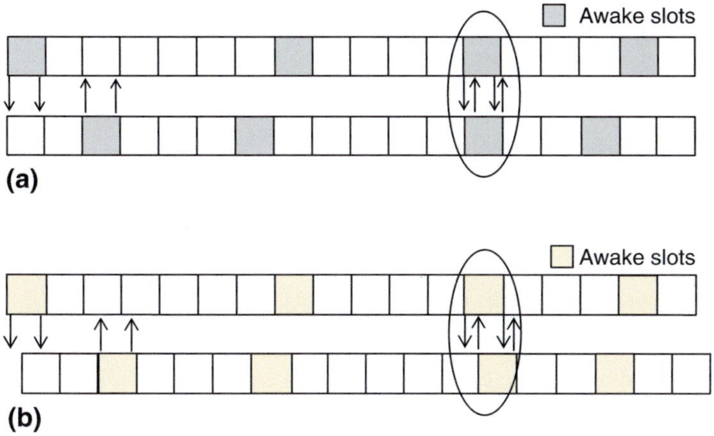

Fig. 7.5 **a** Duty cycle for discovering neighboring SN. **b** Asymmetric duty cycle for discovering neighboring SN

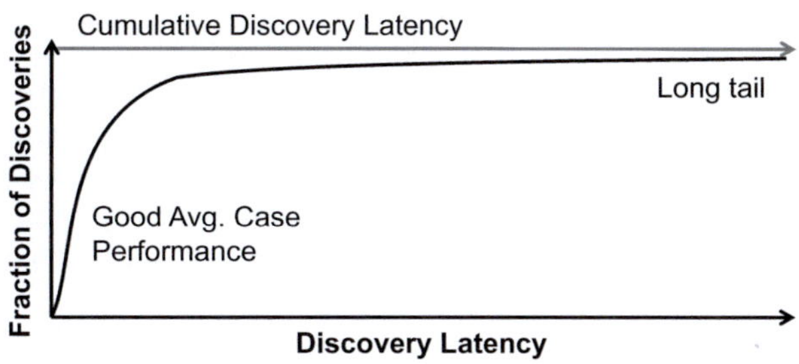

Fig. 7.6 Fraction of discovering neighboring SN as discovery time in Birthday algorithm [15]

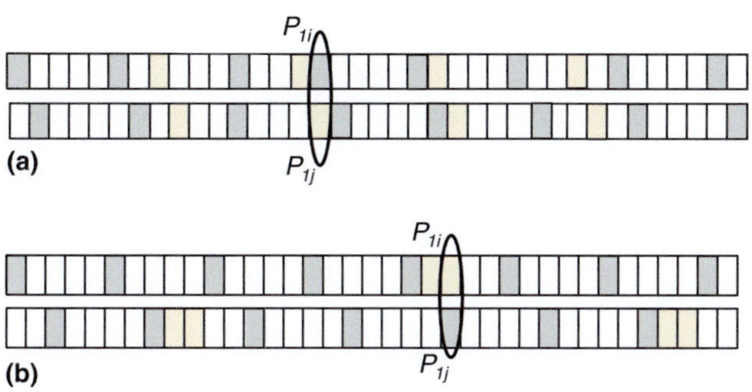

Fig. 7.7 **a** Symmetric duty cycle in discovering neighboring SN, **b** asymmetric duty cycle [15]

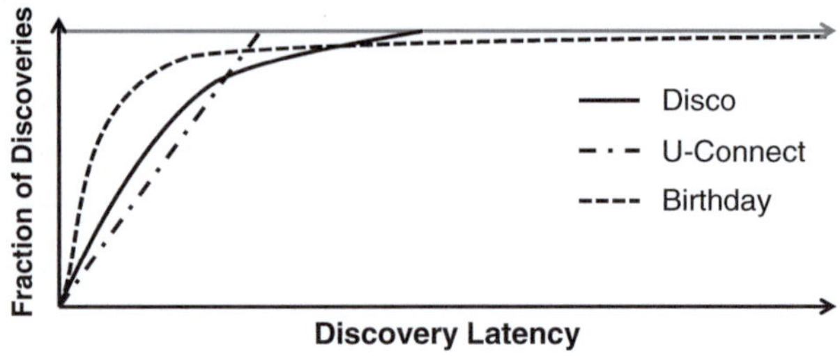

Fig. 7.8 Comparing three schemes for discovering neighboring SN

with fixed period remains fixed, with SN B awake in the first $\lceil t/2 \rceil$ slots of SN A's cycle, and SN A remain awake within the first $\lceil t/2 \rceil$ slots of SN B's cycle. A systematic approach (Fig. 7.9c) selects a fixed period t (does not need to be prime) and keep one slot fixed (anchor slot) [1]. The objective is to meet the fixed/anchor slot of the other SN and only need to search in the range 1 to $\lceil t/2 \rceil$ and no need to probe all $\lceil t/2 \rceil$ slots (Fig. 7.10).

In Sequential probing scheme [4], two slots per period t are used. The anchor slot of SN A keeps one slot fixed at slot 0, and the probe slot at SN B moves around the other slot sequentially that guarantees overlap in $t * t/2$ slots. This provides an improved bound over existing protocols, and time needed to ensure probe-anchor overlap also lead to discover neighboring SNs. In randomized probing [5], no specific patterns are followed. Instead, the probe slot randomly (A: 1–3–2; B: 3–1–2) in a pseudo-random instead of totally random scheme, and each node randomly

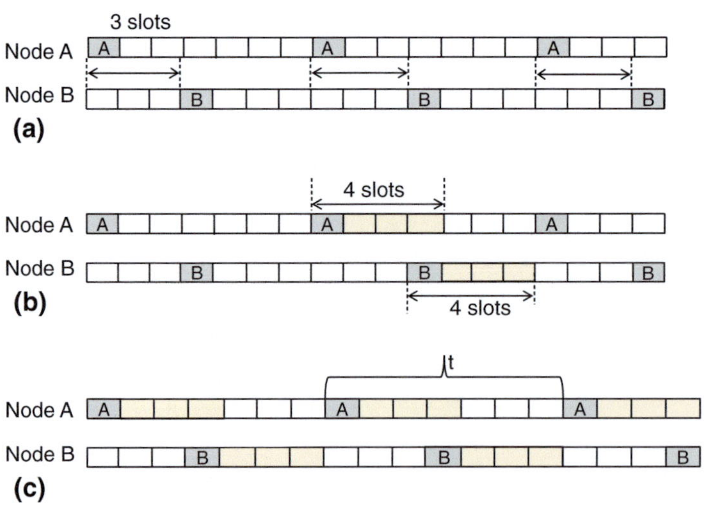

Fig. 7.9 Searchlight schemes for discovering neighboring SN **a** simple neighbor discovery **b** with deterministic approach, and **c** with pseudo-random component

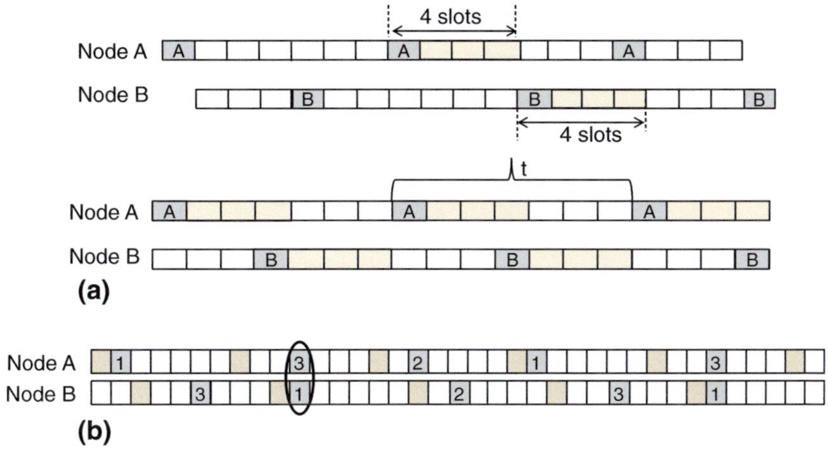

Fig. 7.10 **a** Sequential scheme for discovering neighboring SN [4] and **b** randomized probing scheme

chooses a schedule for its probe slot that repeats every (t * t/2) slots. The schedules of two SNs A and B appear random to each other, while retaining the same worst-case bound an average case performance is improved. As anchor slots no longer have constant distance, striping cannot be used. So, a new approach is to restrict value of the bigger period to an integer multiple of the smaller period only primes are allowed by Disco and U-Connect and a comparison is shown in Fig. 7.11.

All neighbor discovery schemes guarantee given time frame while keeping SNs awake for minimum number of time slots to save energy. This requires matching SNs' awake-sleep schedules to prolong overall battery lifetime. Both HEDIS (HEterogeneous DIScovery) and TODIS (Triple-Odd based DIScovery) [6] guarantee asynchronous neighbor discovery when each SN operates at different duty cycle. For example, in Fig. 7.11, two discovery schedules are A = {0, 0, 0, 0, 0, 1} and B = {0, 1, 0, 0, 0, 0, 0, 0, 1}. Without clock drift, SNs A and B can discover each other every 18 time slots since lcm (Ta, Tb) is = 18. Figure 7.12 shows how two SNs A and B with different duty cycles can achieve successful neighbor discovery, even any amount of clock drift. TODIS is based on creating wake-up schedules for SNs based on numbers co-prime to each other. A comparison of different schemes is given in Table 7.1.

7.3 Delivering Packets to BS/Sink

In a WSN, there could be a *single receiver* SN or m*ultiple receivers* SNs, and they could be in the close vicinity or could be spread out. The routing scheme that can be used could be based on unicast routing along a *single path,* or multiple *paths* can

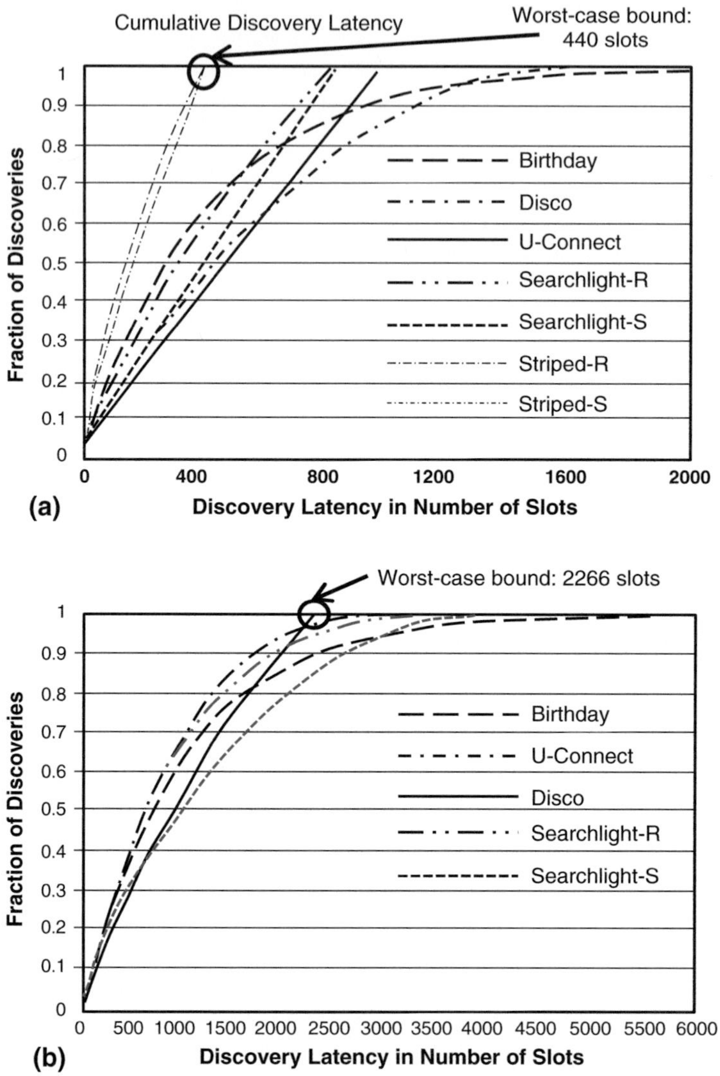

Fig. 7.11 **a** Comparing symmetric schemes with 5% duty cycle. **b** Comparing asymmetric schemes with 1–10% duty cycle

be used between source/destination pair SNs. There is no support for routing, and one has to rely on *flooding and/or gossiping*. Multiple paths could be disjoint or braided alternative routes (Fig. 7.13) for fault tolerance and could be used simultaneously. The send same packet could be sent along multiple paths or redundant fragments could be employed. One example is ReInForM (Reliable Information Forwarding Using Multiple Paths in a WSN).

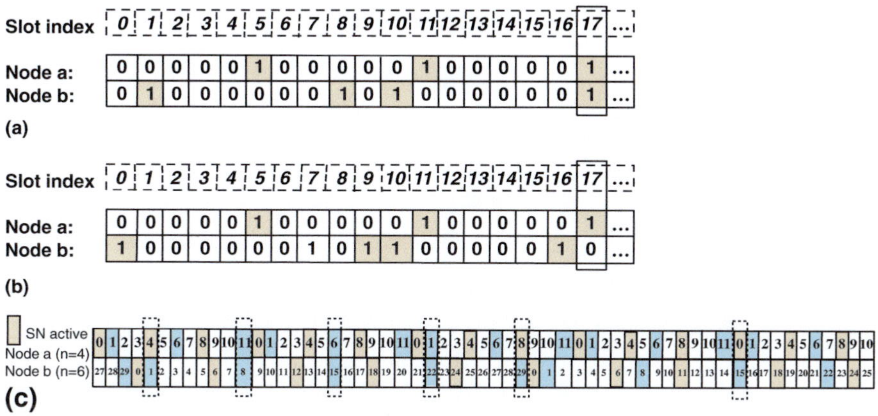

Fig. 7.12 HEDIS scheme for discovering neighboring SNs [6] **a** no clock drift **b** drift by 1 tile **c** 2 SNs with different active/sleep cycle

Table 7.1 Comparison of different protocols for locating Neighboring SNs

Protocol name	Parameter restriction	Average dis. delay	Supported duty cycles
Disco	Prime p_1, p_2, p_3, p_4	$O(\min\{p_1p_3, p_1 p_4, p_2 p_3, p_2p_4\})$	$1/p_1 + 1/p_2$
U-Connect	Prime p_1, p_2	$O(p_1p_2)$	$(3p_1 + 1)/2p_2$
Searchlight	Power-multiple t_1, t_2	$O(t_1t_2)$	$2/t_1^2$
HEDIS	Same parity n,m	$O(nm)$	$2/n$
TODIS	Odd n,m	$O(nm)$	$3(n^2 - n - 1)/\{n(n^2-4)\}$

7.4 Reliability of Delivered Packets

In a WSN, destination is usually the BS or sink node. and routing requires determination of each SN to BS. Reliable Multi-Segment Transport (RMST) [7] provides end-to-end data packet transfer reliability between SN and BS (Fig. 7.14) as each RMST SN caches the packets and is NACK is sent after predetermined timeout period, the packet is retransmitted. In-network caching causes significant overhead, and a transport layer protocol employs diffusion for routing based on selective NACK. It provides guaranteed delivery of all packet fragments, even though delivery is not guaranteed. At the BS, fragmentation reassembly is needed. RMST covers 3 layers of MAC, transport, and application, even though emphasis is on MAC and transport. No ARQ is used in RMST and no RTS/CTS or ACK. All transmissions are broadcasted , and reliability is deferred to upper layers. There are no control overheads, and no erroneous path selection is involved. At MAC layer, there is an option to use ARQ always as all transmissions are unicast and RTS/CTS

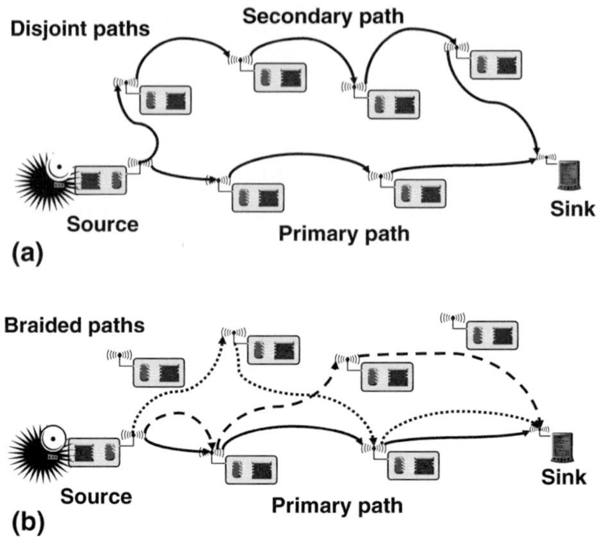

Fig. 7.13 **a** Multiple paths between SNs. **b** Multiple braided paths between SNs

and ACKs are used. One-to-many communication is achieved by retaining multiple unicasts as packets traveling on established paths have higher probability of delivery for good signals along the path. Selective ARQ is broadcast for one-to-many and unicast for one-to-one packet transfer. Data and control packets traveling on established paths are unicast, while route discovery is based on broadcast. End-to-End Selective Request NACK is employed, and loss is detected only at the BS. Any repair requests travel in the reverse direction in multi-hop fashion from BS to source SNs. Hop-by-hop selective request NACK is adopted, and each node along the path caches the data. The loss detection occurs at each intermediate SN along the path. Repair requests are only sent to immediate neighbors, and if data is not found in the local caches, NACKs are forwarded to next hop SN toward the source SN.

Fig. 7.14 Source SN to the BS data transfer

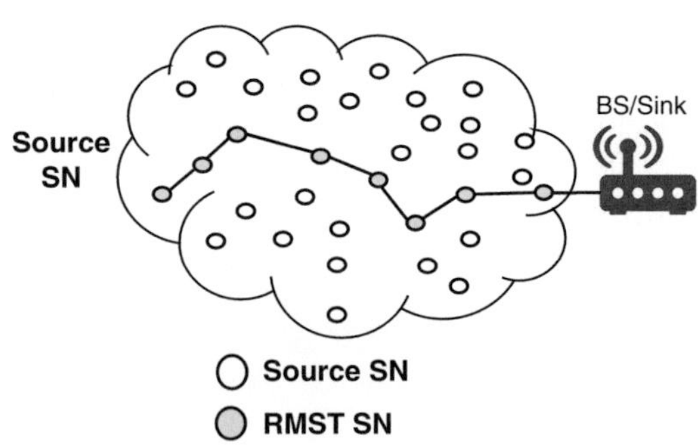

At application layer, RMST provides end-to-end positive ACK as BS requests a large data entity, and source data could be fragmented. BS keeps sending interests until all necessary fragments have been received at the BS and does fragmentation reassembly. To ensure guaranteed delivery at BS, intermediate receiver SNs are accountable for fragment retransmission . Caching or non-caching mode determines classification of a SN and NACKs are triggered by a gap in sequence number. A watchdog timer occasionally examines fragment map for gaps that have stayed for too long. NACKs are propagated from BS to the source SNs. NACK is forwarded only if segment not found in local cache, and back channel is used to deliver NACKs to upstream SNs.

7.5 BS to SNs Reliability

Reliability semantics are different for information flow from BS to SNs, and various schemes have been devised to provide reliable communication. One of the schemes of Pump Slowly, Fetch Quickly (PSFQ) [8], the data from a source SN is obtained at a relatively low speed that allow intermediate nodes to fetch missing data segments from their neighbor SNs. It is assumed that no congestion and losses due only to poor link quality are present, and hop-by-hop recovery is possible. The objective is to recover from losses locally and ensure data delivery with least support from transport setup, and signaling overhead due to detection/recovery operations is minimized. It is assumed that the links function correctly in poor link quality settings. This provides loose delay bounds for data delivery to all intended receivers. Three basic operations are *pump*, *fetch*, and *report* and change between multi-hop forwarding with low error rates and store-and-forward when error rates are higher.

For *Pump* operation, SN broadcasts a packet to its close by SNs every $T_{\min}$ until all the data fragments have been sent out. Adjacent SNs who receive the packet, check their local cache to discard any duplicates. If it is just a new message, the packet is buffered and the Time-To-Live (TTL) field in the header is decreased by 1. If TTL is not zero and there is no gap in the sequence number, then the packet is scheduled for transmission within a random delay time T_{tx}, where $T_{\min} < T_{\text{tx}} < T_{\max}$ (Fig. 7.15a). The random delay allows a downstream SN to recover missing segments before the next segment arrives from a SN. It also reduces the number of redundant broadcasts of the same packet by neighbors.

A node goes into *Fetch* mode when a sequence number gap is detected and NACK messages to its immediate SNs are sends out. Since it is very likely that consecutive packets are lost because of diminishing conditions, a “window” is used to define the range of missing packets. A SN receiving a NACK message checks its cache. If the packet(s) are found, they are scheduled for transmission at a random time in (0, T_r). Neighbors cancel a retransmission when the same segment is transmitted by some other SNs. NACK messages are not spread to avoid message implosion. Single multi-hop path routing needs to consider which node to go

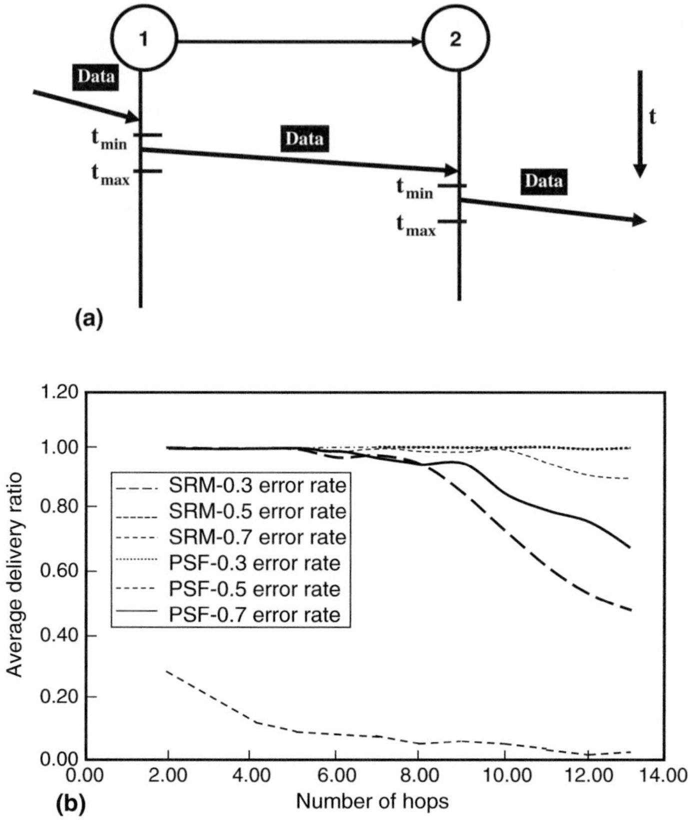

Fig. 7.15 **a** Pump slowly, fetch quickly data transfer. **b** Comparison of pump slowly, fetch quickly data transfer with multicasting [8]

through, how to detect losses, how to initiate retransmissions, and which SN carries out retransmissions. Detection of loss of a *single packet* is involved as only positive acknowledgments (ACK) are feasible and NACK is not an option. The question is which SN sends ACKs at each intermediate SN at MAC/link level. As link layer retransmissions are done and that too only a bounded number of attempts. If this is done at the destination SN, the transport layer retransmission creates an issue of timer selection. Link layer acknowledgments can be done by neighboring SNs, and at the transport layer, acknowledgment is done back to source SN, representing end-to-end ACK.

The next question is how large an inter-packet gap can be? It should be big enough to accommodate at least one, or several fetch operations, so that repairs are done before the next packet arrives. An in-sequence packet is forwarded only after a random wait time, only after 3 neighboring SNs have already forwarded the packet. Out-of-order packet is not forward as it is critical to fill the gap first. As the fetch request is broadcast, it arrives at multiple SNs. It is better to use a slotted resend

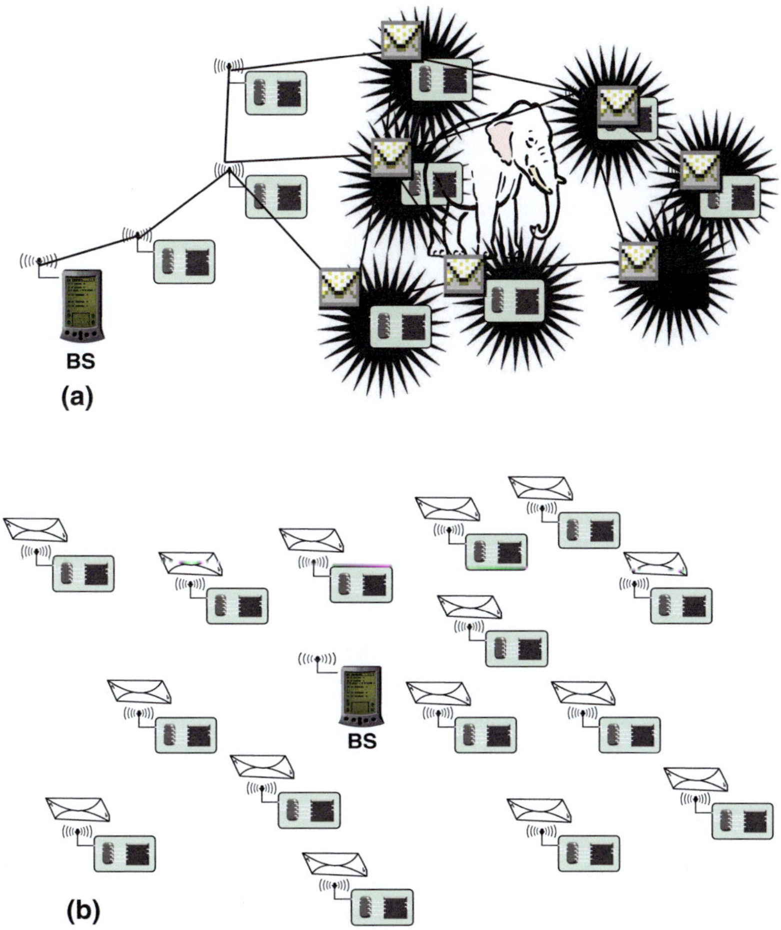

Fig. 7.16 **a** An event detected by multiple SNs. **b** An event causes flooding around the BS

mechanism for requested packets as each one corresponds to a time slot and can be filled by SN if requested packet is available in the buffer. A comparison of this scheme with reliable multicasting is shown in Fig. 7.15b [7]. The main difference is that the in-sequence packets are not enforced, and end-of-block treatment differs. PSFQ is observed to work for higher error rates.

7.6 Congestion Control

When several SNs detect an event and try to report it intermittently (Fig. 7.16a), congestion could be present, as many SNs send data to BS [9], congestion happens around the BS which could lead to shocking significances (Fig. 7.16b). Continuous data broadcasting can reduce goodput and accuracy, leading to increased packet loss and reduced WSN lifetime. In wired network, TCP detects congestion by missing acknowledgments and is not applicable to WSNs, if no ACKs are used. Congestion can be detected *locally* if buffer at SN fills up above threshold. It is important to take growth rate into account as well as only occupancy is not an adequate indicator (Fig. 7.17). CSMA-type MAC interprets high channel utilization as congestion and is easy to detect. In TDMA -type MAC, high channel utilization does not create any problem for throughput , as congestion is more difficult to detect and many schemes have been introduced.

GARUDA (Achieving Effective Reliability for Downstream Communication) [9] is used only for first packet reliability, as short duration pulses are employed in a single radio and an incoming packet is advertised. Negative ACK is used with forced pulses that sense the carrier with extensive use of piggybacking. A sink sends WFP (Wait-for-First-Packet) pulses periodically for a deterministic period before the first packet is sent. WFP prevents ACK implosion with small overhead and addresses the single packet or all packets lost problem, and less energy is consumed. This provides robustness to wireless errors or contentions. GARUDA also supports other reliability measures that might be required for BS–SN

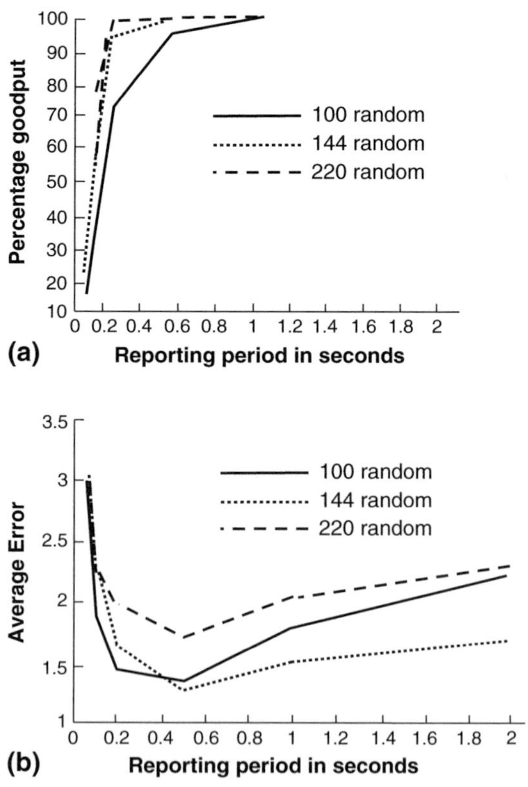

Fig. 7.17 **a** Effect of reporting period on goodput. **b** Effect of reporting period on accuracy [9]

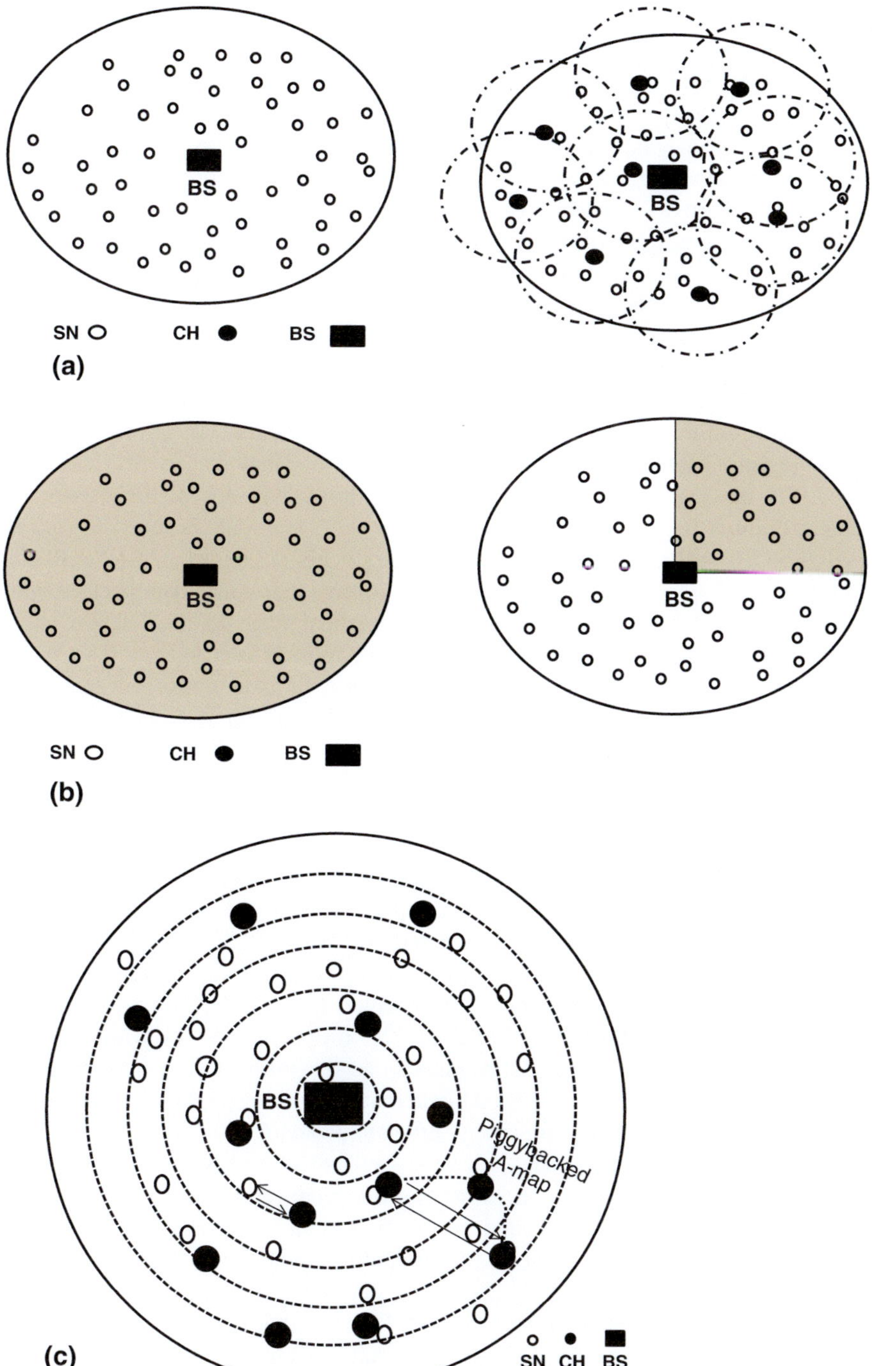

Fig. 7.18 **a** WSN with SNs and CHs and their coverage area. **b** Broadcasting within WSN or segment. **c** Piggybacking in a WSN

communication such as reliable delivery to all SNs within a subregion of WSN, reliable delivery to minimal number of SNs required to cover entire sensing area of WSN, and reliable delivery to a probabilistic subset of the SNs in the network (Fig. 7.18).

MOAP [10] provides an energy and memory efficient code distribution mechanism specifically targeted for Mica2 motes where multi-hop operation is achieved through recursive single-hop broadcasts and data rippled from neighborhood-to-neighborhood to produce multi-hop ripple. Very few source SNs are expected at each neighborhood, and a receiver SN becomes source SN when it has the *entire* area. SNs are prevented to become source if another source is present. If data transmission is in progress, a source will *always* be one hop away that allows local repairs, increased latency, and loss responsibility lies with the receiving SN. It is a NACK-based scheme, and there is no need to route NACKs as that saves energy and minimizes its complexity. Ultimately, all nodes will have data for the whole region. Mote is attached to PC becomes original source SN and sends PUBLISH message. All receiving SNs 1 hop away will accept the message, if version number is greater than their own, and once area data is received, the SN sends a PUBLISH message. Bitmap of few segments are kept in RAM with each SN. Retransmissions of packets have higher priority than other data packets while duplicate requests are blocked as SNs keep track of their sources' action with a timer. If the source SN dies, a broadcast repair request is triggered. The system requires all SNs to sporadically advertise version of their data. The late joiner mechanism allows mote SNs to participate in code transfer if recently recovered from failure.

Congestion Detection and Avoidance (CODA) [11, 12] mainly aims to detect and avoids congestion on the forward path via receiver-based congestion detection, open-loop hop-by-hop backpressure signaling to inform the source about the congestion, and closed-loop multi-source regulation for persistent and larger-scale congestion conditions. Congestion control performed at the SNs without considering the reliability impairs end-to-end reliability between SNs and BS, and three mechanisms of congestion detection, open-loop hop-by-hop backpressure, and closed-loop multi-source regulation are involved (Fig. 7.19). The work watches average energy tax as the ratio of total packets dropped in WSN to total packets received at BS, and average fidelity penalty as the ratio of measures difference between average number of packets delivered at a BS using CODA and using ideal congestion scheme. CODA is an energy efficient protocol and period can deal with persistent and transient hotspots.

In a typical WSN application, BS wants to collect data and location of the SNs and not just an individual SN data, and accuracy of data received at BS need to be measured. The BS is assumed to do reliability evaluation using application-dependent parameters such as decision time interval τ. BS computes a reliability indicator r_i based on the reports from the SNs. If R is the packets required for an event detection, then $r_i > R$ is needed for a reliable event detection, where r_i is the number of packets received by BS during decision interval. There is a need to have an event ID. The Event-to-Sink Reliable Transport (ESRT) [13, 14] protocol aims to

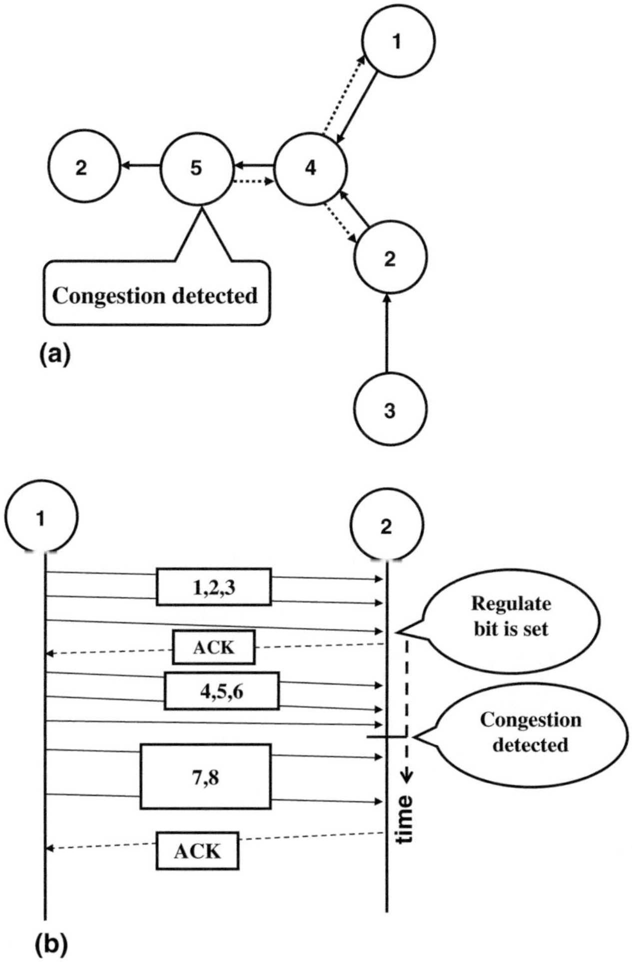

Fig. 7.19 **a** Open-loop hop-by-hop backpressure in a WSN. **b** Closed-loop multisource regulation in a WSN

dynamically adjust the reporting rate to realize required reliability R at the BS (Fig. 7.20). The algorithms mainly run at the BS as the SNs listen to broadcast from the BS to update their reporting rates. The BS computes normalized reliability parameter $\eta_i = r_i/R$ and updates reporting rate. The congestion decisions based on feedback reports from the source SNs. Congestion is detected by looking at local buffer level of SNs and that leads a simple congestion detection mechanism with energy savings. Analytical performance evaluation and simulation results show that the system converges to the state OOR (Optimal Operating Region) regardless of the initial state and is very valuable for random and dynamic topologies. The open c events and how to make BS–SNs transfer also reliable (Tables 7.2 and 7.3).

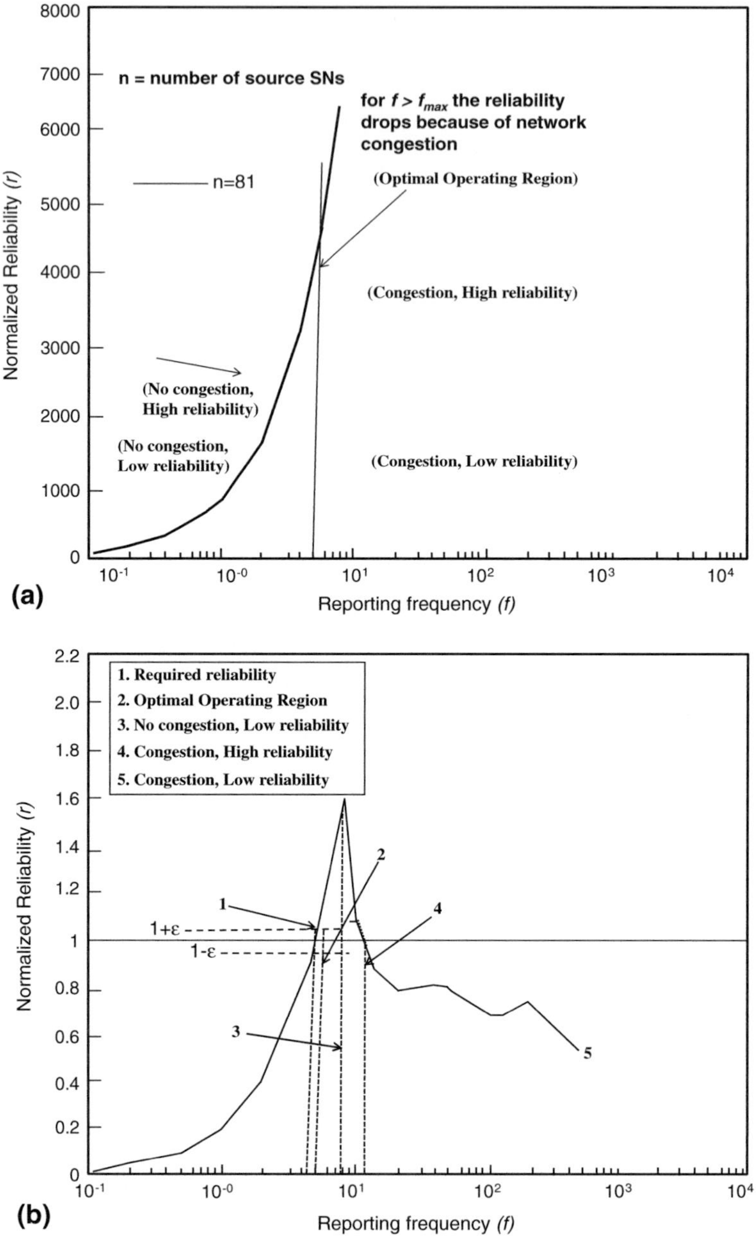

Fig. 7.20 **a** Reliability r versus reporting frequency f based on simulation [13]. **b** Normalized reliability $\eta = r/R$ versus reporting frequency f [13]

Table 7.2 Frequency of update *f* and corresponding action

State	Frequency of update *f*	Action
No congestion, low reliability	$f_{i+1} = \frac{f_i}{\eta_i}$	Multiplicative increase *f* to achieve required reliability as soon as possible
No congestion, high reliability	$f_{i+1} = \frac{f_i}{2}\left(1 + \frac{1}{\eta_i}\right)$	Decrease *f* conservatively, reduce energy consumption, and not to lose reliability
Congestion, high reliability	$f_{i+1} = \frac{f_i}{\eta_i}$	Aggressively decrease *f* to relieve congestion as soon as possible
Congestion, low reliability	$f_{i+1} = f_i^{(\eta_i/k)}$	Exponential decrease. *k* is the number of successive decision intervals spent in state (C, LR)
Optimal operating region	$f_{i+1} = f_i^{(\eta_i/k)}$ $f_{i+1} = f_i$	Unchanged

Table 7.3 Comparing different protocols for reliable transfer [16, 17]

Protocol	Error recovery	Reliability direction	Congestion ctrl	MAC/routing requirement	Sim	OS
PSFQ	h–h	Sink to sensors	–	Broadcast	ns-2	Tiny os
GARUDA	h–h/e–e	Sink to sensors	–	Broadcast	ns-2	–
MOAP	h–h	Sink to sensors	–	Broadcast/ unicast	Em star	Tiny os
RMST	h–h/e–e	Event to sink	–	DF	ns-2	–
ESRT	–	Collective event to sink	Event report frequency	CSMA/CA	ns-2	–
CODA	–	Collective event to sink	Event report frequency	CSMA	ns-2	Tiny os
Protocol	Error recovery	Reliability direction	Congestion Ctrl	MAC/Routing Requirement	Sim	OS
PSFQ	h–h	Sink to sensors	–	Broadcast	ns-2	Tiny os

h–h hop by hop; *OOR* optimal operating region; *e–e* end to end; *DF* directed diffusion

7.7 Impact of Lower Layers on TCP

MAC layer is intended for providing an efficient shared broadcast channel through which the involved SNs can communicate, and RTS/CTS handshake in IEEE 802.11 is only employed when the DATA packet size exceeds some predefined threshold. Each of these frames carries the remaining period for the communication, so that other SNs in the neighborhood can hear it and defer their transmission . The SNs must await an IFS (inter frame space) interval and then struggle for the

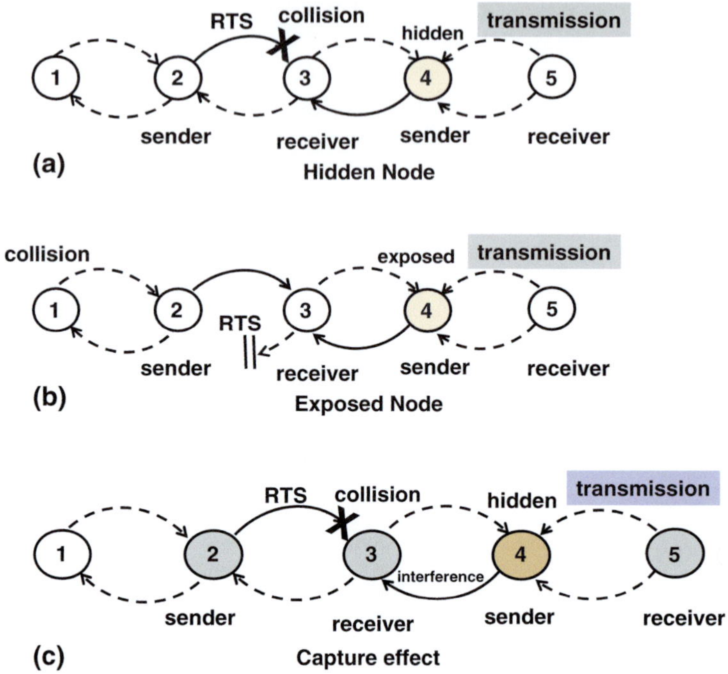

Fig. 7.21 **a** Effect of hidden SN. **b** Effect of exposed SN. **c** Capture effect in WSN

medium again. The contention is carried out by means of a binary exponential back-off scheme which imposes a further random interval. At every unsuccessful attempt, this random interval increases. Consider a linear topology of Fig. 7.21 in which each SN can only communicate with its adjacent neighbor SN. In addition, consider that in Fig. 7.21a, b, there exist a single TCP connection running between nodes 1 and 5. In Fig. 7.21c where there are two independent connections of 2–3 and 4–5.

Assuming that connection between SNs 2–3 experiences collision due to the hidden node problem caused by the active connection of SN 4, SN 2 will back-off and retransmit the lost frame. At every retransmission, the binary exponential back-off mechanism imposes an increasingly back-off interval, and implicitly, this is actually decreasing the possibility of success for the connection between SNs 2–3 to send a packet as connection SNs 4–5 will "dominate" the medium access once it has lower back-off value. In reality, the connection SNs 2–3 will hardly obtain access to the medium as connection SN 4–5 will capture it.

7.8 Conclusions

TCP is widely accepted in wired networks as there is end-to-end data transfer and handshaking mechanism. But, in wireless, transfer of data from SN to BS occurs in multi-hop fashion and follows store-and-forward mechanism. So, there is no end-to-end communication. This is true for data packets from SNs to BS as well as explicit ACK signal back from BS to SNs. Similarly, congestion in WSN and reliable transfer of data need to be handled in different way. There are some specific problems in WSN due to its underlying characteristics.

7.9 Questions

Q.7.1. Can you keep SNs ON all the time to avoid delay in neighborhood discovery?
Q.7.2. How can you make TCP/IP viable for wireless sensor networks?
Q.7.3. Why do you need multiple paths for a single SN to BS data transfer?
Q.7.4. What is the impact of exponential back-off in a WSN?
Q.7.5. How is ACK sent back from BS to source SN?
Q.7.6. How can you improve distributed TCP caching for wireless sensor networks?
Q.7.7. What are the characteristics of a WSN that make end-to-end delay in a WSN difficult to predict?
Q.7.8. How can you detect congestion in a WSN when data being transferred from a SN to BS?
Q.7.9. How can you make data transfer from SN to BS reliable?
Q.7.10. Can you improve TCP Performance over wireless sensor networks by exploiting cross-layer information awareness?
Q.7.11. What are the congestion detection strategies in wireless sensor networks?
Q.7.12. Do wireless sensor networks need to be completely integrated into the Internet?
Q.7.13. How can you handle prioritized heterogeneous traffic in a WSN?

References

1. Mehedi Bakht, Matt Trower, and Robin Hilary Kravets, "Searchlight: won't you be my neighbor?," Proceedings of the 18th annual international conference on Mobile computing and networking, Mobicom 2012, pp. 185–196.
2. Prabal Dutta and David Culler, "Practical asynchronous neighbor discovery and rendezvous for mobile sensing applications," Proceedings of the 6th ACM conference on Embedded network sensor systems, SenSys 2008, pp. 71–84.

3. Arvind Kandhalu, Karthik Lakshmanan, and Ragunathan (Raj) Rajkumar, “U-connect: a low-latency energy-efficient asynchronous neighbor discovery protocol,” Proceedings of the 9th ACM/IEEE International Conference on Information Processing in Sensor Networks, pp. 350–361.
4. Tao Shu and Marwan Krunz, “Throughput-efficient sequential channel sensing and probing in cognitive radio networks under sensing errors,” Proceedings of the 15th annual international conference on Mobile computing and networking, pp. 37–48, Mobicom 2009.
5. Mehedi Bakht and Robin Kravets, “SearchLight: A Systematic Probing-based Asynchronous Neighbor Discovery Protocol,” https://www.ideals.illinois.edu/bitstream/handle/2142/17351/paper.pdf?sequence=2.
6. Lin Chen, Ruolin Fan, Kaigui Bian, Lin Chen, Mario Gerla, Tao Wang, and Xiaoming Li, “On Heterogeneous Neighbor Discovery in Wireless Sensor Networks,” http://campuspress.yale.edu/lchen/.
7. Qixiang Pang, Vincent W.S. Wong and Victor C.M. Leung, “Reliable data transport and congestion control in wireless sensor networks,” Int. J. Sensor Networks, vol. 3, no. 1, 2008.
8. C. -Y. Wan, A. T. Campbell, and L. Krishnamurthy, “Pump-slowly, fetch-quickly (PSFQ): a reliable transport protocol for sensor networks,” IEEE Journal on Selected Areas in Communications, vol. 23, no. 4, pp. 862–872, September 2006.
9. Chonggang Wang, Mahmoud Daneshmand, Bo Li, and Kazem Sohraby, “A Survey of Transport Protocols for Wireless Sensor Networks,” http://citeseerx.ist.psu.edu/viewdoc/download?doi=10.1.1.84.1185&rep=rep1&type=pdf.
10. H. Kim, “Comparison of Network Reprogramming Protocols for Wireless Sensor Networks,” http://www.academia.edu/433777/Comparison_of_Network_Reprogramming_Protocols_for_Wirel.
11. Chieh-Yih Wan, Shane B. Eisenman, and Andrew T. Campbell, “CODA: Congestion Detection and Avoidance in Sensor Networks,” http://www.cs.dartmouth.edu/~campbell/papers/coda-2011.pdf.
12. Valerie Galluzzi and Ted Herman, “Survey: Discovery in Wireless Sensor Networks,” International Journal of Distributed Sensor Networks, vol. 2012, 12 pages.
13. Yogesh Sankarasubramaniam Özgür B. Akan Ian F. Akyildiz, “ESRT: Event-to-Sink Reliable Transport in Wireless Sensor Networks,” http://www.gta.ufrj.br/wsns/Transport/ESRT.pdf.
14. Wei Sun, Zheng Yang, Keyu Wang, and Yunhao Liu, “Hello: A Generic Flexible Protocol for Neighbor Discovery,” http://www.greenorbs.org/people/liu/infocom14-SunWei.pdf.
15. Michael J. McGlynn and Steven A. Borbash, “Birthday protocols for low energy deployment and flexible neighbor discovery in ad hoc wireless networks,” Proceedings of the 2nd ACM international symposium on Mobile ad hoc networking & computing, MobiHoc 2001, pp. 137–145.
16. Chonggang Wang, Kazem Sohraby, Bo Li, and Weiwen Tang, “Issues of Transport Control Protocols for Wireless Sensor Networks,” http://www.academypublisher.com/jnw/vol03/no06/jnw0306698l.pdf.
17. Hongchao Zhou, Xiaohong Guan, and Chengjie Wu, “Reliable Transport with Memory Consideration in Wireless Sensor Networks,” http://www.cwc.oulu.fi/~carlos/WSNPapers/ST03.pdf.

Chapter 8
Sensor Nodes (SNs), Camera Sensor Nodes (C-SNs), and Remote Sensor Nodes (RSNs)

8.1 Introduction to Sensor Nodes

A wireless sensor network (WSN) consists of a large number of sensor nodes (SNs) that transmit a volume of data to a central station, commonly known as a base station (BS) or sink node in a multi-hop fashion. From an Internet point of view, computation is primarily done at the BS which is rich with computing and storage resources. So, the real question is how to describe the deployment of SNs, how to discover neighboring SNs, how to integrate distributed processing done by SNs, and all these considerations constitute a GSN (Global Sensor Network). For example, temperature can be measured by many SNs and instead of sending each of them individually; it may be desirable to aggregate these value before reaching BS. There is no well-accepted standard for transducers and sensors while a lot of resources are needed in developing them. Besides, they are not portable and their compatibility ought to be examined carefully. Moreover, the SNs need to be deployed as soon as there is a need for their utilization for ever-increasing application domain. With the advances in technology, the price of SNs and transducers is decreasing day by day and innovative schemes need to be devised to having a flexible environment. To make a friendly situation, a wrapper can be employed which can extract desired parameter and create a combined representation that could reflect them in an effective way. Typically, a wrapper transforms and translates information from SNs to an understandable form [1]. This may be in plain text or in HTML format and conversion could be time-consuming. Different type of wrappers such as for TinyOS, RFID, and UDP could be used for different level of abstractions and lines of code for different wrappers are shown in Table 8.1.

To understand the effect of the network size in a heterogeneous environment, experiment was conducted [3] to determine processing time from different types of transducers when the number of devices and query is varied. The system shown in Fig. 8.1 consists of 5 desktops, 10 Mica2 motes with light and temperature

D.P. Agrawal, *Embedded Sensor Systems*,
DOI 10.1007/978-981-10-3038-3_8

Table 8.1 Coding efforts for different types of wrappers [2]

Wrapper type	Lines of code
Tiny OS	120
WiseNode	75
Generic UPD (Universal Printer Driver)	45
Generic serial	180
Wired camera	300
Wireless camera (HTTP)	60
RFID reader (TI)	50

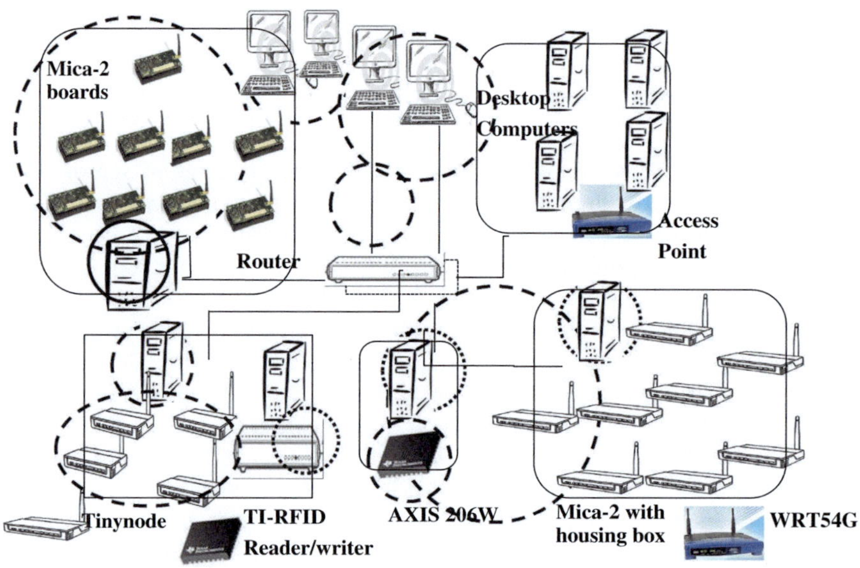

Fig. 8.1 Multi-hop communication from a SN to the BS

transducers (message size 15 Bytes); 8 Mica2 motes each equipped with light, temperature, acceleration, and sound sensors (packet size of 100 Bytes); 4 motes with TinyOS with one light and two temperature sensors (packet size of 29 Bytes); 15 Wireless AXIS 206W 15 cameras (which can capture 640 × 480 JPEG pictures with a rate of 30 frames per second and varying degree of compression); and Texas Instruments RFID reader with three different RFID-tags (each tag can store up to 8 KB of binary data). Each mote generates a burst of size R with probability B > 0. Figure 8.2 shows the results for a stream element size (SES) of 30 Bytes and the average time to process a query if 500 clients issue queries is less than 50 ms. The spikes in the graph are due to bursts and the system comes back to normal behavior.

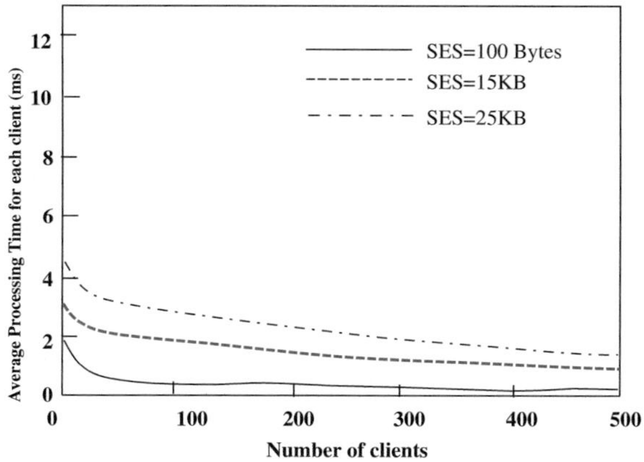

Fig. 8.2 Processing time per client

The spikes in the graphs are bursts as described above. Basically this experiment measures the performance of the database server under various loads which heavily depends on the used database. As expected the database server's performance is directly related to the number of the clients as with the increasing number of clients more queries are sent to the database and also the cost of the query compiling increases. Nevertheless, the query processing time is reasonably low as the show that the average time to process a query if 500 clients issue queries is less than 50 ms, i.e., approximately 0.5 ms per client. If required, a cluster could be used to improve the query processing times which is supported by most of the WSNs.

8.2 Camera Sensor Nodes (C-SNs)

A camera acts as sensors as the image or picture is converted into a form that a computer can recognize as bits and bytes. A digital picture is just a long string of pixels; 1's and 0's and all of these pixels make up the image. A digital camera has different lenses that help focus the light to create the image of a scene and is recorded electronically. A lens focuses light on to a film and other points in the image are projected in the image as a "circle of confusion" as illustrated in Fig. 8.3 [5].

Operation of a digital camera is drastically enhanced by the signal processing schemes [4] and one such scheme is shown in Fig. 8.3. A camera could be doing analog or digital recording and could be printed or optical image recorded on magnetic tape or stored in a RAM. A part of scene satisfying the following relation is focused correctly and is called circle of confusion:

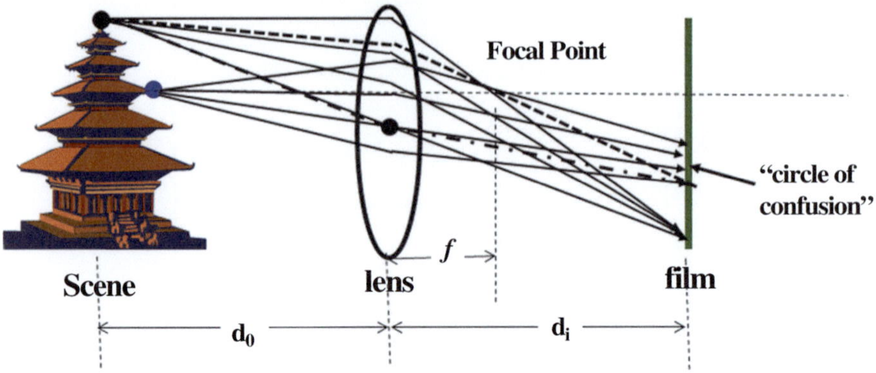

Fig. 8.3 Projection of a scene by a C-SN lens

$$\frac{1}{d_o} + \frac{1}{d_i} = \frac{1}{f}. \tag{8.1}$$

The data captured by a camera can be processed in many ways and different levels of representation are shown in Table 8.2 [6]. The "raw" information measured by the SNs is transmitted by radio-signal and are called "Level 0" data. Level-1 data contain all the Level-0 data, and calibration and navigation information is appended. Each pixel of Level-2 data contains geophysical values by applying the calibration and atmospheric corrections from Level-1 raw data. Level-3 contains geophysical parameters observed during a certain period and interpolated on a global grid such as one day or 8 days with spatial resolution in degrees.

Raw data received by a camera goes through many steps of processing and are summarized in Table 8.3 [7].

Typically, a sensor converts light into electrical charges and digital cameras use either CCDs (Charged Couple Devices) or CMOS (Complementary Metal Oxide Semiconductor) as both convert light into electrons. The value of each pixel (picture intensity) of the image is read and convert light into a readable form [9] (Table 8.4).

Although numerous differences do exist between these sensors, the same role is played in the camera of turning light into electricity. For all practical purpose, both types of digital cameras work nearly identically. The important characteristic of a

Table 8.2 Different levels of C-SN data and the corresponding representation

Level	Details of operation
Level 0	Raw data received in standard binary form
Level 1	Image data in C-SN coordinates
Level 2	Derived geophysical variables
Level 3	Variable mapped on grid scales
Level 4	Model output

Table 8.3 Digital photography, computational steps and C-SNs [8]

	Digital photography	Computational processing	Computational imaging/optics	Computational sensor
Process	Image processing applied to captured images and create "better" images	New image created by processing a captured image	Optically coded images are Captured and computationally decoded leading to "new" images	Detectors combine sensing and processing and create "smart" pixels
Example operations	Interpolation, filtering, enhancement, dynamic range compression, color management, morphing, hole filling, artistic image effects, image compression, watermarking	Mosaicking, matting, super-resolution, multi-exposure HDR, light field from multiple view, structure from motion, shape from X	Coded aperture, optical tomography, diaphanography, SA microscopy, integral imaging, assorted pixels, cat dioptric imaging, holographic imaging	Artificial retina, retinas sensors, adaptive dynamic range sensors, edge detect chips, focus of expansion chips, motion sensors

Table 8.4 Comparing CCD and CMOS technologies [9]

	CCD	CMOS
Pixel signal	Electron packet	Voltage
Chip signal	Voltage (analog)	Bits (digital)
Uniformity	Very high	Low to moderate
Noise	Low	Moderate
Complexity	Low	High
Cost	Low	High
Camera cost	Depends on design	Depends on design
Dynamic range	High	Moderate
Speed	Moderate to high	High
Power consumption	High (up to 100 times of CMOS)	Low to moderate

camera is the resolution that represents details of a camera and is measured in pixels. More pixels a camera has, more detail it can contain and there will be no blur in larger pictures. Size of pixels is summarized in Table 8.5 [10].

As photosets can only keep track of striking light intensity, most sensors use filtering scheme to show three primary color components.

Table 8.5 Pixel numbers and camera sensor node (C-SN) specifications

Size of pixels	Type of camera	Total # pixels	Comments
256 × 256	Cheap	65,000	Unacceptable picture quality
640 × 480	Low end		Ideal for e-mailing
1600 × 1200	High resolution	1,109,000	4 × 5 inch print
2240 × 1680	Good quality	4 megapixel	For prints up to 16 × 20 inches
4064 × 2704	Top-of-the-line	11.1 megapixels	13.5 × 9″ prints possible with no loss

8.3 Digital Images Using CCDs and CMOS Sensors

A simple mechanism for a color photo in a still camera is to separate the three basic color components of red, blue, and green and combine them at the receiver end. But, to obtain superior image quality through enhanced resolution and lower noise [12], it is better to detect only one-third of the color information for each pixel and interpolate other two-thirds with a demosicing algorithm to "fill in the gaps" from other pixels, resulting in a much lower effective resolution. Such consideration of multiple pixels together is due to Dr. Bryce E. Bayer [11] who invented Red, Green and Blue color filters that could capture color information from multiple adjacent pixels. An alternating Red/Green and Blue/Green arrangement of 4 pixels is shown in Fig. 8.4 and is called an RGBG filter that is trapped on the silicon sensor substrate with tiny cavities, like wells (pixels). This bonded color filter on the substrate records colors based on light photons it receives, with each pixel having only two of three colors. As human eyes are more sensitive to green color, there are twice as many green squares as either blue or red. As illustrated in Fig. 8.5, in each 2 × 2 square of four pixels, each pixel contains a single color; either red, green or blue and are called G1, B1, R1, G2. If a single pixel G1 is considered, then it finds out how many blue photons B1 has got and adds them to its green. In a similar way, G1 gets information from R1 and G2. In this way, G1 gets a complete set of primary color data, and can use it at its place on the sensor. When G1 acquires data to close by pixels, it also shares its value. In a similar way, a 3 × 3 grid can be analyzed as shown in Fig. 8.5b where the center pixel decides the constituent pixel

Fig. 8.4 Bayer arrangement of color filters on the pixel array of an image sensor

R	G	R	G	R	G
G	B	G	B	G	B
R	G	R	G	R	G
G	B	G	B	G	B
R	G	R	G	R	G
G	B	G	B	G	B
R	G	R	G	R	G

R: Red; G: Green; B: Bule

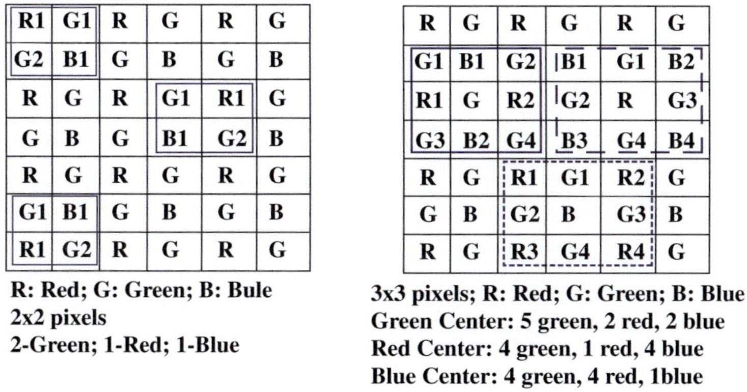

Fig. 8.5 **a** 2 × 2 pixel array in an image sensor. **b** 3 × 3 pixel array in an image sensor [13]

colors. If the center pixel is green, it will have 4 blue and 4 red pixels. A red pixel in the center will have 4 green and 4 blue pixels. A blue center pixel will have 4 red and 4 green pixels. In this way, every single pixel is used by 8 other pixels in the neighborhood and is known as "demosicing" to provide smoothness of pictures.

CCD sensors and a camera circuit board are shown in Fig. 8.6 [13] where after capturing the image, sensor send built-up charges row by row to an output register and amplified before sent to A/D converter. The corresponding digital file is then stored for display and further processing. A similar arrangement for CMOS is illustrated in Fig. 8.7 [13] which is cheaper to produce than a CCD scheme. It consumes less power with more complex circuitry. Each pixel is accessible independently and has more noise than CCD due to larger space requirement (Fig. 8.8).

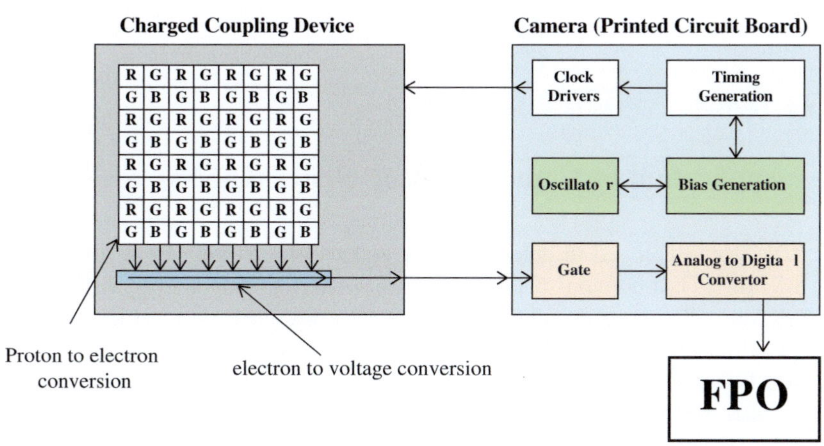

Fig. 8.6 CCD sensors and camera circuit board [14]

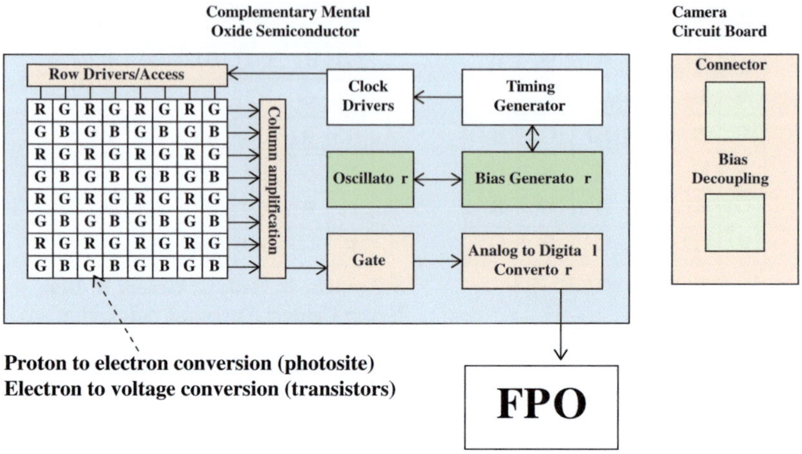

Fig. 8.7 CMOS sensors and camera circuit board [15]

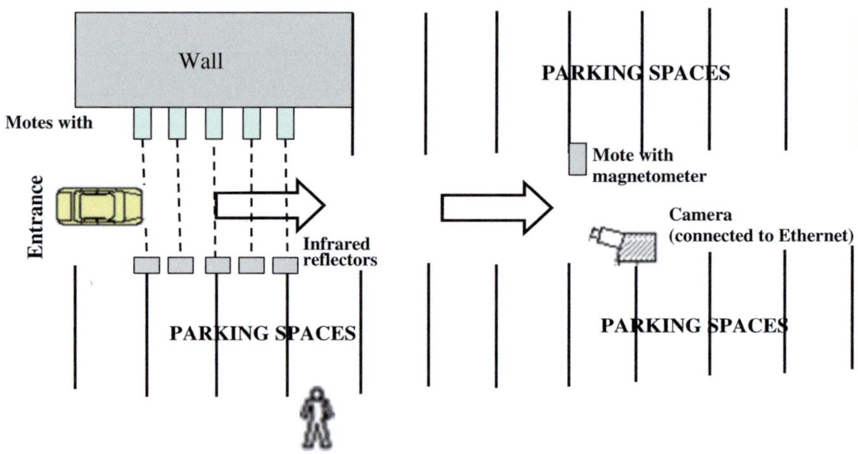

Fig. 8.8 Deployment of SNs in a parking garage

8.4 Application of Camera Sensor Nodes (C-SNs)

SNs have been deployed in detecting open spaces in a parking lot as illustrated in Fig. 8.3 [16, 17]. It contains 5 infrared beam sensors to detect any passing vehicle and determine their speed and length. To ensure presence of a vehicle, a magnetometer is placed at an appropriate place and a camera takes photos of speeding vehicles. Figure 8.9a shows systematic conversion of signal by infrared sensors at 256 Hz rate, magnetometer at 16 Hz rate and the camera in detecting presence of a vehicle, its length, speed, and its assigned unique ID. Variation of signals is shown in Fig. 8.9b (Figs. 8.10 and 8.11).

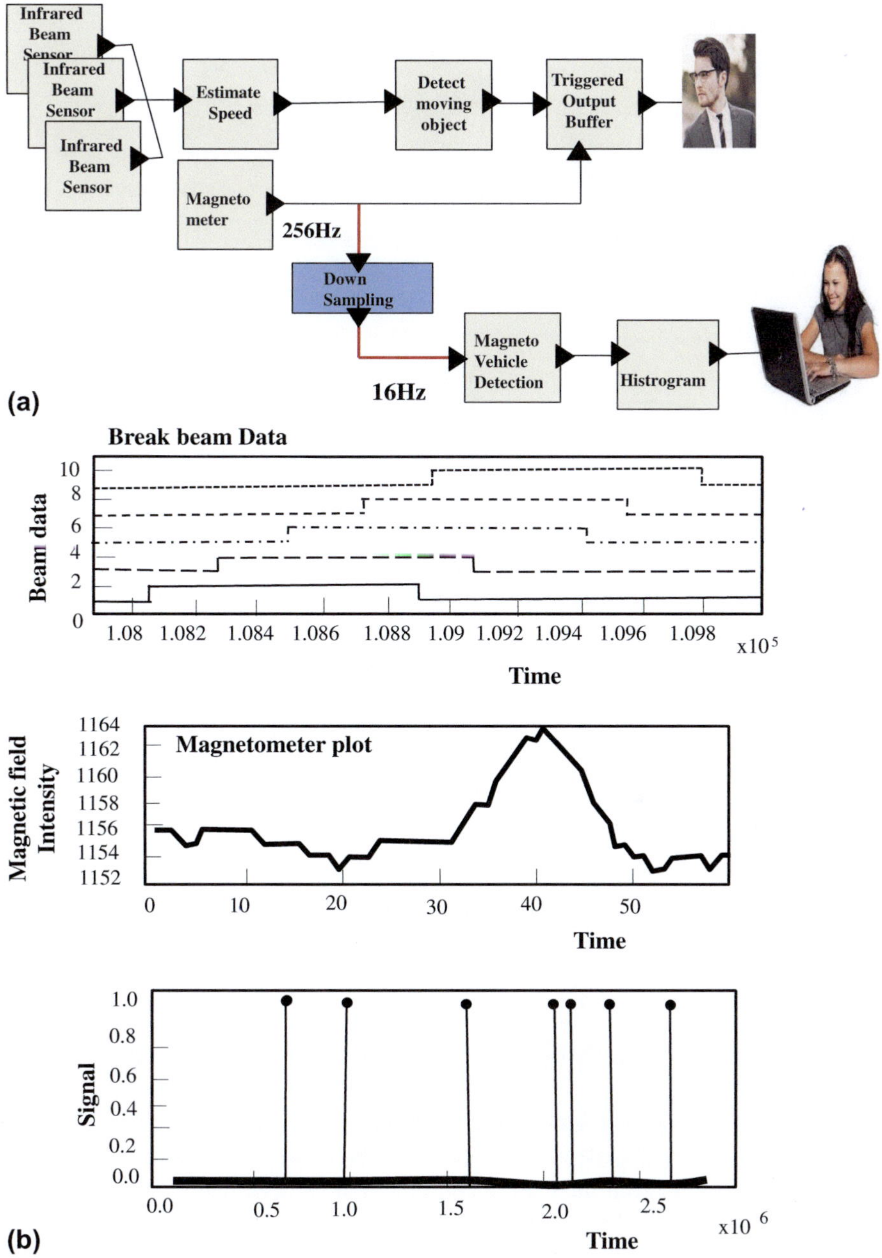

Fig. 8.9 **a** Vehicle detection steps in a parking lot. **b** Signals for vehicle detection

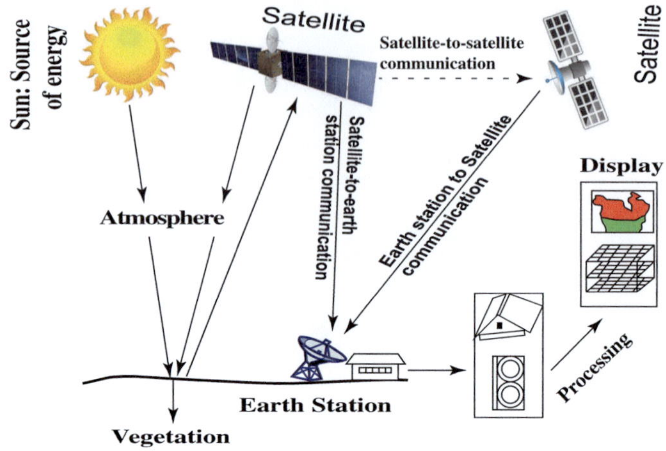

Fig. 8.10 Illustration of remote senor nodes (R-SNs) scheme

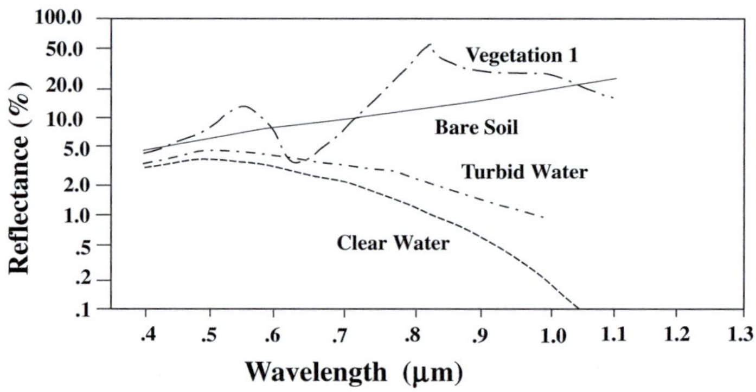

Fig. 8.11 Reflectance from Water and vegetation

8.5 Remote Sensor Node (R-SNs) Applications

It is interesting to note that camera-based techniques can be used in many different areas and remote sensing (RS) is a good example of extending the usefulness. In fact, RS can provide both qualitative and quantitative information about detached objects without approaching in direct contact. RS has numerous applications [18] such as meteorology (weather forecast), environmental studies (pollution effect), agricultural engineering, physical planning (scenario studies), hydrology (water and energy balance), soil science (vegetation mapping), nature conservation (vegetation mapping) forestry (such as fire detection), and land surveying (typography). The geographical information system includes topography, soils, geology, precipitation, land cover, vegetation, remote sensing data, surface temperature, hydrology, population, nature conservation, environment, digital terrain model, and topological map [18]. Selection of a remote sensing system depends on the primary sources of EM energy, possible atmospheric windows, and spectral characteristics of surface being sensed and spectral sensitivity of sensors available. This is due to fact that the reflection from the surrounding surface affects the reading.

RSN provide a regional view (large areas) as remote sensors in all seasons, understand a broader portion of the spectrum and can simultaneously focus in on a number of bandwidths and can provides digital data. A generic remote sensing (RS) scheme is shown in Fig. 8.12 [19] where aerial data is received from both airplane and satellite, and such geographical information system (GIS) provides useful information about the surface of earth, and possible signal processing operations are summarized in Table 8.6 [20].

The GIS system provides synopsis of the area as an overview of the region in terms of differences and coherence. Such flexibility is due to variety of sensors available, the underlying techniques used and processing algorithms used and reproducible analysis adopted. In practice, the actual results depend on the conventional mapping of specified data for given application and the process of updating the information. The interactivity depends on cooperation of human knowledge and machine operations. The dynamic capability of monitoring depends on the time series of data that reveals changes. Moreover, the invisible could become visible as it depends on the objective, quantitative data, extrapolation of

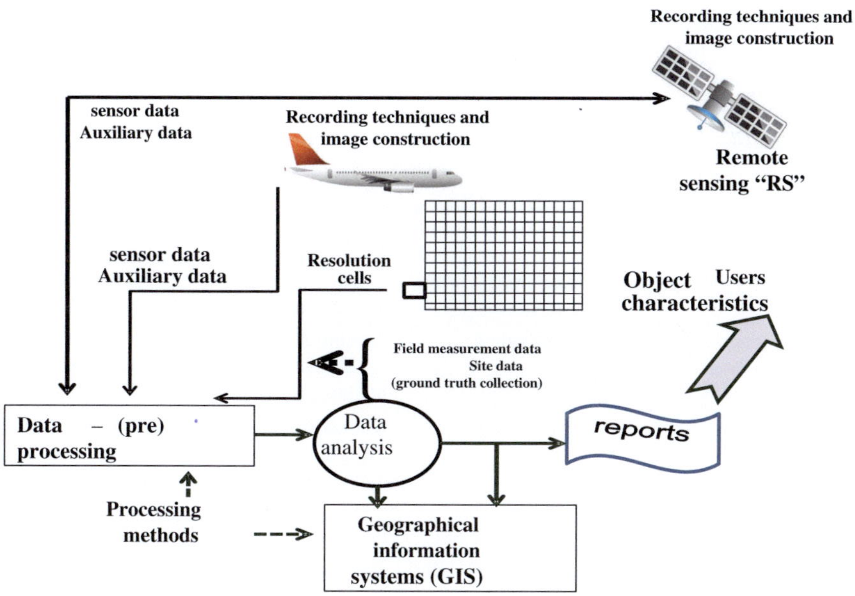

Fig. 8.12 Remote sensor nodes scenario

Table 8.6 GIS applications using RSNs and signal-processing operations [21]

Applications of GIS	Various signal processing operations
Topography	Spectral band selection
Soils	Filtering
Geology	Displaying band composites
Precipitation	Band rationing
Land cover	Contrast modification
Vegetation	Vegetation index computation
Remote sensing data	Histogram equalization
Surface temperature	Classification
Hydrology	Grayscale enhancement
Population	Texture analysis
Nature conservation	Color assignment
Environment	Segmentation
Digital terrain model	Density slicing
Topographical map	Geometric correction
–	Principal components
–	Transformation

point measurements and opening of inaccessible regions. The RS provides a regional view of large areas, and enables repetitive view of the same area. This enables a broader share of the spectrum than human eye. Moreover, SNs can use a number of bandwidths instantaneously and many remote sensors operate 24 × 7 (Table 8.7).

Table 8.7 Processing levels in R-SNs and detailed steps [22]

Level	Processing done	Explanation
Level 0	Raw binary data received from satellite	"Raw" transmitted information from SN on satellite and received by ground station
Level 1	Image data containing individual calibrated channels	Contain Level-0 data, appended calibration and navigation data, and instrument
Level 2	Derived variable, atmospherically corrected and geo-located	Derived from Level-1 raw data by applying the sensor calibration, atmospheric corrections, and geophysical value
Level 3	Composite images of derived variable resampled and averaged over a certain time period	Geophysical parameters interpolated on a global grid
Level 4	Image representing a variable averaged within each grid cell	Objective achieved

8.6 Conclusions

The area of SNs is much wider than just SNs with just few transducers measuring physical quantities. But, a camera lens is more or less equivalent to having a large number of SNs in a single lens, measuring pixel intensity of the whole area that could be focused by the lens and can be termed as C-SNs. Thus, a camera has become an important SN component in providing useful information and monitoring larger areas. Digital color images can be obtained using CCDs and CMOS devices. Remote sensor nodes (R-SNs) take one step further in the coverage area by capturing information from airplanes and satellites and fusing information together. It is really challenging to think of all these different types of SNs in a given application as these technologies are complimentary to each other. Future integration of these devices appears very promising and could lead to a very powerful and versatile system.

8.7 Questions

Q.8.1. How do you compare sensing area and power consumption in a SN versus a camera, versus an aerial camera unit versus a satellite camera? Use both qualitative and quantitative measures to understand associated implications.

Q.8.2. A WSN consists of 160 randomly deployed SNs in a large parking lot of size 2000×2000. The parking lot has an area of 6×9 for parking each car. If each SN can sense an object within 10 units of radius, what is the probability that cars at two selected spots can be detected by the SNs?

Q.8.3. The lot is divided into 16 equal parts, with each part having 10 randomly deployed SNs, what is the probability that a car can be detected?

Q.8.4. If one-fourth of SNs in Q.8.2 is replaced by a camera that can cover 4 times the area of a SN, what is the probability that a vehicle will be detected either by a camera or a SN?

Q.8.5. If one-half of SNs in Q.8.2 is replaced by a camera that can cover 8 times the area of a SN, what is the probability that a vehicle will be detected either by a camera or a SN?

Q.8.6. How can you place several cameras so that sensing area is equivalent to a: (a) Triangle and (b) hexagon.

References

1. http://gsn.sourceforge.net.
2. http://gsn.sourceforge.net.
3. Ali Salehi, "Design and Implementation of an Efficient Data Stream Processing System," D. Sc. Thesis, EPFL, Lausanne, 2010.

4. Clement Farabet, Cyril Poulet, Jefferson Y. Han, Yann LeCun, "CNP: An FPGA-Based Processor for convolutional networks," www.cs.nyu.edu/~yann/research/…/index.html.
5. http://www.phy.ntnu.edu.tw/java/Lens/lens_e.html.
6. http://en.wikipedia.org/wiki/Remote_sensing.
7. S. K. Nayar, "Computational Cameras: Redefining the Image," IEEE Computer Magazine, Special Issue on Computational Photography, Aug. 2006, pp. 30–38 also http://www1.cs.columbia.edu/CAVE/projects/what_is/.
8. http://www1.cs.columbia.edu/CAVE/projects/what_is/.
9. Petr Vodnak Photography - CMOS vs. CCD, in http://www.petrvodnakphotography.com/Articles/CMOSvsCCD.htm
10. http://www.gyes.eu/photo/sensor_pixel_sizes.htm
11. http://en.wikipedia.org/wiki/Bayer_filter.
12. http://en.wikipedia.org/wiki/Three-CCD_camera.
13. Losson Olivier Losson, Ludovic Macaire, Yanqin Yang, "Comparison of color demosaicing methods," March 28 Mar 2012, at web site https://hal.archives-ouvertes.fr/hal-00683233/document.
14. http://sensorcleaning.com/
15. http://sensorcleaning.com/
16. Jie Liu and Feng Zhao, "Towards Semantic Services for Sensor-Rich Information Systems," Proceedings of the 2nd International Conference on Broadband Networks, 2005, Boston, vol. 2, Oct. 7, 2005, pp. 967–974.
17. Jie Liu and Feng Zhao, "Composing Semantic Services in Open Sensor-Rich Environments," IEEE Network magazine, July-Aug. 2008, vol. 22, no 4, pp. 44–49.
18. https://www.google.com/?gws_rd=ssl#q=Applications+of+remote+sensing +wageningen&start=10.
19. https://www.wageningenur.nl/.
20. http://www.geo-informatie.nl/courses/gima_rs/Introduction/GIMA%20Intro.pdf.
21. http://www.geo-informatie.nl/courses/gima_rs/Introduction/GIMA%20Intro.pdf.
22. https://www.eeb.ucla.edu/test/faculty/nezlin/RemoteSensingOfTheSea.htm.

Part II
Random Topology

Chapter 9
Sensor Node Coverage and Connectivity for Random Deployment

9.1 Introduction

A comprehensive sensor node (SN) has been developed by the University of California at Berkeley and marketed by Crossbow, and the resulting product Mica mote [1] is shown in Fig. 9.1. It employs an Atmel Atmega 103 microcontroller running at 4 MHz, with a transceiver radio operating at 916 MHz band for bidirectional communication at a bandwidth of 40 kbps. It runs with a pair of AA batteries and includes thermopile sensors (for local temperature difference or temperature gradient), a photoresistor (incident light intensity), a barometer (for atmospheric pressure), and a humidity transducer. A/D Converter and 8 × 8 power switch on the sensor board further enhances its usefulness.

Typically, a wireless sensor network (WSN) consists of a large number of SNs that transmit a volume of data to a central station, commonly known as a base station (BS) or sink node in a multi-hop fashion.

SNs are protected from variable weather conditions by packaging in an acrylic enclosure, without affecting its sensing or radio communication functionalities. The MICA 2 mote employs one of the three RF frequency bands of 915, 433, or 315 MHz, and miniaturized version of Mote is also available as MICA2DOT. Another version of MICA 2 jointly developed by Crossbow and MIT can use either ultrasound or RF signal and has a flexibility to be used either as a listener or a beacon transmitter. Details of sensor transducers could be obtained from different Web sites.

As all functional components of a Mica mote and SN boards get energy from the batteries, it is critical to minimize energy consumption and has become a challenging task in the design of a WSN. The energy consumption in SN involves three different components of sensing transducer, A/D converter, and data transceiver. The SN consumes only 3.1 μW and 41 pW in standby mode. The transducer is in charge for catching the physical parameters of the surrounding area. The lower bound on energy per sample is roughly $E_{min} = C_{total}V_{ref}^2$, where C_{total} is net system

D.P. Agrawal, *Embedded Sensor Systems*,
DOI 10.1007/978-981-10-3038-3_9

Fig. 9.1 Mica mote SN unit [2]

capacitance and V_{ref} is input voltage. The energy to transmit k-bit message to distance d can be given by:

$$E_{\mathrm{Tx}}(k,d) = E_{\mathrm{Tx-elec}}(k) + E_{\mathrm{Tx-amp}}(k,d) = E_{\mathrm{elec}} * k + * k * d^2, \qquad (9.1)$$

where $E_{\text{Tx-elec}}$ is the transmission electronics energy consumption and $E_{\text{Tx-amp}}$ is the transmit amplifier energy consumption. An example value can be given as $E_{\text{Tx-elec}} = E_{\text{Rx-elec}} = E_{\text{elec}} = 50$ nJ/bit; the receiver energy $E_{\text{Rx}}(k) = E_{\text{Rx-elec}}(k) = E_{\text{elec}}{*}k$; and computing/processing unit $E_{\text{switch}} = C_{\text{total}}V_{\text{dd}}^2$. In order to conserve energy, some SNs may go to sleep mode to minimize energy consumption.

A Mica mote SN is equipped with an embedded microcontroller, such as the Atmel or the Texas Instruments MSP 430 that can be put into various operational and sleep modes. Associated radio transceiver includes RFM TR1001 or Infineon or Chipcon device which engages amplitude-shift keying or *frequency shift keying* while the Berkeley Pico SN employs *on–off keying* modulation. TinyOS is the popular operating system as an open-source event-driven platform. If each subarea can be covered by more than one subset of SNs all the time, then some of the SNs can be placed into sleep mode while satisfying full coverage. There are many versions of Mica mote boards with varying parameters and characteristics and have been summarized in [3] of wireless SNs.

SNs can be deployed in two different ways before they can send data to the BS or sink node and depends on whether the area to be monitored is physically inaccessible or is easily accessible. Many such possible schemes are summarized in

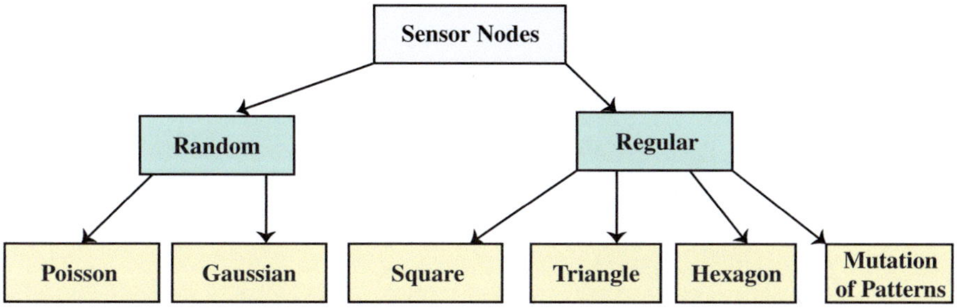

Fig. 9.2 Deployment of SNs

Fig. 9.2. Poisson allows SNs to be distributed in a random way within the area of interest and is achieved by low-flying airplanes or drones. The Gaussian has variable concentration of randomly deployed SNs, with increased density near the BS to take care of any possible energy-hole problem. Regular placement is feasible by placing SNs at regular intervals, and the basic arrangement basically characterizes the pattern used.

9.2 Coverage and Placement of SNs

Transducer of a SN can sense physical parameter of the surrounding area and Fig. 9.3 employs a disc model with a radius of r_s to represent the area being covered by a single SN_i. Data from a single SN are not adequate to make any useful decision and need to be collected from a set of SNs. There could be some areas not covered by any SN, or there could be some overlapped areas by multiple SNs. The question that comes is where to place the second sensor at location SN_j'. That creates an overlap between SN_i and SN_j'. An alternative is to place it at SN_j'', leaving a gap between SN_i and SN_j''. Both these issues can be addressed if the second sensor is placed in location SN_j'''. An appropriate place for the third and fourth SNs are, respectively, at SN_j'''' and SN_j''''', even though they leave a small gap in between.

Transmission between adjacent SNs is feasible if there is at least one SN within the communication range of each SN. Therefore, not just the sensing coverage, but the communication connectivity is equally important. As shown in Fig. 9.3, the wireless communication range r_c of a SN's wireless transmitter must be at least twice the sensing distance r_s.

As considered earlier, a single SN is not adequate to provide adequate information to an actuator to make a useful decision. In fact, one SN can sense only small area and several SNs need to be utilized to monitor a given area. This is shown in Fig. 9.4 by utilizing 9 random SNs to cover the area A with corner points given by WXYZ. The sensing area is exemplified by circles around each SN, assuming a homogeneous system is used with all SNs having the same sensing area

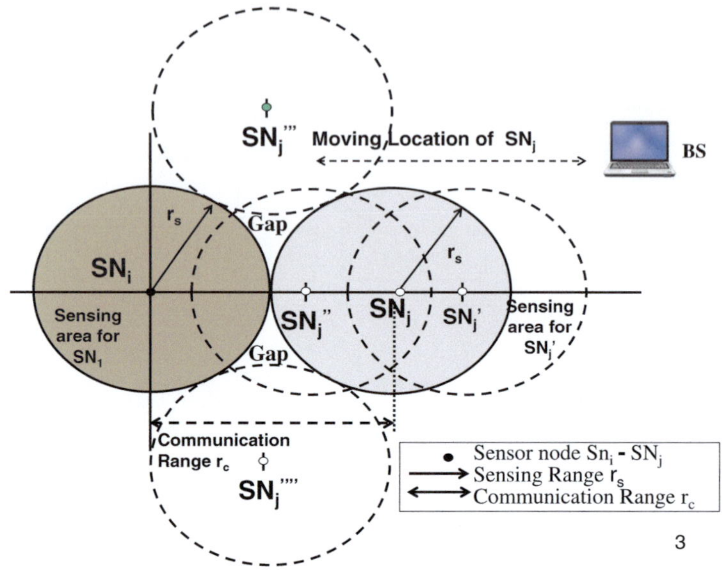

Fig. 9.3 Coverage and placement of second and successive SNs

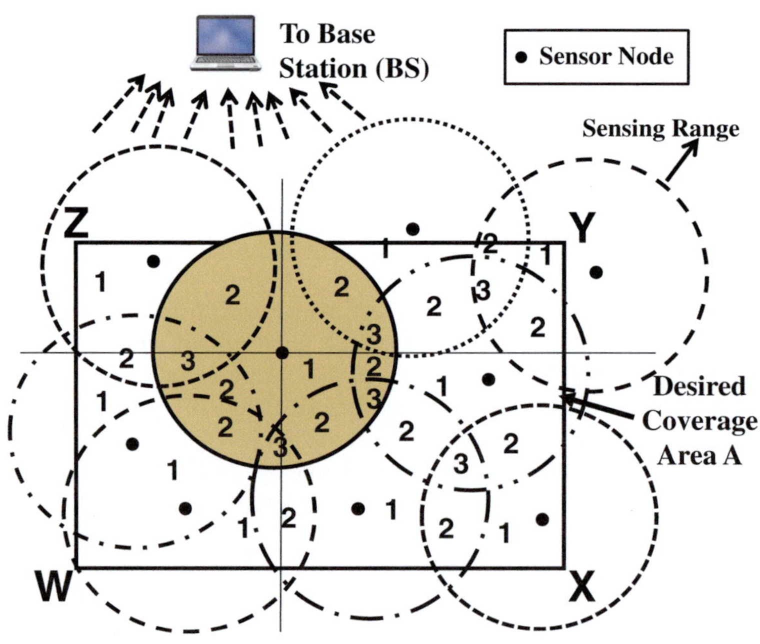

Fig. 9.4 SNs covering an area

r_s. The parameter in some small areas can be indicated by reading from a SN and is marked by region 1 in Fig. 9.4. Some areas are marked by 2, signifying that reading from one of the two adjacent SNs can be used to exemplify the parameter. As the physical parameter of the surrounding environment is measured, two SNs will provide almost the same value and reading from any of the two SNs is acceptable.

If N SNs are randomly placed within an area A, then the SNs density can be given by:

$$\lambda_s = \frac{N}{A}. \tag{9.2}$$

The whole space A has to be covered with N SNs, with adjacent SNs located at most at a distance of $2r_s$ from each other. As the disk shape covered by single sensor $a = \pi r_s^2$, the probability that point p is located within an arbitrary sensor's sensing range $= a/A = \pi r_s^2/A$.

Probability that point p is not located at an arbitrary sensor = $1 - a/A$

Probability that point p is not located at any N sensors = $(1 - a/A)^N$

$$(1 - \pi r_s^2/A)^N = (1 - \lambda_s \pi r_s^2/N)^N \tag{9.3}$$

When A increases to infinity N also goes to infinity and this probability

$$\mathrm{e} -^{a\lambda} s = (\mathrm{e}-^{\lambda_s \pi r_s^2}). \tag{9.4}$$

$$\text{The fraction of Area being covered} = f_a = (1 - \mathrm{e}-^{\lambda_s \pi r_s^2}) \tag{9.5}$$

So, required density d and # SNs can be given by:

$$\lambda_s = -\frac{\ln(1 - f_a)}{\pi r_s^2} \tag{9.6}$$

The sensing model is known as Boolean sensing model as any event occurring outside the sensing radius r_s *is assumed to be unknown.* The probability that monitored area is covered by at least one SN is given by:

$$f_{1-\text{covered}} = 1 - f(0) = 1 - \mathrm{e}^{-\lambda_s} \tag{9.7}$$

where $f(0)$ is coverage by zero SN.

In many situations, a given parameter need to be sensed by at least k close-by SNs for a cooperative decision, then concurrent sensing by k SNs can be expressed as:

$$f_{k-\text{covered}} = 1 - \sum_{m=0}^{k-1} f(m) = 1 - \sum_{m=0}^{k-1} \frac{(\lambda_s)^m}{m!} \mathrm{e}^{-\lambda_s}. \tag{9.8}$$

9.3 Individual Sensor Coverage and Area Coverage by SNs

Sensing area by each SN remains the same assuming homogeneous environment. As the SNs are deployed randomly, the area covered by two or more SNs could overlap and net overall area covered by SNs could be smaller. This is illustrated in Fig. 9.5 where SNs are randomly deployed in 10,000 × 10,000 pixels area. Each SN has sensing range r_s of 10 pixels, while the SN density is changed from 0 to 0.025 per unit pixel area (SNs varied from 0 to 2,500,000). Variation in SNs coverage and effective area is shown in Fig. 9.6.

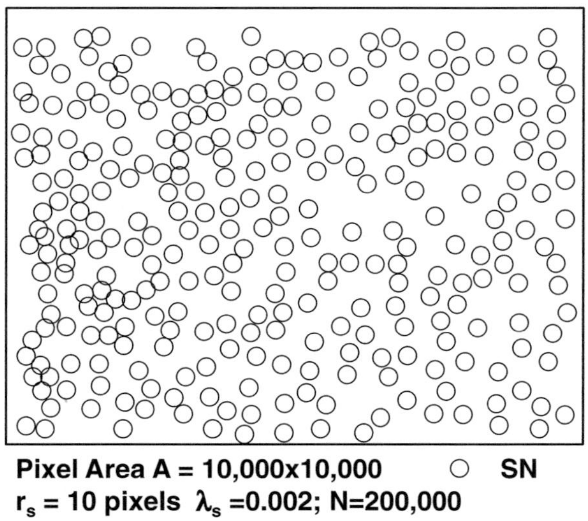

Fig. 9.5 Example randomly deployed SNs in an area of 10,000 × 10,000 pixels [4]

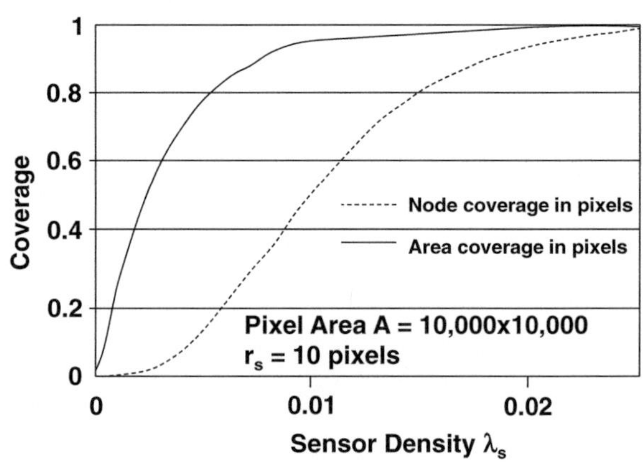

Fig. 9.6 Coverage by randomly deployed SNs in an area of 10,000 × 10,000 pixels [4]

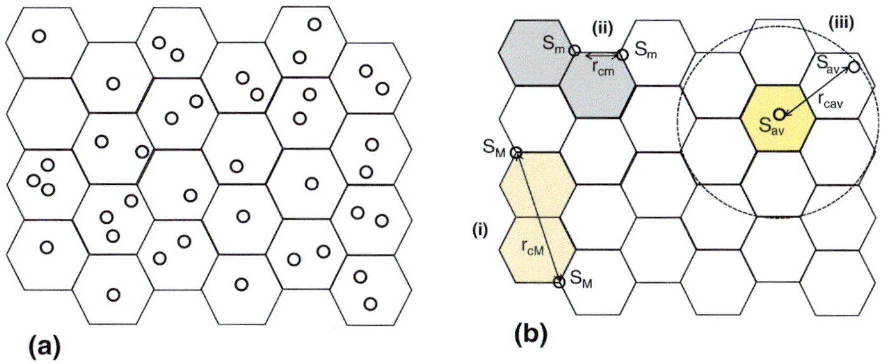

Fig. 9.7 **a** Given area subdivided into hexagons, with at least one SN inside each hexagon [5]. **b** Connectivity probability of randomly deployed SNs in three extreme cases [5]

The coverage indicated by circles implies overlapped areas that could be eliminated if a hexagon is used to represent both coverage and connectivity areas. In order to ensure connectivity to BS from SNs, a given area can be subdivided into hexagons as illustrated in Fig. 9.7a. As SNs are deployed randomly, as long as there are at least one or more SNs inside each hexagon, the whole area will be covered by SNs and they could remain connected. In order to determine connectivity in terms of communication distance, three extreme cases [5] are considered as shown in Fig. 9.7b. In case (i), SNs of two adjacent hexagons are located at far extreme as possible at distance $\mathrm{r_{cM}}$; minimum distance of r_{cm} is used in case (ii); while case (iii) depicts an average distance of r_{cav} between two SNs where seven hexagons are approximated by a circle covering equal area $R_s \geq 2R_c/\sqrt{13}$. By geometry, each side of hexagon [5] can be expressed by $l = \frac{r_{cM}}{\sqrt{13}}$, $l = \mathrm{r_{cm}}$, or

$$l = \sqrt{\frac{2\pi}{21\sqrt{3}}}\, r_{cav}. \tag{9.9}$$

These three possibilities of placing SNs require different densities of SNs λ_s, and they lead to different probabilities of SNs being connected together. A detailed analysis has been performed in [4], and the use of r_{cM} requires deployment of fewer SNs, while r_{cm} leads to upper bound of SNs density. Final results of variation in connectivity for four different models are illustrated in Fig. 9.8a. For a given SN connectivity of 95%, variation in connectivity as a function of SNs density for four different values of hexagons is shown in Fig. 9.8b [4]. It is clear from Fig. 9.8 that the connectivity increases drastically as a given SN density changes between 1 and 1.3, and such threshold density ought to be used if coverage and connectivity need to be ensured for a randomly deployed WSN. As a typical WSN application does not require 100% sensing coverage and 100% connectivity, Fig. 9.8 can be used as a generic guideline for selecting density of SNs. It is obvious that with increasing SNs, the number of neighbors for each SN also increases. Simulations are done [5] to

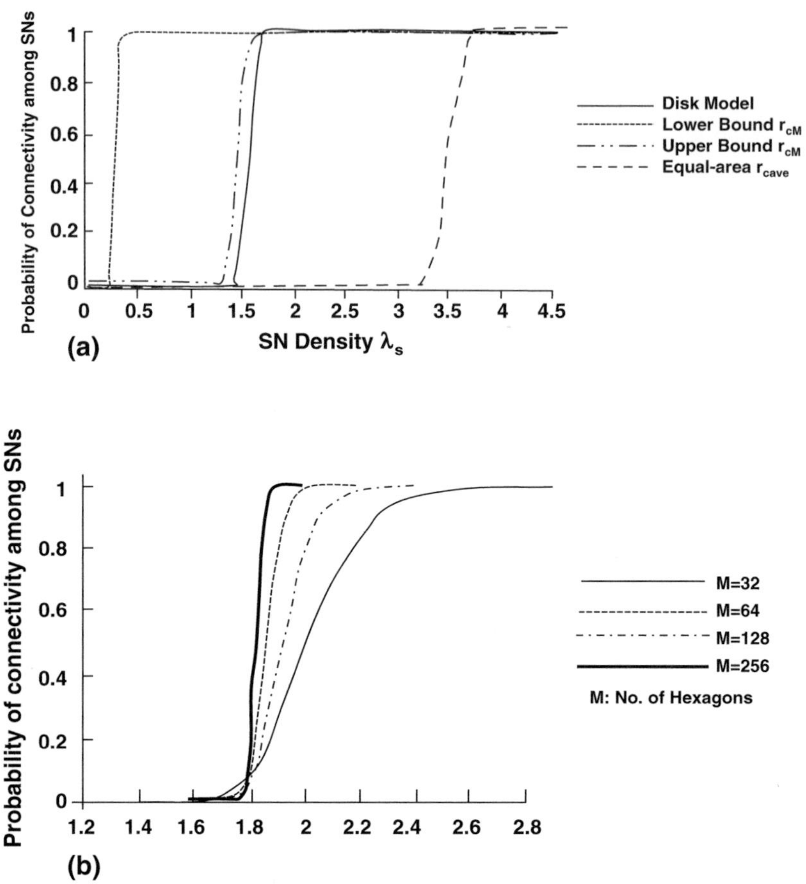

Fig. 9.8 **a** Probability of SNs connectivity for four different models as a function of SN density λ_s [5]. **b** Probability of SNs connectivity for four different values of hexagons as a function of SN density λ_s for given SNs connectivity of 95% [5]

study the effect of adding SNs by partitioning 50 × 50 area into 80 hexagons and is given in Fig. 9.9a. The impact of number of neighbors for each SN on the probability of connectivity of SNs is given in Fig. 9.9b [5]. It is clear that a SN connectivity of 6–8 is desirable in keeping SNs connected in a randomly deployed WSN.

9.4 Energy-Hole Problem in a Randomly Deployed WSN

Energy consumption in a SN depends on the number of transmissions and receptions performed by a given SN. Let us have a closer look at a randomly deployed WSN of Fig. 9.10a, assuming that packets are forwarded to the BS in multi-hop

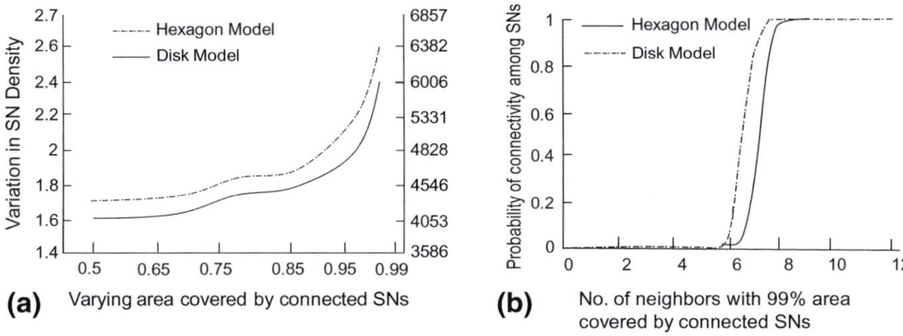

Fig. 9.9 **a** No. of neighbors for each SN with 95% area covered by connected SNs [5]. **b** Probability of connectivity among SNs as a function of no. of neighbors for each SN [5]

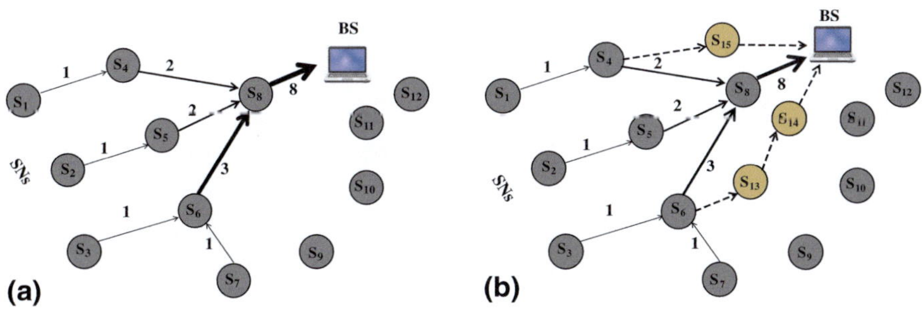

Fig. 9.10 **a** Energy-hole problem in a randomly deployed WSN. **b** Possible solution to the energy-hole problem

fashion without making any changes to the packet. The number of packets sent by each SN is given in Table 9.1. As SNs closer to the BS need to receive and transmit a larger number of packets, they could run out of energy at a much faster rate than far away SNs. Once the SNs near the BS runs out of energy, no information can be transmitted to the BS. Such an unbalance of residual energy in SNs causes a phenomenon commonly known as Energy Hole problem. So, what can be done to resolve this issue? A simpler approach is to deploy larger number of SNs near the BS so that packets could be forwarded to BS via different SNs and is illustrated in Fig. 9.10b. The numbers of packets from S_8 to BS is reduced from 8 to 2 in this manner.

One such scheme that utilizes more sensors near BS is to employ SNs in a Gaussian-distributed manner *f(x, y) given by* Eq. (9.1) and is shown in Fig. 9.11. Gaussian is a continuous probability distribution that has a bell-shaped probability density function; parameters x_i, y_i are the *mean* and *expectation,* and σ^2 is the variance (σ is known as the standard deviation) and given by:

Table 9.1 SNs and number of transmissions/receptions in WSN of Fig. 9.10a

SN	# Transmissions without aggregation	# Receptions
S_1	1	0
S_2	1	0
S_3	1	0
S_4	1	0
S_5	3	2
S_6	2	1
S_7	2	1
S_8	8	7

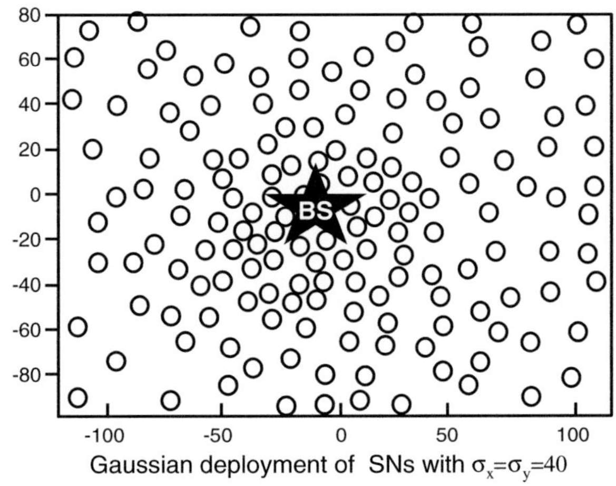

Fig. 9.11 Gaussian-Distributed SNs in a randomly deployed WSN

$$f(x, y) = \frac{1}{2\pi\sigma_x\sigma_y} e^{-\left(\frac{(x-x_i)^2}{2\sigma_x^2} + \frac{(y-y_i)^2}{2\sigma_y^2}\right)} \tag{9.10}$$

Standard normal distribution is given by $x_i = y_i = 0$ and $\sigma^2 = 1$ and $f(x, y)$ is reduced to:

$$f(x, y) = \frac{1}{2\pi\sigma_x\sigma_y} e^{-\left(\frac{x^2}{2\sigma_x^2} + \frac{y^2}{2\sigma_y^2}\right)} \tag{9.11}$$

If multiple BSs are to be used, then the area can be subdivided into many subregions and can have Gaussian distribution in each subregion as illustrated in Fig. 9.12.

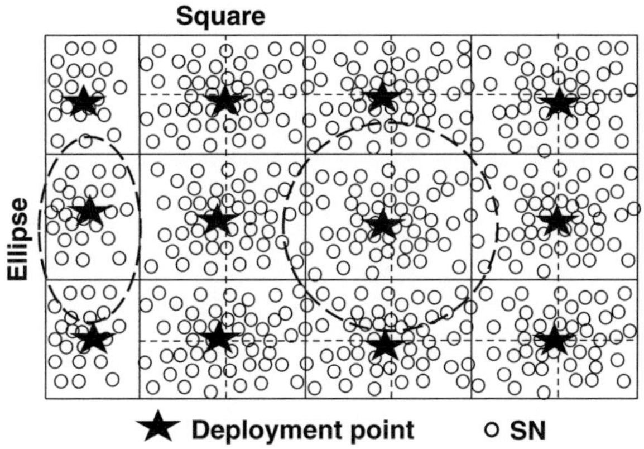

Fig. 9.12 Gaussian distributed in a randomly deployed WSN with several deployment points

9.5 Conclusions

Both SN coverage and connectivity are important as the coverage represents what portion of the area is being monitored by SNs while connectivity indicates whether information from a SN can be forwarded successfully to BS along multi-hop path from a SN to BS. Energy is consumed in transmitting a packet as well as in receiving a packet. As multi-hop is used to minimize energy consumption, SNs closer to BS have to receive more packets and retransmit to BS and could run out of power at much faster rate than SNs far away from the BS. Gaussian distribution has been suggested to avoid such energy-hole problem.

9.6 Questions

Q.9.1. How do you compare sensing area and power consumption in a SN versus a camera, versus an aerial camera unit versus a satellite camera? Use both qualitative and quantitative measures to understand associated implications.

Q.9.2. What is the role of sensing radius on the performance of a WSN?

Q.9.3. Assuming that energy consumption in transmission is proportional to square of distance you need to cover, what is skip value that optimizes the product of delay and consumed power?

Q.9.4. Can you have full coverage and connectivity in a three-dimensional WSN?

Q.9.5. What are the advantages of having full-coverage with multiple connectivities in three-dimensional networks?

Q.9.6. How can you have four-connectivity and full coverage in a randomly deployed WSN?

Q.9.7. What is the role of Gaussian distribution in addressing energy-hole problem in WBNs?

Q.9.8. How does higher connectivity help in minimizing energy-hole problem?

Q.9.9. What kind of mechanism you can utilize to position SNs in a randomly deployed WSN?

Q.9.10. What is the role of frequency of SN data transmission on energy hole?

Q.9.11. Can you employ high-power SNs to minimize the effect of energy-hole problem?

References

1. www.memsic.com/wireless-sensor-networks/.
2. http://computer.howstuffworks.com/mote4.htm.
3. http://en.wikipedia.org/wiki/List.
4. Benyuan Liu and Don Towsley, “A Study of the Coverage of Large-scale Sensor Networks,” University of Massachusetts—Amherst, Computer Science Department Faculty Publication Series Computer Science, 2004.
5. Haoyan Cai, Xiaohua Jia, and Mo Sha, “Critical Sensor Density for Partial Connectivity in Large Area Wireless Sensor Networks,” Proceedings of the IEEE INFOCOM 2010, 25 pages.

Chapter 10
Medium Access and Routing

10.1 Introduction

A wireless sensor network (WSN) consists of a large number of sensor nodes (SNs) that collect data at BS or sink node in a multi-hop fashion. The SNs use a single channel and could have collision present under many different scenarios as the hidden terminal problem was discussed in earlier chapter. Similarly, when SNs send beacon signals to determine neighbours, collision could be present due to asynchronous nature of SNs. SNs sense and provide data from the surrounding area, and the presence of any event is deduced based on the values provided by many SNs.

10.2 Collision Avoidance in a WSN

Important primary attributes of MAC protocols in WSN are collision avoidance, energy efficiency, and scalability in terms of SNs density. The secondary attributes are latency, fairness, throughput, bandwidth utilization, always sensing versus transceiver sleep–awake cycle, and data aggregation. Each SN has limited sensing range of r_s and communication range of r_c. These allow SN to cover a given area and transmit data to BS in multi-hop fashion. So, the sensing range allows SN to determine neighboring SNs. But, SNs far apart do not know the presence of each other, and hidden terminal problem was discussed earlier. This can be avoided by using small hand-shaking packets of RTS-CTS.

When RTS (request-to-send) packet is sent by SN A to SN B as shown in Fig. 10.1, the signal is received by all SNs in the communication range of SN A. RTS also contains the length of packet to B, and all devices in communication range of A knows how long the medium will be kept busy by A and the time period is known as NAV (network allocation vector). If device B is ready to receive, it sends CTS (clear-to-send) packet back to SN A. This message is heard by all

D.P. Agrawal, *Embedded Sensor Systems*,
DOI 10.1007/978-981-10-3038-3_10

devices in communication range of SN B, including the period of follow-up data from A to B. This NAV allows SN C to understand ongoing communication and thus learns the presence of SNs B and A. In a similar way, using TRC packet by SN B and CTS by SN A could avoid exposed terminal problem as shown in Fig. 10.1b. All devices within SN B's RTS packet and handshaking CTS packet by A allow all devices to keep quiet for NAV period. SN C knows the ongoing transmission from B to A and will be able to send data to SN D, thereby avoiding exposed terminal problem (Fig. 10.2).

SNs are distributed randomly in a WSN, and simplest scheme for collision avoidance is Aloha where SN ready to transmit packet immediately as shown in Fig. 10.3a. As packets are of different size and slots are not synchronized, collision could occur either at the beginning of a transmitted packet or at the end of the packet. An improvement occurs when slot size is fixed and slots are synchronized. Throughput is increased in slotted Aloha protocol and is illustrated in Fig. 10.3b.

In a WSN, packets are transmitted as fixed size packets, and time is basically slotted. As many distributed sender SNs may want to communicate at the same time, the question is how to avoid or minimize collision. TDMA schedule can be used if WSN is partitioned into clusters, and CH of each cluster dictates the TDMA schedule. The other option before forming clusters is to use contention-based scheme. CSMA/CA (carrier sense multiple access/carrier avoidance) shown in Fig. 10.2a is somewhat similar to IEEE 802.11 ad hoc mode CSMA/CA where SN senses the medium if some other SN is using; if so, then wait till it becomes free and then wait for random period between 1 and CW-1 (Contention window-1). A collision between transmissions from two or more SNs is possible and is indicated by the absence of ACK message from destination SNs if the same random delay is selected by transmitting SNs. In such a situation, the CW is doubled and the same process is repeated with the hope that colliding SNs will generate different random delays with increased window size. Collision is known only to colliding SNs as they do not receive ACK message, and the remaining SNs will keep the initial value of CW. The whole process is known as CSMA/CA.

In CSMA/CA, after sensing the medium, the SN can start transmitting immediately after the delay counter becomes zero and that is called persistent CSMA/CA protocol. If the SN waits for one more slot with probability $(1 - p)$ before starting the transmission of a packet, it is called non-persistent CSMA/CA (Fig. 10.2b).

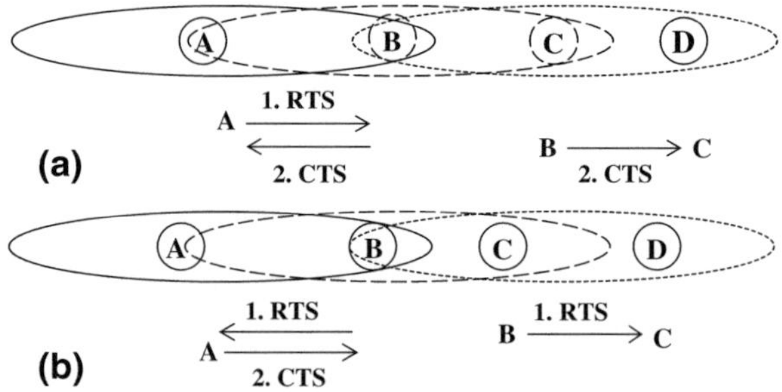

Fig. 10.1 **a** Avoiding hidden terminal problem using RTS/CTS, **b** avoiding exposed terminal problem using RTS/CTS

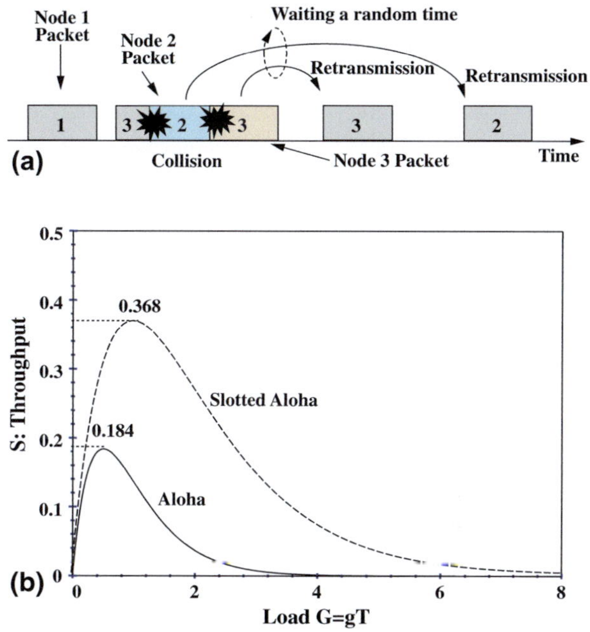

Fig. 10.2 **a** Collision in Aloha protocol, **b** throughput of Aloha/slotted Aloha protocols

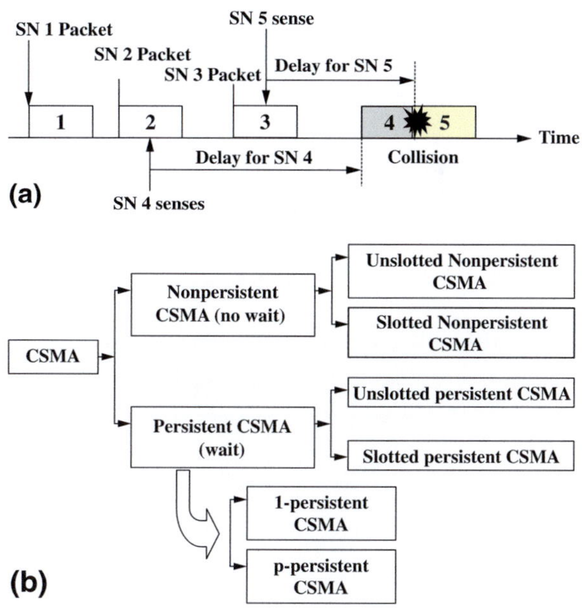

Fig. 10.3 **a** Collision in CSMA/CA, **b** classification of CSMA/CA schemes

Here, p indicates the probability of persistent wait for a slot by a SN. The use of random delay is shown in Fig. 10.4a. The counter for second device is frozen as soon as the first SN starts transmitting. In addition, DIFS (distributed interframe space) is also added to ensure better functioning of CSMA/CA as shown in Fig. 10.4b. In order to completely avoid collision, similar to ad hoc networks, RTS/CTS-based handshaking mechanism can be used (Fig. 10.4c) between a sender and receiving SNs before transmitting actual data as summarized in Table 10.1. A good trade-off remains between non-persistent and 1-persistent CSMA, and throughputs are compared with basic Aloha and slotted system in Fig. 10.5.

The SNs provide their location as well as event number. Thus, ID of SNs is not very critical and what is important is the values provided and WSN is considered as "data-centric" as it may require a huge ID due to large number of deployed SNs. In addition to location, SNs provide associated attribute values. So, unlike ad hoc networks, SN-to-SN routing is not required and data need to be routed from SNs to BS or sink node. Once SNs are deployed randomly, each SNs determine their adjacent neighbors by sendingbeacon signals. Then, the SNs could be grouped together to form clusters, and all SNs of a cluster could elect a cluster head (CH) that would be responsible to communicate with the BS/sink node. SNs measure physical parameter from the surrounding area; the reading provided by

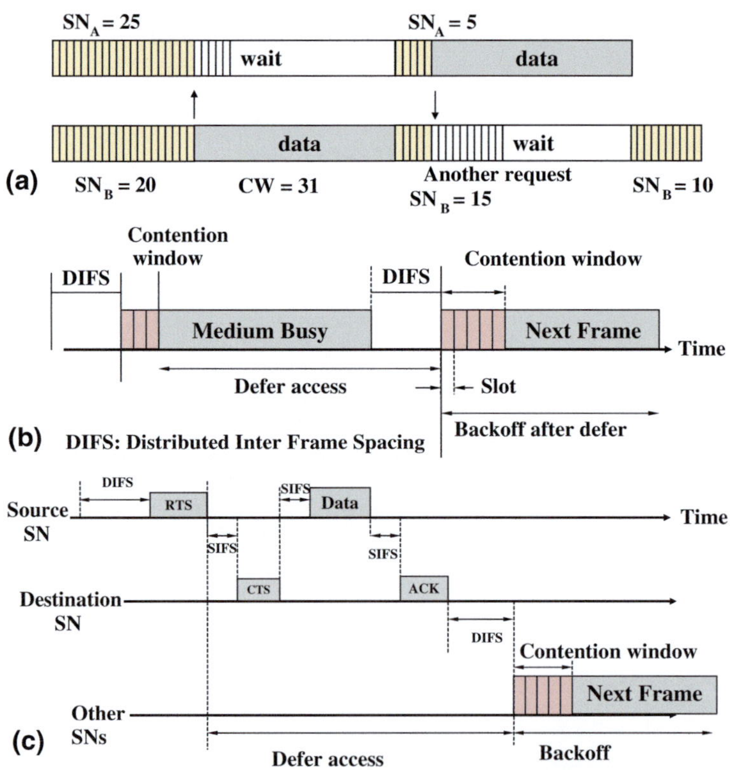

Fig. 10.4 **a** CSMA/CA illustrated, **b** CSMA/CA with DIFS, **c** CSMA/CA with DIFS and RTS/CTS

Table 10.1 Steps for *p*-persistent CSMA protocol

Step 1: If the medium is idle, transmit with probability *p*, and delay for worst case propagation delay by one packet with probability $(1 - p)$
Step 2: If the medium is busy, continue to listen until medium becomes idle, then go to Step 1
Step 3: If transmission is delayed by one time slot, continue with Step 1 $p = 0$: non-persistent and $p = 1$, 1-persistent CSMA

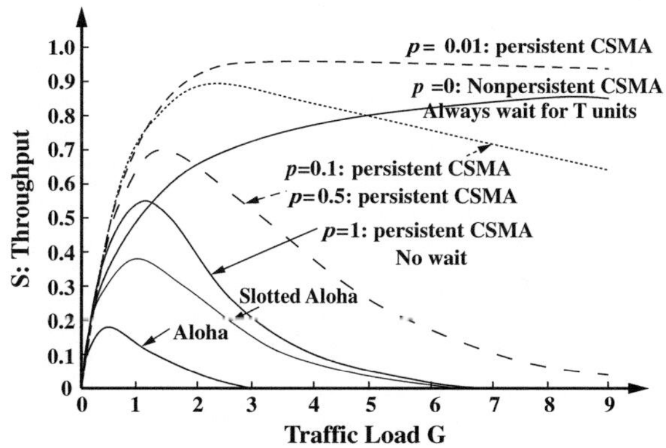

Fig. 10.5 Throughput comparison between persistent and non-persistent CSMA/CA

adjacent SNs could be similar or close to each other, and it is desirable to combine such values together, commonly known as data aggregation. This process can be done at each CH as it can collect data from cluster member SNs, and in this way, the energy dissipation through WSN is also controlled and is discussed later.

10.3 Routing in a WSN

In a WSN, data need to be sent from SNs to BS/sink node in multi-hop fashion. BS/sink node needs to broadcast a query to all SNs, indicating which physical parameters need to be forwarded and what frequency they need to be sent. You may have CHs if a WSN is partitioned into multiple clusters and each cluster elects a CH (Cluster Head). Then, the packet could travel to CH from associated SNs and then CH can combine (aggregate) data to the BS. In brief, routing is needed in a WSN, and many results obtained for ad hoc networks can be used effectively in a WSN. As discussed earlier, beacon signals are used to determine neighbors and 2-hop neighbors can be obtained by piggybacking 1-hop neighbor information. In the previous section, we discussed how to utilize CSMA/CA effectively to schedule transmission from SNs with sleep–awake cycle.

In a WSN, two different routings are needed: one dissemination of query information from the BS/sink node to all SNs, and routing of sensed data from responding SN to the BS. BS could broadcast the query to all SNs in a single high power transmission as shown in Fig. 10.6. In another scheme, BS is assumed to be far away and query from BS cannot reach all SNs in a single transmission. Hence, query in directed diffusion protocol [1] has to be diffused throughout WSN in multi-hop fashion between SNs (Fig. 10.7a). While the query is diffused from BS/sink node of the WSN, a primary path is selected from each SN to BS that requires minimum number of hops and is termed as gradient determination. If there are multiple paths with the same hop-length from a SN to BS, then other criterion such as packet delivery ratio can be used to select among them as detailed in Fig. 10.7b. It is interesting to note that an event could start from some specific points and move outwards to reach the requesting SN. So, a small number of SNs can be reinforced to prevent further flooding in the WSN.

Three types of queries are possible and are either historical queries with analysis of data collected over time, one-time queries giving snapshot view of the network, or persistent queries that involves periodic monitoring of data at regular intervals for a long period of time. Historical query is mainly used for the analysis of historical data stored at the BS, e.g., "What was the temperature 2 h back in the northwest quadrant?" One-time query gives a snapshot view of the network, e.g., "What is the temperature in the northwest quadrant?" Persistent query is mainly used to monitor a network over a time interval with respect to some physical parameters, e.g., "Report the temperature in the northwest quadrant for the next 2 h."

The routing required to respond to a query is application-specific and data-centric, while data aggregation capability with clustered WSN is desirable so as to minimize energy consumption. The query could be to respond either proactive or reactive way. In proactive approach, the periodic-sensed value is sent by SN, e.g., every 2 min for 20 h. In reactive scheme, sensed value is sent if it crosses a threshold, e.g., whenever temperature is beyond 35 °C. Hybrid combines both proactive and reactive approaches, e.g., reactive and proactive with increased reporting time (e.g., every 2 h). In proactive protocol, the SNs in a WSN periodically switch on their SN's transceiver, *sense the environment, and record the data of interest and transmit it to the during* allocated time slot by the CH (Fig. 10.8a). WSNs in responding to a query could follow either flat or clustered topology architecture and classification of different routing protocols is summarized in

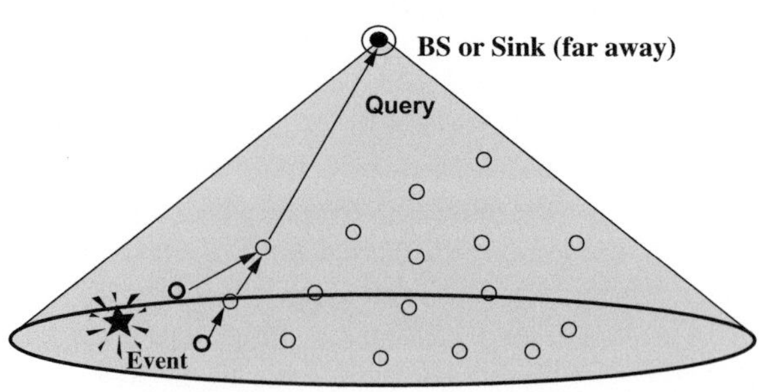

Fig. 10.6 Broadcasting of a query from BS/sink to SNs in a WSN

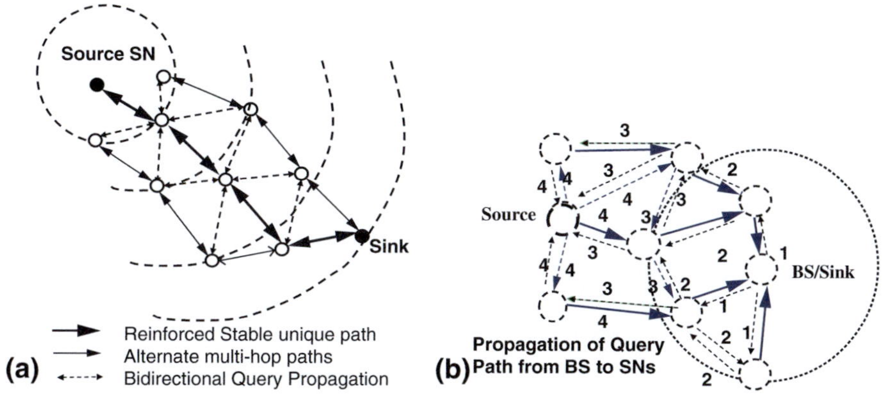

Fig. 10.7 **a** Directed diffusion of query from BS/sink node to SNs, **b** propagation of query in directed diffusion from BS/sink node to SNs

Table 10.2. Since the SNs switch off their transceiver at all times except the report times, the energy of the WSN is conserved. Report time T_R is the time period between successive reports sent by a transceiver of the SN and attributes (A) is a set of physical parameters which the BS is interested in obtaining data about. It is possible that the time-critical data may reach the BS only after report time.

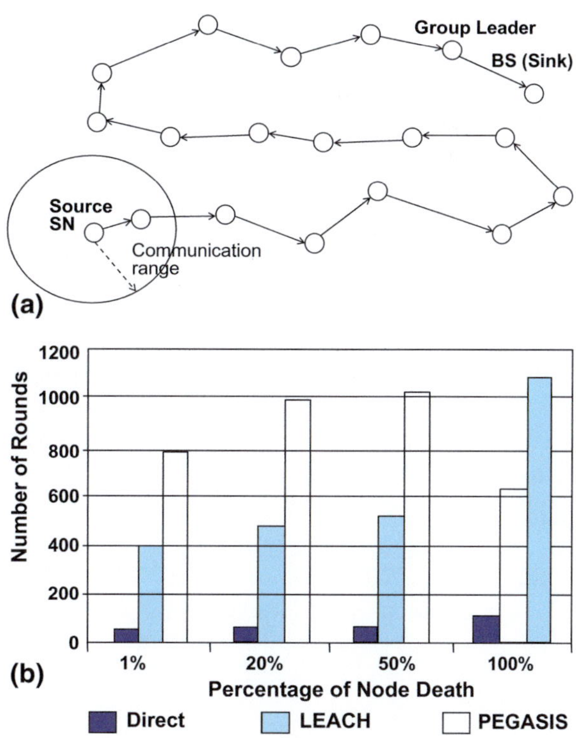

Fig. 10.8 **a** PEGASIS scheme, **b** comparing energy consumption of PEGASIS with direct scheme

As more energy is consumed by SN in transmitting and receiving data and SNs near BS face the problem of energy-hole problem, efforts ought to be made in making the number of transmissions/reception per SN equal. To achieve equalization of energy consumption in a WSN, a novel scheme known as PEGASIS [2] has been introduced where in a continuous traversal between SNs and BS is determined as shown in Fig. 10.8a. By having a single path, all SNs have to transmit one packet and receive one packet (except the source SN at one end transmits a packet and BS at the other end receives a packet). This balances energy consumption in a WSN and performs much better than direct transmission to BS (Fig. 10.8b).

In principle, clustering has been suggested to limit the number of transmissions to BS and conserve energy. Minimum transmission energy (MTE) scheme selects path [3] between SNs and BS so as to minimize energy consumption in transmission and reception by the transceivers. In LEACH (low energy adaptive clustering hierarchy) protocol [3], clusters are formed and CH is selected and changed dynamically based on residual energy of SNs. The advantage is that SNs within a cluster and from neighboring area need to communicate only with the CH which can aggregate received similar data and send a single packet to the BS/sink node. LEACH works in rounds, each with short setup and long steady state. In this proactive protocol, the entire WSN is divided into a number of clusters; all SNs transmit only to their immediate CH (Fig. 10.9b) and CHs at increasing levels in the hierarchy need to transmit data over relatively longer distances. The set-up phase is subdivided into self-advertising as a CH, SNs opting for a cluster, and TDMA schedule assigned by the CH to each SN. In steady state, TDMA is used for data transmission as everyone uses the same channel. CHs use different CDMA codes to communicate with BS/sink node and the hierarchy can be extended to hierarchical clustering. The algorithm is totally distributed, and any SN having larger residual energy can be CH.

LEACH, being proactive, transmits data periodically and any event detected by a SN has to wait for allocated time slot and in many applications that may not be acceptable. TEEN (threshold-sensitive energy-efficient sensor network) has been introduced [4] as a reactive protocol which uses LEACH-based clustering and SNs dynamic reconfiguration capability that suits for time-critical applications. In TEEN , a SN transmits only if sensed value is greater than a hard threshold (H_T),

Table 10.2 Classification of different routing protocols

Broadcast	Unicast			Multi-cast
	Proactive	Reactive	Hybrid	
– Direct	– Pegagus – Leach	–TEEN – Minimum cost forwarding – Rumor routing – Random walk	– APTEEN – Two-tier data dissemination – Directed diffusion	– SPIN

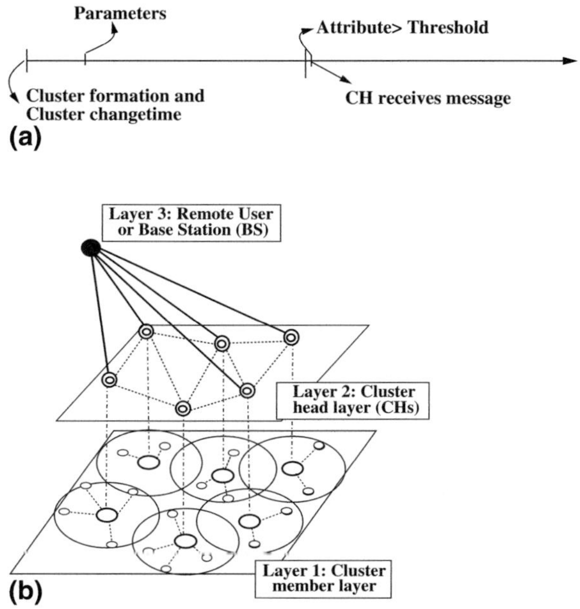

Fig. 10.9 **a** Periodic cluster formation, **b** clustering of SNs in a WSN

and it transmits data if the sensed value differs from last transmitted value by more than a soft threshold (S_T). An important feature of this scheme is that time-critical data reaches the BS almost instantaneously. As message transmission consumes much more energy than data sensing, energy consumption can be controlled by changing the threshold values and it offers flexibility by allowing the user to set the threshold values for the attributes (Fig. 10.10a). Attributes can be changed as every cluster change. The only drawback of this scheme is that if threshold is not reached, then the user never gets to know about the SNs. But, it is useful for time-critical environment-like intrusion detection, etc.

If the threshold is never crossed, TEEN will not provide any value and one would start wondering whether SNs are working correctly or not. To address this shortcoming, APTEEN (adaptive periodic threshold-sensitive energy-efficient sensor network protocol) [5] was introduced that combines features of reactive and proactive networks by having reactive element as it is and drastically increasing the sampling period for proactive component that could indicate health of SNs. It offers a lot of flexibility by allowing the user to set the time interval and the threshold values for the attributes. Similar to TEEN, the energy consumption can be controlled by changing periodic interval as well as the threshold values, and the hybrid network can emulate a proactive network or a reactive network, based on the application, by suitably setting the periodic interval and the threshold values as it combines both proactive and reactive policies (Fig. 10.10b).

Sensor protocol for information via negotiation (SPIN) [6] protocol allows network-wide broadcast by limiting negotiation and using local communication. Blind flooding causes excessive resources consumption, and this problem is limited

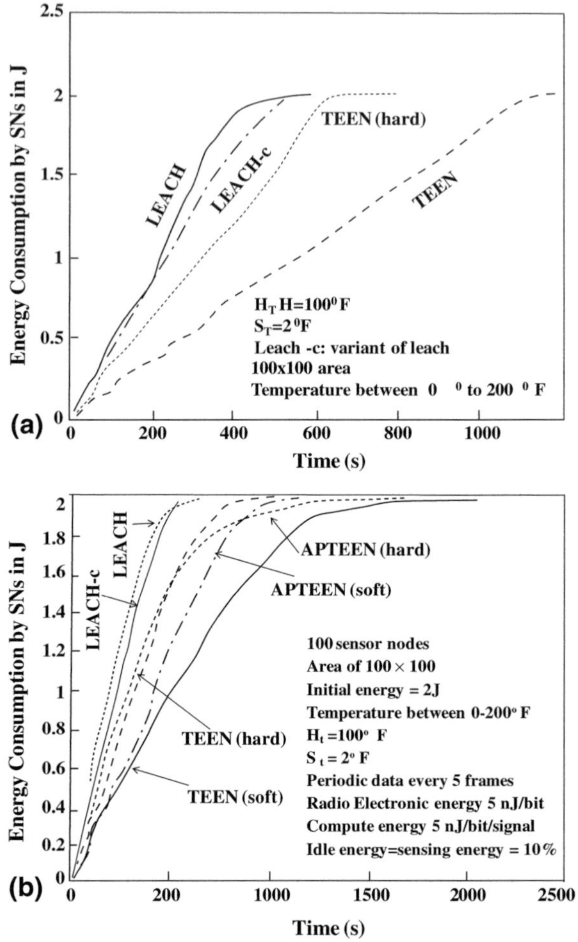

Fig. 10.10 **a** Energy consumption by TEEN, **b** comparison of energy consumed by APTEEN

by getting localized information and detecting overlapping regions as the same data available from many neighbors. Broadcast is limited in a WSN with *n*SNs, and every SN handles O(*n*) messages. Two versions of protocols SPIN-PP (point to point) and SPIN-BC (local broadcast communication) [7, 8] require data to be updated throughout WSN. In SPIN-PP, SN with data advertises (ADV) to all its neighbors (Fig. 10.11a) and receiving SNs requests for data. The receiving SN follows the same sequence by advertising the presence of data to its neighbors. This process continues around the network, and SNs may aggregate their data to ADV. In a lossy network, ADV may be repeated periodically and REQ if not answered. SPIN-BC is a variant that supports broadcast operation in a WSN so as to reach all neighbors.

Random walk scheme depends on finding a random walk over a grid with little state information, and SNs are assumed to be located at cubic grid junctions. With

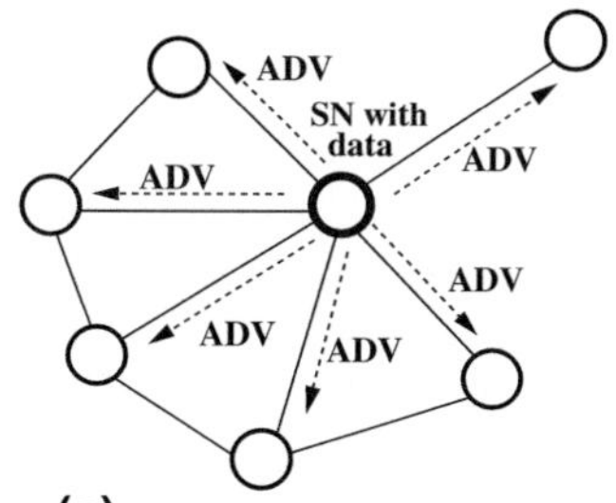

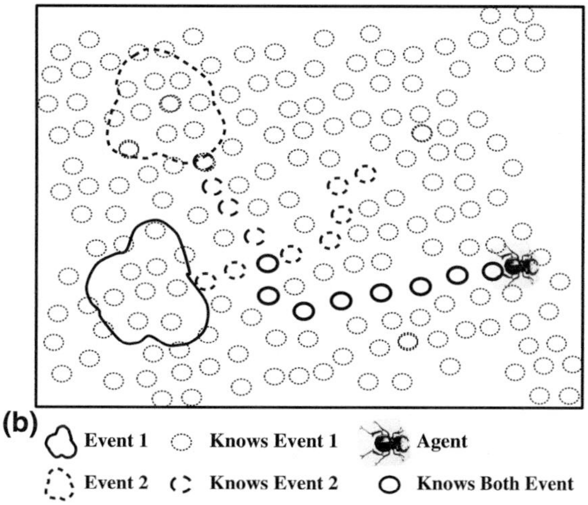

Fig. 10.11 **a** ADV message sent to neighbors by SN having data in SPIN-PP, **b** rumor routing in a WSN

localized communication, coordinates differences (D_x, D_y) using Distributed Bellman Ford using local communication. Every SN computes the probability of moving on *X* and *Y* directions and in each move to an adjacent SN on *X* or *Y* using computed probability. By carefully adjusting the probability, load balancing can be achieved by selecting diverse routes at different time. If one adjacent SN is missing, then go to the other SN.

If both are missing, then go to a neighbor whose breadth-first distance to the destination is strictly smaller than the current node. In rumor routing (Fig. 10.11b), the network follows a dynamic graph [9] as SNs may sleep and wake. For many applications, any arbitrary path will be able to do the job and there is no need for the shortest path as SNs are densely distributed. The scheme is specifically attractive when the ratio of events with queries is within a threshold where it is not attractive to flood the network. Optimal parameters heavily depend on topology as it does not guarantee delivery. The movement on the network is done by several agents, randomly trying to walk straight as every SN maintains a list of neighbors and events. An agent coming from an event is updating SNs it visits.

10.4 MAC Challenges for a WSN

Traditionally, there are many challenges in a WSN at the medium access control level, including fairness, latency, and throughput. For a WSN, both power efficiency and scalability are important as WSNs are designed to operate unattended for long time as it is rather impractical to replenish the batteries. However, SNs are in idle state for most time when no transmission of sensed data occurs.

Measurements have shown that a typical radio transmitter of a SN consumes a similar level of energy in idle mode as in receiving mode. Therefore, it is important that SNs are able to operate in low duty cycles as idle listening consumes significant energy and it is better to have periodic listening and turn off the radio when sleeping most of the time. For example, duty cycle can be reduced to $\sim 10\%$ (200 ms on and 2 s off). Sleep/awake cycle can be used in a randomly deployed WSN and only one set is active at a time among 3 sets (Fig. 10.12a). This is done by organizing available SNs into mutually exclusive sets and they periodically listen and sleep rest of the time. Both *transmitter and receiver need to be awake simultaneously to have data transfer. It is preferable for the* neighboring SNs to have the same schedule for easy broadcast and maintaining low control overhead.

As both *transmitter and receiver SNs need to be awake simultaneously to have data transfer, it is better if* neighboring SNs have the same schedule. Therefore, it is not only required to synchronize SNs but also desirable to remember neighbors' schedule to help when to send data to them (Fig. 10.12b). Each SN broadcasts its schedule few periods of its sleeping and listening cycle so that adjacent SNs are resynchronized with receiving schedule update. The scheduled packets also serve as beacons for new SNs to join neighborhood. A SN wakes up, indicating high signal level while the channel is free and it simply wastes energy contributing as false positive. SN fails to wake up indicating low signal level while the channel is busy and could miss important data and is termed as negative false wakeup. An important problem is how RSSI (receive signal strength indicator) varies and how can RSSI be used to detect presence of a signal. The real question is what is the right wake-up threshold value. In a WSN, BS sends command and collects data while SN side can send or receive tone, and very little work has been done on physical layer protocols for WSNs. Most of the existing MAC works make SNs sleep as long as possible and include at least some aspects of TDMA. The channel could be allocated either as static or dynamic. In static channel allocation with N SNs, available bandwidth is divided into N equal portions and used in FDMA,

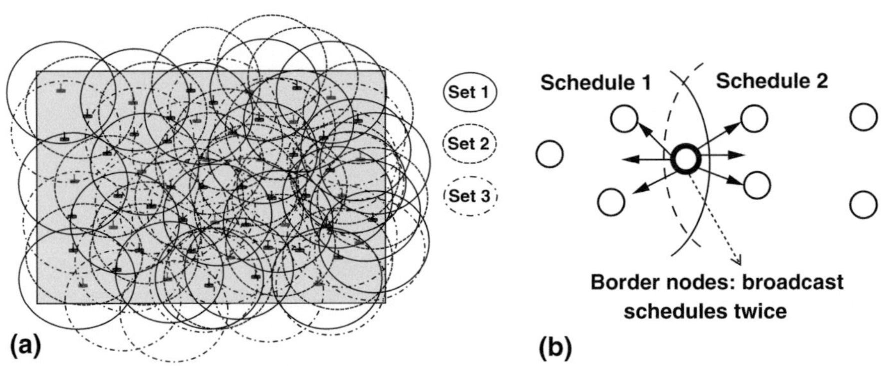

Fig. 10.12 **a** Three set of SNs covering the area for sleep–awake cycle in a WSN, **b** sleep–awake cycle between 2 groups in a WSN

TDMA, CDMA, SDMA or OFDM or ultra-wide band. In dynamic channel allocation, there is no fixed assignment of bandwidth as the number of active SNs changes dynamically, and contention-based allocation scheme is followed. In a WSN with clusters, the number of SNs per cluster is kept small, encouraging the use of static channel allocation. TDMA is suitable for either proactive or reactive type of WSN. In reactive networks, a sudden change could be noticed immediately by many SNs and that could possibly lead to collisions, and it is better to employ TDMA. The use of CDMA by CHs avoids collisions due to the use of the same channel by different clusters and use of TDMA/CDMA avoids intra/inter cluster collisions. When SN fails, MAC and routing protocols must accommodate the formation of new links and routes to other SNs and the BS. The traditional MAC protocols have negative impact in terms of collisions, overhearing, control packets overhead, and idle cycle time. The issue is if an entire message is process in-network in a WSN, it requires *entire* message. So it is advisable not to interleave different messages, and rather a long message is fragmented and sent in burst as RTS/CTS mechanism can reserve medium for entire message and fragment-level error recovery is possible by extending transmitting time and re-transmit immediately if error. Other SNs sleep for whole message time. Overhearing can be avoided when SNs r*eceive packets destined to other SNs. This can be done if SNs sleep when neighbors talk.* So, all immediate neighbors of sender and receiver SNs should sleep *as the* interval *informed by* each packet to other SNs to remain in the sleep mode.

10.5 S-MAC Protocols with Sleep–Awake Cycles

SNs periodically sleep in a WSN and energy efficiency is traded for lower throughput and higher latency as SN has to sleep during other SNs transmissions. The sensor-MAC (S-MAC) [9] protocol explores design trade-offs for energy conservation in the MAC layer. It reduces the radio energy consumption from

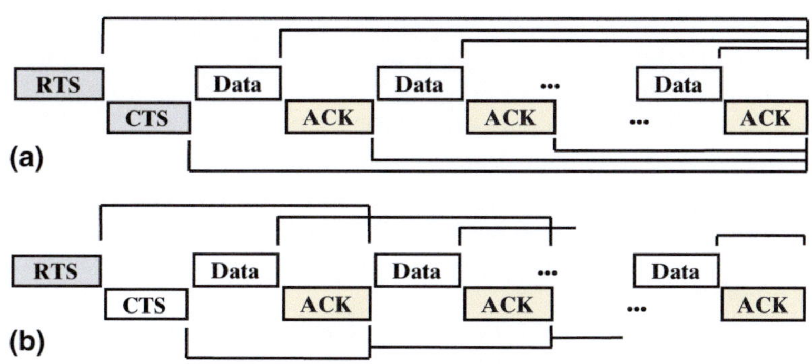

Fig. 10.13 **a** S-MAC message passing, **b** fragmentation in fixed data size

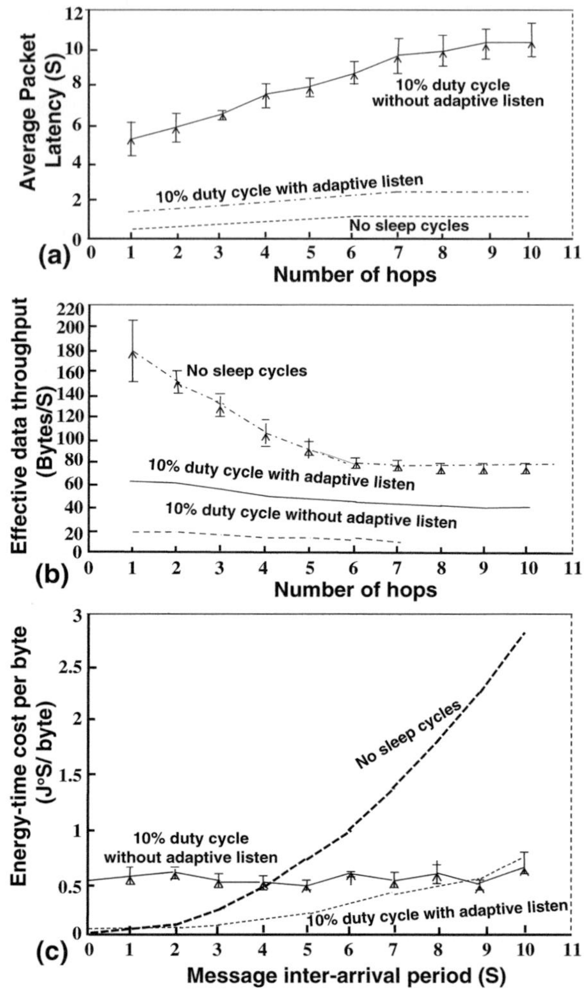

Fig. 10.14 **a** Average packet latency in S-MAC, **b** throughput in S-MAC, **c** energy savings by S-MAC

collision, control overhead, overhearing unnecessary traffic, and idle listening. The basic scheme of S-MAC is to put all SNs into a low-duty-cycle mode of listen and sleep periodically. When SNs are listening, they follow a contention rule to access the medium, which is similar to the IEEE 802.11 DCF. The major components in S-MAC are periodic listen and sleep, collision avoidance, overhearing avoidance, message passing, and synchronize every 10 s. In S-MAC, SNs exchange and coordinate on their sleep schedules rather than randomly sleep on their own and before each SN starts the periodic sleep, it needs to choose a schedule and broadcast it to its neighbors.

To prevent long-term clock drift, each SN periodically broadcasts its schedule as the SYNC packet. To reduce control overhead and to simplify broadcasting, S-MAC encourages neighboring SNs to choose the same schedule, but it is not a

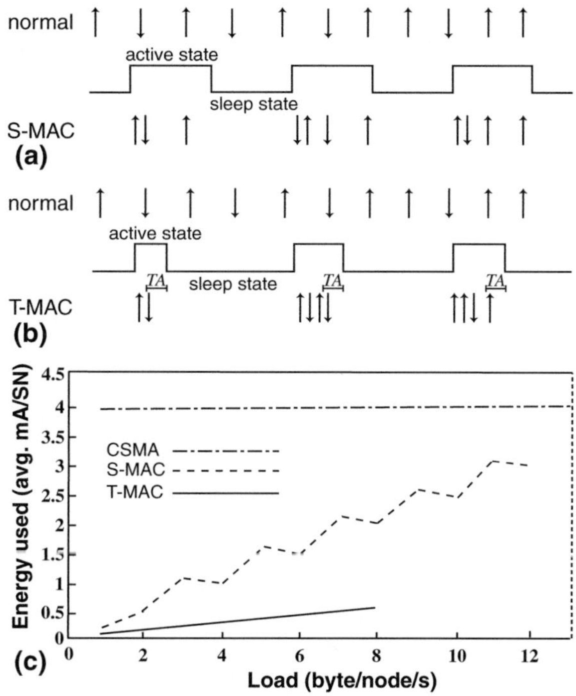

Fig. 10.15 **a** S-MAC, **b** T-MAC, **c** energy saving in T-MAC

requirement. A SN first listens for a fixed amount of time, which is at least the period for sending a SYNC packet. If it receives a SYNC packet from any neighbor, it will follow that schedule by setting its own schedule to be the same (Fig. 10.13a). Otherwise, SN will select an independent schedule after the initial listening period. Such listening period considerably longer than the clock drift. The neighboring SNs exchange SYNC messages while exchanged timestamps are relative rather than absolute. Utilization of RTS/CTS avoids hidden terminal and contains expected duration of message (Fig. 10.13b). Every packet is acknowledged, and adaptive listening can be used so that potential next hop SNs wake up in time for possible transmissions. Performance of S-MAC is given in Fig. 10.14.

T-MAC (timeout MAC) is a RTS/CTS/ACK-based scheme and transmits all messages in bursts of variable length and sleep between bursts. The synchronization is similar to S-MAC (Fig. 10.15a, b). T-MAC is observed by simulation to save energy compared to S-MAC (Fig. 10.15c), while the "early sleeping problem" limits the maximum throughput.

The objective behind B-MAC (Berkeley MAC) [10] is to have low power operation, effective collision avoidance, simple implementation (small code), to be efficient at both low and high data rates, reconfigurable by upper layers, tolerant to changes in the network, and scalable to a large number of nodes. In B-MAC, RTS/CTS, ACKs, etc. are considered higher layer functionality and low power listening (LPL) using preamble sampling (periodic channel sampling) and hidden

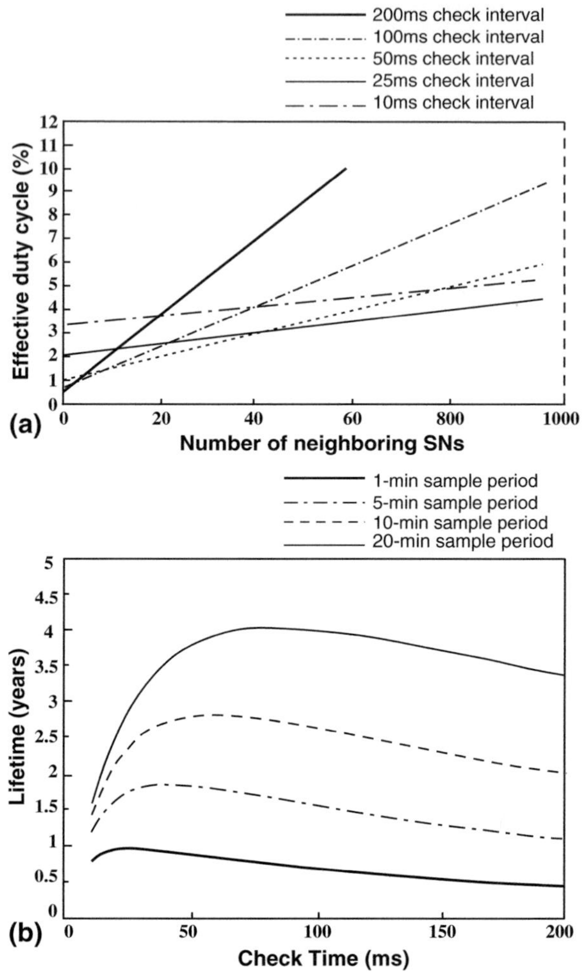

Fig. 10.16 **a** Duty cycle of B-MAC. **b** WSN lifetime in years [10]

terminal and multi-packet mechanisms not provided, should be implemented, and can be implemented by higher layers if needed. Initial and congestion back-off is considered on a per packet basis. Clear Channel Assessment (CCA) could be on or off. The protocol establishes noise floor and then a transmitting SN starts monitoring RSSI of the channel. If a sample has considerably lower energy than the noise floor during the sampling period, then the channel is free and no one is using it. The goal is to minimize listen cost (Fig. 10.16a). The SN periodically wakes up, turns radio on, and checks channel—if energy is detected, node powers up in order to receive the packet. Then, the SN goes back to sleep. If a packet is received after a timeout, then the preamble length matches channel checking period. No explicit synchronization is needed. The energy consumption can be minimized by varying check time/preamble if constants, sample rate, and neighboring nodes are known

(Fig. 10.16b). B-MAC performs better than the other studied protocols in most cases.

P-MAC (Pattern MAC) [11] is based on scheduling received slots as every SN announces its sleeping pattern for the next frame; n-sleep/1 awake cycle. The transmission time is equal to contention window + RTS + CTS + DATA + ACK time and simulated results are observed to be better than SMAC. This protocol is good for relatively stable traffic conditions as adaptation to changes in traffic might be slow. Loose synchronization of time is required.

10.6 Conclusions

Self-organization is a must in any WSN. Address of a SN is not critical as compared to the location of SN where an event has occurred. Routing from SNs to BS is needed, and load balancing in energy consumption of a WSN is important as it is desirable to maintain the same residual energy among SNs. Sleep cycle is useful for enhancing lifetime of a WSN. Collision avoidance is desirable for Beacon signals as well as simultaneous transfer of data while data aggregation could prove to be very useful. How to merge heterogeneous data is still an open question.

10.7 Questions

Q.10.1. What is the goal of localization in a WSN?

Q.10.2. Why do you need SNs to self-organize in a WSN?

Q.10.3. What do you do in self-organization?

Q.10.4. Three beacons are located at $a = (1, 1)$, $b = (1, -1)$, and $c = (-1, 1)$. The received power from nodes a, b, and c is 1.2, 1.5, and 1.7, respectively. Calculate the unknown position of the receiver through a weighted centroid computation.

Q.10.5. What is meant by 1-hop and 2-hops neighbors?

Q.10.6. What is the use of piggybacking 1-hop neighbor information?

Q.10.7. Why do you need routing from SN to BS?

Q.10.8. Do you need routing of data from SN to another SN? Justify your answer.

Q.10.9. What is the role of sleep–awake cycle in life-time of a WSN?

Q.10.10. A WSN is given in the following figure; SN_1 needs to send data to the BS. In the second of routing, a path can be selected either passing through SN_4, SN_5, or SN_6. Under what condition, a particular path will be followed?

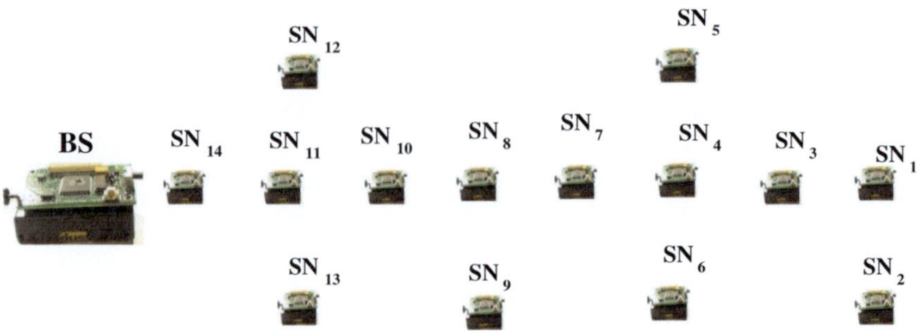

Q.10.11. Explain the following terms:

(a) Sleep–awake cycle
(b) Data aggregation
(c) One-hop neighbors
(d) Anchor nodes
(e) Directed diffusion
(f) Synchronization
(g) Intrusion detection using WSN

References

1. Chalermek Intanagonwiwat, Ramesh Govindan, and Deborah Estrin, "Directed Diffusion: A Scalable and Robust Communication Paradigm for Sensor Networks," www.isi.edu/div7/publication_files/directed_diffusion_scalable.pdf.
2. Stephanie Lindsey and Cauligi S. Raghavendra, "PEGASIS: Power-Efficient Gathering in Sensor Information Systems," http://ceng.usc.edu/~raghu/pegasisrev.pdf.
3. Wendi Rabiner Heinzelman, Anantha Chandrakasan, and Hari Balakrishnan, "Energy-Efficient Communication Protocol for Wireless Microsensor Networks," Proceedings of the Hawaii International Conference on System Sciences, Maui, Hawaii, January 4–7, 2000.
4. Arati Manjeshwar and Dharma P. Agrawal, "TEEN: A Routing Protocol for Enhanced Efficiency in Wireless Sensor Networks," Parallel and Distributed Processing Symposium, International. Vol. 3. IEEE Computer Society, 2001.
5. Arati Manjeshwar, Qing-An Zeng, and Dharma P. Agrawal, "An Analytical Model for Information Retrieval in Wireless Sensor Networks Using Enhanced APTEEN Protocol," IEEE Transactions on Parallel and Distributed Systems, vol. 13, no. 12, pp. 1290–1302, DECEMBER 2002.
6. Jaewan Seo, Moonseong Kim, Sang-Hun Cho, and Hyunseung Choo, "An Energy and Distance Aware Data Dissemination Protocol Based on SPIN in Wireless Sensor Networks," ICCSA 2008, Part I, LNCS 5072, pp. 928–937, 2008.
7. Luis Javier García Villalba, Ana Lucila Sandoval Orozco, Alicia Triviño Cabrera and Cláudia Jacy Barenco Abbas, "Routing Protocols in Wireless Sensor Networks," *Sensors*, vol. *9, no.* 11, pp. 8399–8421, 2009.

8. Sergio D. Servetto and Guillermo Barrenechea, "Constrained Random Walks on Random Graphs: Routing Algorithms for Large Scale Wireless Sensor Networks," Proceedings of the 1st ACM international workshop on Wireless sensor networks and applications, WSNA'02, pp. 12–21, September 28, 2002.
9. Wei Ye, John Heidemann, Deborah Estrin, "An Energy-Efficient MAC Protocol for Wireless Sensor Networks," INFOCOM 2002.
10. Joseph Polastre, Jason Hill, and David Culler, "Versatile low power media access for wireless sensor networks," Proceedings of the 2nd international conference on Embedded networked sensor systems, pp. 95–107, SenSys'04.
11. Tao Zheng, S. Radhakrishnan, and V. Sarangan, "PMAC: an adaptive energy-efficient MAC protocol for wireless sensor networks," Proceedings 19th IEEE International Parallel and Distributed Processing Symposium, 4–8 April, 2005.

Chapter 11
Broadcasting, Data Aggregation, and Opportunistic Forwarding

11.1 Introduction

A large number of SNs are randomly deployed in a WSN, and the question that comes is how each SN route data to BS? In order to conserve energy, usually multi-hop solution is adopted. But, how is the path determined in a WSN? There is no other choice but to broadcast request for route to BS. Broadcasting is a common operation in many applications, e.g., graph-related and distributed computing problems, and is also widely used to resolve many network layer problems.

11.2 Broadcasting

The broadcast in a WSN is spontaneous as little or no local information about neighboring nodes or location of BS may be collected in advance. The broadcast is frequently unreliable as you shoot in the dark, and a broadcast message is distributed to as many neighboring SNs as possible without putting too much effort. A SN may miss a broadcast message because it is off-line, or it is temporarily isolated from the network, or it experiences repetitive collisions. Broadcast acknowledgements may cause serious medium contention. In many applications, 100% reliable broadcast is unnecessary. Moreover, SN can easily detect duplicate broadcast messages. If flooding is used blindly, many redundant messages will be sent and serious contention/collision will be incurred in a WSN. When a SN decides to rebroadcast, all its neighbors may already have the message and this is called redundant rebroadcasts. Transmissions from neighbors may severely contend with each other, and due to the absence of any collision detection mechanism, collisions are more likely to occur and cause more damage. For example, in WSN of Fig. 11.1a, transmission and blind retransmission to look for BS can be defined as follows: Step 1, SN 1; Step 2, SNs 2, 6, 7, and 9; Step 3, SNs 4, 3, 8, and 10; and

D.P. Agrawal, *Embedded Sensor Systems*,
DOI 10.1007/978-981-10-3038-3_11

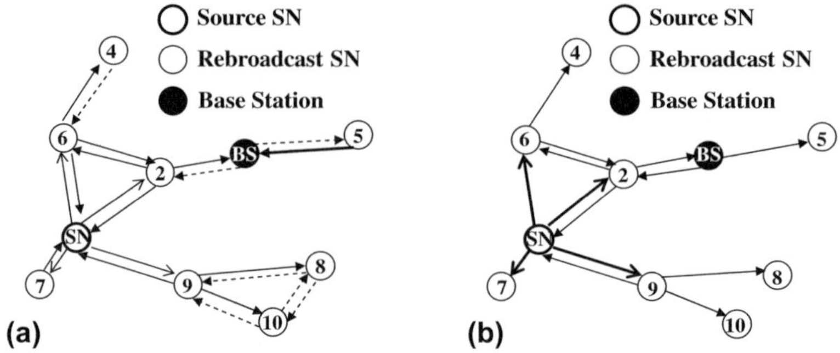

Fig. 11.1 **a** Blind rebroadcast in a WSN with each SN doing this and **b** controlled rebroadcast in a WSN with only 4 rebroadcasts

Step 4, SN 5. This makes many SNs to receive multiple copies of the query, and if controlling is done, this can be done only in 4 rebroadcast steps: Step 1, SN 1; Step 2, SN 2, 6, and 9; and Step 3, SN 3, and is shown in Fig. 11.1b.

Step for rebroadcast in a 13-node WSN is illustrated in Fig. 11.2a. It may be noted that there could be many different collisions present as two close-by SNs rebroadcast to a single neighbor. Many possible collisions are also shown. Figure 11.2b, c shows optimal rebroadcast schemes in two different WSNs. Assume that the total area covered by the radio signal transmitted by a transceiver is a circle of radius r_c. When two SNs are deployed close to each other, then an intersection area of two circles of radio transmission range of r_c whose centers are d apart is defined as INTC(d) (Fig. 11.3a). *An additional coverage* provided by a SN that rebroadcasts the packet is equal to $\pi r_c^2 - \text{INTC}(d)$. When $d = r_c$, the additional coverage is approximately $0.61\pi r_c^2$ and is the maximum value as shown in Fig. 11.3b. On the other hand, the average additional coverage by rebroadcasting from randomly located neighboring SN is $0.41\pi r_c^2$. An *expected additional coverage* provided by a host's rebroadcast after the same broadcast packet received by host k times is denoted by EAC(k) and is given in Fig. 11.3c. Every attempt should be made to minimize the number of retransmissions of a broadcast message, and attempt is made to deliver a broadcast packet to each and every SN in the WSN. Jitter allows one neighbor SN to acquire the channel first, while other neighbor SNs detect that the channel is busy. A random delay time (RDT) allows a SN to keep track of redundant packets received over a short time interval. By keeping track, SN rebroadcasts a given packet no more than one time by caching original source SN ID of the packet and the packet ID.

Broadcasting can be categorized as either simple flooding or probability-based methods. In simple flooding, a source SN broadcasting a packet to all neighbors, the neighbors, upon receiving the broadcast packet, rebroadcast the packet exactly once. Probability-based methods are similar to ordinary flooding except SNs only rebroadcast with a predetermined probability. In dense WSNs, multiple SNs share similar transmission coverage area. In a Counter-Based Scheme, a relationship

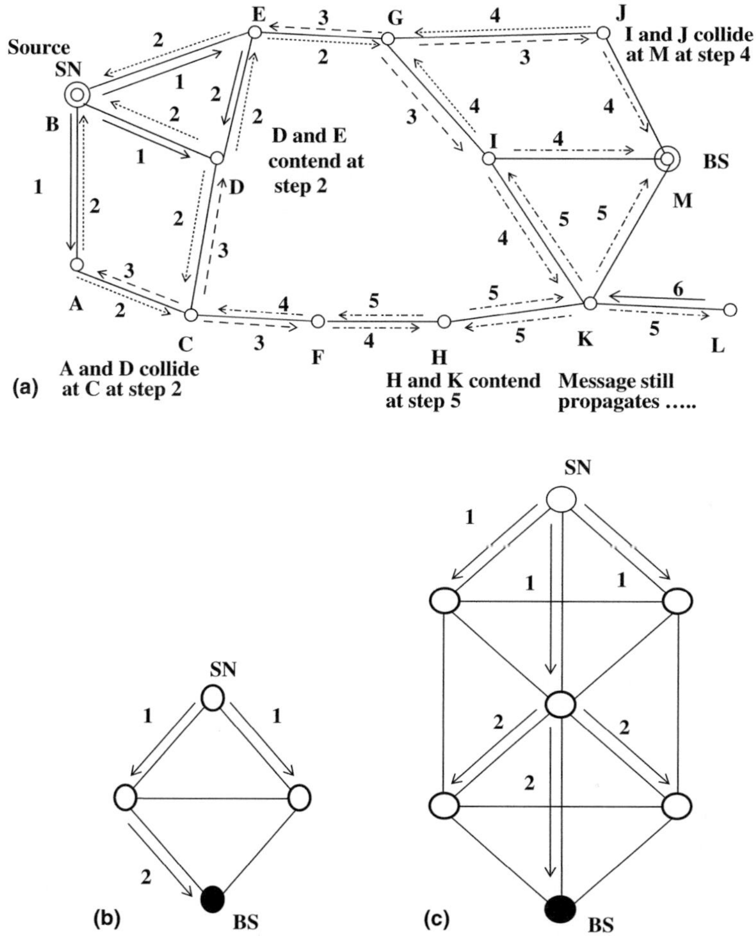

Fig. 11.2 **a** Blind rebroadcast in a WSN and **b**, **c** optimal broadcast steps in 2 WSNs

between the numbers of times a packet is received by the SN and the probability of a SN's transmission to cover additional area on a rebroadcast. One special case is dominating set of SN group is defined such that all remaining SNs are at the most 1-hop away from this group and is called dominating set (Fig. 11.4). Thus, if all members receive at least one copy of the message, then rebroadcasting by dominating set will cover all SNs of the whole WSN. This is true as all remaining SNs are one hop away from a SN of the dominating set. A dominating set of a graph $G = (V, E)$ is a vertex subset $S \subseteq V$, such that every vertex $v \in V$ is either in S or adjacent to a vertex of S and is applicable to any WSN, both randomly deployed and regular WSN. Now, the next question is how the SNs of a dominating set get a

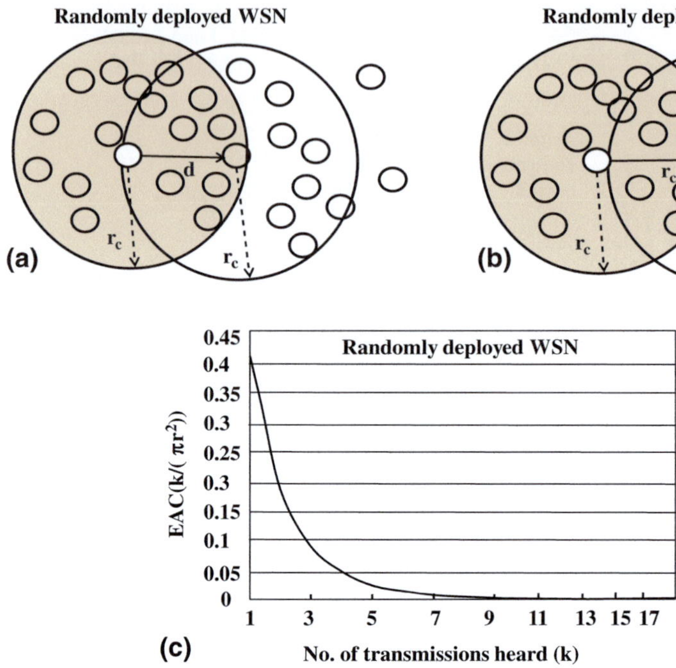

Fig. 11.3 **a** Coverage by second SN in a WSN, **b** maximum possible coverage, and **c** additional area covered as a function of number of transmissions

copy of the query about route to BS. A simple scheme is to determine a connected dominating set (CDS) of a graph G whose induced graph is connected, and this enables any SN to be within communication range of another member SN so that each member gets a copy of query for broadcast to rest of SNs in WSN (Fig. 11.5).

Achieving broadcast in a large WSN is quite involved as shown in Fig. 11.6. A simple observation is that some SNs are highly connected than others, and similarly, some SNs are sparsely connected than others, while some SNs are in between (Fig. 11.7). Determining dominating set is a complex process and needs to have knowledge of global connectivity. A simple scheme has been devised in [1] where local knowledge of 2-hop neighboring SNs is analyzed to determine connectivity of each SN with respect to adjacent SNs (Table 11.1).

If A be the area of the WSN, N be the number of SNs, and r_c be the communication range of each mobile node with a being the fraction of the area covered and a being the fraction of area covered,

$$\alpha = \frac{\pi r_c^2}{A} \tag{11.1}$$

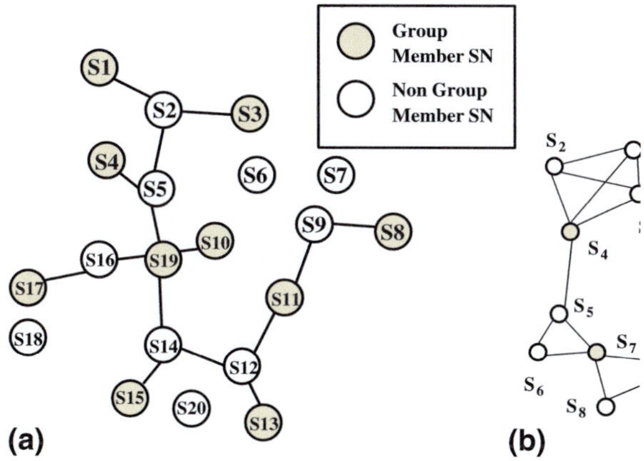

Fig. 11.4 **a** Dominating set group of a WSN and **b** another example

then, the average number of neighboring nodes $N_{\text{neighbors}}$ can be given by:

$$N_{\text{neighbors}} = (N-1)\alpha. \tag{11.2}$$

The probability p_i that a mobile node has i neighbors can be given by:

$$p_i = \binom{N-1}{i}(1-\alpha)^{N-1-i}\alpha^i. \tag{11.3}$$

The probability p_1, p_2, p_3, and p_4 can be given by:

$$P_1 = \sum_{i=1}^{N-1} p_i \binom{N-1}{i}\left(\sum_{j=1}^{i-1} p_j\right)^i\left(1-\sum_{j=1}^{i-1} p_j\right)^{N-1-i} \quad \text{Largest.} \tag{11.4}$$

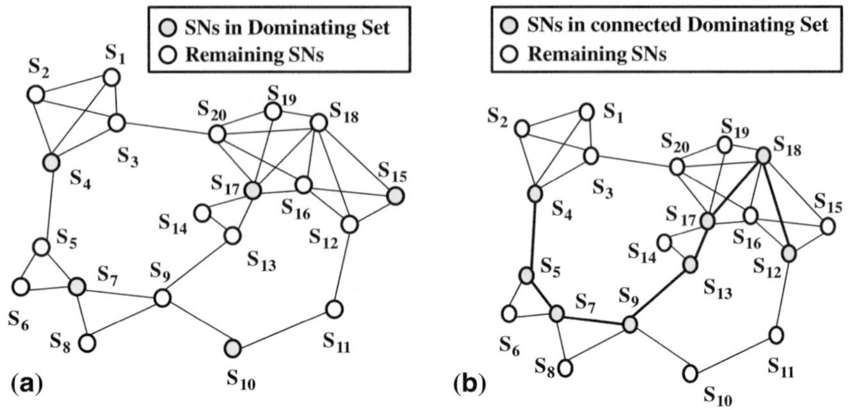

Fig. 11.5 **a** Dominating set group of SNs in a WSN and **b** connected dominating set in a WSN

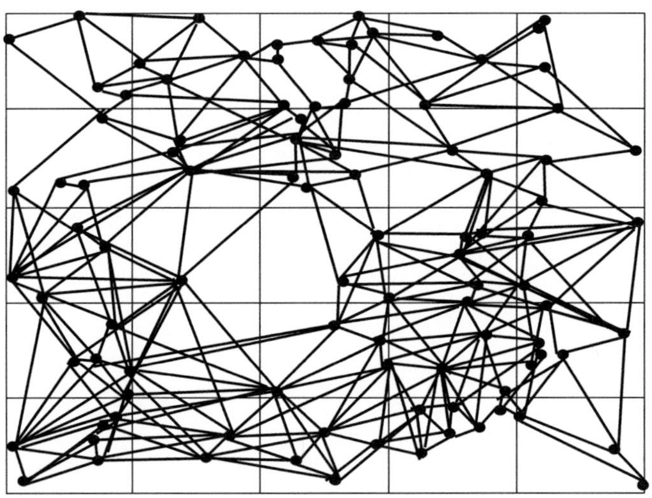

Fig. 11.6 Dominating set of a large WSN

Fig. 11.7 Dominating set and SN connectivity in a WSN

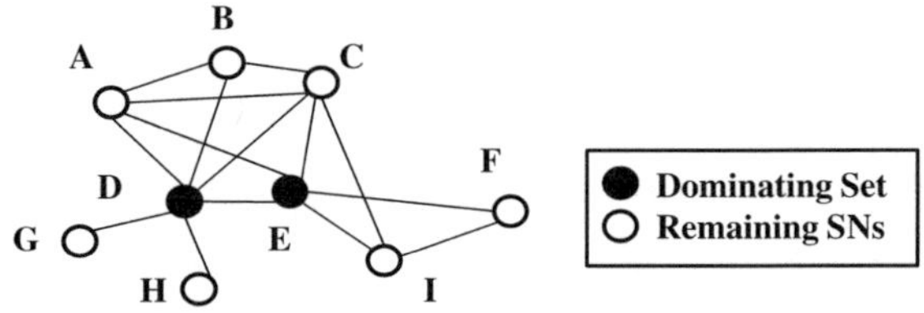

Table 11.1 Connectivity of SNs in a WSN

SN	A	B	C	D	E	F	G	H	I
Degree/# neighbors	4	3	5	6	5	2	1	1	3
Neighbors	B, C, D, E	A, C, D	A, B, D, E, I	A, B, C, E, G, H	A, C, D, F, I	E, I	D	D	C, E, F
Connectivity compared to neighbors	Medium	Low	Medium	High	Medium	Low	Low	Low	Low
Group no.?	2	4	2	1	2	4	3	3	4

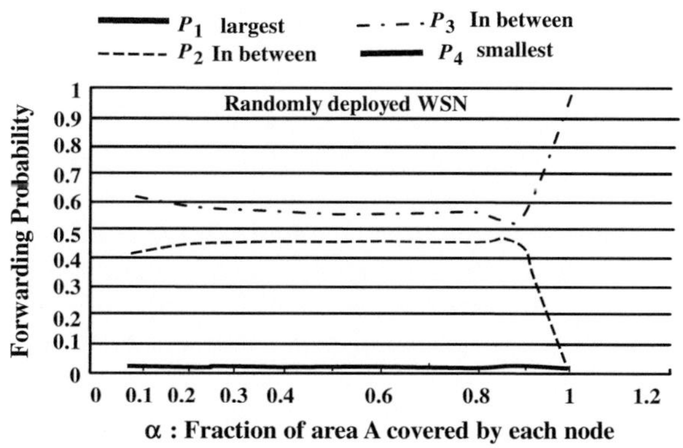

Fig. 11.8 Forwarding probability of SNs

$$P_2 = \sum_{i=1}^{N-1} p_i \left(\sum_{k=1}^{i/2} \left(\binom{i}{k} \left(\sum_{j=1}^{N-1} p_j \right)^k \left(\sum_{j=1}^{i-1} p_j \right)^{i-k} \right) \right) \quad \text{in between.} \tag{11.5}$$

$$P_3 = \sum_{i=1}^{N-1} p_i \left(\sum_{k=i/2}^{i} \left(\binom{i}{k} \left(\sum_{j=1}^{N-1} p_j \right)^k \left(\sum_{j=1}^{i-1} p_j \right)^{i-k} \right) \right) \quad \text{in between.} \tag{11.6}$$

$$P_4 = \sum_{i=1}^{N-1} p_i \binom{N-1}{i} \left(\sum_{j=i+1}^{N-1} p_j \right)^k \left(1 - \sum_{j=i+1}^{N-1} p_j \right)^{N-1-i} \quad \text{smallest.} \tag{11.7}$$

Relations 11.4–11.7 indicate probability that different SNs will rebroadcast the query for locating BS and are mapped in Fig. 11.8. The total number of forwarding steps is given by:

$$N_F = N(\tau_1 P_1 + \tau_2 P_2 + \tau_3 P_3 + \tau_4 P_4), \tag{11.8}$$

where t_i is the forwarding probability for the SN in group i. Even though such a scheme does not provide 100% coverage of all SNs in a WSN, a good coverage and excellent saving are achieved under mobility and about 20% higher goodput is obtained than the conventional AODV protocol [2]. Lightweight and efficient network-wide broadcast relies on two-hop neighbor knowledge obtained from hello packets (beacon signals); however, instead of a SN explicitly choosing other SNs to rebroadcast, the decision is implicit due to connectivity property.

11.3 Lifetime of a WSN

Data is transmitted by SNs to BS in a multi-hop fashion in response to query issued by the BS to all SNs. The types of queries include one-time queries, persistent queries, and historical queries and need to be routed to all SNs from the BS. There is no well-accepted definition of lifetime of a WSN. Some researcher consider when one of the SNs runs out of energy is considered lifetime. A better definition is if 95% of monitored area is covered.

In responding to a query, a shortest path is first determined using broadcasting of needed path from a SN to BS. Such a route is used in forwarding data, and this increases disparity in energy levels of SNs, limiting the life of the network. So, enough energy is consumed by SNs involved in receiving data and forwarding toward BS by retransmission and could run out of energy at a faster rate than remaining part of the WSN (Fig. 11.9).

Better utilization of network resources is feasible if diverse service requirements for different type of data can be taken into account in selecting the route. In a real application, only few real-time data are generated that create alarms or warning, and most of the time, data for periodic monitoring are generated. So, non-critical data can be forwarded from SNs to BS along longer paths, and shorter paths are reserved only for time-critical data. This is illustrated for 2-D mesh-connected SNs in

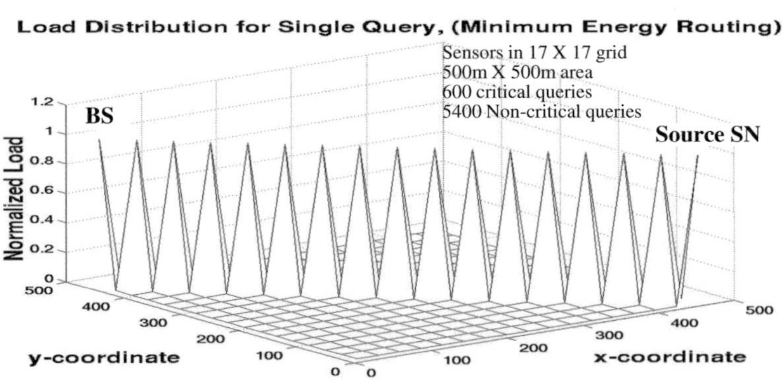

Fig. 11.9 Forwarding along shortest path from a SN to the BS

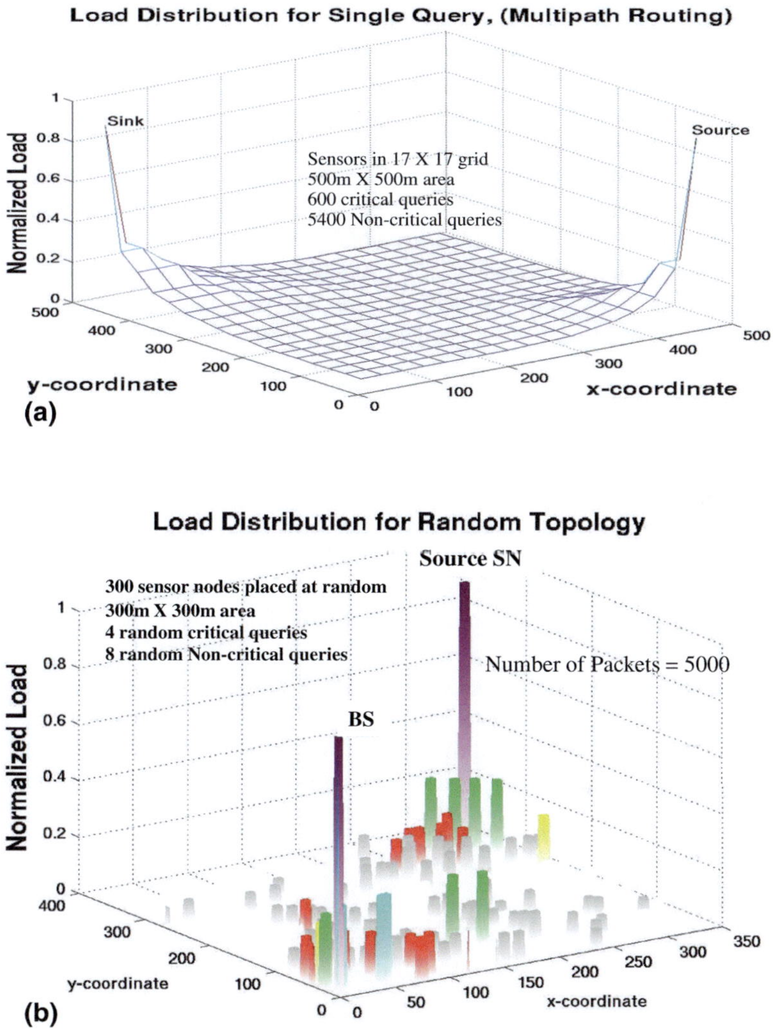

Fig. 11.10 **a** Multi-path routing balances energy consumption in a mesh-connected WSN and **b** multi-path routing balances energy consumption in a randomly deployed WSN

Fig. 11.10, and a quick situational awareness is needed for timely decision making. SNs are yet to become inexpensive to be deployed with any redundancy. A denser deployment of sensor and transmission of sensed data may cause more energy consumption and increased delay due to collisions. However, transmitting data between two far apart SNs may cause increased energy consumption. However, data reduction by aggregation plays a vital role in minimizing the energy consumption.

Data aggregation is needed as individual SN readings are of limited use while delivering large amount of data from all SNs to a central BS consumes lot of energy. Some of the operations that are adopted for aggregation are average, sum, min–max, count, and variance. Such a scheme provides approximate contours of SNs' residual energy assuming Gaussian distribution, and the boundary line between sensing and no-sensing is approximated. This approach conserves limited energy and bandwidth and increases system lifetime. Data aggregation scheme is expected to process data as it flows from SNs to BS (sink node), and there is a trade-off between energy efficiency and data quality. An initial tree can be established using sink as the root and SNs as nodes. Nodes aggregate data from their children and forward result to BS as data are generated periodically (Fig. 11.11a). A tree is established, and BS broadcasts request for data. Nodes send reply to parent, and it is important to discover how many children are there at each level and establish reverse paths to BS.

The nodes set time-outs based on position in the tree, and data wave reaches BS in one period. Optimal data aggregation is known to be NP-hard. In a practical WSN, instead of tree-based approach (e.g., shortest path), it is advantageous to use cluster-based approach as shown in Fig. 11.11b. There is trade-offs between data accuracy versus lossy or loss-free versus delay in computation. In LEACH [3], the role of CH keeps on changing to preserve energy and the network lifetime can be increased further if only one CH sends packet directly to BS. The role of CH can be changed randomly per round based on residual energy. In a distributed kernel regression, even though any node has all kernel regression coefficients, it cannot answer queries involving an entire region. The message size is variable vector corresponding to a square region and BS may need data values for the entire

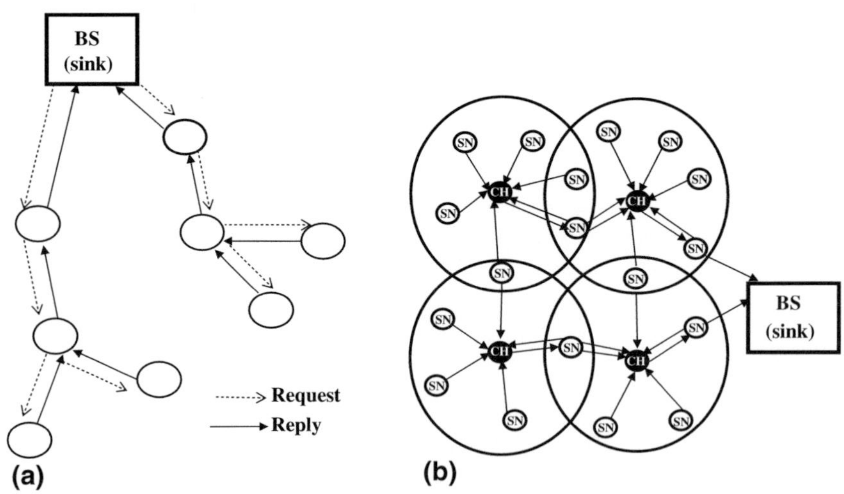

Fig. 11.11 **a** Tree structure with the BS as the root and SNs as sources and **b** tree structure in a WSN with CHs

network. Another approach is rather than each SN forwarding data along selected path to BS in multi-hop fashion, and BS moves around randomly and distributes the query when it is one hop away from a SN. We consider an equivalent reverse problem of a mobile BS or a relay node (RN) pick up data from a SN just one hop away, and we call this random mobility of a BS or RS as shown in Fig. 11.12.

If total m SNs are used in the network of diameter L, BS with communication range l_i, each BS can collect info from N_A SNs as given in Fig. 11.13a. Let N_D be the number of data-centric sensors within area of diameter L, and the question is how many such transmissions are required? For fixed m and L, as l_i is reduced = min(l_i), then asymptotically,

$$\lim_{l_i \to \infty} \frac{N_D}{N_A} = \frac{1}{m}. \tag{11.9}$$

If m and L are fixed and l_i is increased, BS can aggregate their information and send just a single copy, thereby achieving m-fold saving.

PEDAP (Power-efficient Data gathering and Aggregation Protocol) [4] employs *MST (minimum spanning tree)-based routing* using energy as the metric. It optimizes energy only locally, and end-to-end latency is increased. TREG Scheme for data aggregation shown in Fig. 11.13b [5] assumes that NT nodes report their data to the tree node closest to them. Measured attribute varies smoothly in a continuous manner with respect to space. Values (z) stored at each tree node can be considered as function values having $x - y$ inputs and values are stored as tuples (z_m, x_m, y_m), where z_m is the attribute value sensed by a node at location (x_m, y_m) (Fig. 11.13b).

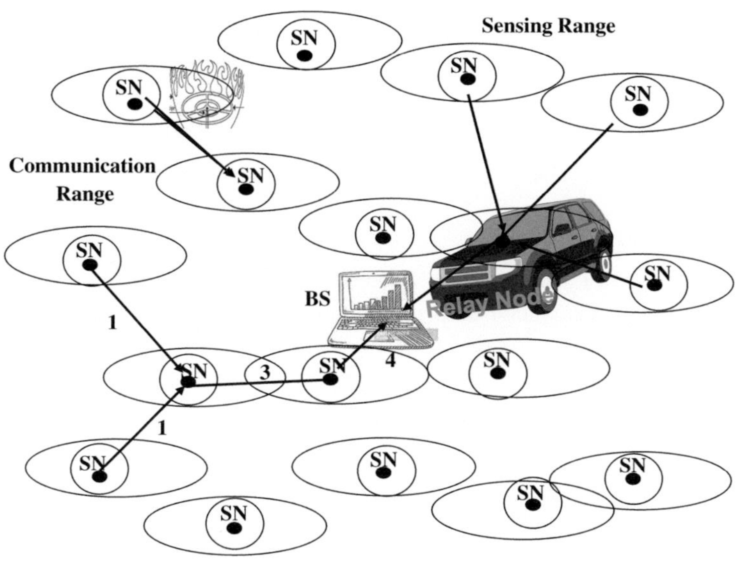

Fig. 11.12 Mobile BS or RN collecting data from SN when one hop away

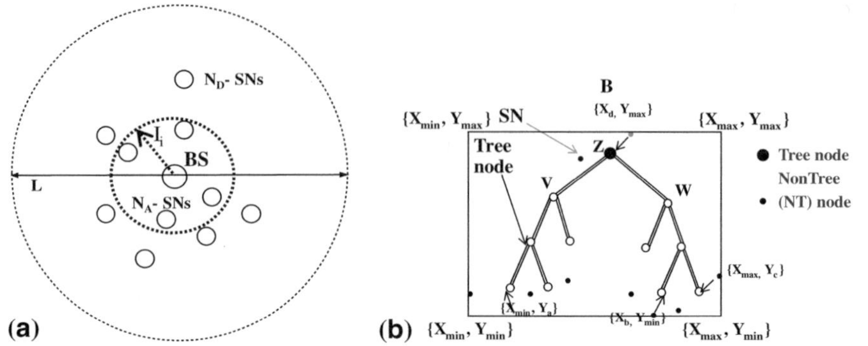

Fig. 11.13 **a** Aggregation with clustering in a WSN and **b** TREG-based aggregation in a WSN

A polynomial equation is generated through function approximation with three input variables (z, x, y) for all the data points in one particular node of the query tree. The goal is to compress redundant data into few β coefficients and transmitted as follows:

$$p(x, y) = \beta_0 + \beta_1 y + \beta_2 y^2 + \beta_3 x + \beta_4 xy + \beta_5 xy^2 + \beta_6 x^2 + \beta_7 x^2 y + \beta_8 x^2 y^2. \tag{11.10}$$

Parent node regenerates values of attribute by using coefficients obtained from their children, and values are recomputed by generating random x's and y's in the range $\{x_{\min}, y_{\min}, x_{\max}, y_{\max}\}$ and substitute them in $p(x, y)$ for temperature distribution on rooftop (Fig. 11.14a). Recomputing data values in entire range helps increase the accuracy of the function approximation process and substituting them in the relation:

$$\begin{aligned} p(x, y) = {} & 26.1429 + 0.0427163y - 0.000167934y^2 + 0.014x \\ & + 0.000249xy - 0.00000009231xy^2 - 0.0000181258x^2 \\ & - 0.000000860054x^2 y + 0.00000000116143x^2 y^2 \end{aligned} \tag{11.11}$$

This leads Fig. 11.14b as values obtained using regression polynomial. Percentage error ε is calculated as the absolute deviation from the true value and percentage error:

$$E = \left(\frac{|z - \bar{z}|}{z} \times 100 \right) \leq \varepsilon_{\mathrm{Th}}, \tag{11.12}$$

where $\varepsilon_{\mathrm{Th}} = 6\%$ is error threshold.

Maximum error of 5.64% is observed with tree of depth 4, while error is mostly in 0–1.68% range. Compression ratio of 50% is obtained at each level of the tree and is independent of depth of aggregation tree. The data field will contain just the 9

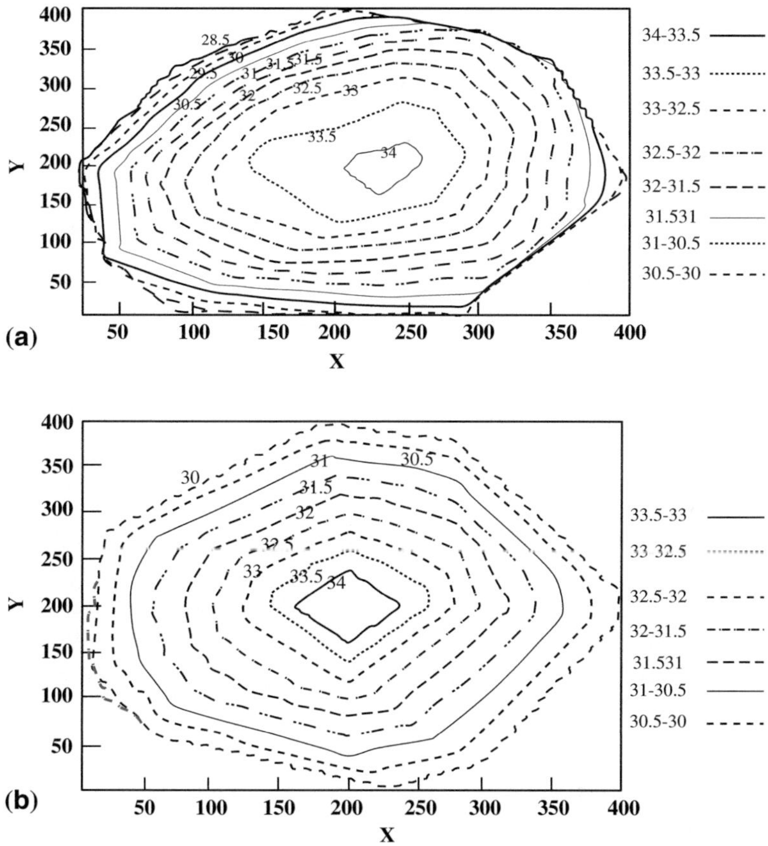

Fig. 11.14 **a** Rooftop temperature distribution using a WSN and **b** temperature distribution using regression polynomial in a WSN

coefficients that lead to fixed data packet size, giving substantial energy savings. A general multi-linear regression model is as follows:

$$z = f(x_1, x_2, \ldots, x_m) = \beta_0 + \sum_{k=1}^{m} \beta_k x_k. \tag{11.13}$$

Squared error needs to be minimized by applying least square criteria given by:

$$F\left(\vec{\beta}\right) = \left(X\vec{\beta} - \vec{z}\right)^{\mathrm{T}}\left(X\vec{\beta} - \vec{z}\right). \tag{11.14}$$

$$X = \begin{pmatrix} 1 & y_1 & y_1^2 & x_1 & x_1y_1 & x_1y_1^2 & x_1^2 & x_1^2y_1 & x_1^2y_1^2 \\ 1 & y_2 & y_2^2 & x_2 & x_2y_2 & x_1y_2^2 & x_2^2 & x_2^2y_2 & x_2^2y_2^2 \\ \vdots & \vdots & \vdots & \vdots & \vdots & \vdots & \vdots & \vdots & \vdots \\ 1 & y_n & y_n^2 & x_n & x_ny_n & x_1y_n^2 & x_n^2 & x_n^2y_n & x_n^2y_n^2 \end{pmatrix} \quad \beta = \begin{pmatrix} \beta_0 \\ \beta_1 \\ \vdots \\ \beta_8 \end{pmatrix} \tag{11.15}$$

$$\vec{\beta} = \left(X^TX\right)^{-1} X^T\vec{z}.$$

For estimating β, a unique inverse of X should exist. Attribute values generated at a particular node is sent to its parent along with $\{x_{\min}, y_{\min}, x_{\max}, y_{\max}\}$, where the minimum and maximum of x's and y's are taken over all the SNs in the subtree under the current parent that report to tree nodes. % error is plotted in Fig. 11.15a for tree of depth 4, while its variation as a function of depth is given in Fig. 11.15b.

A large number of SNs produce data periodically and internal nodes in the data collection tree average data received from downstream nodes and forward the result toward sink. Aggregator *concatenates* multiple data items and transmits as a single packet. So, the question is how long should a node wait to receive data from its children before forwarding data already received? There is a trade-off between data accuracy and freshness. It is quite possible that the coefficients need not be updated every time new data are received. When coefficients are updated every two hours over a 24-h period, % error strictly increases with time due to increased number of approximations and is shown in Fig. 11.16a. The first five updating steps (till 10th hour of observation) has error 5.3% (<6) while approximated maximum and minimum temperature for a 12 h period closely follows the actual data (Fig. 11.16b).

A similar analysis has been done for 3-D mesh-connected WSN (Fig. 11.17a) using the following regression polynomial using 8 coefficients:

$$t = F(x, y, z) = a_0 + a_1z + a_2y + a_3yz + a_4x + a_5xz + a_6xy + a_7xyz. \tag{11.16}$$

The error rate as a function of tree height is given in Fig. 11.17b.

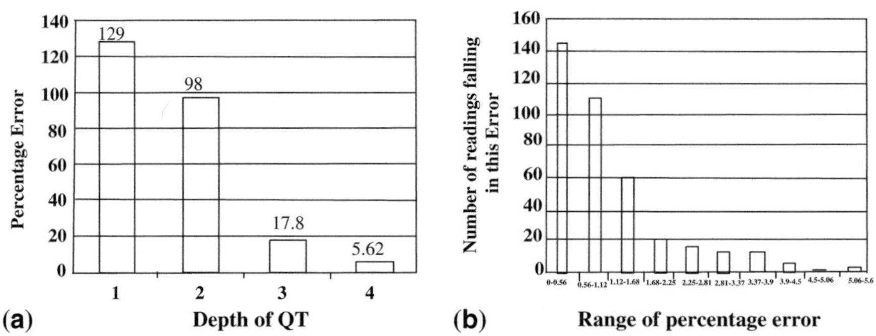

Fig. 11.15 **a** % error with depth 4 using regression polynomial and **b** % error as a function of depth in regression polynomial

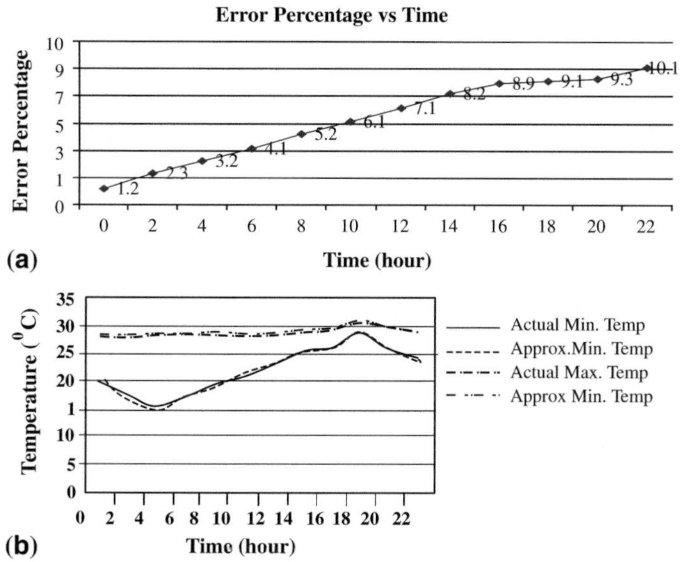

Fig. 11.16 **a** % error when coefficients updated every 2 h and **b** temperature variation when coefficients updated every 2 h

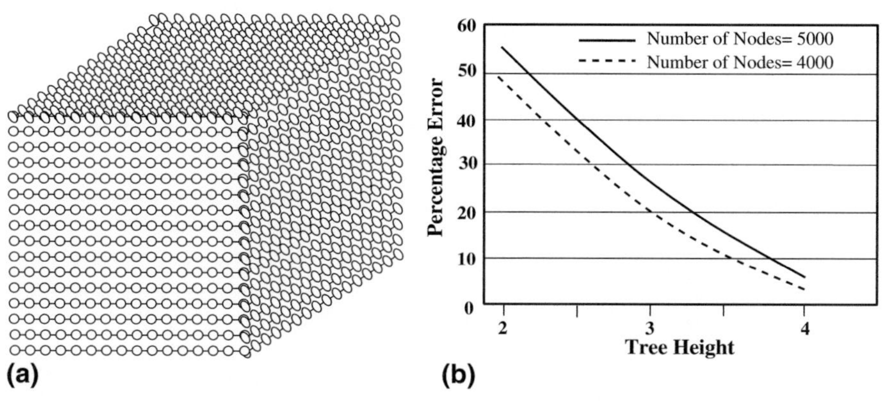

Fig. 11.17 **a** 3-D mesh deployment of SNs and **b** % error in 3-D mesh deployment of SNs

11.4 Query Processing and Data Collection

As discussed earlier, routing from each SN to BS is required to respond to a query from BS and could be application specific and data centric while it is desirable to incorporate data aggregation capability at CHs while minimizing energy consumption. Users can issue declarative queries through BS without having to worry about how the data are generated, processed, and transferred within the network from SNs to BS via CHs, and how SNs/CHs are programmed to satisfy changing user interest to maintain transparency. The BS sends an inquiry packet to CH having data as well as all CHs located between BS and CH to request an update on average of SNs in their clusters. Up on receiving the request from the BS, CHs sends back a reply packet carrying the aggregated value of sensed parameter to the BS. These values then can be used as indexes into table to lookup for the optimal value of value that should be used in response to the query. The BS then sends this value as a query packet to the CH. The WSN should be able to concurrently handle several user query requests through running multiple queries. Given a set of queries Q that have been submitted to the base station, rewrite them into a new query set Q′. The optimal situation is that data requested by queries in Q′ will be just enough to answer queries in Q, and the same data needed for various queries in Q will be acquired only once by queries in Q′. Whenever a query is updated, it is checked against the synthetic query list to see whether it is beneficial to other synthetic queries; if so, the most beneficial pairs are rewritten, and the newly updated synthetic query will be checked against the synthetic query list; this process terminates when there is no further beneficial rewriting (Fig. 11.18a). Long running queries to monitor events occurring in several target regions geographically were separated from each other. Communication architecture supports continuous in-network query processing (Fig. 11.18b). The presence of CH introduces heterogeneity in the network, and few CHs could be used as additional query processor rather than a central storage at the BS. So, the question is which of the CHs could be selected as a query processor in a WSN. In-network query processing could lead to possible reduction in volume of data. One example of query tree is illustrated in Fig. 11.19a. If data tuples of size x is generated by each of n target regions, then data are aggregated at each higher level and are reduced by a factor φ, and then the corresponding data reduction tree is shown in Fig. 11.18b.

Thus, a query is processed in a distributed manner by few CHs acting as aggregator operators. The CHs acting as query operators have additional storage, processing capability, and power to perform effectively. The cost of data transferring data is proportional to distance to be transferred and the cost from L and R to P through X of Fig. 11.20a given by cost function $f(X, Y)$ is given by:

$$f(X,Y) = ||LX||d_l + ||RX||d_r + ||XP||d_p, \tag{11.17}$$

where $||LX||$ is the distance between L and X and d_l is the volume of data from L. Optimal placement the operator X requires minimization of function $f(X)$. A simple nonlinear optimization method of steepest descent can be used to find X such that $f'(X) = 0$. The issue that needs to be addressed is how to translate query tree to energy-aware routing tree. This requires adapting operator placement in a decentralized manner and providing robustness and scalability in a decentralized

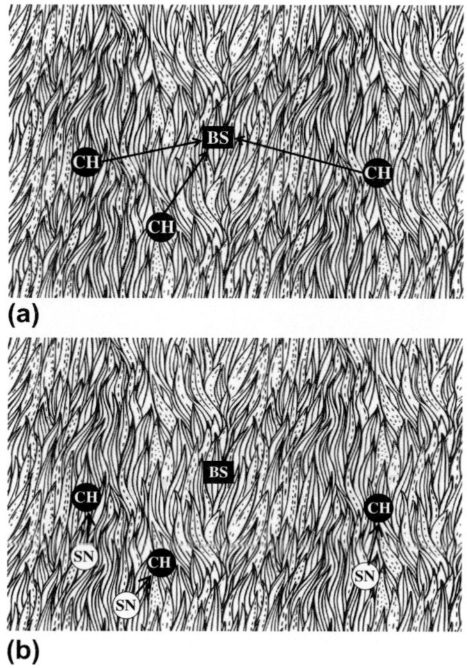

Fig. 11.18 **a** External storage and processing (BS) in a WSN and **b** in-network processing (at CHs) in a WSN

manner so that right set of CHs can be selected as a query operator. Initially, data rates and reduction factor are not known and is determined iteratively using bottom-up starting from leaf CHs and top-down from BS approaches as shown in Fig. 11.20b. After few iterations, results are obtained which are closer to optimal solution as shown in Fig. 11.20b.

11.5 Mobility as an Enabler in WSNs

In an ad hoc network, distributions of node, speed, position, distances, etc., change with time, while in a WSN, the number of SNs are assumed fixed and are static with no mobility. But, in order to minimize energy consumption in SNs, rather than forwarding data to BS in multi-hop fashion, BS itself moves around and collects data from those SNs which are within communication range of BS. This can also be achieved by special nodes called relay nodes (RNs) which move around the field of WSN, collect data from different SNs, and ultimately deliver to the BS. This enables SNs to transmit data that can be collected by RN in one

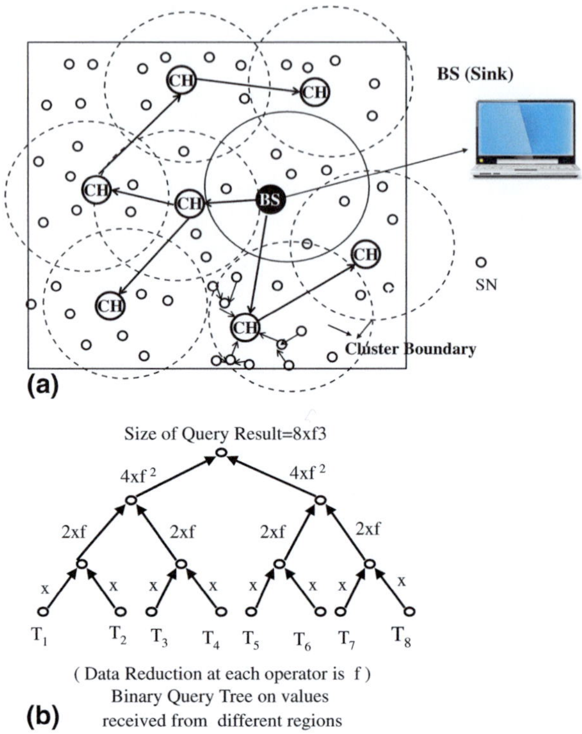

Fig. 11.19 **a** Routing tree corresponding to a query and **b** query tree from different regions in a WSN

hop and conserve energy. This makes the WSN a delay-tolerant system where latency is average time taken for an event to be detected at the BS and total delay E is given by:

$$E(\text{Total Delay}) = E(\text{Forward}) + E(\text{Relay}), \qquad (11.18)$$

where E(Forward) is time taken to forward the sensed data to RNs and E(Relay) is time taken for the RNs to deliver the message to the BS. SN can forward data to RN. If multiple RNs are utilized, a SN can forward data through multiple copies via several RN that helps minimize total delay for data to be delivered at BS. This delay with multiple copies is usually smaller than single copy forwarding. Such a scheme is based on an opportunistic connection for a message forwarding until it reaches the destination (BS). Opportunistic connection also implies that another mobile RN is encountered as RNs are randomly moving and the network does not have any control over the mobility of RNs while the network connectivity to SNs and BS is intermittent. The latency of such networks is such that a message may take very long time to reach the destination.

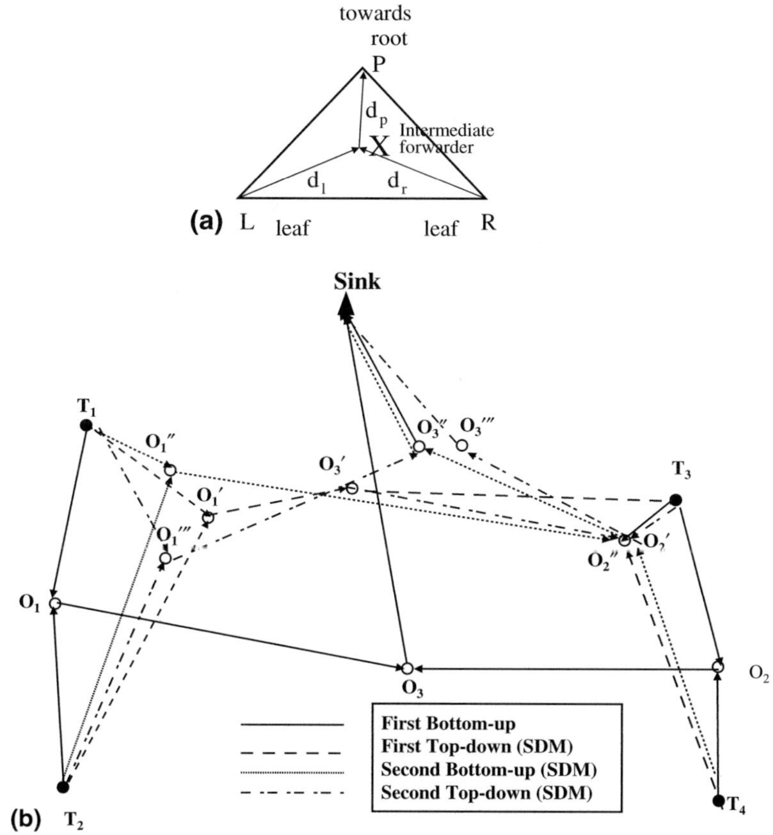

Fig. 11.20 **a** Translating query tree to energy-aware routing in a WSN and **b** top-down and bottom-up iteration to lead close to optimal solution

11.6 Conclusions

It is important to find WSN topology of WSNs, and some way of broadcasting is needed. It is desirable to have an effective broadcasting mechanism of beacon message cs for this purpose. Energy needs to be conserved as SNs are battery powered with limited resources while they still have to perform basic functions such as routing. The method of dominating set and connected dominating set schemes do offer many advantages, but are themselves difficult to implement for a large WSN. Data aggregation allows drastic reduction in volume of data, and such effect ought to be explored as much as possible. In-network processing inside WSN offers many advantages and allows to incorporate effective query processing. Mobility of Relay node makes a WSN delay-tolerant, and its usefulness in conserving energy by opportunistic routing needs to be examined carefully.

11.7 Questions

Q.11.1. What is meant by delay-tolerant WSN?

Q.11.2. What are the important factors in selecting a CH?

Q.11.3. What are the advantages of a dominating set?

Q.11.4. How do you differentiate connected dominating set with a regular dominating set?

Q.11.5. A topology is shown in the following diagram. Can you determine the dominating set for the network? Can you also determine connected dominating set?

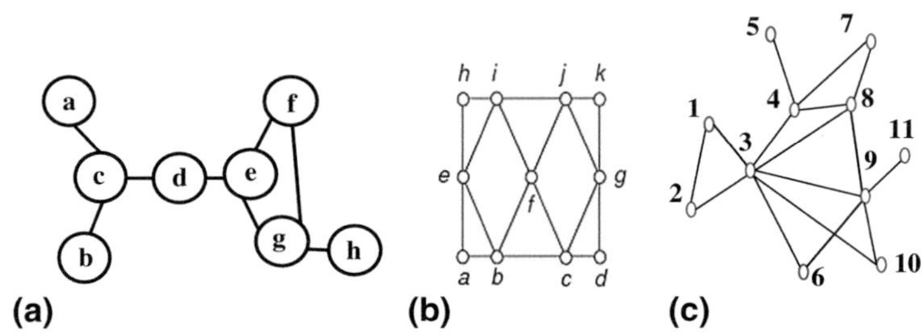

Q.11.6. The following topologies use two or more tiles as base in defining the coverage of a given area. Draw Voronoi diagram for each of them and determine the following:

(a) The number of each type of tile needed in covering an area of size 1000 × 1000.

(b) What is the maximum sensing radius needed for each type of tile?

(c) Draw the Delaunay Triangles and determine the communication requirements.

Q.11.7. What is meant by data aggregation?

Q.11.8. Why so much emphasis is given on data aggregation in a WSN?

Q.11.9. What are the limitations of regression polynomials in representing data of a WSN?

Q.11.10. Can you try other polynomials and determine percentage error for a given application?

References

1. Qi Zhang and Dharma P. Agrawal, "Performance Evaluation of Leveled Probabilistic Broadcasting in MANETs and Wireless Sensor Networks," Special issue of SIMULATION: Transactions of the Society for Modeling and Simulation International. vol. 81, no. 8, Aug. 2005, pp. 533–540.

2. Dharma P. Agrawal and Qing-An Zeng, "Introduction to Wireless and Mobile Systems," 4th edition, textbook published by Brooks/Cole, 650 pages, 2016.
3. Wendi Rabiner Heinzelman, Anantha Chandrakasan, and Hari Balakrishnan, "Energy-Efficient Communication Protocol for Wireless Microsensor Networks," Proceedings of the 33rd Hawaii International Conference on System Sciences – 2000, https://pdos.csail.mit.edu/archive/decouto/papers/heinzelman00.pdf.
4. Huseyin Ozgür Tan and Ibrahim Korpeoğlu, "Power Efficient Data Gathering and Aggregation in Wireless Sensor Networks," http://citeseerx.ist.psu.edu/viewdoc/download?doi=10.1.1.59.7487&rep=rep1&type=pdf.
5. Torsha Banerjee, Kaushik R. Chowdhury and Dharma P. Agrawal, "Using polynomial regression for data representation in wireless sensor networks," International Journal of Communication Systems, vol. 20, no. 7, pages 829–856, July 2007.

Chapter 12
Clustering and Energy Consumption Minimization

12.1 Introduction

A large number of SNs are deployed in a WSN and SNs need to route data to BS. This creates a volume of data at BS and efforts need to be made to reduce the size of data. One effective approach is to explore the use of clustering WSN such that cluster members can send data to selected CH where data are aggregated and a single packet is forwarded by each CH. This leads to drastic reduction in data size.

12.2 Clustering

Topology management protocols for WSNs suffer from at least one of the following problems. It depends on location awareness (e.g., GAF: Geographic Adaptive Fidelity), or converges slowly (i.e., dependent on the network diameter). As energy efficiency is not the main goal of many protocols, e.g., Max–Min clustering maximizes the number of CHs and minimizes distance from each SN to CH and there is no focus on the quality of clustering, such as having CHs well-distributed in the network (e.g., LEACH [1]). For electing CHs, the primary parameter is the residual energy (E_r) with a SN and a secondary parameter being the communication cost that is used to break ties, such that it maximizes energy consumption and minimizes cost. Selecting CH reduces structural complexity in a WSN, and complex/energy consuming activities are delegated to a subset of SNs in the network. It not only reduces routing complexity but reduces wireless interference and preserves network capacity, while maintaining connectivity in a WSN. It also lessens routing complexity, reduces wireless interference, and preserves network capacity.

Besides reducing complexity, topology control approaches used in a WSN involve how to reduce radio power consumption. Most work concentrate on minimizing radio interference and reducing routing complexity while not worrying

D.P. Agrawal, *Embedded Sensor Systems*,
DOI 10.1007/978-981-10-3038-3_12

too much about its capacity. The net effect is routing selectivity is lost and incurs increased topology maintenance overhead. The literature is full of numerous theory/simulation results, and very few experimental results have been explored as algorithms are complicated and underlying assumptions in the algorithm are difficult to realize in practice. Moreover, wireless links usually vary in quality over time as wireless links do not follow binary value (good/bad) in nature and wireless links may be asymmetric. SNs are designed using low-speed CPUs, and it may not be possible to run complex algorithms. So, the overall objective should be to define a clustering approach that is completely distributed, has low message/processing overhead, terminates in O(*l*) iterations with *l* links, creates generates high-energy and well-distributed cluster heads (CHs), and can provide other characteristics such as balanced or dense clusters.

Hybrid, energy-efficient, distributed [2] clustering approach (HEED) (Fig. 12.1a) is proposed that it allows every SN to use information from its 1-hop neighbors only (within cluster range). HEED is distributed and energy-efficient as it

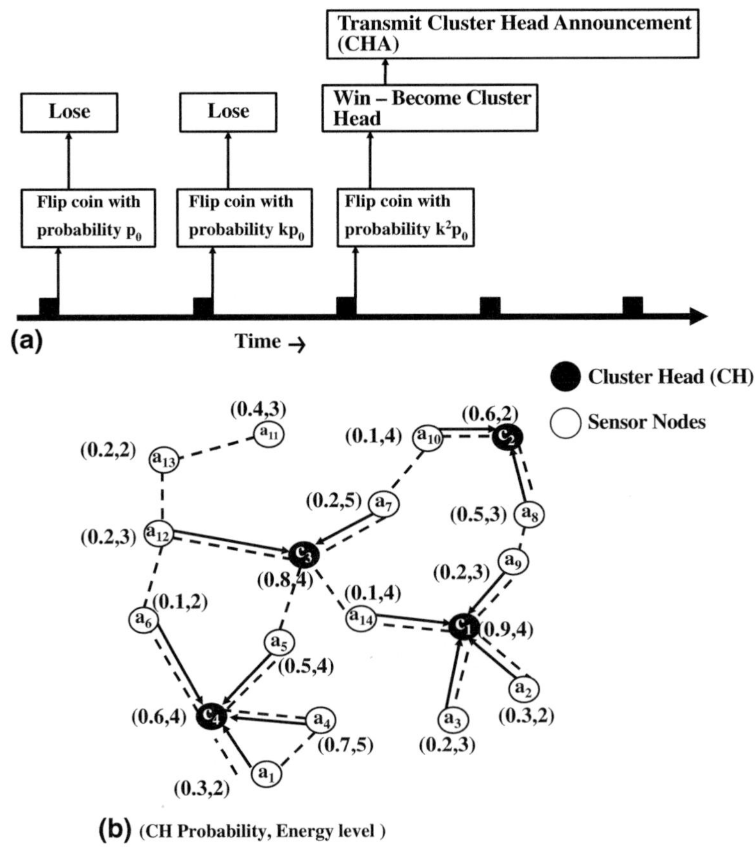

Fig. 12.1 **a** CH election algorithm in HEED, **b** example WSN in electing CH with HEED

elects those SNs as CHs that are rich in residual energy E_r, and reclustering is used to result in distributing energy consumption, with E_{max} being the maximum energy of each SN. The algorithm is implemented in three steps of initialization, main processing, and finalization. In initialization step, neighbors are discovered within the cluster range and initial cluster head probability $CH_{prob} = f(E_r/E_{max})$ is computed as the ratio of residual energy of a SN with maximum energy (Fig. 12.1b). In main processing, if a SN v receives message from few SNs to be a CH, then select one SN as CH that costs minimum energy consumption. If SN v does not have a CH, then elect to become a CH with CH_{prob} and adjust its probability by $CH_{prob} = \min(CH_{prob} * 2, 1)$ and repeat until CH_{prob} reaches 1 and a CH is found. In finalization step, join the CH if found; otherwise, elect SN to be a CH.

Thus, HEED consists of discovering neighbors, computing ($CH_{prob,}$ cost), elect to become CH and resolve ties, and finally select CH. The HEED has the properties like completely distributed, and clustering terminates with iterations as follows:

$$N_{\text{iter}} \leq \left\lceil \log_2 \frac{1}{p_{\min}} \right\rceil + 1, \tag{12.1}$$

where P_{min} is the minimum node degree of SNs. The processing overhead is O (n) per SN where n is # SNs and the message overhead is O(1) per node = # iterations required. The CHs are well-distributed. P_r and there exists at least one CH in any area of size

$$\left(2 + \frac{1}{\sqrt{2}}\right) r_c \times \left(2 + \frac{1}{\sqrt{2}}\right) r_c. \tag{12.2}$$

HEED produces a connected multi-hop CH graph asymptotically, and its performance comparison for a field of size 100 × 100 with 1000 SNs is shown in Fig. 12.2 [2], with initial energy E_r of 2 J, and each round consists of 5 TDMA frames. So, the next question is how the network density and initial bootstrap parameters change the rate of energy consumption and lifetime of WSN. Clustering scheme has been implemented [3] on a simple simulator without modeling contention or message losses due to collisions. A WSN in an 500 × 500 m area is modeled as a Poisson point process with intensity λ varied from 0.0002 to 0.002 with communication range r_c of 90 m, and initial probability of being a CH in round 0 be denoted as p_0. In the ith round, a SN becomes a CH with probability $k^i p_0$. This leads to each SN having an average of 5–50 neighboring SNs. The number of CHs as a function of SN density is shown in Fig. 12.3a while b illustrates the impact of p_0.

Another clustering scheme [3] performs the sequence of time synchronization, neighbor discovery, cluster selection, gateway selection, and routing as a part of setup phase as shown in Fig. 12.4 and needs to be repeated at regular intervals. The clustering algorithm has been tested on 42 Mica2 SNs test bed [3], with the maximum one-hop neighbors of 15. B-MAC is adopted with DSDV-like table-driven proactive query–response approach. Link-level measurements are

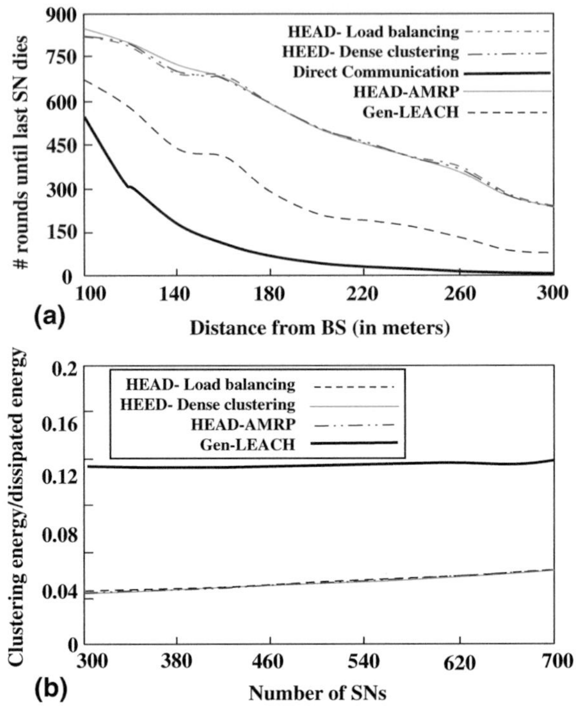

Fig. 12.2 **a** Rounds before SN dies as a function of distance to the BS, **b** energy consumption in forming CH as a function of no. of SNs

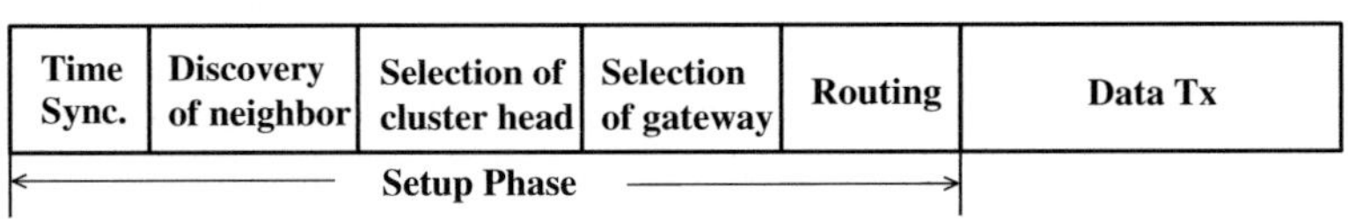

Fig. 12.3 Set-up phase for clustering [3]

used to select routing parents (Fig. 12.5). As CHs consume a lot more energy, it is desirable to recluster occasionally. Using δ as controlling power saving degree, Fig. 12.5c shows the effect of network density and δ on the reclustering period with $p_0 = 0.032$, $k = 2$, and duty cycle $X = 50\%$. To balance energy, frequent reclustering is required as fewer CHs are used here. CHs are tried for varying duty cycles ($X = 2$–45%). Radio is selected as 19.2 Kbps with a packet payload of 36 bytes. Every SN transmits packets with probability $\alpha\%$, and α is varied for two types of scenarios: low data rate experiment, with SN remains idle most of the time with a very brief periods of activity (e.g., earthquake detection) and $\alpha = 0.1$ to 1; high data rate experiment with larger active periodicity (e.g., temperature monitoring) and $\alpha = 10$–100. Experimental default parameters are shown in Table 12.1 and clustering overhead is given in Table 12.2. 11 out of 42 SNs are selected as CHs and

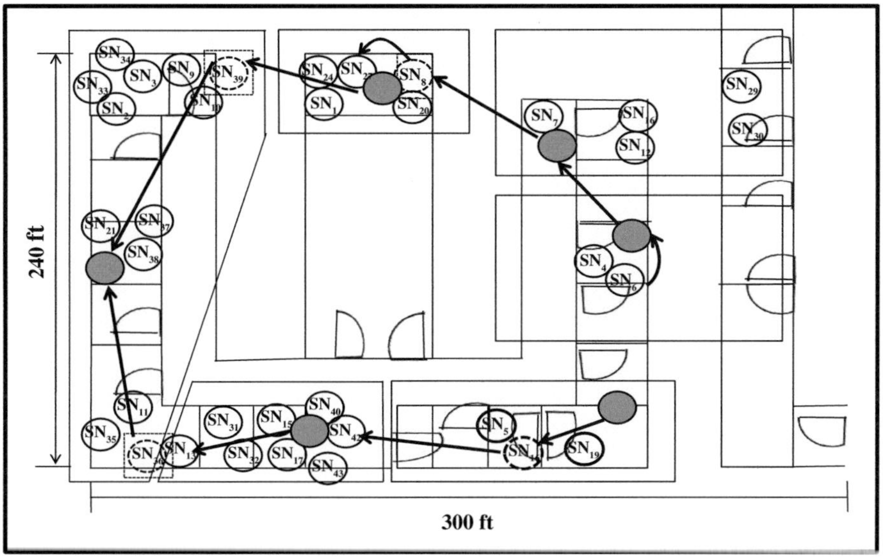

Fig. 12.4 Experimental test bed [3]

achieve roughly 3 times of energy efficiency of B-MAC. About half the overhead is due to routing and clustering consumes about remaining half. With time synchronization of 17 h leads to synchronization error of up to 10 s. During the neighbor discovery phase, SNs periodically broadcast beacon signals to discover neighbors and receiving SNs use this to determine link quality. In cluster selection phase, few SNs adjacent to current CHs are selected as new CHs with probability $k^i p_0$ in the ith round where p_0 is probability of being CH in round 0. When a SN becomes a CH candidate, it broadcasts a *CH Advertisement* (CA) message till the timer expires. In cluster selection phase, SNs select their CHs and two adjacent CHs choose their gateway SN. A proactive routing algorithm is used to determine path to BS via CHs and gateway nodes. Throughput is illustrated in Fig. 12.6; energy efficiency in topology control and routing is illustrated in Fig. 12.7, and the results depend on the values of p_0 and constant k. A parameter δ is used as a controller that reflects power saving desired by the designer, and $\delta = 1$ means all SNs follow the same duty cycle while $\delta = 0$ means all SNs are turned off and only CH maintain a duty cycle.

12.3 Sensor Properties and Resource Constraints

WSNs have lower transmission distances (<10 m), lower bit rates (typically < kbps) and limited battery capacity [4]. Many different SNs have been developed including MIT μAMPS with 59–206-MHz processor, 2 radios, capable

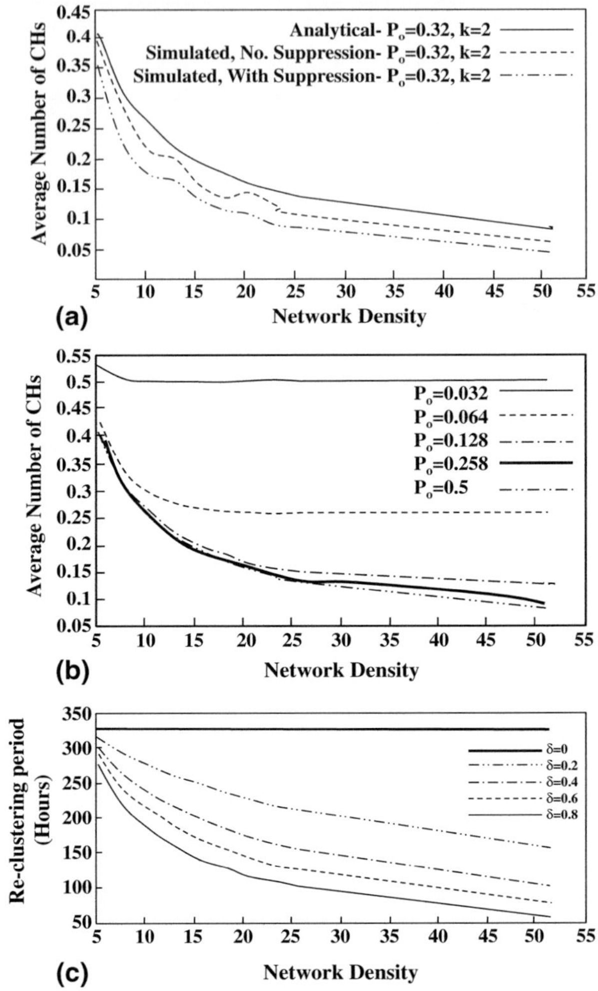

Fig. 12.5 **a** Number of CHs as a function of SN density for $p_0 = 0.032$ and $k = 2$, **b** Impact of p_0 on number of CHs in a WSN, **c** Impact of network density and δ on reclustering period

of transmitting at 1 Mbps, and 4 KB RAM; Berkeley MICA motes with 8-bit, 4-MHz processor; 40-kbit CSMA radio; 4-KB RAM, and TinyOS -based operating system. However, circuit gains are nearing flat as circuit tricks and voltage scaling provided a large part of the gains (Fig. 12.8).

While energy needs functionality, speed, they continue to climb at the rate of 10× increases in gate count every 7 years) and in frequency every 9 years. Speed power efficiency has indeed gone up 10 times every 2.5 years for μPs and DSPs in 1990s and varies from 100 mW/MIP to 1 mW/MIP since 1990. IC processes have provided 10 times improvements every 8 years since 1965. Power consumption is being lowered for a given function and performance, such as DSP is reduced by 1.6 times every year since the early 1980s and most optimistic projections are 60 pJ/op (about 20 times) for a given function and performance (Table 12.3).

Table 12.1 Setting for MICA2 experiments

TinyOS experiment parameters	Default
Initial cluster probability (p0)	0.032
Constant multiplier (k)	2
Cluster head and gateway duty cycle (X)	Varied
Member node duty cycle ($\delta = 0$)	0
Transmission power (full power)	10 dBm
Communication bandwidth (CC1000 radio)	19.2 Kbps
Preamble size	Varied
Payload size	36 bytes
Radio transmission power consumption	60 mW
Radio receiving power consumption	45 mW
Channel sampling power consumption	5.75 mW

Table 12.2 Clustering set up overhead cost

Phase	Average time (s)	Average energy (J)
Time synchronization	10	0.235
Neighbor discovery	20	0.469
Cluster selection	60	1.288
Gateway selection	20	0.429
Routing	120	2.583
Total	230	5.004

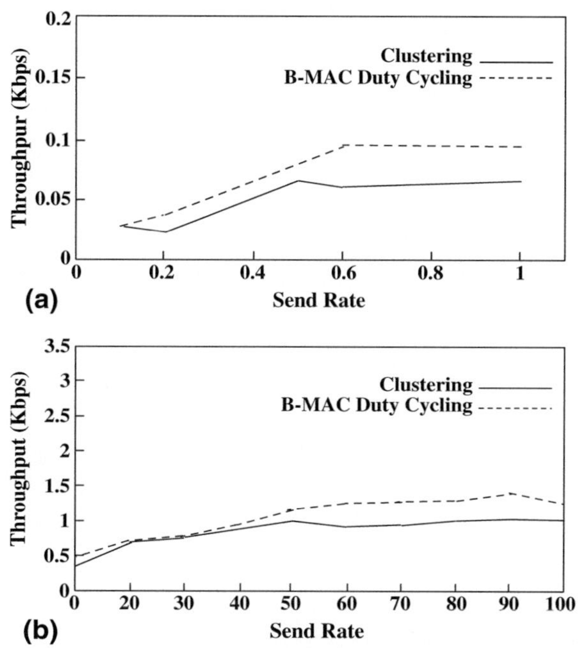

Fig. 12.6 **a** Throughput for low rate SNs, **b** throughput for high rate SNs

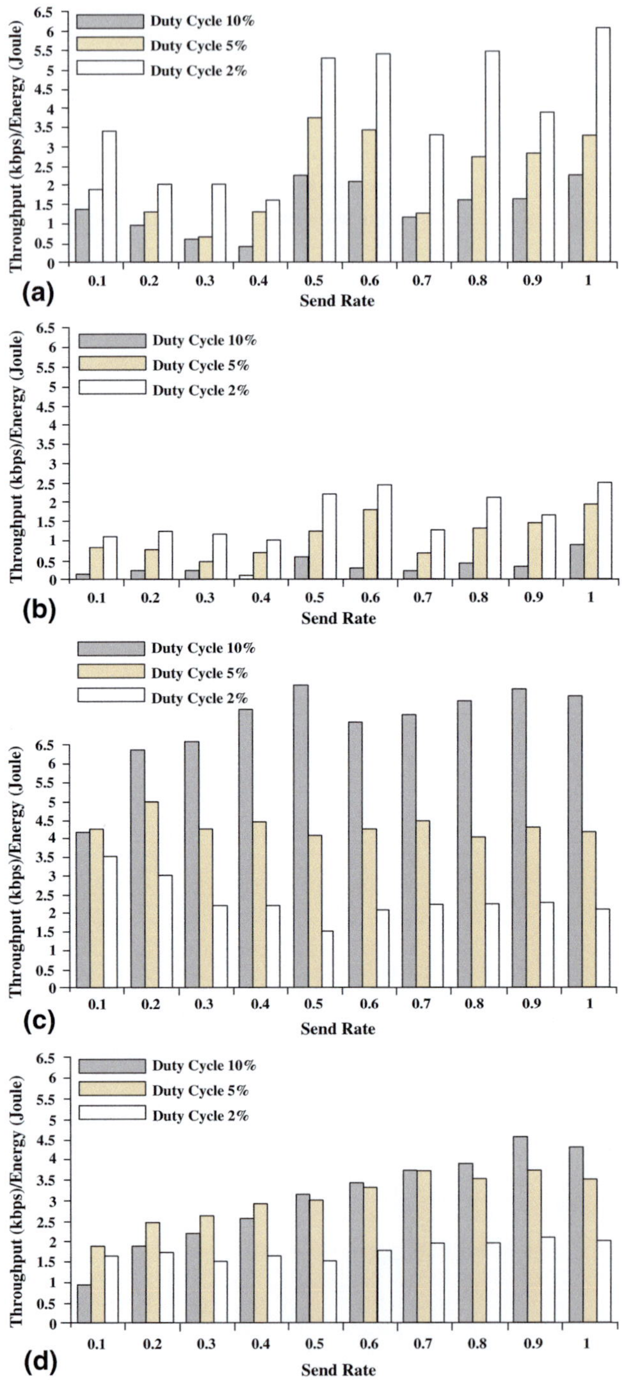

Fig. 12.7 Energy efficiency **a** topology control in low rate, **b** B-MAC in low rate, **c** topology control in high rate, **d** B-MAC in high rate

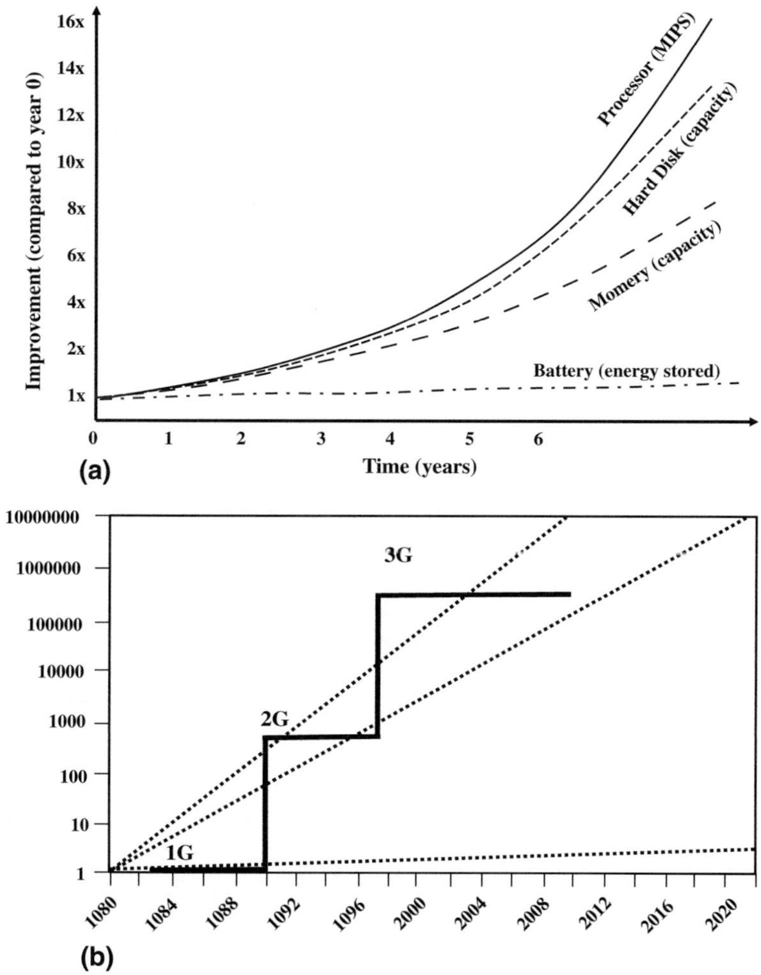

Fig. 12.8 **a** Processor, memory, and battery improvements, **b** processor performance and battery capacity

Table 12.3 Properties of sensor nodes

Radio mode	Power consumption (mw)
Transmit	14.88
Receive	12.50
Idle	12.36
Sleep	0.016

Power efficiency (or energy efficiency) η_P is defined as the ratio of signal energy per bit E_b to noise power spectral density N_0 required at the receiver for a certain BER (bit error rate) and is given by

$$\eta_P = E_b/N_0. \quad (12.3)$$

High power efficiency requires low (E_b/N_0) for a given BER and the bandwidth efficiency

$$\eta_B = \text{bit rate/bandwidth} = R_b/\text{W bps/hz}. \quad (12.4)$$

The ratio of throughput data rate to bandwidth occupied by the modulated signal typically ranges from 0.33 to 5, and there is a trade-off between the two as for a given BER. Adding FEC reduces η_B, but reduces required η_P. Modulation schemes with larger number of bits per symbol have higher η_B, but also require higher η_P. Projected computation cost in 2004 is 60 pJ/op, and minimum thermal energy for communications is 20 nJ/bit transmitting for 100 m at a bandwidth of 1.5 GHz is equivalent of 300 ops and 2 nJ/bit at 1.5 GHz for 10 m is equivalent of 0.03 ops (Fig. 12.9a). There exists a clear trade-off between significant processing energy versus communication cost. A reduction in supply voltage decreases speed (Fig. 12.9b) and the supply voltage ought to be reduced when slower speed can be tolerated.

To combat slow operation, another alternative is to use architectural techniques such as concurrency and pipelining via compiler techniques. Parameter of Rockwell WINS SN of Fig. 1.10a is shown in Fig. 12.10b. The processor consumes 360 mW power doing repeated transmit/receive, and transducer takes 23 mW. Shutdown is attractive for many wireless applications due to low duty cycle of many subsystems, and issues to be addressed are the cost of restarting as latency versus power trade-off remains that increases in latency (response time) and power consumption due to start-up. So, the question is, when to shut down for optimal gain and how to select idle time threshold or use some predictive approach and when to wake up for optimal performance or depend on on-demand service or utilize predictive approach? (Fig. 12.11). Two main approaches are reactive versus predictive. In a reactive approach, the SN goes to reduced power mode after the SN has been idle for a few seconds/minutes and restarts on demand. In predictive mode, computation history is used to predict whether $T_{\text{block}}[i]$ is large enough satisfying $T_{\text{block}}[i] \geq T_{\text{cost}}$. It has been observed that it is better to lower voltage than to complete shutdown in case of digital logic. For example, a task with 100 ms deadline requires 50 ms CPU time at full speed. A normal system gives 50 ms computation, 50 ms idle/stopped time. At half speed/voltage system gives 100 ms computation, 0 ms idle and thus requires the same number of CPU cycles but 1/4 in energy consumption. So, the voltage gets dictated by the tightest (critical) timing constraint both on throughput and latency. So, it is better to change voltage dynamically and use voltage to control the operating point on the power versus speed curve. As the power and clock frequency are functions of voltage, the main

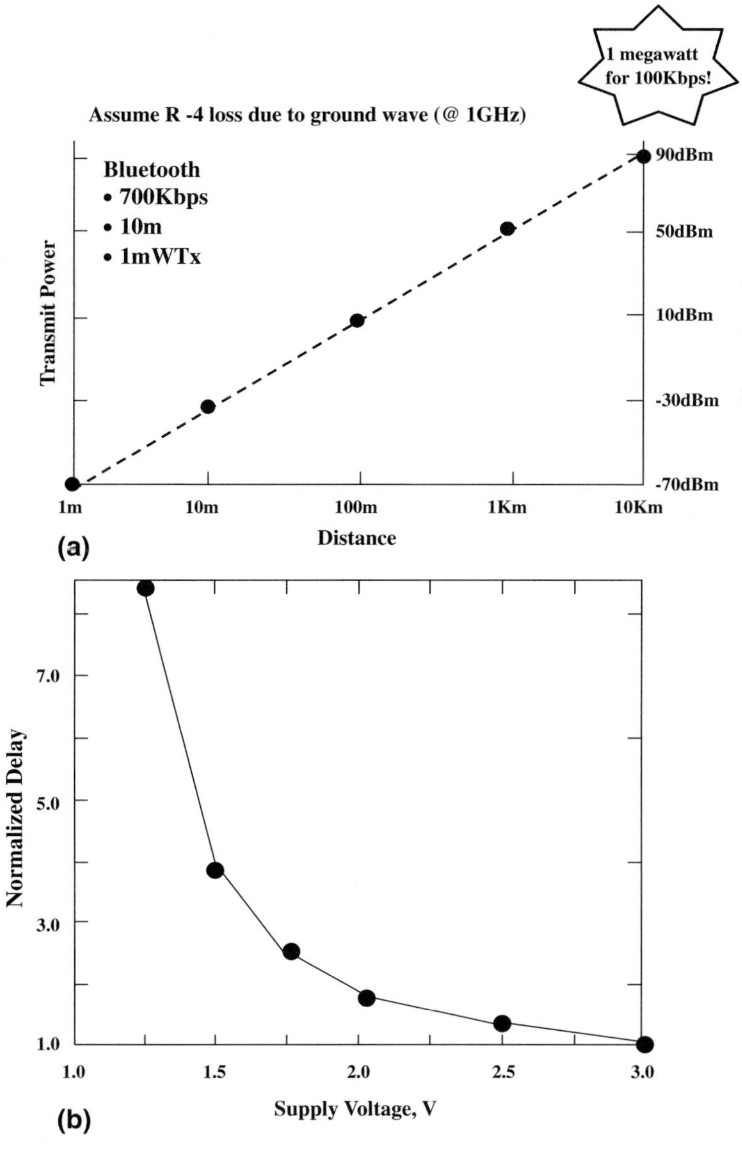

Fig. 12.9 **a** Power consumption as a function of distance, **b** slowdown by reducing supply voltage

challenge is algorithmic and need to schedule the voltage variation as well either through compiler or OS or the hardware (Fig. 12.12).

Advanced Configuration and Power Management Interface (ACPI) allows OS/drivers of SN to be in sync regarding power states, and a standard way is used for the system to describe its device configuration and power control hardware

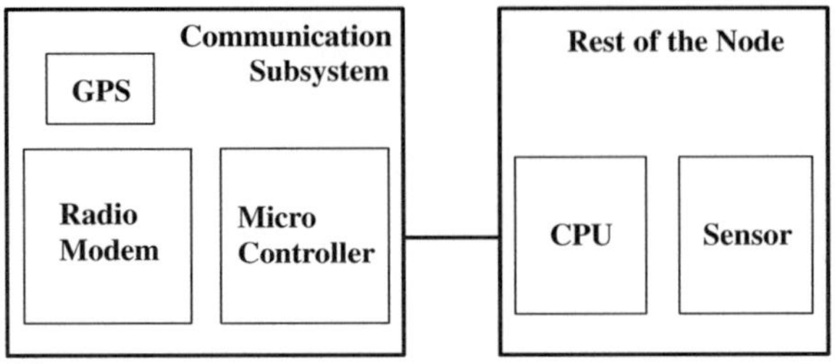

(a)

Processor	Seismic Sensor	Radio	Power (mW)
Active	On	Rx	751.6
Active	On	Idle	727.5
Active	On	Sleep	416.3
Active	On	Removed	383.3
Active	On	Removed	360.0
Active	On	TX (36.3 mW)	1080.5
		TX (27.5 mW)	1033.3
		TX (19.1mW)	986.0
		TX (13.8 mW)	942.6
		TX (10.0 mW)	910.9
		TX (3.47 mW)	815.5
		TX (2.51 mW)	807.5
		TX (1.78 mW)	799.5
		TX (1.32 mW)	791.5
		TX (0.955 mW)	787.5
		TX (0.437 mW)	775.5
		TX (0.302 mW)	773.9
		TX (0.229 mW)	772.7
		TX (0.158 mW)	771.5
		TX (0.117 mW)	771.1

(b)

Fig. 12.10 **a** Power measurements on Rockwell WINS SN, **b** parameters of Rockwell WINS SN

interface to the OS. Common functions are registered through interface and the system controls the events, processor power and clock control, and thermal management. Information on devices, resources, and control mechanisms is described using tables (actually linked to a table of tables) and power management capabilities

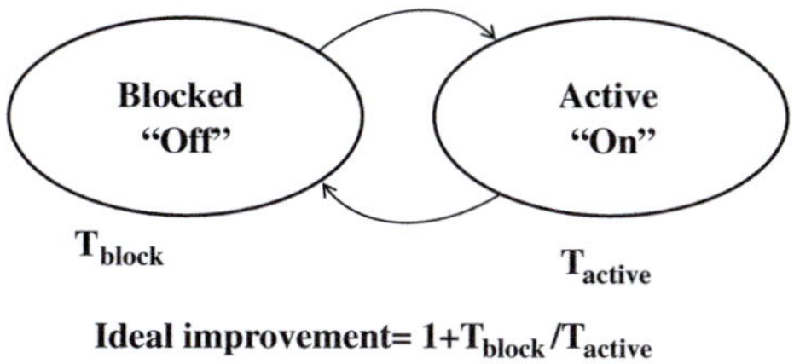

Fig. 12.11 Blocked and active state of Rockwell WINS node

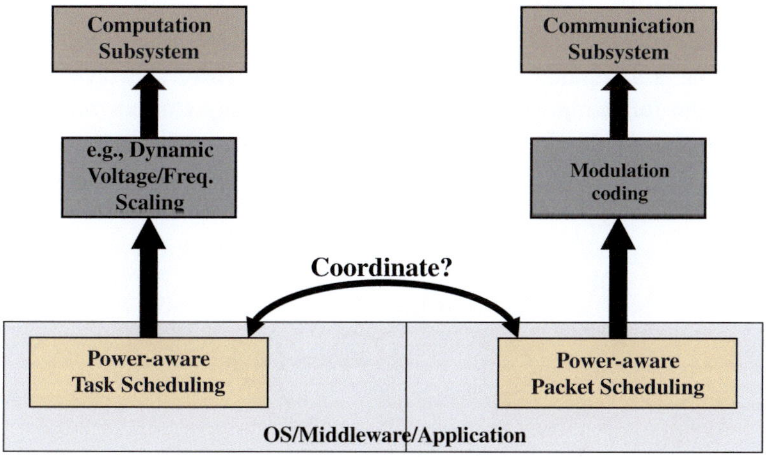

Fig. 12.12 Communication subsystem of Rockwell WINS node

and requirements for each device are described. There are specific methods for setting and getting the power state, and hardware resource settings are managed by appropriate methods. Thus, ways are provided for application, operating system, and hardware to efficiently exchange energy/power and performance related information. This facilitates a continuous dialogue/adaptation between OS and applications, and power-aware OS services are implemented by providing a software interface to low-power devices. A power-aware API to the end user enables one to implement energy-efficient RTOS services and applications. The applications' interface enables the application to pass on RT (round trip) information to OS (period, deadlines), WCET (worst-case execution time), hardness by creating new threads. It also predicts when OS time is expected to finish a given task depending on the conditions of the environment (application dependent and not yet implemented). OS is also able to predict and tell applications the time estimated to finish the task depending on the scheduling scheme used, and a task with hard deadline must be killed if its deadline is missed.

12.4 Conclusions

There are many issues critical to WSNs, and clustering is very useful, especially for randomly deployed WSNs. The question has been addressed as to how to do it effectively with minimal information about neighboring SNs. What is the impact of multi-hop communication and how to keep the process distributed in nature. Reducing supply voltage could be another effective solution in order to minimize energy consumption.

12.5 Questions

Q.12.1. What is an optimal cluster sizes for wireless sensor networks?

Q.12.2. In the following examples of WSNs, how many clusters will be formed using (i) LEACH, (ii) HEED, and (iii) topology control schemes?

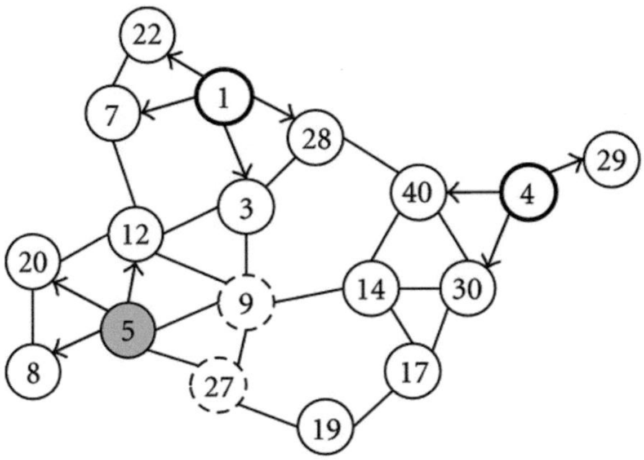

Prob. 12.2(i): Example WSN

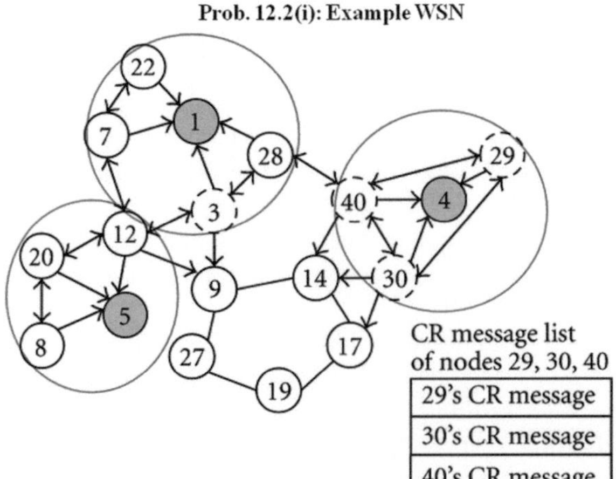

Prob. 12.2(ii): Example WSN

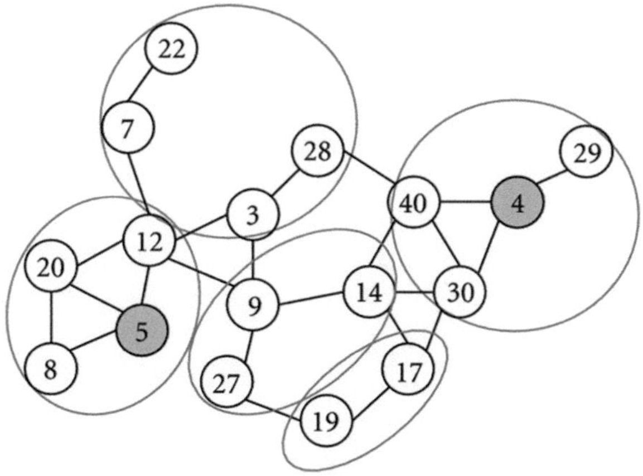

Prob. 12.5(iii): Example WSN

Q.12.3. In Q. 12.2, can you identify CH? How much is the average distance in hops for three cases?

Q.12.4. How many control messages are used in each case, neglecting any collisions?

Q.12.5. How frequently you need to do reclustering if SNs synchronized every 2 min? Assume appropriate parameters for initial energy and power consumption rate.

Q.12.6. What is the difference between two different clustering schemes done for the WSN? Explain clearly.

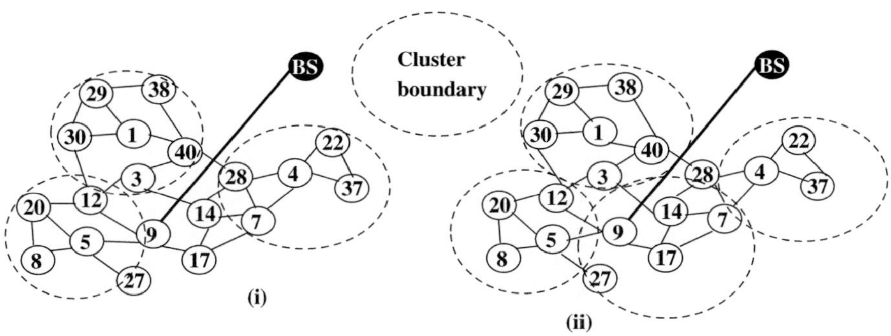

Q.12.6

Q.12.7. In the following WSN, how can you group SNs to form clusters? Explain your answer carefully.

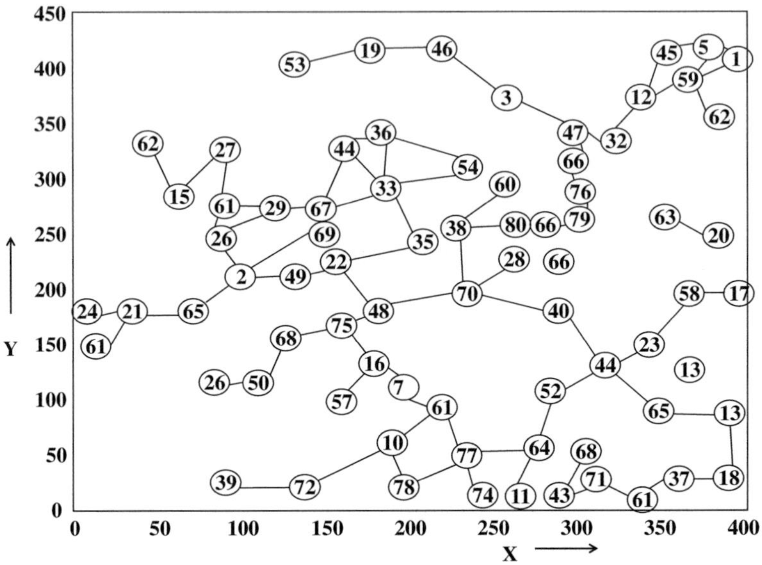

References

1. Wendi Rabiner Heinzelman, Anantha Chandrakasan, and Hari Balakrishnan, "Energy-Efficient Communication Protocol for Wireless Microsensor Networks," Proceedings of the 33rd Hawaii International Conference on System Sciences—2000, https://pdos.csail.mit.edu/archive/decouto/papers/heinzelman00.pdf.
2. O. Younis and S. Fahmy, "HEED: a hybrid, energy-efficient, distributed clustering approach for ad hoc sensor networks," IEEE Transactions on Mobile Computing, vol. 3, no. 4, pp. 366–379, Oct-Dec. 2004.
3. Ajit Warrier, Sangjoon Park, Jeongki Min, and Injong Rhee, "Impact of Topology Control on Energy Saving through Differential Duty Cycling," http://www4.ncsu.edu/~rhee/export/topology.pdf.
4. http://www.ics.uci.edu/~quasar/tutorial/hipc.ppt.

Chapter 13
Intrusion Detection Using WSNs

13.1 Introduction

A wireless sensor network (WSN) is a collection of small, cheap, and low-powered sensors which can dynamically form a network without any infrastructure and is supported by wireless communication while the objective is to monitor environmental changes in a given area.

A pulsed laser beam is reflected by a laser scanning system [1] from a rotating mirror. A detection field of approximately 50 mm deep in a scanning angle of up to 190° is produced by using a rotating mirror to reflect a pulsed laser beam (Fig. 13.4a). A maximum of 80 m scanning radius can be achieved with angular resolution of 0.25° and 1°. Each complete scan takes about 10–50 ms, and the distance is calculated as $D = c(t/2)$, where D is the distance, c is speed of light, and t is the total propagation time. To recognize an object, at least two neighboring laser pulses must be intersected sequentially (Fig. 13.4b) and the use can stipulate which parts of a scene to be secured. Precise area can be specified where unsolicited invasion has happened. False alarm could occur due to bad weather, and birds flying can be eliminated by defining minimum size of the object. Laser scanning is appropriate for detecting the unofficial existence of vehicles 80–100 m away, as well as finding personnel in 25 m range. Reflective property of the object influences the effective range as intrusion detection is a tool that could detect unsuitable, improper, or inconsistent attacks from outside of the WSN. A large number of SNs can be deployed to monitor the area, and by placing more SNs or increasing the sensing range r_s of SNs, intruder can be noticed in shorter distance of its mobility.

This is illustrated in Fig. 13.1. A fraction of area covered by SNs and area covered by SN A is given by $f_a = \pi r_s^2$, where r_s is the sensing radius of each SN using disk model. The probability of detectability $P_d(x)$ is the probability SNs detect a moving object. If x is the length moved by an external object, then the detection

D.P. Agrawal, *Embedded Sensor Systems*,
DOI 10.1007/978-981-10-3038-3_13

Fig. 13.1 **a** Detection of an event by SNs when object moves in a *straight line*. **b** Detection of an event by SNs when object moves randomly

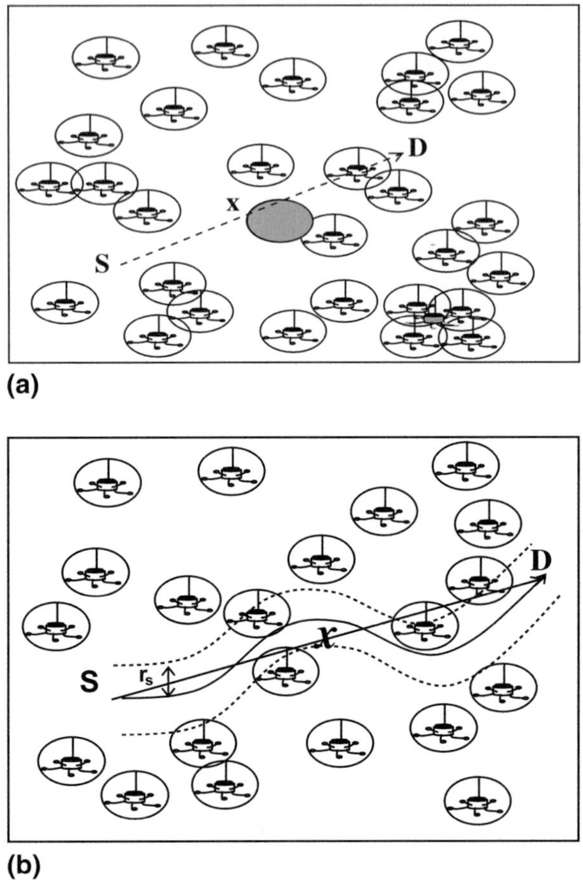

zone is $2r_s x + \pi r_s^2$. To understand the impact of external objects on received signal strength (RSS), an experiment was conducted as follows.

Three MICA2 motes (Fig. 13.2a) are used [2]: one is used as the transmitter *T*, another is used as the receiver *R*, and the third one is used as the jammer (*J*), whose transmission is synchronized with that of the transmitter to generate possible interference (Fig. 13.2b, c). All experiments are conducted in an open parking lot late at night to avoid possible influence from people and moving cars. Transmitter *T* and receiver *R* are fixed in positions, and jammer *J* moves. When the distance between *T* and *S* is 16.2 ft (Fig. 13.2b), only 80% of packets are correctly delivered due to weak link. In Fig. 13.2c, the receiver *R* is *put close to the transmitter T* (8.5 ft) with a stable 100% packet delivery ratio and is called a strong link. When the link is weak, the transmitter's signal power level (RSS Indicator) is very low and can be interfered with distant jammer, and when the link is strong, the transmitter's signal power level is very strong and cannot be interfered. *T* first broadcasts a high-power detection (HD) packet including transmitter's ID and followed by a

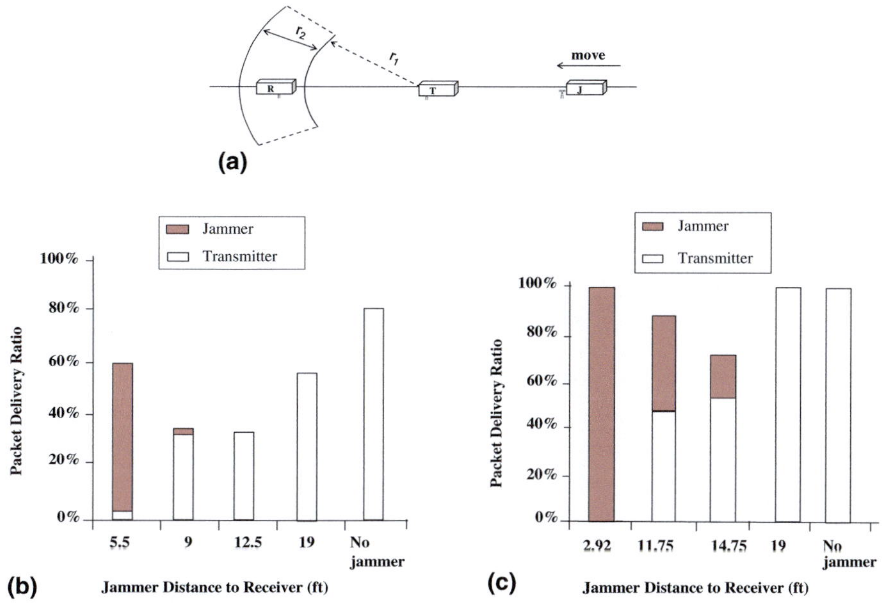

Fig. 13.2 **a** RSSI from transmitter to receiver. **b** Packet delivery ratio for weak link. **c** Packet delivery ratio for strong link

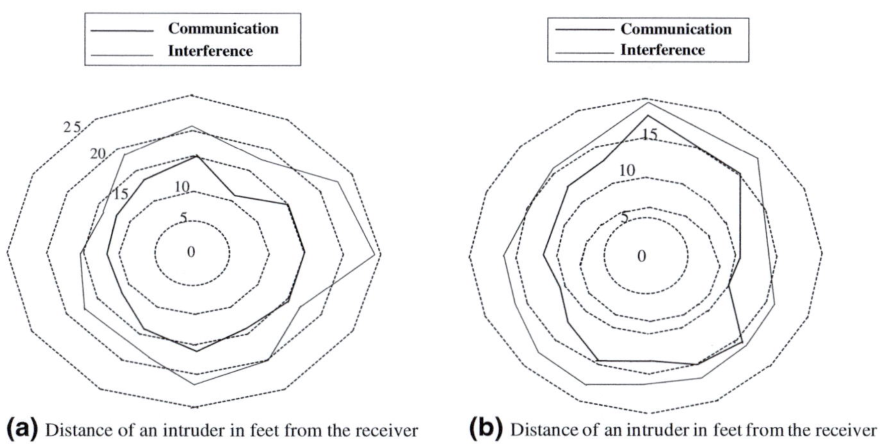

Fig. 13.3 **a** Interference with weak link. **b** Interference with strong link

normal-power detection (ND) packet and is termed as an HD-ND detection sequence that is used to guess *T*'s interference strength (Fig. 13.3). Each SN begins to exchange the interference information detection (RID) in its locality to estimate collisions within the WSN. Three steps of RID are HD-ND detection, information sharing, and interference calculation (Table 13.1).

Table 13.1 Simulation parameters

Terrain	$(144\ \text{m} \times 144\ \text{m})^2$
Node number	144
Node placement	Uniform
Application	Many-to-one CBR streams
Payload size	32 bytes
Routing layer	GF
MAC layer	NAMA/NAMA-RID-B
Radio layer	RADIO-ACCNOISE
Radio bandwidth	250 Kb/s
Radio range	25 m

NAMA Node activation multiple access
RID-B Radio interference detection-basic

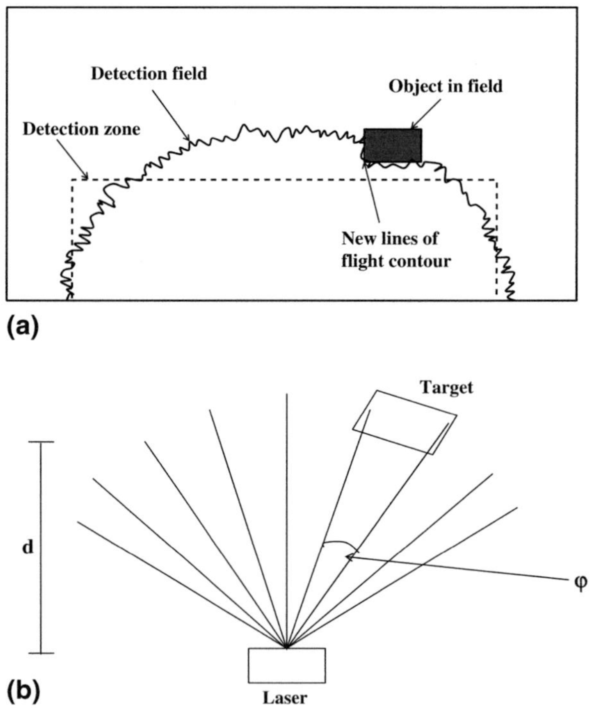

Fig. 13.4 **a** Real-time field monitoring [1]. **b** Target intersecting two pulses [1]

Real-time field monitoring is covered in Fig. 13.4a and is done by having laser beams intersect with each other (Fig. 13.4b).

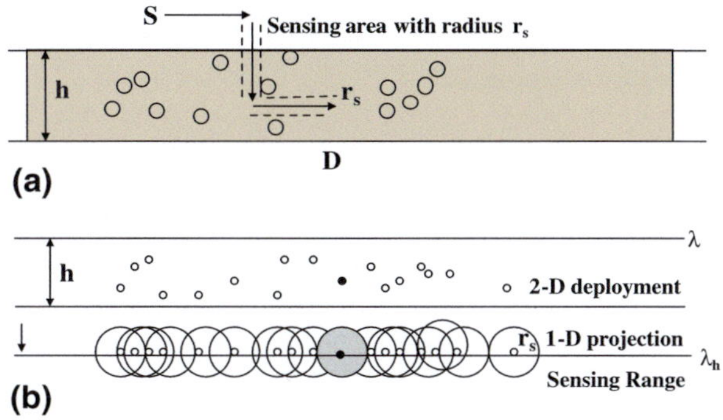

Fig. 13.5 **a** Detectability in 2-D strip with locations of SNs unknown. **b** Detectability in 2-D strip with locations of SNs known

13.2 Intrusion Detection Schemes

The intrusion distance is defined as the position where an intruder goes inside WSN to the point where it is detected [3]. A SN can individually detect the intruder and is commonly known as *single-sensing detection* . Most of the time, collaboration of multiple sensors is required to detect an intruder and is known as *k*-sensing detection by *k*-SNs. A SN can be treated as a barrier and detect any intruders along any paths by at least k distinct SNs. A very important concept for intrusion detection sensitivity analysis is that there is no knowledge of SN locations (Fig. 13.5a). Consider the case *S*, *D* on opposite sides and detection zone minimized if crossing strip perpendicular to edge

$$P_{\mathrm{d}}(2-\dim)=1-\mathrm{e}^{-2\lambda r_s h}. \tag{13.1}$$

If locations of SNs are completely known, then 2-D deployment of SNs can be projected on 1-D plane to check if there are any holes (Fig. 13.5b), and if so, additional darkened SN can fill the gap to ensure that any intruder is detected.

Such a projection suppresses any existing path in 2-D deployment if

$$P_{\mathrm{d}}(2-\mathrm{Dim})<P_{\mathrm{d}}(1-\mathrm{Dim}), \tag{13.2}$$

where P_{d}(2-Dim) is the minimum distance between any two SNs in 2-D and similar value for 1-D. The 1-D projection can detect any crossing objects if *distance between adjacent SNs* < $2r_{\mathrm{s}}$.

A uniform deployment (Fig. 13.6a) has SNs spread out uniformly across the whole area. That means the region covered by the WSN has a distribution of SNs that appears randomly but is spread across the area at a particular density and this density determines the cost of the network. The main advantage of uniform deployment is it works well for intruder that attacks from the boundary of the area under protection. Since the sensors are randomly deployed at a particular density, it is easier to deploy as compared to Gaussian distribution (Fig. 13.6b). The disadvantage is that a uniform deployment performs badly (as compared to Gaussian)

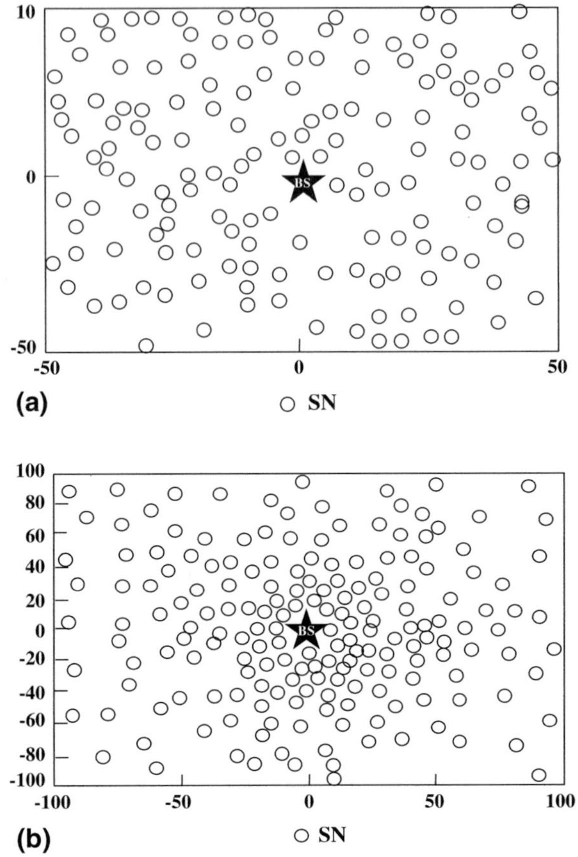

Fig. 13.6 **a** Uniform deployment of WSN. **b** Gaussian deployment of WSN

when intrusion is from within the network. This is because the density is uniform overall and there is no added protection near to the BS located at the center of the network as the BS has a few SNs surrounding it. Since all the traffic is directed toward the BS, these SNs tend to overwork and subsequently die out quickly due to energy-hole problem. This leads WSN to become disconnected quickly and thus decreases its lifetime. In two-dimensional Gaussian distribution [4], the BS or sink node is situated at the central point (0, 0) with same sigma. Thus, we rewrite the 2-D Gaussian distribution PDF as (Fig. 13.7)

$$g(x,y) = \frac{1}{2\pi\sigma_x\sigma_y} e^{-\left(\left(\frac{x-\mu_x}{2\sigma_x^2}\right)^2 + \left(\frac{y-\mu_y}{2\sigma_y^2}\right)^2\right)}. \quad (13.3)$$

where (μ_x, μ_y) is the deployment point. When the BS or sink node is situated at the central point (0, 0), we rewrite the 2-D Gaussian distribution PDF as

Fig. 13.7 **a** PDF of uniform deployment in a WSN. **b** PDF of Gaussian deployed WSN with $\sigma_x = \sigma_y = \sigma$

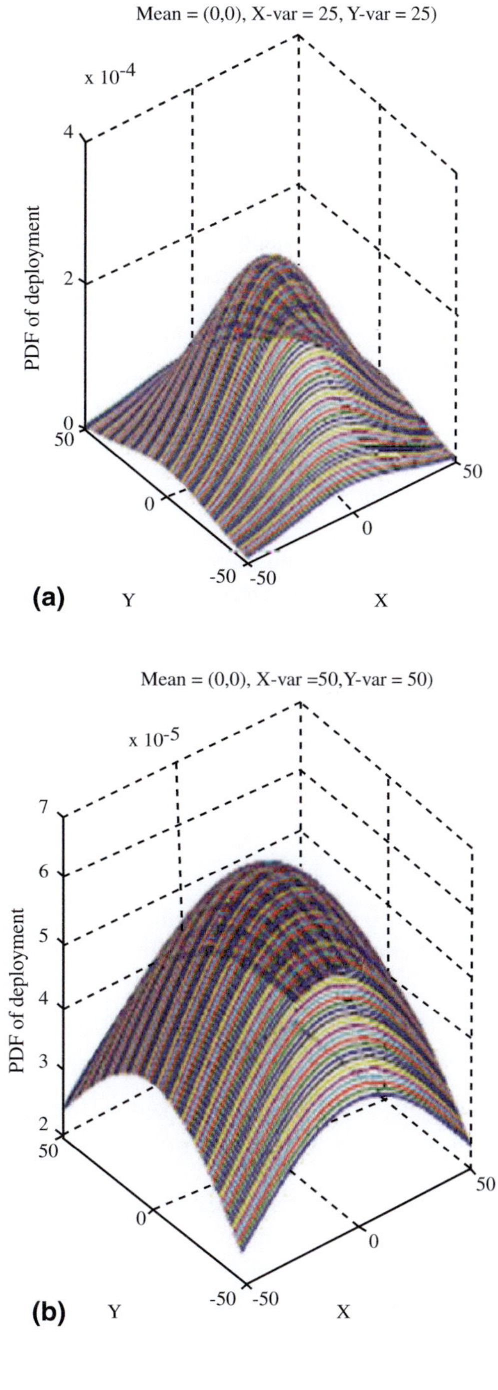

$$
\begin{aligned}
g(x,y) &= \frac{1}{2\pi\sigma^2} \mathrm{e}^{-\left(\frac{x^2+y^2}{2\sigma^2}\right)} \quad \text{and} \quad S = \pi r_{\mathrm{s}}^2 \\
P(m) &= \frac{(S\lambda)^m}{m!} \mathrm{e}^{-S\lambda} \quad S = \frac{4\pi r_{\mathrm{s}}^3}{3}.
\end{aligned}
\tag{13.4}
$$

To date, several deployment strategies for intrusion detection have been proposed and they have their own advantages over others, as per their specific scenarios. In Gaussian deployment, the SNs' density is very high at the center and then decreases as we move away from the center and toward the border. The disadvantages of the uniform deployment led to the proposal of a new scheme which has sensor nodes deployed according to a 2-D Gaussian distribution with high density of SNs around the BS (center) and this density gradually decreases as we move away from the center toward the border. The Gaussian deployment can address the "energy-hole" problem, can provide differentiated intrusion detection capabilities around the network, and can provide additional protection to key areas within the network. The disadvantage of Gaussian deployment is that it has insecure (sparsely populated) borders and immediate detection of intrusion from border area is very hard, especially when more than one detection is necessary to confirm intrusion. It also assumes that the intruder will not cause any harm even if the distance traveled by it before detection is significant, which is not acceptable is the critical applications such as military bases. With uniform sensor distribution, the intrusion detection probability is the same for any point in a WSN.

The expected intrusion distance $E(D)$ is

$$
E(D) = \int_0^{\sqrt{2}L} 2\xi\lambda\, r_{\mathrm{s}} \mathrm{e}^{-\lambda\left(2\xi\lambda r_{\mathrm{s}} + \frac{\pi r_{\mathrm{s}}^2}{2}\right)} d(\xi), \tag{13.5}
$$

where $\xi > 0$ is the maximum allowable distance to be traveled by intruder before it is detected. A problem occurs if

$$
\sqrt{\left[(x_i - x_t)^2 + (y_i - y_t)^2\right]} \leq E(D), \tag{13.6}
$$

where the target can be attacked no matter how large the area of the uniform deployed WSN. Different degrees of intrusion detectability at different places are required in some WSNs. WSNs with Gaussian-distributed SNs can provide differentiated node density in different locations. Uniform and Gaussian deployments have their own advantages and disadvantages. Numerous studies have analyzed many realistic probabilistic deployment sensors such as Uniform AND Gaussian. However, these strategies always focus on protecting a single facility or an individual facility of system. They rarely considered the system or monitored area as a whole unit as multiple facilities within a monitored area could be desired and system is supposed to either detect an intruder approaching any facility or alert the

BS once any intrusion occurs toward a facility. An intruder might cause severe damage to the whole system by cutting off communication between two facilities. Probability that there are m nodes within the area or space of S is Poisson distributed.

The probability that the monitored area or space is one covered can be expressed as

$$p_{\text{cover}-1} = 1 - P(0) = 1 - \mathrm{e}^{-\lambda S}. \quad (13.7)$$

Coverage implies that the probability that any target point in the WSN is within the sensing range of any nearby SNs. The probability that the monitored area or space is k-covered is

$$p_{\text{cover}-k} = 1 - \sum_{m=0}^{k-1} P(m) = 1 - \sum_{m=0}^{k-1} \frac{(\lambda S)^m}{m!} \mathrm{e}^{-\lambda S}. \quad (13.8)$$

least on given by e.

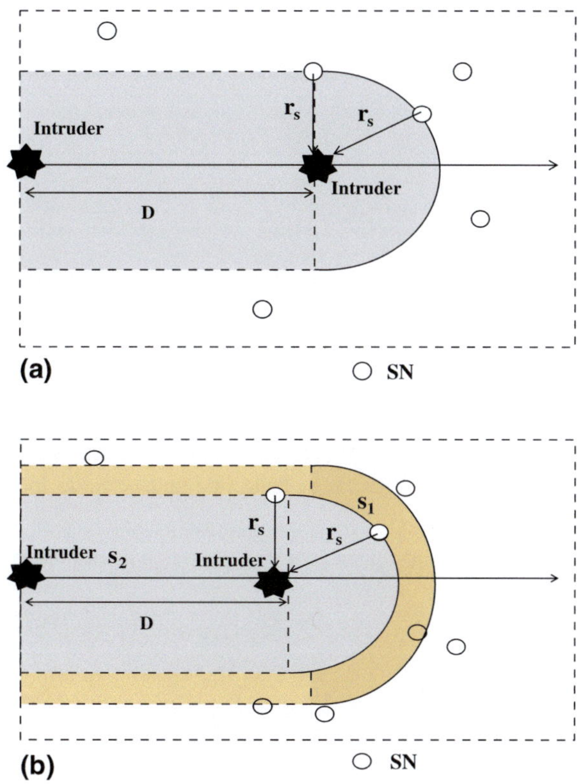

Fig. 13.8 **a** One-covered intrusion detection. **b** k-covered intrusion detection

As illustrated in Fig. 13.8, the intruder is detected before it has moved in a distance D inside the monitored area, and there must be at least one SN within the area of S_D and it can be given by

$$S_D = 2Dr_s + \frac{\pi r_s^2}{2}. \tag{13.9}$$

Thus, the probability that the intruder is sensed at an intrusion distance D can be derived from the differential of its CDF as follows

$$p_1[D = 0] = 1 - P\left(0, \frac{\pi r_s^2}{2}\right) = 1 - e^{-\rho\frac{\pi r_s^2}{2}}, \tag{13.10}$$

where $P(m, S) = \frac{(S\rho)^m}{m!} e^{-S_\rho}$.

The probability that the intruder can be sensed immediately once it enters the monitored area can be expressed by

$$P_k[D = 0] = 1 - \sum_{i=0}^{k-1} P\left(i, \frac{\pi r_s^2}{2}\right). \tag{13.11}$$

The probability that the intrusion distance is less than ξ is equivalent to the probability that there is at least k sensor node located within the area of where $S_\xi = 1 - 2\xi\lambda r_s + \frac{\pi r_s^2}{2}$.

$$P_k[D = 0] = 1 - \sum_{i=0}^{k-1} P\left(i, \frac{\pi r_s^2}{2}\right). \tag{13.12}$$

Thus, the average intrusion distance can be derived as:

$$E(D) = \frac{k\sum_{i=0}^{k-1} P\left(i, \frac{\pi r_s^2}{2}\right)}{2\rho r_s}. \tag{13.13}$$

The probability that the intrusion distance is less than ξ is equivalent to the probability that there is at least one sensor node located within the area of

$$\begin{aligned} p_1[D \le \xi] &= 1 - P(0, S_\xi) = 1 - e^{-\rho\left(2\xi r_s + \frac{\pi r_s^2}{2}\right)} \\ E(D) &= \frac{k\sum_{m=0}^{k-1}\left[\sum_{j=0}^{m} P_1\left(j, \frac{\pi r_{s1}^2}{2}\right) P_2\left(m - j, \frac{\pi r_{s2}^2}{2}\right)\right]}{2\rho\, r_s} \end{aligned} \tag{13.14}$$

Thus, the average intrusion distance can be derived as

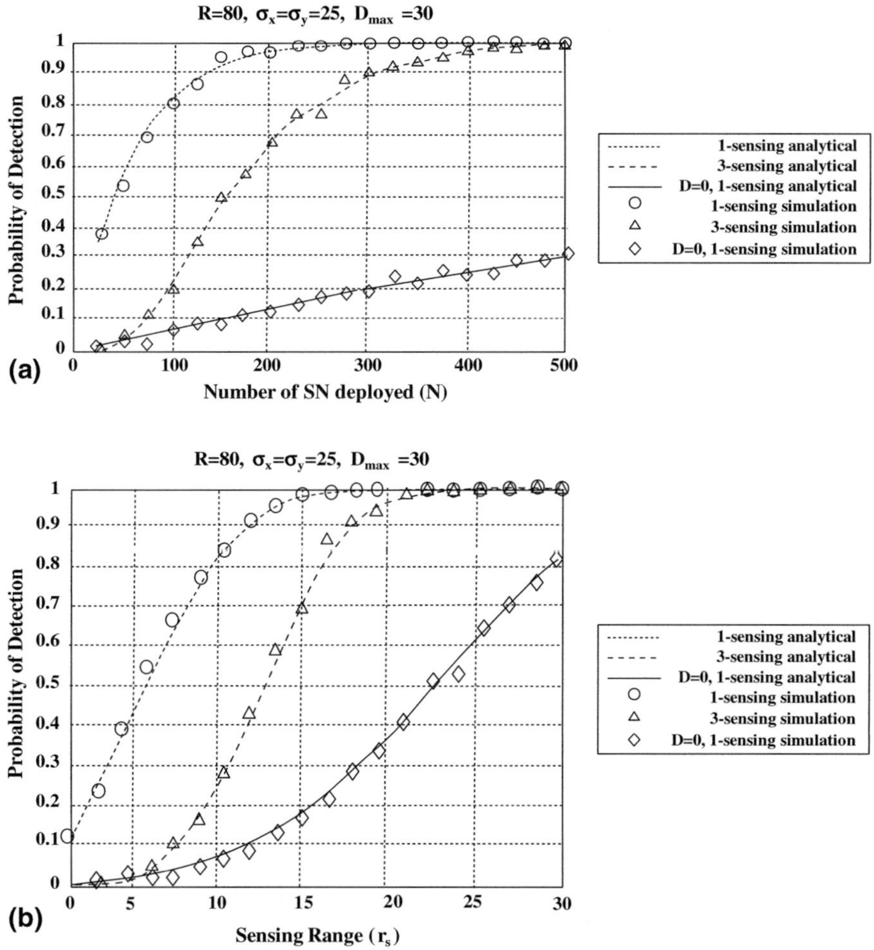

Fig. 13.9 **a** Detection probability with number of SNs in Gaussian-distributed SNs. **b** Detection probability with sensing range r_s in Gaussian-distributed SNs

Detection probability is shown in Figs. 13.9 and 13.10.

The monitored area has multiple facilities (Fig. 13.11a) as a whole unit and focuses on security of the whole region. Theoretically, we have a random distribution satisfying lower probability at the center and higher probability at the border, which is called reverse Gaussian distribution. The PDF of 2-D reverse Gaussian distribution can be derived from the upside-down curve of the 2-D Gaussian distribution. We consider a circular monitored area *A in which the WSNs* are deployed. The BS is located at the center, and facilities are randomly deployed within the area *A under protection. The* intruder attack type is only considered invading from a

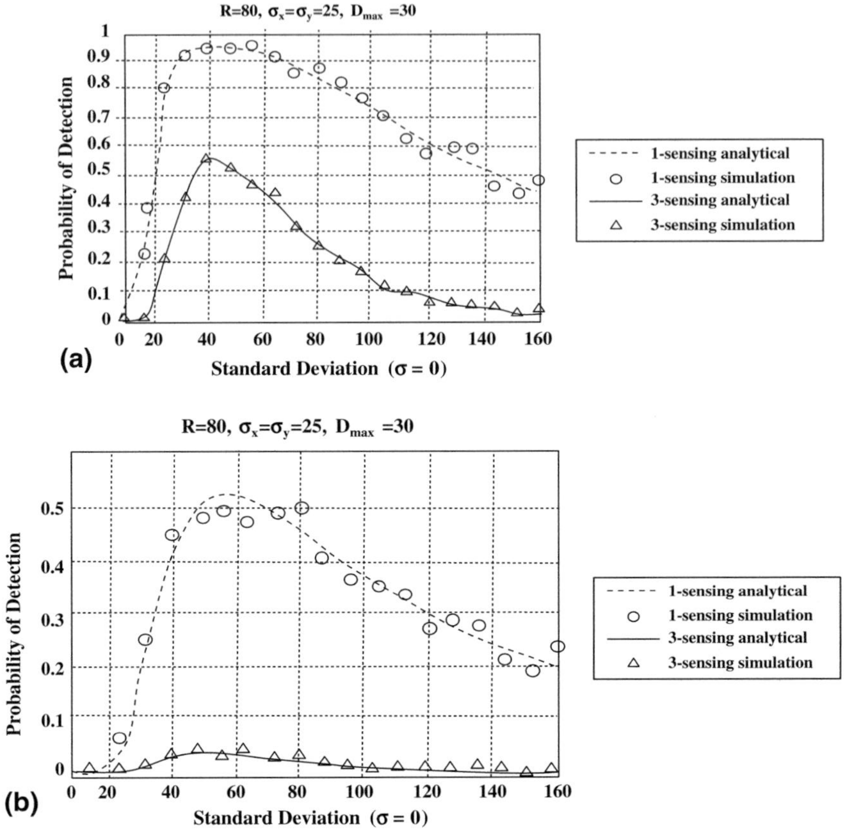

Fig. 13.10 **a** Three-sensing detection probability with deployment deviation in Gaussian distribution of SNs. **b** Three-sensing detection probability with deployment deviation in Gaussian distribution of SNs

random point on the border of the monitored region. The simulation is run for 3 different deployment schemes such as Gaussian, uniform, and our reverse Gaussian distribution deployments. Two types of attacking are simulated: *Border to Center, where the intruder enters through a random* point on the border and heads straight to the center (BS) of the circle; and second, *Border to Facility, where the intruder enters through a* random point on the border and tries to reach to a certain facility using shortest path. Java is utilized to write the simulation, and 1000 runs are used to determine the metrics of *intrusion distance and the number of SNs* triggered by the intruder. Reverse Gaussian scheme performs better than the other two schemes when intrusion is at the center as it quickly performs intrusion detection. When facility is located near the border, the reverse Gaussian scheme could still detect intrusion very quickly. Uniform scheme has worse performance than reverse

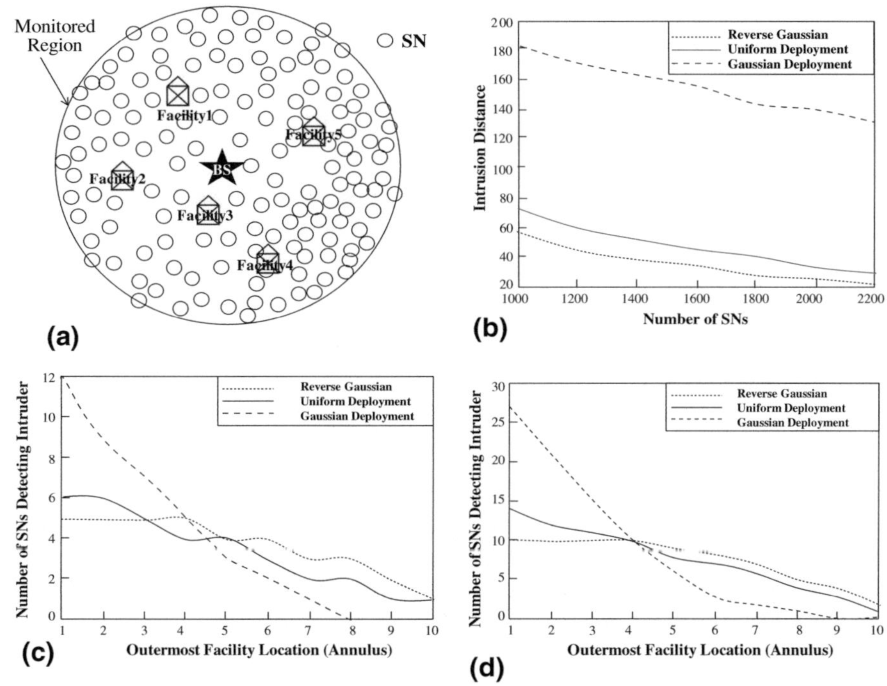

Fig. 13.11 **a** Reverse Gaussian deployment of SNs for intrusion detection. **b** Intrusion detection distance from border. **c** Intrusion detection with facility location with 1000 SNs. **d** Intrusion detection with facility location with 2200 SNs

Gaussian scheme in terms of border region protection. Gaussian scheme can rarely provide any protection for facility near edge area. 1000 SNs are deployed into monitored region with three schemes, and reverse Gaussian performs better in the region from annulus 5 to annulus 10. For all inner annuli, uniform and Gaussian distribution schemes work better. When 2200 SNs are deployed, reverse Gaussian scheme performs better in annuli 4–10 *than the other two schemes.* Reverse Gaussian does better than both Gaussian and uniform in terms of protecting facilities located in the border region.

13.3 Intrusion Detection Based on Hybrid Gaussian Deployment

The disadvantages in both uniform and Gaussian deployments and their obvious strengths led to a hybrid deployment scheme called as hybrid Gaussian-ring deployment [5] where a "ring" of uniformly deployed sensors over an underlying Gaussian deployment combines the strengths of both uniform and Gaussian

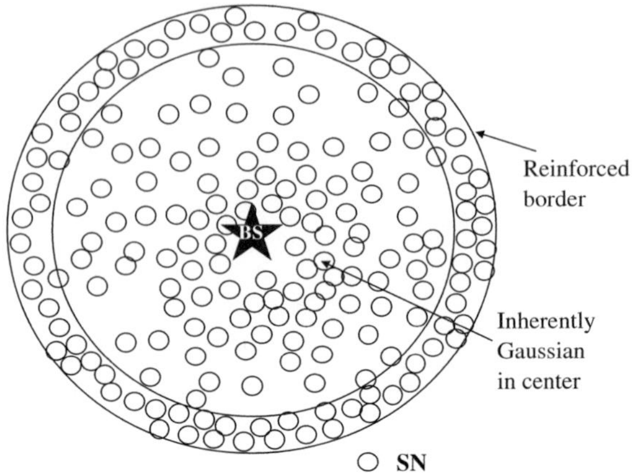

Fig. 13.12 Hybrid Gaussian-Ring deployment

distribution schemes (Fig. 13.12). A quick detection of an intruder in a critical application led to this scheme. It gives high importance to immediate detection, i.e., the distance traveled by intruder before detection should have very low delay such that $D \sim 0$. This model ensures that the borders are well protected and at the same time ensures that the network does not have an energy-hole problem as it has the inherent properties of a Gaussian distribution. The primary challenges in modeling this hybrid deployment are the connectivity of the network and good coverage in the outer ring for early detection. The modeling problem can be thus divided into two parts such as determining the number of sensors required in the Gaussian part using the connectivity requirement (P_c) and determining the number of sensors required in the uniform part using the coverage requirement (P_s). The important point to be noted is that the Gaussian distribution network should have at least one connectivity with the border to ensure that the border is not disconnected from the base station. In the underlying Gaussian deployment, the outermost annulus has the lowest density and ensuring that the minimum connectivity (P_c) is achieved here will ensure the whole network has P_c connectivity. This is because every inner annulus has higher density than the outermost annulus in a Gaussian distribution. Thus, minimum number of nodes that should be present in the outer annulus can be calculated and this is further used to calculate the total number of nodes needed in the Gaussian network for a particular connectivity requirement (P_c). This means a dense border leads to immediate intrusion detection. The outer ring should be well connected within itself and also dense enough to detect any intrusion. Therefore, a ring of SNs can be deployed manually or using uniform or Gaussian distribution.

A comparison of the performance of these 3 deployments of uniform, Gaussian, and Hybrid Gaussian-Ring is done in Fig. 13.13a. An intruder is simulated to be entering the protected area, and if it moves through the sensing range of any node, then that is considered as an intrusion. The metrics used are the number of SNs deployed, the intrusion distance (distance traveled by intruder before first detection), and total SNs that detected the intruder while it is in the area. The types of attacks are edge to center (The intruder enters through a random point on the

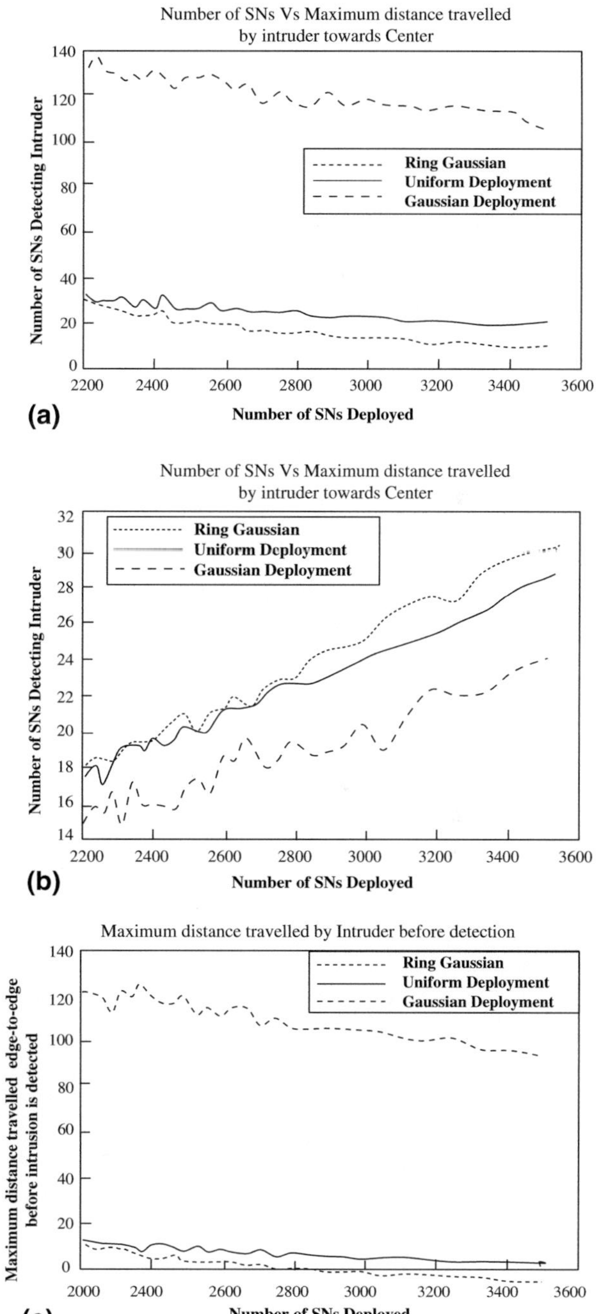

Fig. 13.13 **a** Distance traveled by intruder from border in a hybrid Gaussian-Ring deployment. **b** No. of SNs detecting intruder in a hybrid Gaussian-Ring deployment. **c** Distance traveled by intruder from/to random point in a hybrid Gaussian-Ring deployment

boundary and heads straight to the center of the area) and edge to edge (The intruder enters through a random point on the boundary and leaves through another random point, traveling through the area in a straight line). 100 SNs are deployed in an area of 500 unit radius, and sensing range is set to 10 units. A plot of Fig. 13.13b gives an idea of how many sensors detect an intruder as it moves from a point on the boundary toward the center where the base station is situated. It can be observed that the hybrid deployment performs very well when the intruder just enters the network. This is because of the "reinforced" border that is present. When compared to the other two schemes, this means that the intruder is detected by a higher number of sensors as soon as it enters the WSN. A plot in Fig. 13.13 shows the maximum distance the intruder travels before it is detected while moving from the border to the center of the network. It can be seen that, as expected, Gaussian performs very badly because its borders are very sparse. At the same time, our hybrid scheme shows to have a slight advantage over uniform distribution as the scheme has a reinforced border and that is an advantage over uniform distribution too. A plot of Fig. 13.13c shows the total number of detections of the intruder by the SNs during the time it takes to travel from the boundary to the center. Even in this case, both uniform deployment and hybrid Gaussian-Ring deployment have similar results while the Gaussian deployment performs badly. A plot of Fig. 13.13 shows the distance traveled by an intruder, before being detected, form a random point of the boundary to another random point on the boundary. It can be seen that Gaussian deployment performs very badly in this scenario when compared to uniform and hybrid deployments due to an "edge-to-edge" scenario as the intruder has a high chance of moving in and out of the network without actually coming near the center. Once again, the stronger boundary of the hybrid deployment proves to give a slightly better detection time than uniform deployment.

13.4 Maintaining Anonymity

There could be different levels of anonymity in a WSN (Fig. 13.14) [6]. Beyond suspicion means that an attacker can see evidence of a sent message, but the sender appears no more likely to be the originator than any other potential sender in the system. Probable innocence implies that the sender may be more likely the originator than any other potential sender, but the sender appears no more likely to be the originator than to not be the originator. Possible innocence considers that the sender appears more likely to be the originator than to not be the originator, but

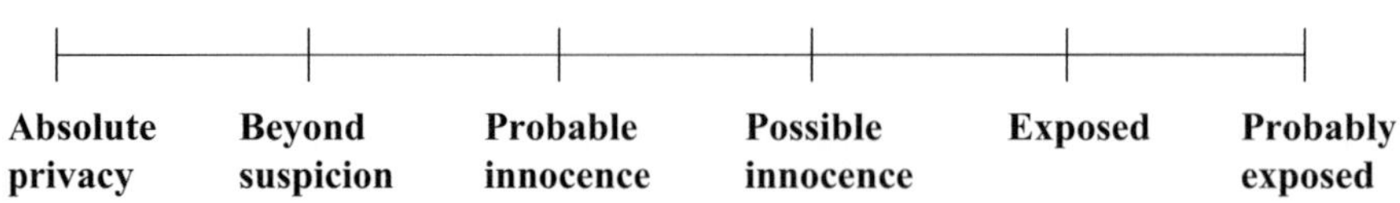

Fig. 13.14 Anonymity levels in a WSN

there is still a non-trivial probability that the originator is someone else. Local eavesdropper implies that it is possible to observe communication to and from the user SNs. Collaborating crowd members means that the crowd members can pool their information and deviate from the protocol. An end server directs to a Web server to which the transaction is directed. A local eavesdropper can see that the user originated a request. When a sender SN is exposed, it typically cannot see the target of the request. Requests are encrypted unless they are submitted to a target server. If the request is encrypted, each end server appears for the attacker equally likely to be the target of the request and leads to beyond suspicion anonymity. If the user's own system submits the request, then the target is exposed and the probability of this is 1/*N,* where *N* is the number of SNs. P(receiver/beyond suspicion) tends to 1 as $N \rightarrow$ infinity. When end server is the target of the request, receiver anonymity is not possible. If anonymity for the originator is strong, user's system always forwards the request to a random member of the crowd ($\sim$hides user identity with a one-time pad), and the end server is equally likely to receive the request from any member of crowd. From the end server perspective, each SN is equally likely to be the originator and sender's beyond suspicion anonymity is guaranteed. In one-hop key establishment system, two more keys are then set (Fig. 13.15a) as follows:

Data encryption key: K^0AB-enc = H(KAB ⊕ C1), where C1 is a constant
MAC function key: K^0AB-mac = H(KAB ⊕ C2), where C2 is a constant
The two keys will change dynamically

$$\text{Data encryption key: } \mathrm{K^{i+1}AB - enc = H\big(K^{i}AB - enc\big)}. \tag{13.15}$$

$$\text{MAC function key: } \mathrm{K^{i+1}AB - mac = H\big(K^{i}AB - mac\big)}. \tag{13.16}$$

Hidden identities (His) are bidirectional as shown in Figs. 13.16 and 13.17.

$$\mathrm{HI_{A\rightarrow B}^{Seq} = H(K_{AB} \oplus ID_B \oplus Seq * rn)}, \text{where rn is a random number.} \tag{13.17}$$

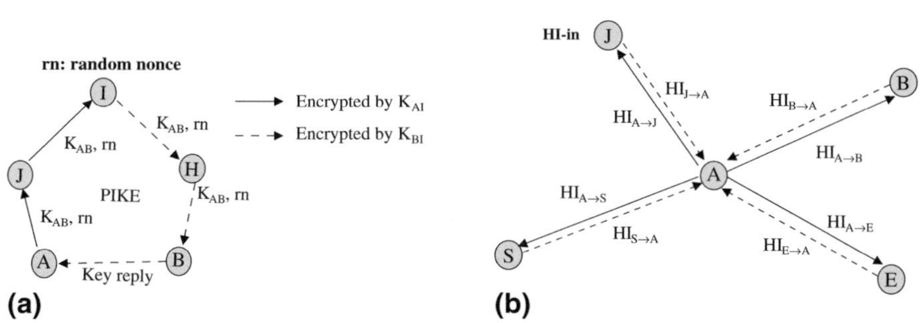

Fig. 13.15 **a** One-hop key establishment. **b** Hidden identity establishment

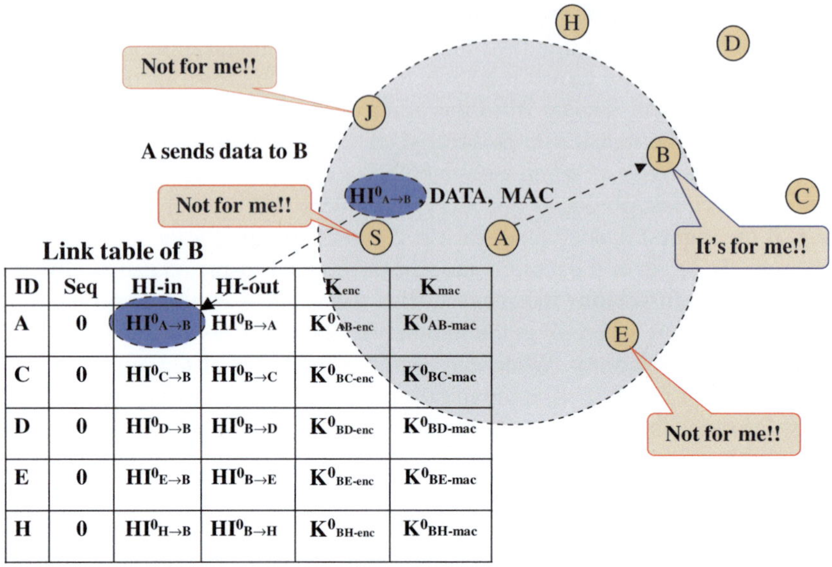

ID	Seq	HI-in	HI-out	K_{enc}	K_{mac}
A	0	$HI^0_{A\rightarrow B}$	$HI^0_{B\rightarrow A}$	$K^0_{AB\text{-}enc}$	$K^0_{AB\text{-}mac}$
C	0	$HI^0_{C\rightarrow B}$	$HI^0_{B\rightarrow C}$	$K^0_{BC\text{-}enc}$	$K^0_{BC\text{-}mac}$
D	0	$HI^0_{D\rightarrow B}$	$HI^0_{B\rightarrow D}$	$K^0_{BD\text{-}enc}$	$K^0_{BD\text{-}mac}$
E	0	$HI^0_{E\rightarrow B}$	$HI^0_{B\rightarrow E}$	$K^0_{BE\text{-}enc}$	$K^0_{BE\text{-}mac}$
H	0	$HI^0_{H\rightarrow B}$	$HI^0_{B\rightarrow H}$	$K^0_{BH\text{-}enc}$	$K^0_{BH\text{-}mac}$

Fig. 13.16 One-hop communication for hidden identity

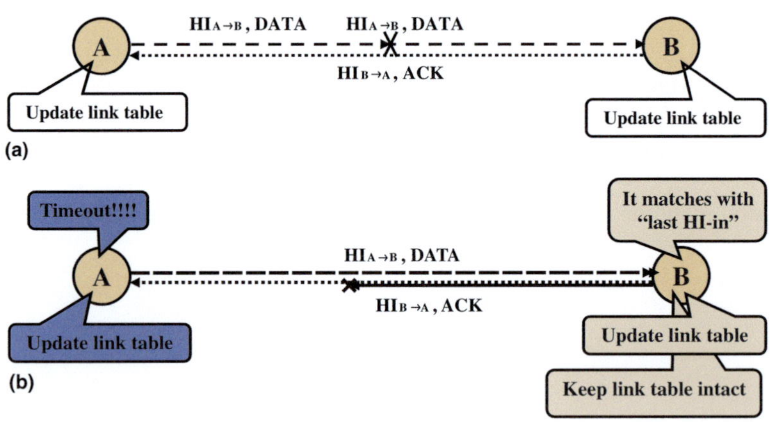

Fig. 13.17 **a** 1-hop acknowledgement. **b** ACK lost

$$HI^{Seq}_{B\rightarrow A} = H(K_{BA} \oplus ID_A \oplus Seq * rn), \text{ where rn is a random number.} \quad (13.18)$$

A sends data to SN B. To solve packet loss problem, B updates sequence number and hide identity (HI). But, A does not, making sequence numbers and HIs different. A simple solution is to store last HI-in. Two more pseudonyms are HIPs (hidden identity for routing path) and PathIDs. HIPs are established for any possible source SN and stored in HIP table for each path (A path is represented by

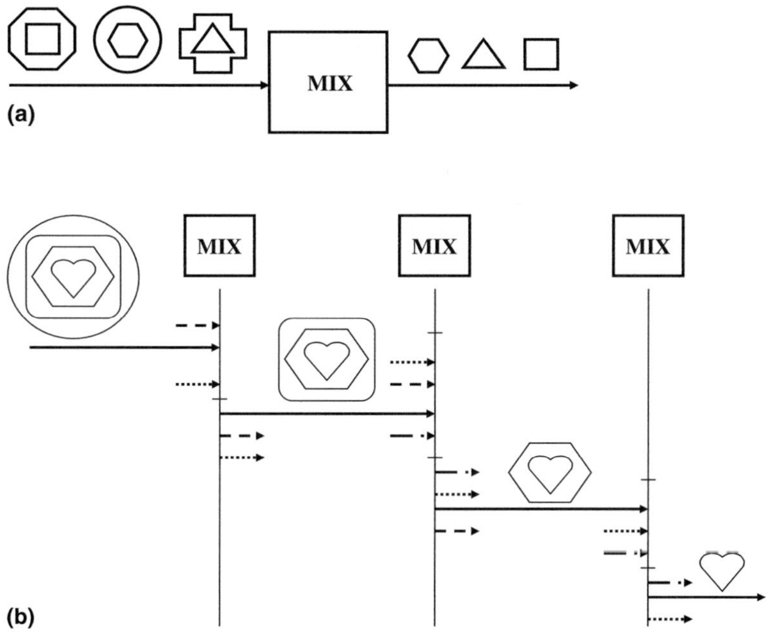

Fig. 13.18 **a** Chaum mixing. **b** Mix chaining

two-end SNs of the path having the source SN and the destination SN). PathIDs are established and used in the routing table. Two types of messages such as anonymous path routing request (APR-REQ) and anonymous path routing reply (APR-REP) are used. Two cases for the source and destination SNs are as follows: one with a predistributed pairwise key as shown in Fig. 13.18a and without predistributed pairwise key.

Defense against colluding compromised to use MIXes as illustrated in Fig. 13.18b, and if a single MIX behaves correctly, unlinkability is still achieved. Onion routing [7] is used to have several anonymous connections that is to be multiplexed on a link as shown in Fig. 13.19a. Connection is identified by a connection ID (ACI) and is unique on a link, but not globally. The application is configured to connect to the application proxy instead of the real destination, and upon a new request, the application proxy decides whether to accept the request, opens a socket connection to the onion proxy, and passes a *standard structure* to the onion proxy. The standard structure contains application type (e.g., HTTP, FTP, and SMTP, retry count (number of times the exit funnel should retry connecting to the destination), format of address that follows (e.g., NULL-terminated ASCII string), address of the destination (IP address and port number), and waits response from the exit funnel before sending application data.

Upon reception of the standard structure, the onion proxy decides whether to accept the request, establishes an anonymous connection through some randomly selected onion routers by constructing and passing along an *onion,* and sends the

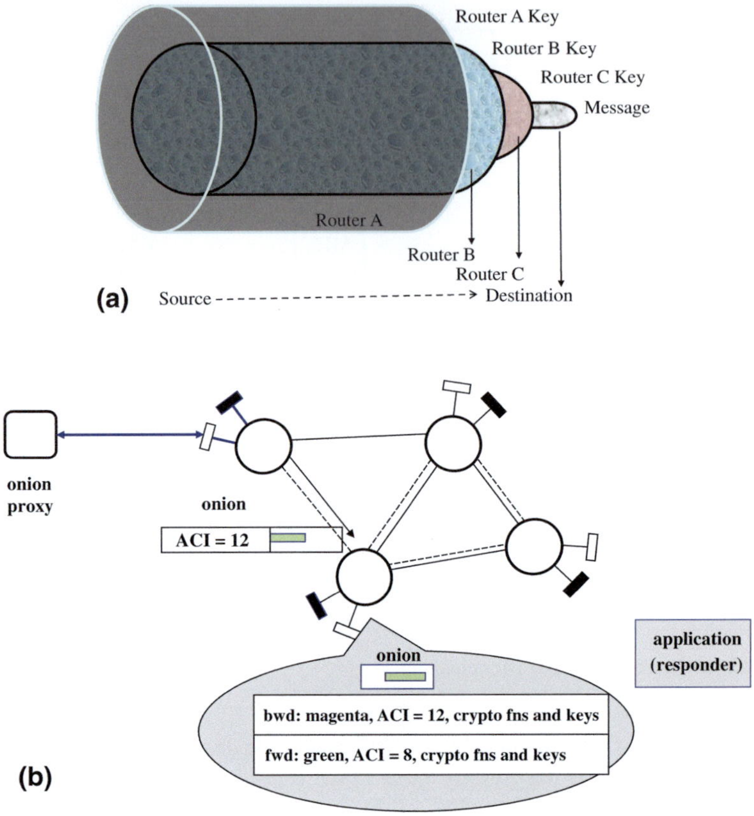

Fig. 13.19 **a** Onion routing . **b** Anonymous connection setup using onion routing

standard structure to the exit funnel of the connection. After that, it relays data back and forth between the application proxy and the connection. Upon reception of the standard structure, the exit funnels and tries to open a socket connection to the destination. It sends back a one-byte status message to the application proxy through the anonymous connection (in backward direction) (Fig. 13.19b). If the connection to the destination cannot be opened, then the anonymous connection is closed; otherwise, the application proxy starts sending application data through the onion proxy, entry funnel, anonymous connection, and exit funnel to the destination. Private information is processed and stored extensively by various individuals and organizations. The location of user is not known. In this way, complete and meaningful profiles on people can be created and information technology makes this easier as there is no compartmentalization of information. The cost of storage and processing (data mining) is decreasing as technology is available to everyone. The proxy (Fig. 13.19b) encrypts the data with 5's public key followed by 3 and then 4. Thus, an onion is created which looks like

$$E_{4pu}\left(3'\text{s IP address}, E_{3pu}\left(\left(5'\text{s IP address}, \left(E_{5pu}(\text{recipient's IP address, data})\right)\right)\right)\right). \tag{13.19}$$

The proxy then sends the onion to the first onion router, i.e., 4. Onion router 4 peels the outer layer of the onion using its private key, and it forwards the onion to 3 which now looks like

$$E_{3pu}\left(\left(5'\text{s IP address}, \left(E_{5pu}(\text{recipient's IP address, data})\right)\right)\right). \tag{13.20}$$

Onion router 3 peels the outer layer of the onion using its private key and it forwards the onion to 5 which now looks like (E_{5pu} (recipient's IP address, data)). Onion router 5 now peels the outer layer of the onion using its private key. It finds plain data and the destination address and forwards it to the destination. This allows the source node S to establish a path up to the destination using intermediate nodes. The beauty of this phase is that none of the intermediate nodes can discover the identity of any of the participating nodes except its neighbors. The source S creates a *path discovery* packet and broadcasts it as private data are protected from abuse by unauthorized entities. So, what do we want to hide? Basically, we want to preserve sender's anonymity so that an attacker cannot determine who the sender of a particular message is and the receiver anonymity so that an attacker cannot determine who the intended receiver of a particular message is. The attacker may determine senders and receivers but not the associations between them. We want to hide this from the communication partner (sender anonymity) and external attackers. Local eavesdropper (sniffing on a particular link (e.g., LAN)), global eavesdropper (observing traffic in the whole network), and internal attackers (colluding) compromise with system elements (e.g., routers). In the forward direction, the onion proxy adds all layers of encryption as defined by the anonymous connection, each onion router on the route removes one layer of encryption, and responder application receives plaintext data. In the backward direction, the responder application sends plaintext data to the last onion router of the connection (due to sender anonymity, it does not even know who is the real initiator application), each onion router adds one layer of encryption, the onion proxy removes all layers of encryption, and anonymous connections are terminated by the initiator, the responder, or one of the onion routers in the middle.

A special DESTROY message is propagated by the onion routers. If an onion router receives a DESTROY message, it passes it along the route (forward or backward) and sends an acknowledgement to the onion router from which it received the DESTROY message. If an onion router receives an acknowledgement for a DESTROY messages, it frees up the corresponding the connection ID. Every message of 48 bytes is encrypted with DES. If the payload is already encrypted (e.g., it carries (part of) an onion), then only the cell header is encrypted. Cells of different connections are mixed, but order of cells of each connection is preserved. The onion routing (TOR) [8] is a distributed overlay network designed to anonymize TCP-based applications and provides circuit-based, low-latency anonymous communication service. TOR provides benefits over previous onion routing protocols, as it provides perfect forward secrecy with telescoping path-building design and keys refreshed periodically. It possesses better deploy ability, and there is no mixing, padding, or traffic shaping. In fact, many TCP streams can share one circuit with less overhead. Its leaky-pipe circuit topology allows traffic to depart anywhere in circuit, and traffic shaping and volume attacks become harder. There is no need for directory servers, and no more flooding state is needed through the network and has variable exit policies. Different users are happy with different levels of involvement, and end-to-end integrity checking is done. An overlay of onion routers' routes is used to route traffic and is maintained with long-term identity key and short-term onion key.

13.5 Base Station Location Anonymity

With a wide range of applications such as military and health, WSN is prone to attacks and security becomes a concern such as monitoring a persons' health, WSN monitoring, and a hostile area in military applications. Restricted power source and limited computation capability of sensors make it more difficult to defend WSNs from security attacks. Energy efficient systems are desirable that require less computation to defend WSNs. Security in WSNs can be classified as follows: (1) content-oriented privacy (can be secured through encryption algorithms) and (2) context-oriented privacy (protecting the contextual information in a WSN that refers to the physical locations of SN in WSN).

By knowing the physical location of SNs, the attacker can introduce DoS or physical attacks. For example, in a contextual attack, an adversary can launch a DOS attack in a military application, or can cause physical damage to the BS to which all the SNs account their events, or the attacker/adversary can capture the animals in an environmental monitoring application, if the physical location of the SNs reporting action is known.

We concentrate on protecting the location of BS. Since the base station forms the bridge between the sensor and the user, a complete sensor network or a part of it can be rendered useless if the base station or the sink node is taken down. Challenges in devising a scheme/technique protect contextual privacy as it is important to take care of the energy consumption and keep the delay as low as possible to maintain usefulness of data. The adversary can be assumed to have different capabilities such

Fig. 13.20 **a** Power control for security. **b** BLAST scheme for security

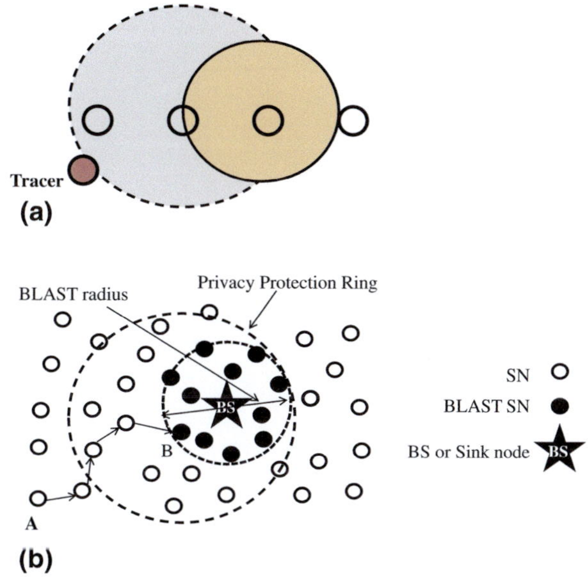

as local and global adversary. Local adversary with a single mobile attacker has to move around to listen to packets, estimate location of the transmitter from the angle of arrival and signal strength and interpret the direction of traffic to determine location of BS. A global adversary can hear all the transmissions in the network by deploying an adversarial network over an existing WSN. The adversary detects the traffic rate at each location, SNs nearer to the BS generate higher traffic rates than SNs far away from the BS, and the adversary can find the physical location of the BS based on traffic rates. Packet tracing in promiscuous mode allows the adversary not to wait at a location to analyze the traffic as it can find the next transmitter of the packet and move along. The adversary can move hop by hop from source SN to BS and is more efficient than the traffic analysis attack as the adversary keeps moving closer to BS. Other approaches that have been used include injecting and routing fake packets in different directions to diverge the attacker from an actual path. Another option is to have multiple BSs as fake BS creates high traffic areas at different locations in the network to mislead the attacker. Another alternative is to route packets along a random path to obscure attacker in choosing the next hop SN in the path. Another approach is to employ random transmission timing in forwarding a packet in place of forwarding it as soon as it arrives. A generic scheme could utilize a combination of these techniques (Fig. 13.20).

In a recent work [9], power control has been employed by varying transmission range of the SNs in maintaining location privacy. SNs have the ability to vary their

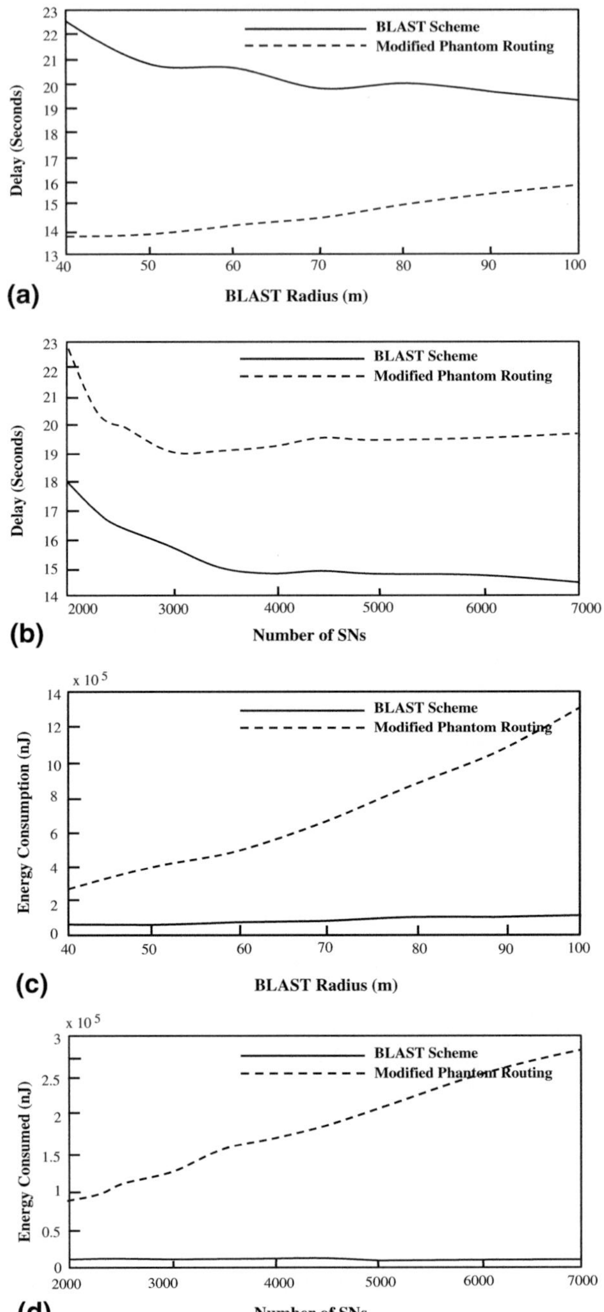

Fig. 13.21 **a** Delay with BLAST radius. **b** Delay with SNs. **c** Energy consumption with BLAST radius. **d** Energy consumption with SNs

transmission power as lesser the transmission power shorter is the distance to which data can be transmitted and lesser is the energy consumed. Each SN has a transmission range of t_c to $K \times t_c$, where K can be any value such that the product stays under the maximum possible transmission range of the SN. This describes a sensor having K level transmission capability. In BLAST [10], the network is divided into two sets of nodes, namely ordinary nodes and blast nodes, and each ordinary node has a transmission range of t_c. The blast nodes are in a ring or circular region encapsulating the real BS. Each blast node has a transmission range of $K \times t_c$. When a SN has a packet to send, it routes the packet along shortest path to a SN which is randomly picked from the blast SNs, and once the packet reaches a blast node, the node transmits it with a range of $K \times t_c$ to the BS. This transmission is received by all SNs in the ring including the real BS. It is possible to use a random directed path to increase the privacy. As the blast SN is randomly chosen, the path is going to be different each time. A local adversary may estimate the presence of ring after many transmissions and have a picture of the ring with its physical location. The adversary cannot make any further progress toward the real BS as any of these special/blast SNs can be the real BS. A global adversary can trace the ring quicker than a local adversary as it simultaneously gets the information from all over the network. Even the global adversary cannot predict the actual location of the base station within the ring. This technique is compared with modified version of phantom routing [11] (a well-known location privacy technique for protecting the source location by reaching one of the special SNs in a circular area and then the packet is flooded in a controlled manner). In this simulation, the motes can vary their transmission range from 10 to 100 m and the WSN area is 500 m × 500 m with the transmission range $t_c = 20$ m for the non-blast sensor nodes. For simplicity, the actual BS is placed at the center of the ring and the results are averaged of 1000 iterations for 3000 SNs and varied the blast radius. BLAST shows 40% less delay than the MPR (modified phantom routing). Blast radius represents the ring size in case of MPR. Though the curves seem to cross each other if blast radius is extended, it is not practically possible as the maximum transmission range is 100 m. The blast radius is fixed at 100 m while varying the number of SNs. As the density increases, the shortest path routing becomes easier which can be seen. The privacy of the technique [9] increases with the blast radius and the node density (Fig. 13.21).

13.6 Conclusions

Intrusion detection with WSNs is an emerging new area, and various deployment alternatives need to be fully explored to deploy sensors so as to effectively detect activities near interest. The hybrid Gaussian-ring deployment makes use of the advantages of uniform and Gaussian-distributed sensor networks. This provides a better border intrusion detection mechanism while minimizing the number of SNs needed. Power control provides location privacy in WSNs. Modified BLAST scheme leads to location privacy of the BS while keeping the delay and energy low.

13.7 Questions

Q.13.1. How can you make data transfer from SN to BS reliable?
Q.13.2. Why wireless sensor networks are difficult to protect?
Q.13.3. How can you utilize hierarchical architecture for intrusion detection in WSN?
Q.13.4. How can you provide border intrusion detection and surveillance using a wireless sensor network?
Q.13.5. What are the requirements of privacy preserving protocols for wireless sensor networks?
Q.13.6. What will be impact on BS determination if transmission power is varied between max and min values?
Q.13.7. In problem 13.7, can you select power randomly at each forwarding SN toward BS?
Q.13.8. In onion routing, if the destination SN is known, what changes do you have to make it more effective?
Q.13.9. Can you apply onion routing to WSN designed with regular topology?
Q.13.10. How can you discriminate between multiple object classes using a network of binary sensors?
Q.13.11. What are the limitations of modified BLAST scheme and how can you address them?

References

1. P. Hosmer, "Use of Laser Scanning Technology for Perimeter Protection," IEEE A&E Systems Magazine, pp. 13–17, August 2004.
2. Gang Zhou, Tian He, John A. Stankovic, and Tarek Abdelzaher, "RID: Radio Interference Detection in Wireless Sensor Networks," www.alice.virginia.edu/~stankovic/psfiles/RIDPaper.pdf.
3. Yun Wang, Xiaodong Wang, Bin Xie, Demin Wang, and Dharma P. Agrawal, "Intrusion Detection in Homogeneous and Heterogeneous Wireless Sensor Networks," IEEE Transactions on Mobile Computing. Vol. 7, no 6, pp. 698–711, July 2008.
4. Y. Wang, W. Fu, and D. P. Agrawal, "Intrusion detection in gaussian distributed wireless sensor networks," in *6th International Conference on Mobile Adhoc and Sensor Systems, MASS. IEEE, pp. 313–321, 2009.*

5. Narendranad Katneni, Vaibhav Pandit, Hailong Li, and Dharma P. Agrawal, "Hybrid Gaussian-Ring Deployment for Intrusion Detection in Wireless Sensor Networks," IEEE International Conference on Communications (ICC), pp. 6678–6682, 2012.
6. J. -P. Sheu, J. -R. Jiang, and C. Tu, "Anonymous Path Routing in Wireless Sensor Networks," Communications, IEEE International Conference on Communications, pp. 2728–2734, ICC 2008.
7. https://en.wikipedia.org/wiki/Onion_routing.
8. https://en.wikipedia.org/wiki/Tor_(anonymity_network).
9. V. P. V. Gottumukkala, V. Pandit, Hailong Li, and Agrawal, D.P., "Base-station Location Anonymity and Security Technique (BLAST) for Wireless Sensor Networks," ICC, 2012.
10. http://www.webopedia.com/TERM/B/blast.html.
11. Yun Li, Leron Lightfoot, Jian Ren, "Routing-Based Source-Location Privacy Protection in Wireless Sensor Networks," http://www.egr.msu.edu/~renjian/pubs/05189579.pdf.

Part III
Regular Topology

Chapter 14
Coverage and Connectivity for Regular Deployments

14.1 Introduction

Up till now, we have considered WSN to have a large number of randomly deployed SNs, primarily useful for defense applications and natural disasters. But, they can be utilized using a regular topology pattern if the area is accessible and SNs can be placed at any point within the area. This is especially useful if cost is to be minimized as the area can be covered with minimum number of SNs. Moreover, location of adjacent SNs is known and there is no need for self-organization or localization. In addition, the geometric symmetry allows clustering can be done in a systemic way without much effort. That allows TDMA schedule to utilize easily. Consider deployment of 9 randomly placed SNs in a given area in Fig. 8.4 which can be easily done by 6 SNs if you have freedom to place SN at any desired location (Fig. 14.1). The sensing range is kept the same, and there is no magic. The main reason for reduction is that the overlapped sensing area between two adjacent SNs is minimized and the cost of SN is not low enough to use them in a redundant way. Deployment of homogenous sensors in 2-D matrix with a sensing range of r_s disk model is shown in Fig. 14.2a. Packing circles like this creates small holes in the area, and these holes basically represent small area where physical parameter cannot be predicted. As deployment of non-overlapping sensors, with communication distance of $2r_s$ leaves some areas uncovered and deployment of full coverage.

Given the SNs' sensing and communication ranges and bounded monitored field, SNs need to be placed at $\sqrt{3}r_s$ apart that can fully cover the area as shown in Fig. 14.2b. So, given square region can be filled with circles without overlap (with holes) or with overlap (no holes) and needed sensing range r_s and the corresponding fraction of area covered is given in Table 14.1 while Table 14.2 shows required sensing range with overlaps in sensing area of each SN and no gaps.

So, the objective should be how to deploy SNs for full coverage, with minimum overlapping sensing areas. One approach is to deploy SNs as needed to have

D.P. Agrawal, *Embedded Sensor Systems*,
DOI 10.1007/978-981-10-3038-3_14

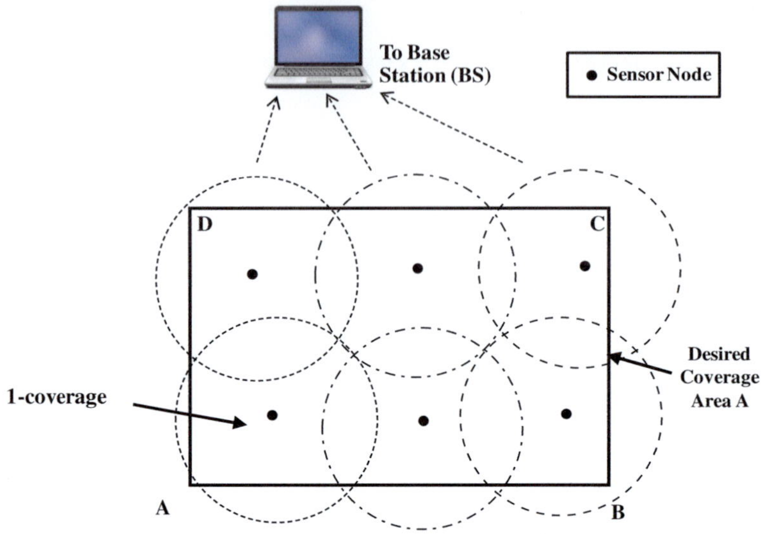

Fig. 14.1 Regular deployment of SNs

maximum coverage and minimum overlapping areas, and such packing of circles results in circles of different sizes. Such a sequential packing-based deployment algorithm (SPDA) is given in Table 14.3, and one example is shown in Fig. 14.3.

14.2 Mesh and Other Topologies

Examples of circle packing-generated embedding of expansion complexes of 2-D mesh commonly have 4 neighbors, while many other alternatives of 3, 6, and 8 neighbors are possible as illustrated in Fig. 14.4 [3]. Simulation of different 512 SNs 32 × 16 2-D and 8 × 8 × 8 3-D mesh topologies has been done to understand the effect of different parameters [3]. Power consumed in transmitting/receiving a message is taken as 50 nJ/bit, while the sender consumes extra 100 pJ/bit/m^2 to strengthen the transmitting signal with interference present and denoted by E_{amp}. The distance between the two neighbors is kept as 0.5 m, and packet length is of 512 bits. Energy consumed in transmission and reception is given by (Table 14.4):

$$\begin{aligned} E_{Tx}(k,d) &= E_{\text{elec}} * k + \varepsilon_{\text{amp}} * k * d^2 \\ E_{Rx}(k,d) &= E_{\text{elec}} * k. \end{aligned} \tag{14.1}$$

Three topologies of rectangular, triangular, and hexagonal tiles allow the repeated use to cover additional areas and are shown in Fig. 14.5. Voronoi diagram is drawn for each case to determine how much non-overlapping area is covered by each module and is shown in Table 14.5. Clearly, triangular topology has less

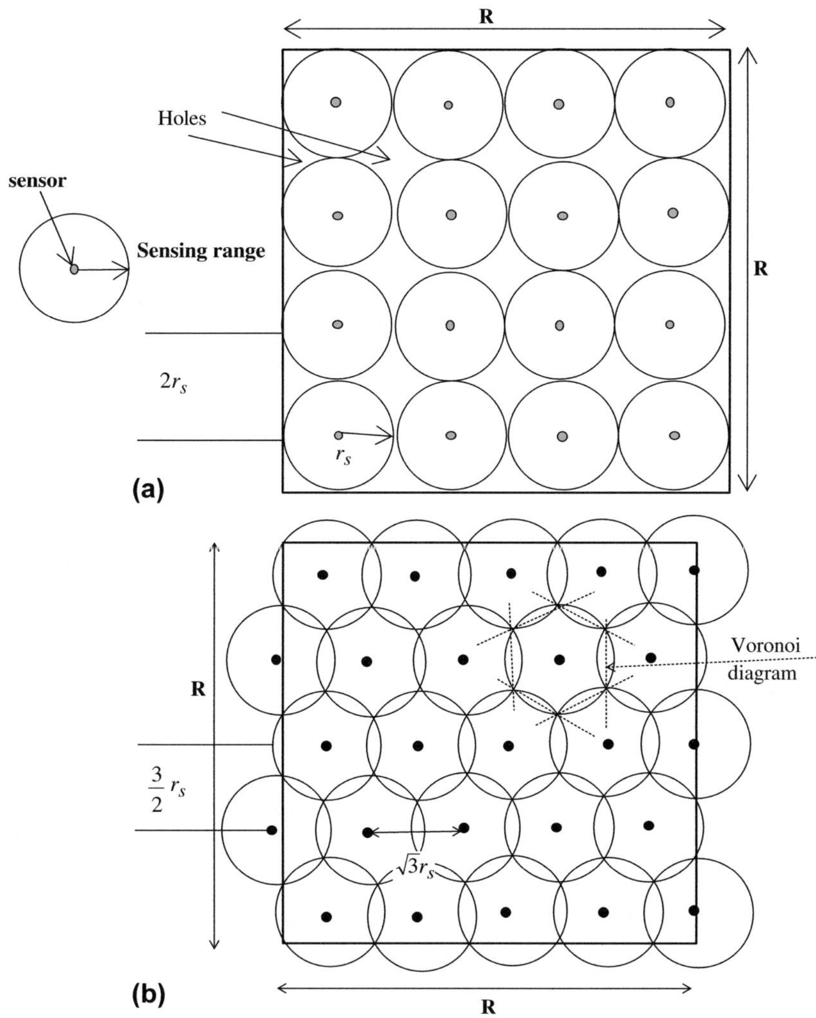

Fig. 14.2 **a** 2-D matrix deployment with holes. **b** 2-D matrix deployment with no holes

Table 14.1 Deployment of non-overlapping SNs [1]

# SNs	1	2	3	4	5	6	7
Sensing range r_s/R	0.500000000000	0.292893218813	0.292893218813	0.292893218813	0.207106781187	0.207106781187	0.207106781187
Fraction of $R \times R$ area sensed	0.785398163397	0.785398163397	0.785398163397	0.785398163397	0.785398163397	0.785398163397	0.785398163397

# SNs	8	9	14	16	25	36
Sensing range r_s/R	0.170540688701	0.170540688701	0.170540688701	0.170540688701	0.170540688701	0.170540688701
Fraction of $R \times R$ area sensed	0.730963825254	0.730963825254	0.730963825254	0.730963825254	0.730963825254	0.730963825254

Table 14.2 Deployment of overlapping SNs with no gap [2]

# sensors	Required sensing range r_s/R	# SNs	Required sensing range r_s/R
1	0.70710678118654752440	16	0.16942705159811602395
2	0.55901699437494742410	17	0.16568092957077472538
3	0.55901699437494742410	18	0.16063966359715453523
4	0.35355339059327376220	19	0.15784198174667375675
5	0.32616058400398728086	20	0.15224681123338031005
6	0.29872706223691915876	21	0.14895378955109932188
7	0.27429188517743176508	22	0.14369317712168800049
8	0.26030010588652494367	23	0.14124482238793135951
9	0.23063692781954790734	24	0.13830288328269767697
10	0.21823351279308384300	25	0.13354870656077049693
11	0.21251601649318384587	26	0.13176487561482596463
12	0.20227588920818008037	27	0.12863353450309966807
13	0.19431237143171902878	28	0.12731755346561372147
14	0.18551054726041864107	29	0.12555350796411353317
15	0.17966175993333219846	30	0.12203686881944873607

Table 14.3 Sequential packing-based deployment algorithm (SPDA) [2]

SPDA algorithm
1: Select a potential SN s_i to be deployed
2: IF (i = 0) then
3: Deploy the SN at one of the field's corners such that it touches 2 boarders
4: ELSE
5: Computer the potential placement points such that the current SN is connected to one of deployed SNs and
6: The SN has to touch at least 2 items (2 SNs, 2 boarders, one boarder and one SN)
7: If no potential point satisfies step 5 and 6
8: Place the SN at a point that gives minimum overlapping
9: ELSE
10: Computer d_y between each potential point for the current SN and the untouched deployed SNs
11: Computer d for each point.
12: Place the SN at the point that has the minimum value.
13: End IF
14: IF there is no more SNs and/or the field is totally covered go to 18.
15: ELSE go to 1.
16: End IF
17: End IF
18: Compute the coverage contribution of each SN
19: Stop.

overlap as sensing area is still considered as circular in nature (disk model). This is once again verified by Tables 14.6, 14.7 and 14.8 that show fewer SNs required if they are placed in a triangular pattern. That confirms that overlap in the sensing area is minimum, leading to fewer SN requirements (Fig. 14.6; Table 14.10).

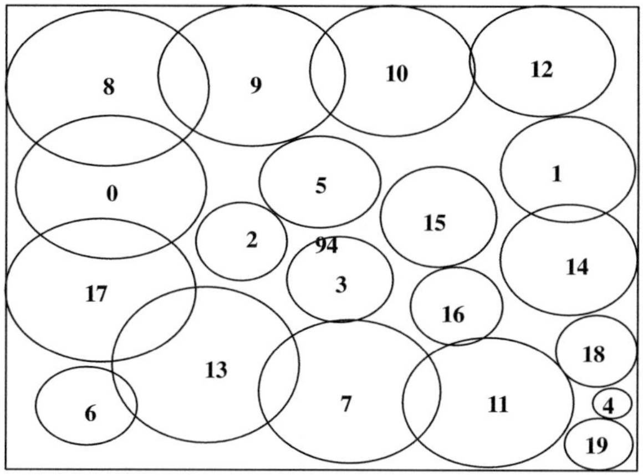

Fig. 14.3 Sequential packing-based deployment of SNs

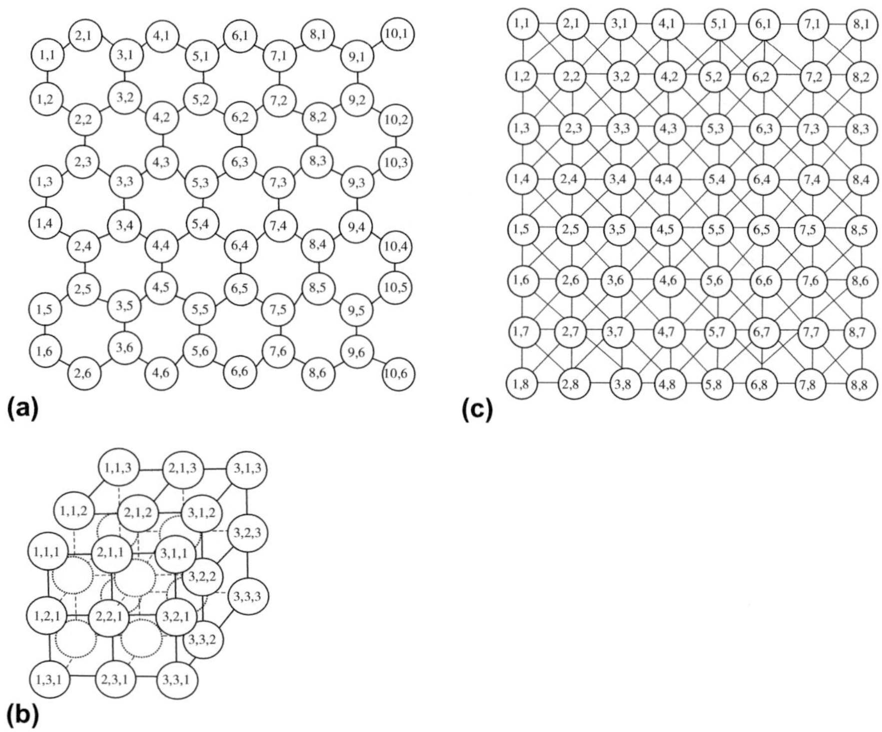

Fig. 14.4 **a** 2-D mesh with 2 neighbors. **b** 2-D mesh with 6 neighbors. **c** 2-D mesh with 8 neighbors

Table 14.4 Power consumption in transmission, broadcasting, and time delay [3]

Topology/ connectivity	Power consumption in transmission			Power consumption in broadcasting			Time delay in broadcasting
	Transmit	Receive	Consumption in Joules	Transmit	Receive	Consumption in Joules	
2-D-3	255	765	2.61×10^{-2}	301	798	2.81×10^{-2}	46
2-D-4	170	680	2.18×10^{-2}	208	714	2.36×10^{-2}	45
2-D-8	102	816	2.35×10^{-2}	143	895	2.66×10^{-2}	31
3-D-6	124	744	2.22×10^{-2}	167	815	2.51×10^{-2}	20

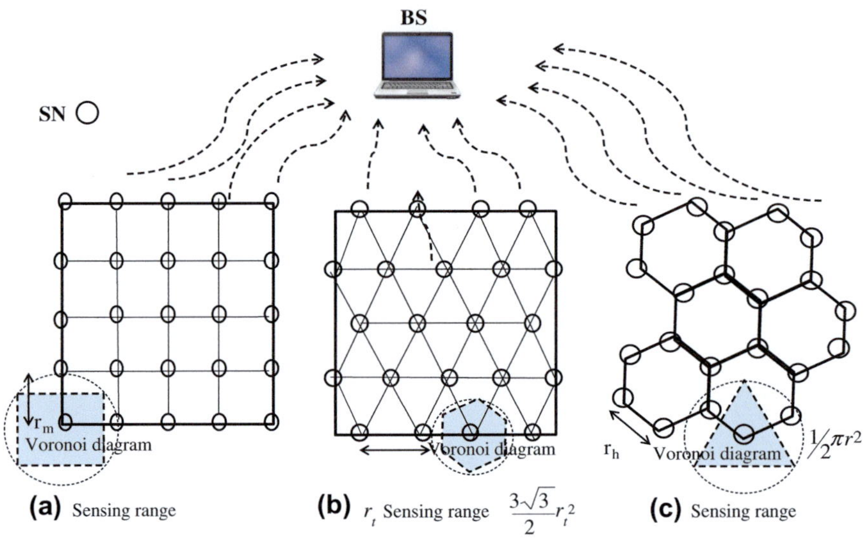

Fig. 14.5 3 regular topologies [4]

Table 14.5 Covered sensing area by three placement topologies

Placement	Distance between adjacent sensors	Area of the topology	Sensing area to be covered by each SN based on Voronoi diagram
Rectangular	rm	rm^2	rm^2
Triangular	r t	$\frac{\sqrt{3}}{2}r_t^2$	$\frac{3\sqrt{3}}{2}r_t^2$
Hexagon	r h	$\frac{3\sqrt{3}}{2}r_h^2$	$\frac{\sqrt{3}}{2}r_h^2$

Deployment efficiency versus ratio of r_s to r_c is given in Table 14.9 that gives a general guideline about which topology to use for different values of sensing range r_s and communication range r_c.

Table 14.6 Required SNs for the **full coverage** to cover area A = 100 * 100 units

r_s	0.5	1	1.5	2	2.5	3	3.5	4	4.5	5
Square	20,000	5000	2222	1250	800	556	408	313	247	200
Hexagon	30,792	7698	3421	1925	1232	855	628	481	380	308
Triangle	15,396	3849	1711	962	616	428	314	241	190	154
r_s	5	6	7	8	9	10	11	12	13	14	15
Square	200	139	102	78	62	50	41	35	30	26	22
Hexagon	308	214	157	120	95	77	64	53	46	39	34
Triangle	154	107	79	60	48	38	32	27	23	20	17

Table 14.7 Required SNs for the **full connectivity** to cover area A = 100 * 100 units

r_c	0.5	1	1.5	2	2.5	3	3.5	4	4.5	5
Square	40,000	10,000	4444	2500	1600	1111	816	625	494	400
Hexagon	30,792	7698	3421	1925	1232	855	628	481	380	308
Triangle	46,188	11,547	5132	2887	1848	1283	943	722	570	462
r_c	5	6	7	8	9	10	11	12	13	14	15
Square	400	278	204	156	123	100	83	69	59	51	44
Hexagon	308	214	157	120	95	77	64	53	46	39	34
Triangle	462	321	236	180	143	115	95	80	68	59	51

Table 14.8 Required SNs for the **full coverage and connectivity** to cover area A = 100 * 100 units

Lattice type	$r_s = 5$	$r_s = 6$	$r_s = 7$	$r_s = 8$	$r_s = 9$	$r_s = 10$	$r_s = 11$	$r_s = 12$	$r_s = 13$	$r_s = 14$	$r_s = 15$
Square	200	139	102	79	62	50	42	35	30	26	23
Hexagonal	924	642	472	361	286	231	191	161	137	118	103
Triangular	154	107	79	61	48	39	32	27	23	20	18

Table 14.9 Deployment efficiency versus ratio of r_s to r_c

Condition	Best deployment pattern
$\frac{r_s}{r_c} \leq \frac{\sqrt{3}}{3}$	Triangle
$\frac{\sqrt{3}}{3} \leq \frac{r_s}{r_c} \leq \frac{\sqrt{43}}{2}$	Square
$\frac{\sqrt{43}}{2} \leq \frac{r_s}{r_c} \leq \frac{\sqrt{2}}{2}$	Triangle
$\frac{\sqrt{3}}{3} \leq \frac{r_s}{r_c} \leq \frac{\sqrt{43}}{2}$	Square
$\frac{\sqrt{43}}{2} \leq \frac{r_s}{r_c} \leq \frac{\sqrt{2}}{2}$	Hexagon
$\frac{r_s}{r_c} \geq 1$	Hexagon

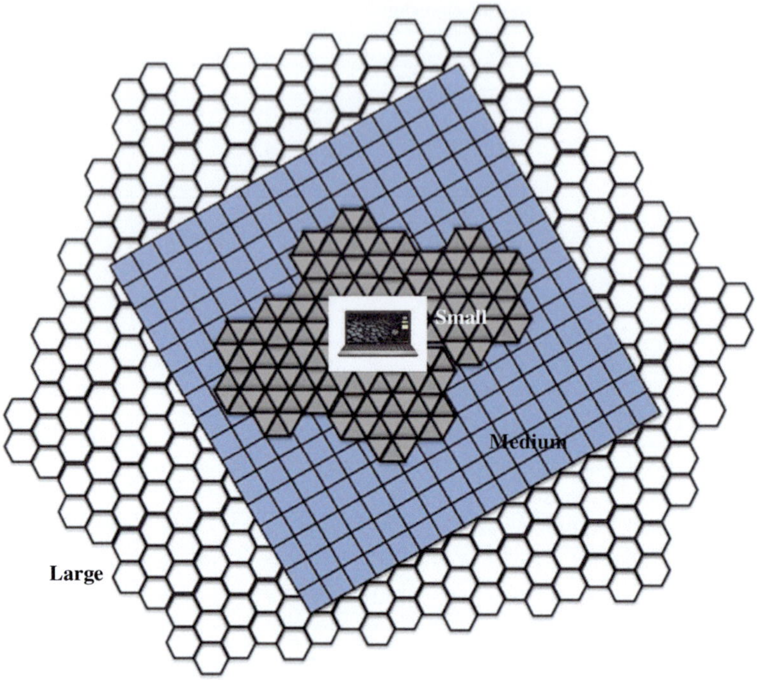

Fig. 14.6 Use of 3 regular topologies for different sizes

Table 14.10 Time delay of different topologies

Covered area	1000 × 100 m	Delay
Length	150 m	100 data packets sent to the goal node with transmission gap of 0.2 s
Sensors in hexagon topology	40	0.0350 s from No. 20 node to node no. 24
Sensors in 2-D grid topology	49	0.0437 s
Sensors in triangle topology	56	To be determined

Throughput and delay in 2-D grid topology are obtained (Fig. 14.7) *using s*imulation on ns-2, which was run for 50 times, and each run lasted for 300 s. *BER is taken as* 0.0000005. It is observed [5] that the optimal number of neighboring SNs for achieving the best throughput is 4. When the number of neighbors is greater than 4, maximum throughput is achieved when some nodes drop interests, i.e., the percentage of nodes not dropping interest is less than 100% as level of interference is reduced and higher throughput is achieved. As the number of neighbors is decreased, optimal percentage of nodes that do not drop interests for

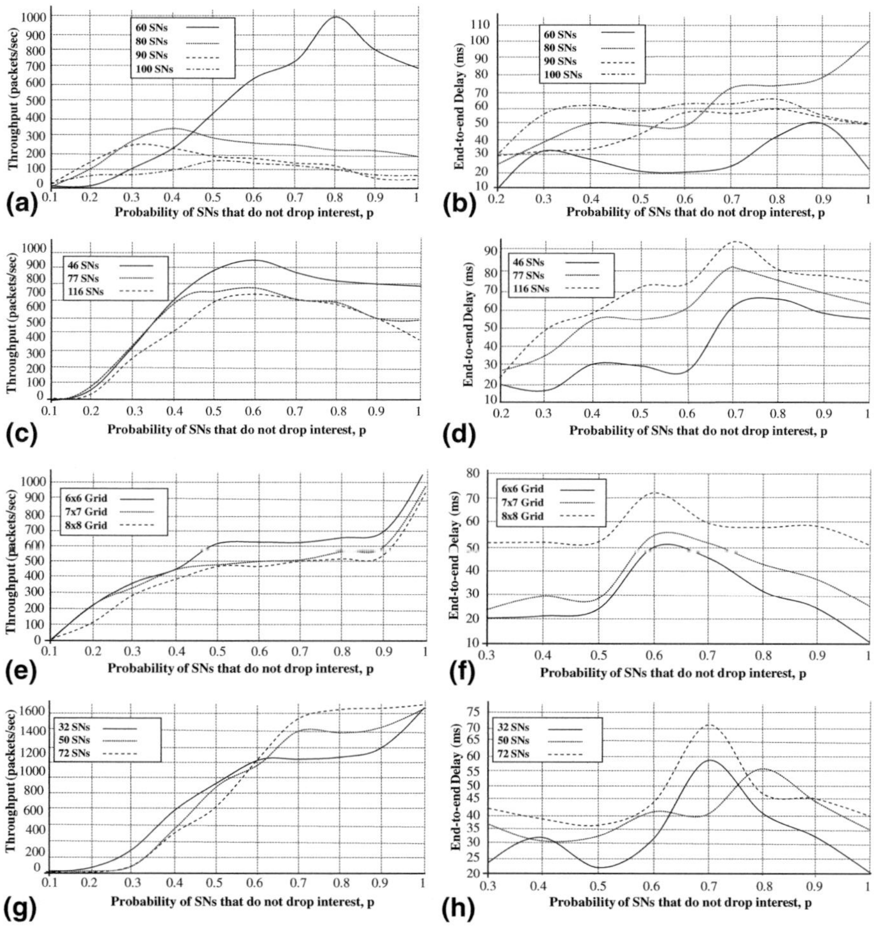

Fig. 14.7 Throughput and delay in different topologies. **a** Throughput in random topology. **b** Delay in random topology. **c** Throughput in triangular topology. **d** Delay in triangular topology. **e** Throughput in 2-D grid topology. **f** Delay in 2-D grid topology. **g** Throughput in hexagon topology. **h** Delay in hexagon topology. **i** Throughput in four topologies. **j** Delay in four topologies. **k** Throughput in four topologies with flooding. **l** Delay in four topologies with full flooding

achieving maximum throughput moves toward 100%. The delay is highest when the percentage of nodes not dropping interest is optimal and random topologies perform worse than topologies with regular tessellation for all densities of SNs. So, a general guideline for comparing random and regular topologies is summarized as in Table 14.11.

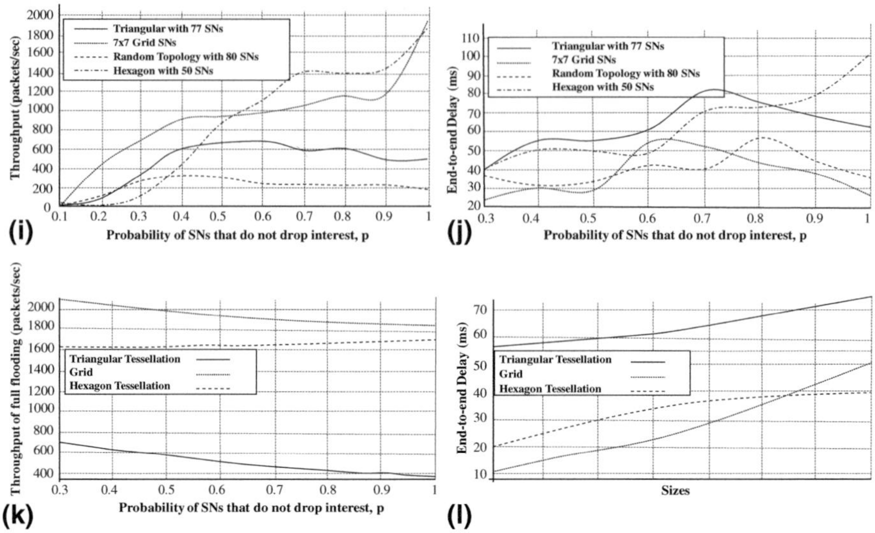

Fig. 14.7 (continued)

Table 14.11 Comparison of random and regular WSNs topologies

WSN topology	Randomly deployed	Regular topology
SN distribution	Random	Uniform
Neighborhood discovery	Use beacon signals to find the nearest neighboring SNs within its communication range and is a very complex process	As SNs are regularly placed, neighboring SN is known ahead of time
Routing	Need to determine path to BS along with intermediate SN neighborhood data	Routing is easy as each SN is placed at a location based on the topology governing rule
Voronoi diagram	Fairly elaborate steps	Easy as SNs are regularly placed

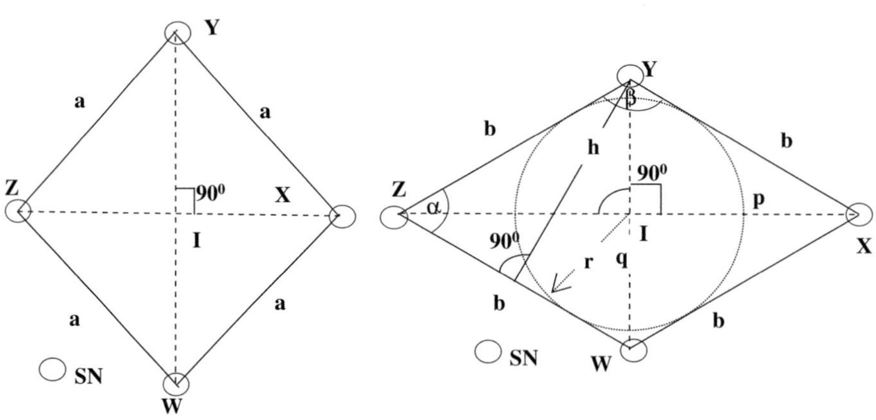

Fig. 14.8 Rhombus topology

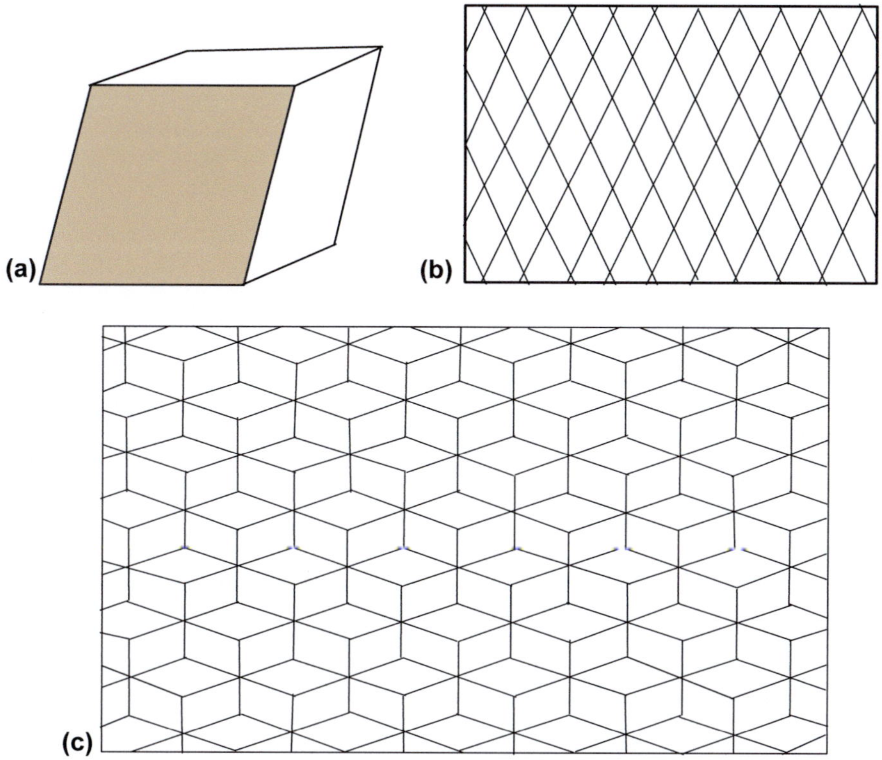

Fig. 14.9 Different rhombus topologies. **a** Rhombohedron. **b** 2-D Tiling. **c** 3D 30°–60° rhombille tiling

Table 14.12 Optimal regular deployment pattern according to the radius proportion

Optimal deployment pattern	Ratio of sensing radius and communication radius
Regular hexagon	$0 < r_c/r_s \leq \frac{1}{2}3^{\frac{3}{4}}$
Square	$\frac{1}{2}3^{\frac{3}{4}} < r_c/r_s \leq \sqrt{2}$
Rhombus	$\sqrt{2} < r_c/r_s < \sqrt{3}$
Regular triangle	$r_c/r_s \geq \sqrt{3}$

Table 14.13 Network lifetime of various 3-D topologies [11]

3-D topology	Size of r_c	Network lifetime compared to theoretical upper bound (%)
Tetrahedron	$\frac{r}{\sqrt{6}} \leq r_c$	60.1
Hexahedron	$r \leq r_c$	34.7
Octahedron		30.1
Dodecahedron		24.1
Icosahedron		21.8

r is the size of each side of polyhedral

14.3 Rhombus and Irregular Topologies

Rhombus is a simple quadrilateral (Fig. 14.8), all of whose four sides have the same length, with an area (Fig. 14.9):

$$A = b \cdot h = b^2 \cdot \sin\alpha = b^2 \cdot \sin\beta = pq/2 = 2br. \tag{14.2}$$

$$\text{Leading to } r = \frac{p \cdot q}{2\sqrt{p^2 + q^2}}. \tag{14.3}$$

This leads to the topology preference as shown in Table 14.12.

Irregular hexagonal pattern can also be used as shown in Fig. 14.9 where (Fig. 14.10) $L = \min\{2\cos(\frac{\theta}{2})r_s, r_c\}$, giving the area A of irregular hexagon as follows:

$$S_p = \left(2\sin\left(\theta - \frac{\pi}{2}\right)L + L + L\right)x\cos\left(\theta - \frac{\pi}{2}\right)L. \tag{14.4}$$

So, the general guideline for selecting a topology is modified as in Table 14.13.

14.4 More Complex Topologies

The number of SNs needed to cover square area of 1000 × 1000 m^2 with $r_s =$ 30 m, 24 m $\leq r_c \leq$ 57 m, and 0.8 $\leq r_c/r_s \leq$ 1.9 is given in Fig. 14.9b, c [6]. Many other patterns are possible and are given in Fig. 14.11 [7]. The corresponding coverage is shown in Fig. 14.12. More complex tiles can be generated and is shown in Fig. 14.13, which can be obtained by subdividing the rectangles as shown in Fig. 14.14 (Fig. 14.15).

14.5 Topologies with K-Connectivity

The number of SNs needed when deployed in 10003 m^3 for the best topology for full coverage and larger connectivity is given in Fig. 14.16 [9]. The sensing range r_s is considered 30 m, and communication range r_c is varied from 15 to 60 m. Most researchers consider hexagon-based pattern for a range of r_c/r_s, and the recent trend is to utilize a collection of squares, hexagons, and octagons. The number of SNs needed by new patterns and hexagon pattern for various values of r_c/r_s when

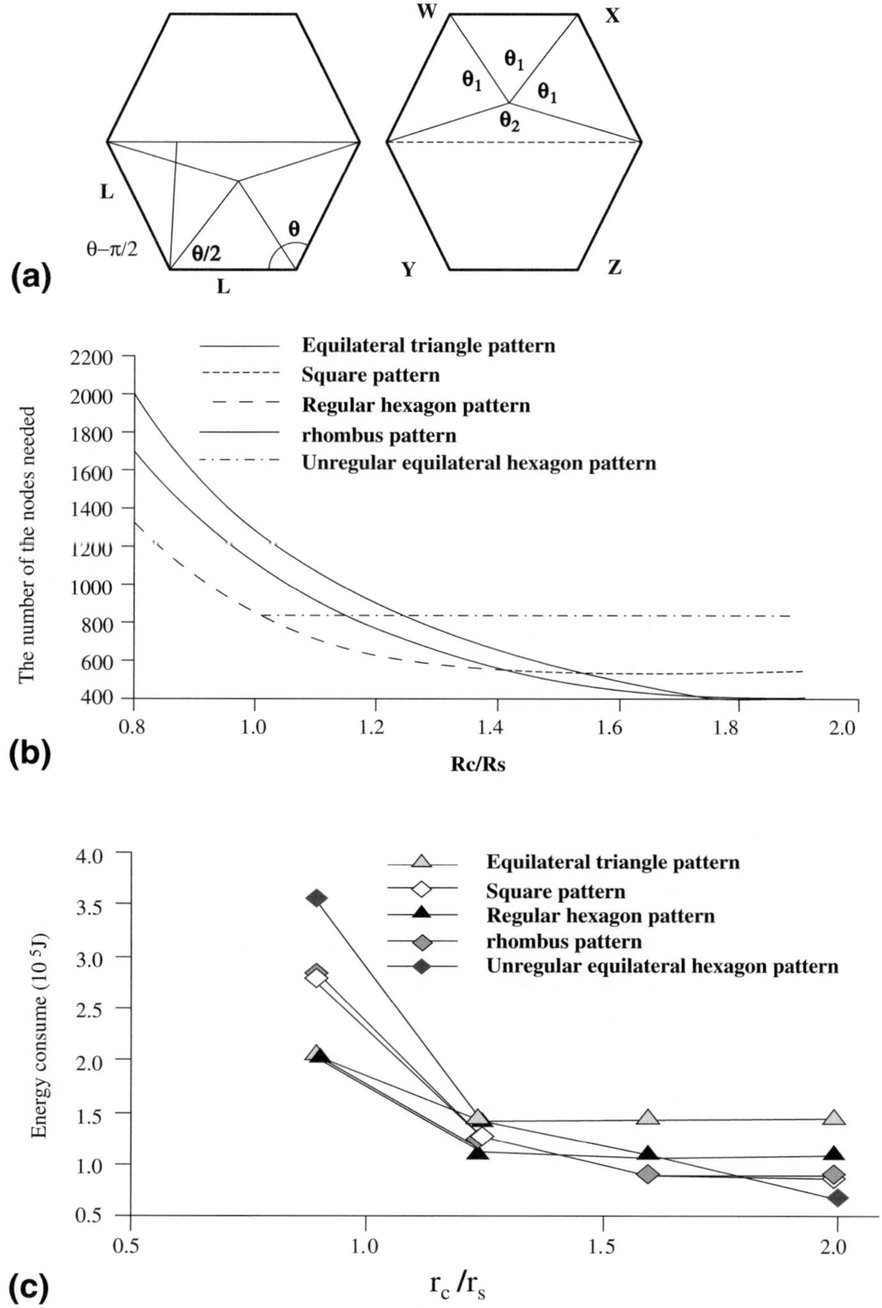

Fig. 14.10 **a** Irregular hexagons. **b** Comparing different topologies as a function of r_c/r_s. **c** Energy consumed [6]

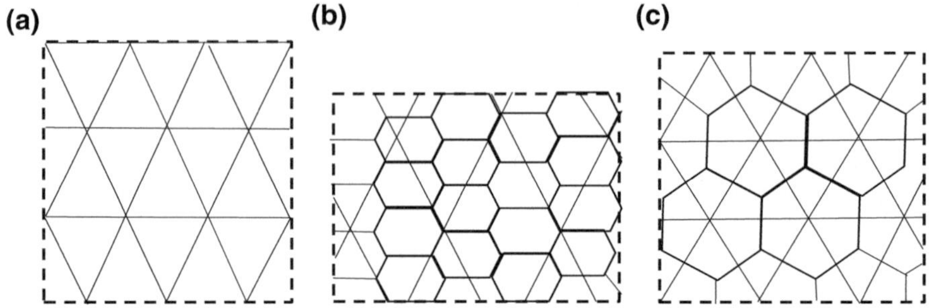

Fig. 14.11 Various topologies possible with SN placed at each cross point [10]

sensors each with r_s = 30 m are deployed over a 1000 m × 1000 m region is given in Fig. 14.17. The mutation in the patterns that achieve the full coverage and 3-connectivity is when r_c/r_s = 1.0459. Similar result for mutating a triangular pattern is given in Fig. 14.18. A different pattern for the full coverage and 3-connectivity is shown in Fig. 14.19. Optimal deployment pattern to achieve the full coverage and 3-connectivity for a full range of r_c/r_s *is* shown in Fig. 14.20. Optimal deployment pattern to achieve the full coverage and 5-connectivity for a full range of r_c/r_s is given in Fig. 14.21. The performance of mutated patterns is given in Fig. 14.22, with each SN with r_s = 30 m deployed over a 1000 m^2 region. r_c is varied from 0.3 r_s to $\sqrt{2}r_s$.

The number of SNs needed for 3-, 4-, and 5-connectivities is shown in Fig. 14.22. Other possible mutations are shown in Fig. 14.23 for larger connectivity.

This can be further extended to 3-D topology, and the network lifetime is given in Table 14.13 [11]. The number of SNs needed to achieve the 4-connectivity and full coverage when r_c/r_s is varied from 1.3 to 1.8 is shown in Fig. 3.24a. Similar figures are shown in Fig. 14.25b when r_c/r_s is varied from 0.9 to 1.43. The number of SNs needed to achieve the 4-connectivity and full coverage when r_c/r_s is varied from 0.3 to 1 is shown in Fig. 14.24c. The impact of coverage from sensing irregularity with 4-connectivity on optimal deployment patterns is illustrated in Fig. 14.25d, with the region size is 1000 m × 1000 m, the average r_s over all directions is 30 m and the standard deviation is taken as 2, 5, 10, and 20, respectively. r_c varied from 10 to 54 m.

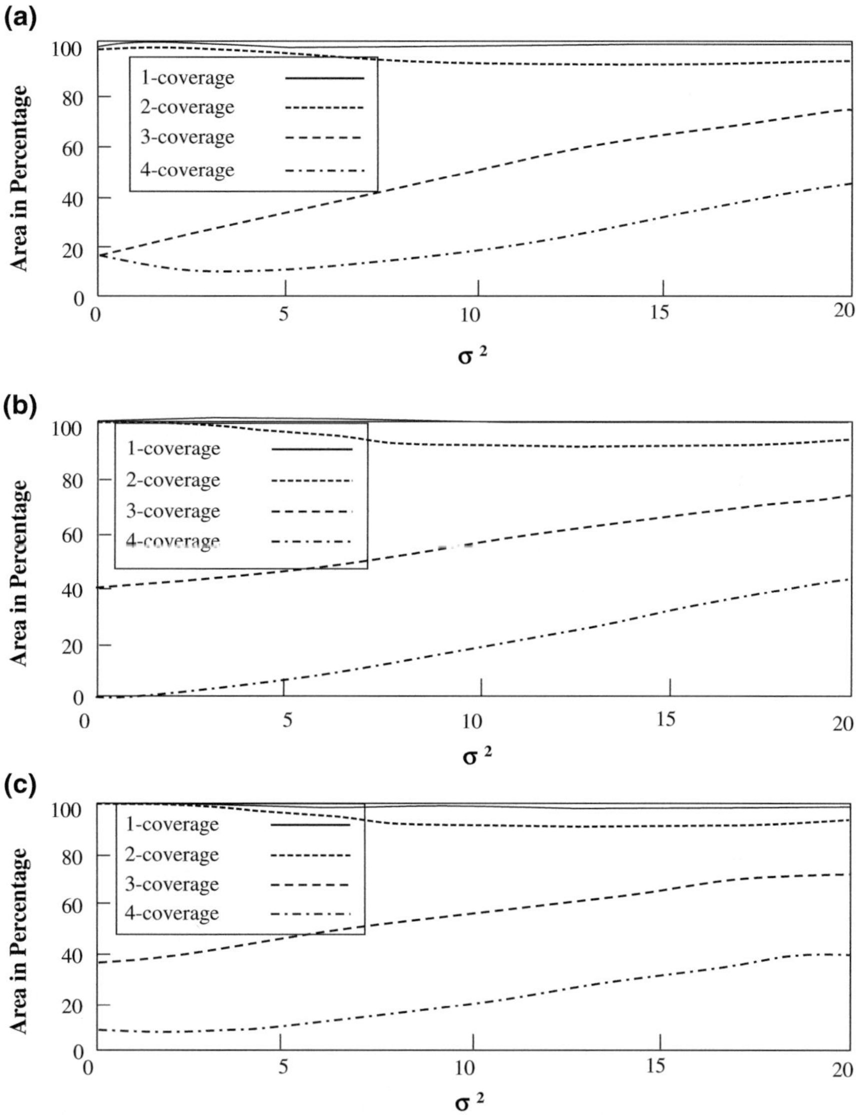

Fig. 14.12 Coverage under sensing irregularity for Fig. 14.10a–c where σ is the coverage density

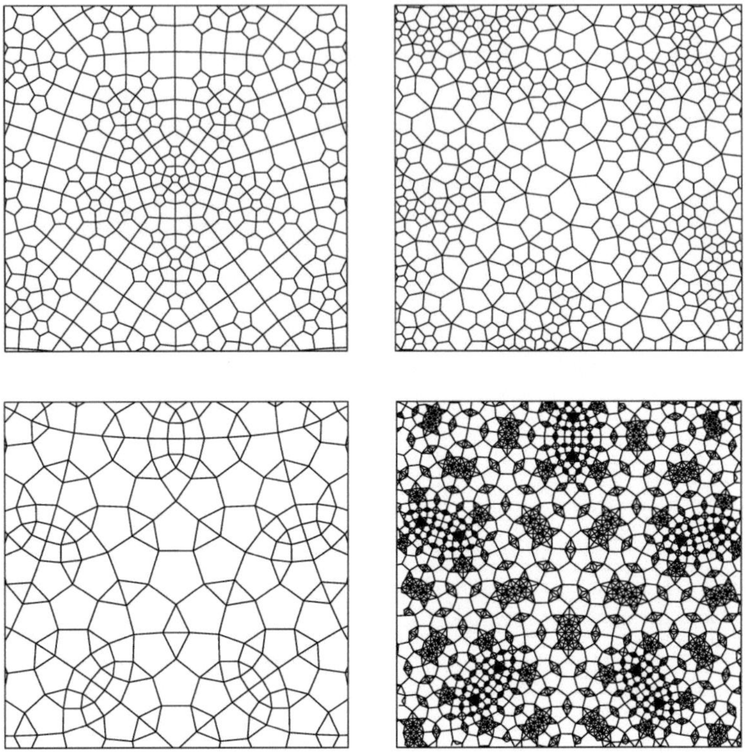

Fig. 14.13 More circle packing [8]

14.6 Connected Coverage with Directional Antennas and C-SNs

Figure 14.26 shows the pattern used by directional antennas. The dotted circles with radius r_s are the sensing range, the solid shaded sectors with radius r_c are the communication range, and the solid lines in Fig. 14.26b, d are wireless propagation of EM waves. Figure 14.26a, b shows the wave shapes which are optimal for nodes with large communication angles, namely $\alpha_c \geq \pi/3$. Two neighboring lines of SNs form a connected and full covering band. Figure 14.26c, d shows the line shape which is optimal for SNs with small communication angles. Every line of SNs forms a linked and full covering band. Connected bands in both the wave and line patterns are connected by SNs spotted on each side. It may be noted that camera sensor nodes (C-SNs) do have a very similar characteristics and could be applied there as well. For simplicity, we will only talk about directional antennas.

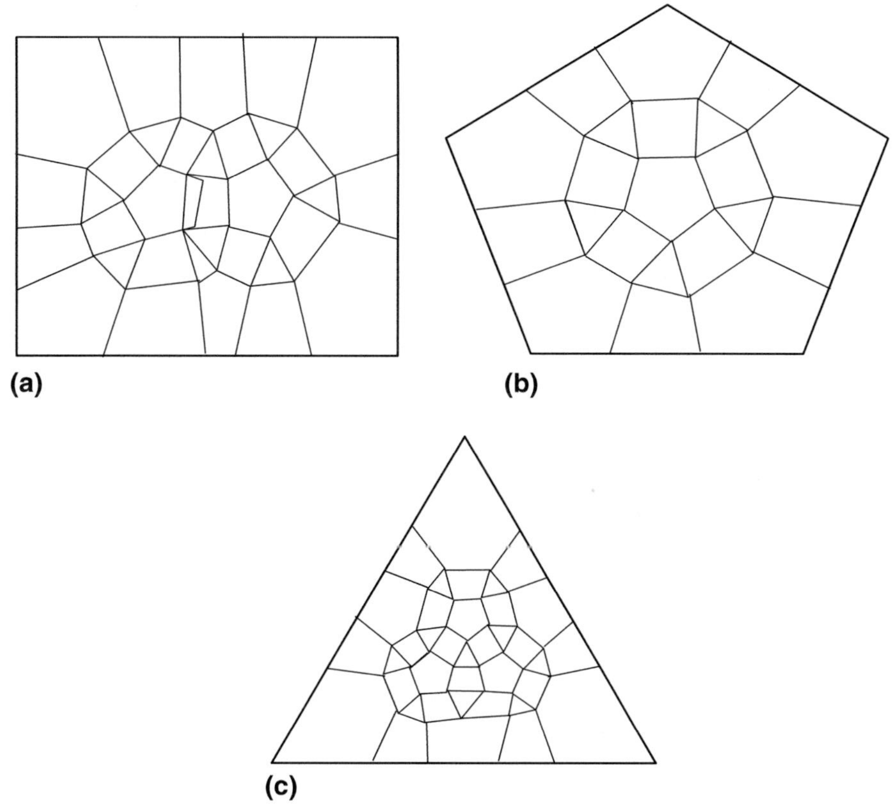

Fig. 14.14 Circle packing of Fig. 14.13 illustrated [8]

In Fig. 14.27a, nodes A and B can communicate with each other. However, in Fig. 14.27b, A may connect to B, but not vice versa. In Fig. 14.27c, nodes A and B point directly to each other, establishing a bidirectional link. In Fig. 14.27d, a circulation chain of connections enables A and B to communicate bidirectionally. Figure 14.28a shows the sector model of the main lobe of the directional antenna, while Fig. 14.28b gives the radiation pattern of a directed antenna. Figure 14.28c shows a doorknob-shaped 3-D radiation model, and Fig. 14.28d illustrates 2-D projection as the doorknob model.

Fig. 14.15 Hyperbolic pattern [8]

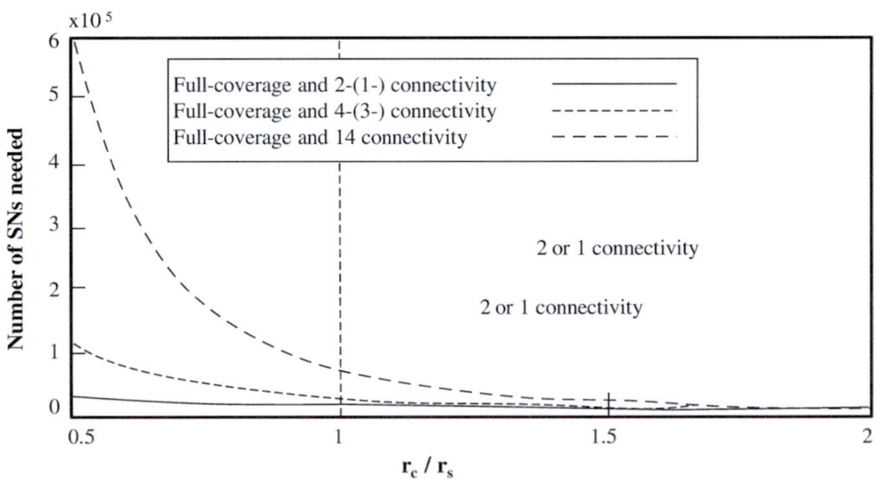

Fig. 14.16 SNs needed for various r_c/r_s to achieve 2-(1-), 4-(3-), and 14-connectivity by optimal lattice patterns [9]

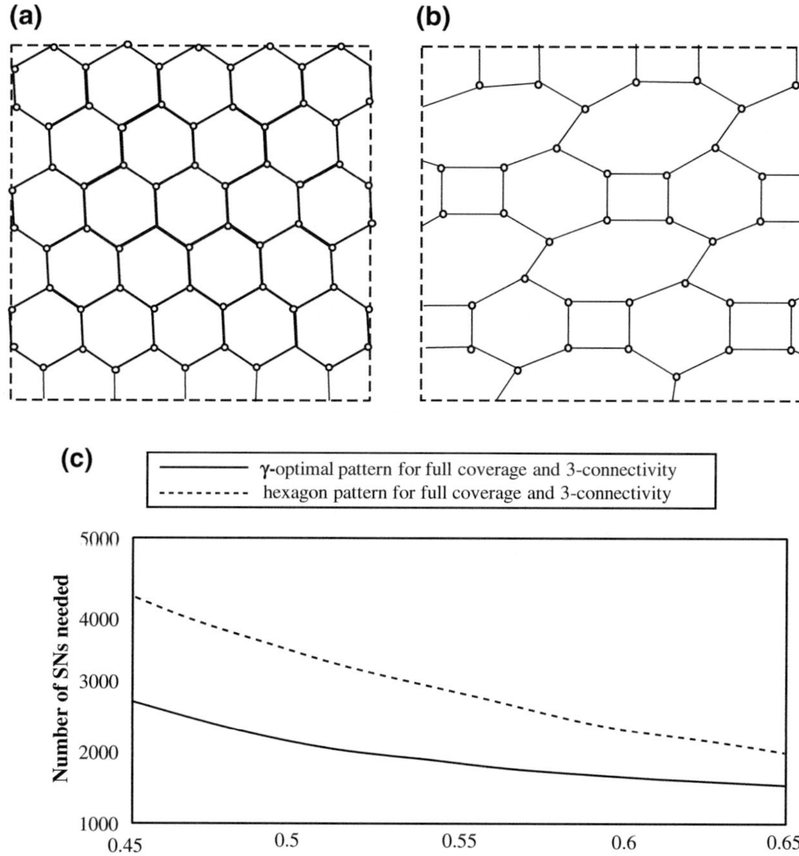

Fig. 14.17 **a** Hexagon pattern. **b** Mutation pattern. **c** # SNs needed for the full coverage and 3-connectivity by hexagon and mutated patterns [10]

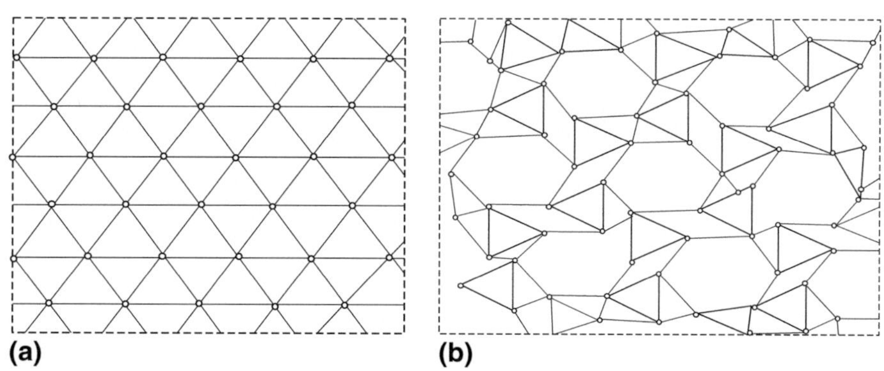

Fig. 14.18 **a** Triangular pattern. **b** Mutation pattern [10]

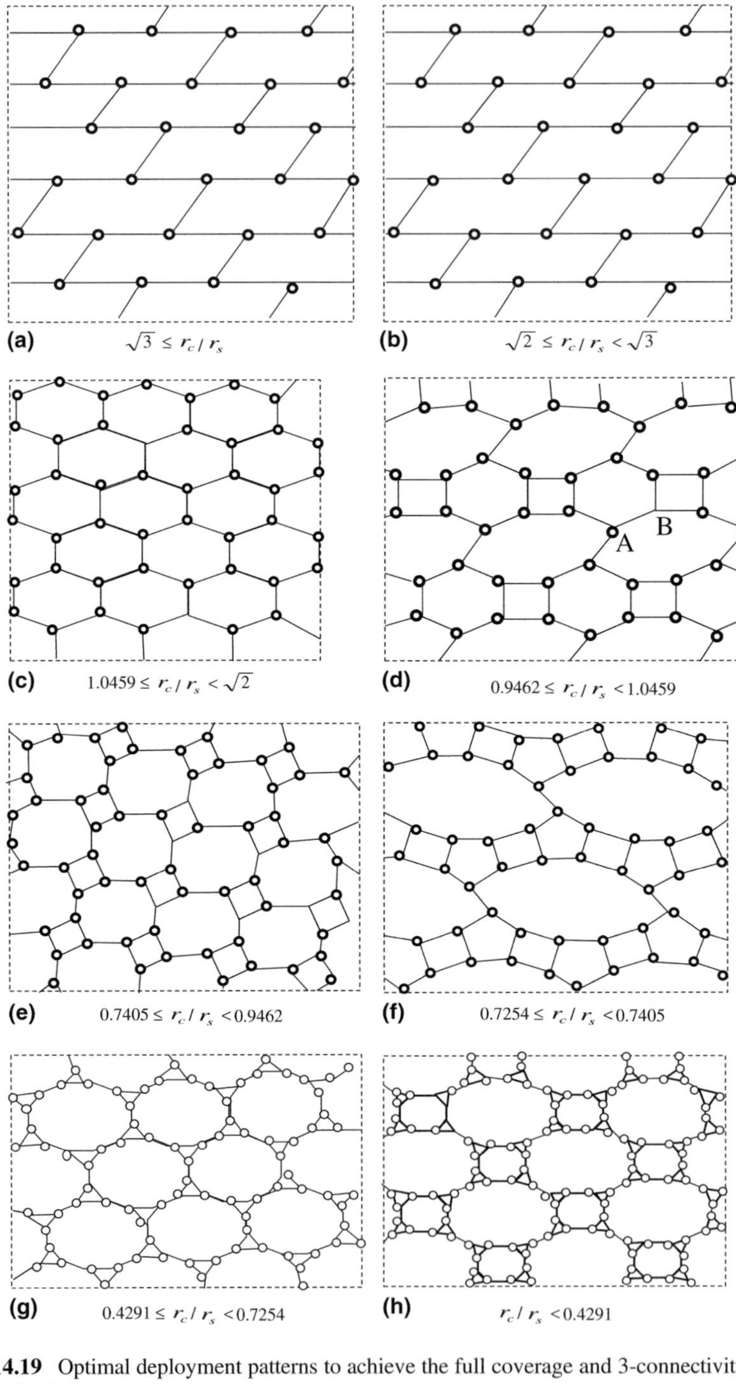

Fig. 14.19 Optimal deployment patterns to achieve the full coverage and 3-connectivity for a full range of r_c/r_s

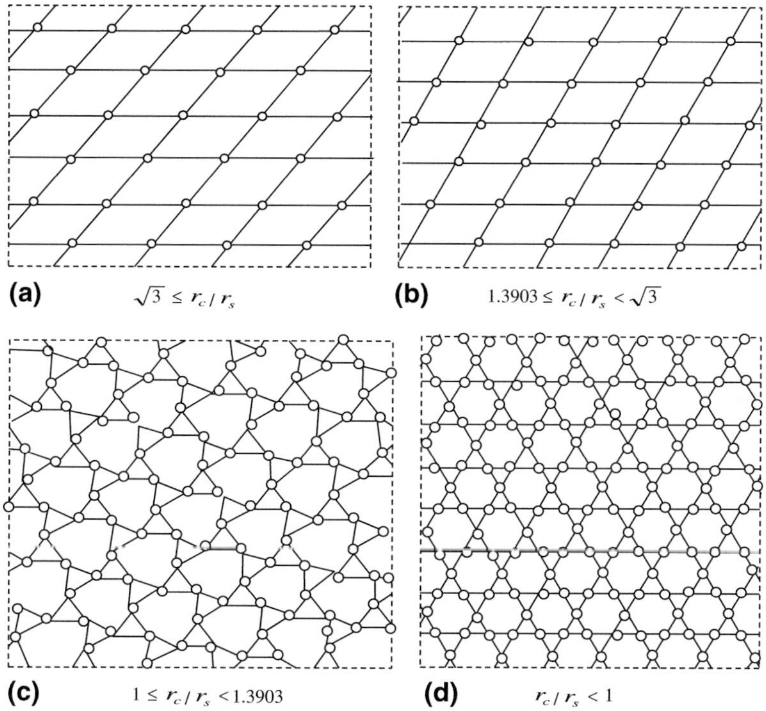

Fig. 14.20 Optimal deployment patterns to achieve the full coverage and 3-connectivity for a full range of r_c/r_s

Figure 14.29a shows the sector communication range, while (b) shows the EM wave design. The light-filled dots show the SN locations that form the horizontal strip, while the dark-filled dots form the boundary strips. Lines with arrows show the directed connection edges between SNs, and sectors illustrate the communication range of nodes. Figure 14.29c, d give a close-up of the area circled in covered eclipse in (b). Figure 14.30 shows two different SN connections that are outlined in a different style. The shaded zone below each letter defines the communication range. Figure 14.31 gives the coverage under anisotropic sensing model and random node placement error. Figure 14.32 shows the coverage in percentage by generating 105 points within the square and then checking how many of them are covered. Assume that each SN with $\mu = 30\ m$ is deployed in a 10,002 m^2 square as per proposed patterns.

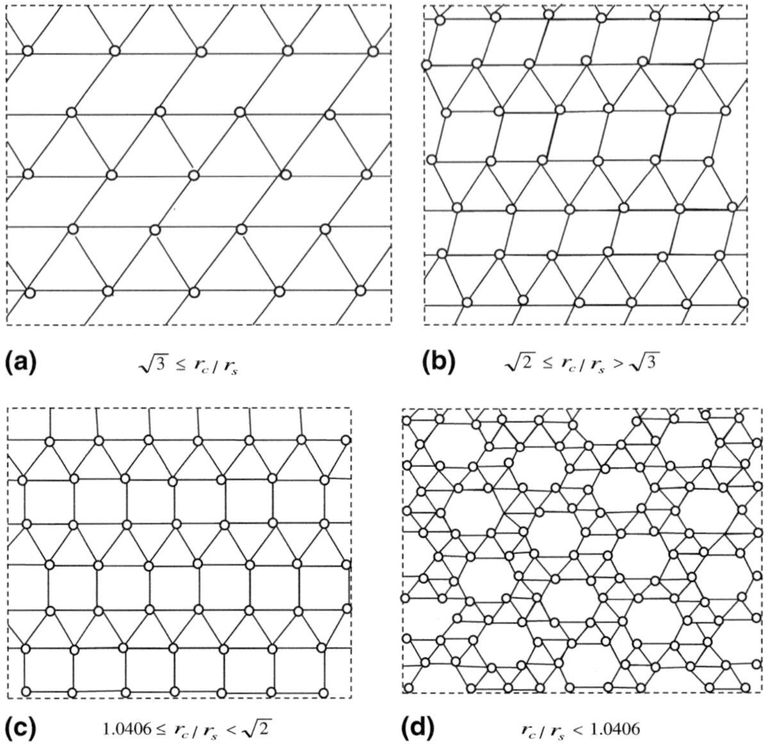

Fig. 14.21 Optimal deployment patterns to achieve the full coverage and 5-connectivity for a full range of r_c/r_s

14.7 Conclusions

Regular topologies are appropriate for numerous civilian applications, and all such future areas are expected to employ some kind of regular WSNs. There are many possible topologies, and the best option is to minimize the overlap in sensing areas of adjacent SNs so that desired area can be covered with minimum number of SNs. Complex schemes could prove to be appropriate for specific area and sensing and communication ranges and can be easily adopted. But, there is no clear winner, and more research is desirable to determine which scheme is optimal. More work is needed for the coverage by SN with directional antennas, and the future appears quite promising.

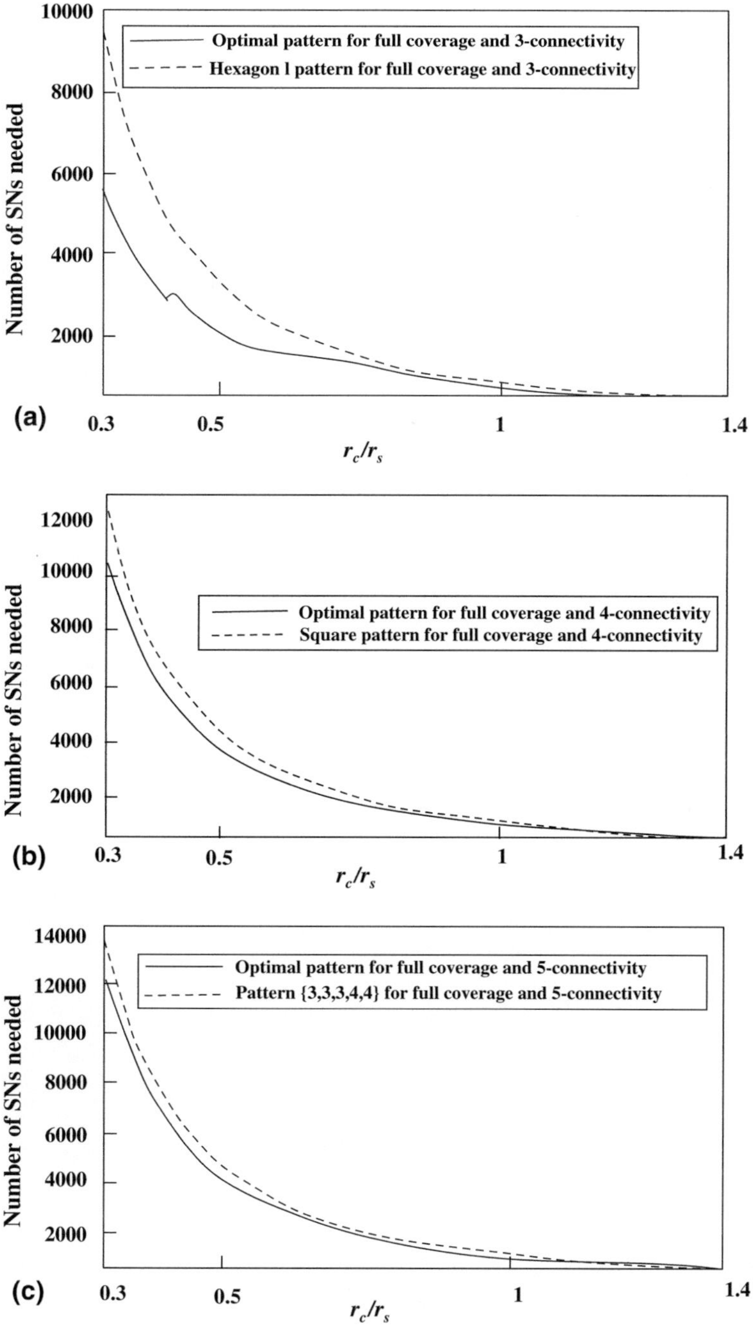

Fig. 14.22 No. of SNs needed for 3-, 4-, and 5-connectivity

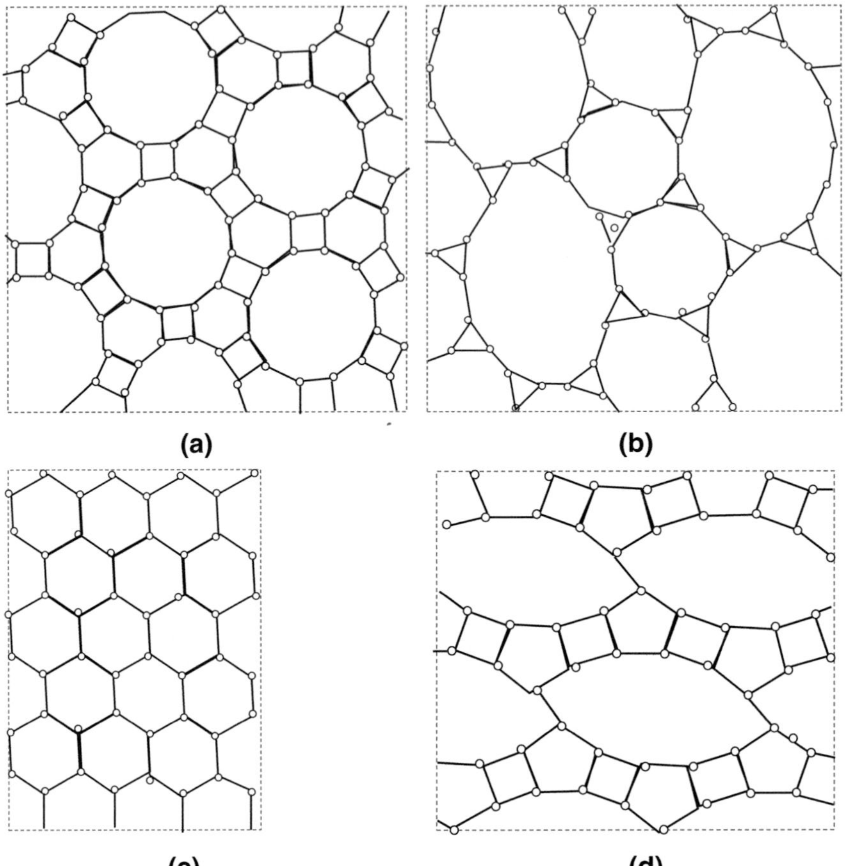

Fig. 14.23 Other possible mutations

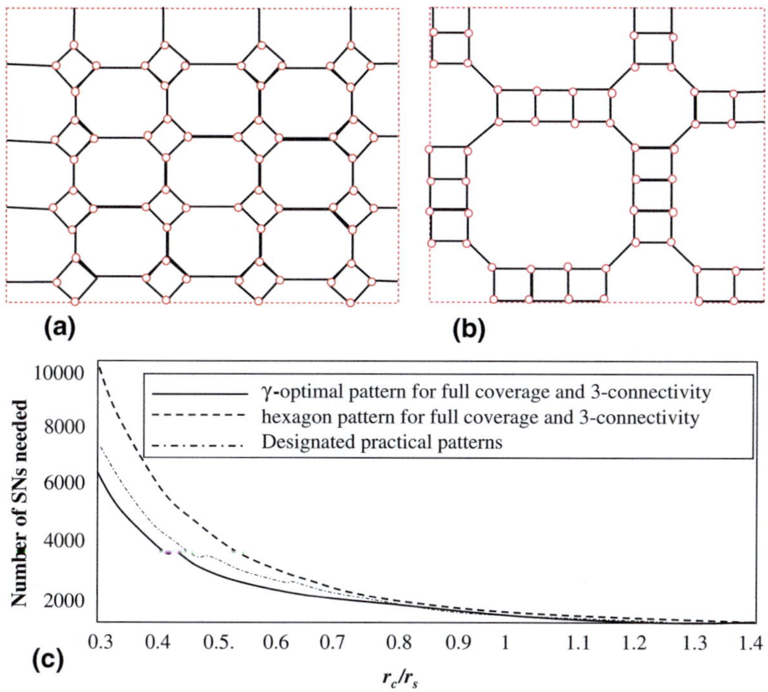

Fig. 14.24 (**a, b**) Patterns for the full coverage and 3-connectivity. **a** For $0.6184 \le r_c/r_s < 0.8089$. **b** For $r_c/r_s < 0.4813$. **c** # SNs needed in optimal patterns with 3-connectivity

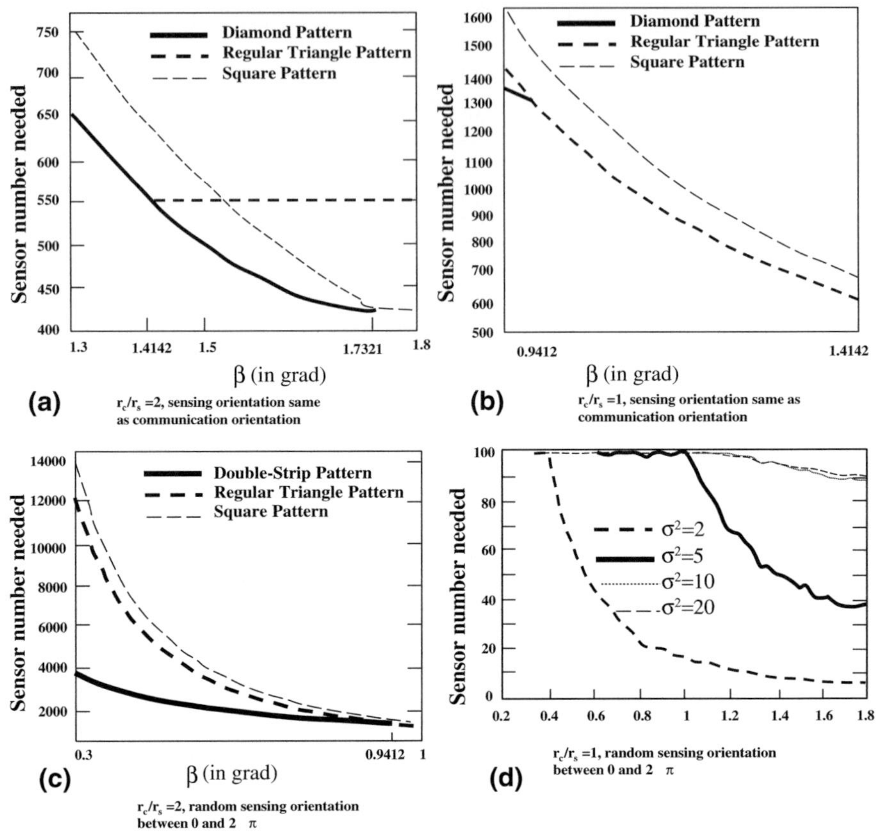

Fig. 14.25 # SNs needed to achieve 4-connectivity and the full coverage. **a** r_c/r_s varied from 1.3 to 1.8. **b** r_c/r_s varied from 0.9 to 1.43. **c** r_c/r_s varied from 0.3 to 1. **d** r_c varied from 10 to 54 m

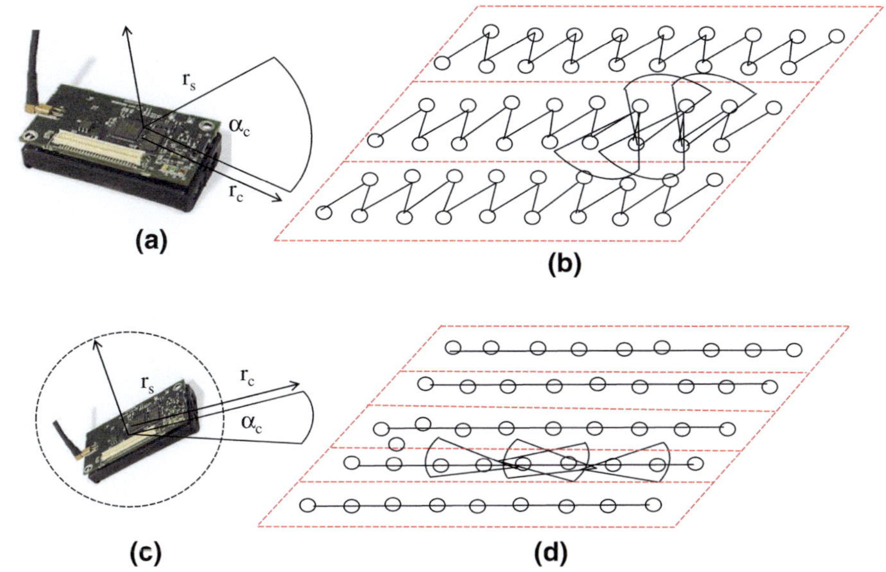

Fig. 14.26 Directional antennas and EM wave propagation [12]

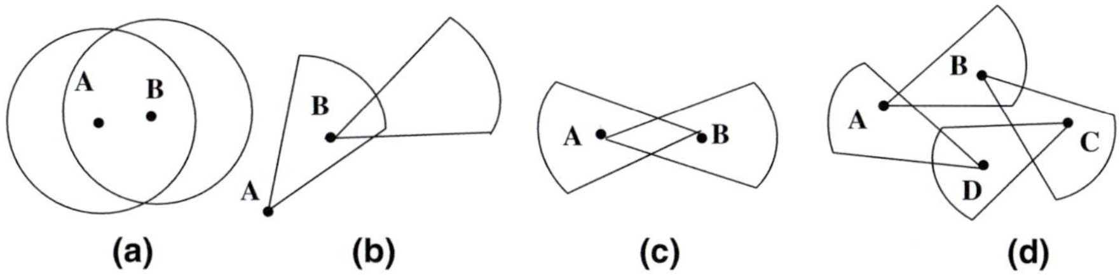

Fig. 14.27 Communication in (**a**) directional antenna, (**b**) directional antenna sets, (**c**) bidirectional connection using directional antennas, and (**d**) diverse directional antennas [12]

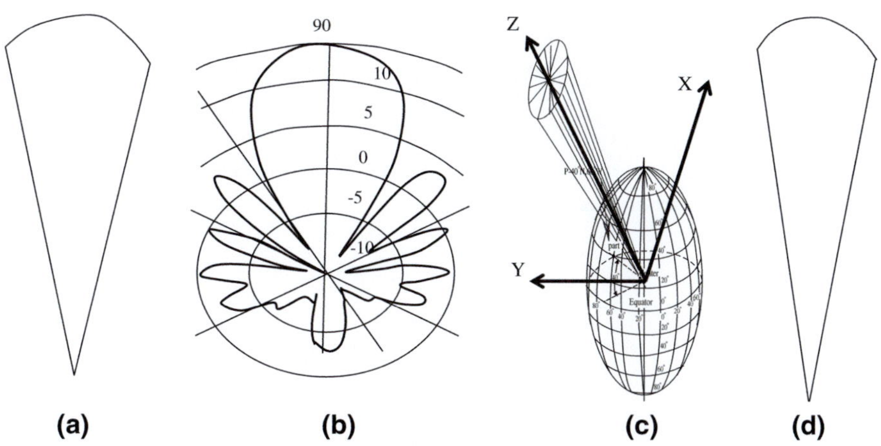

Fig. 14.28 Communication range of directional antenna [12]

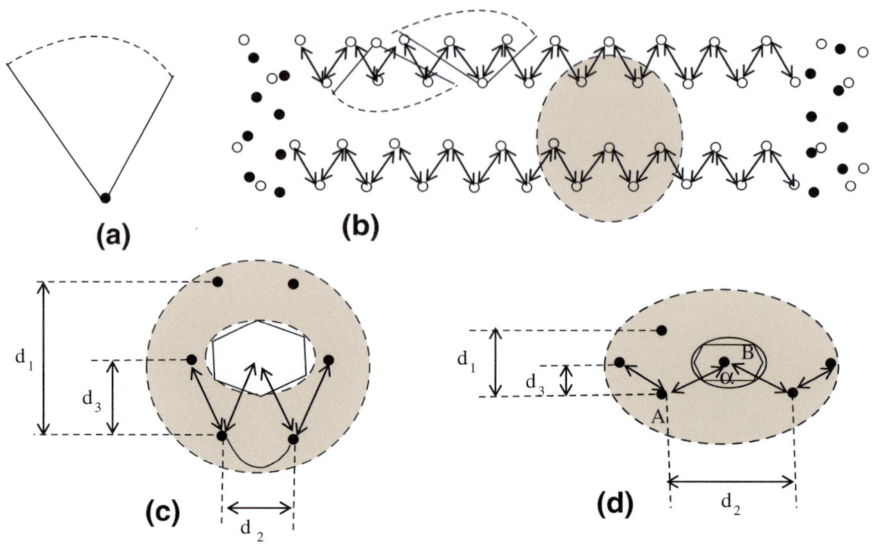

Fig. 14.29 Connection edges and sensing Voronoi polygon for the two variant wave patterns of directional antenna [12]

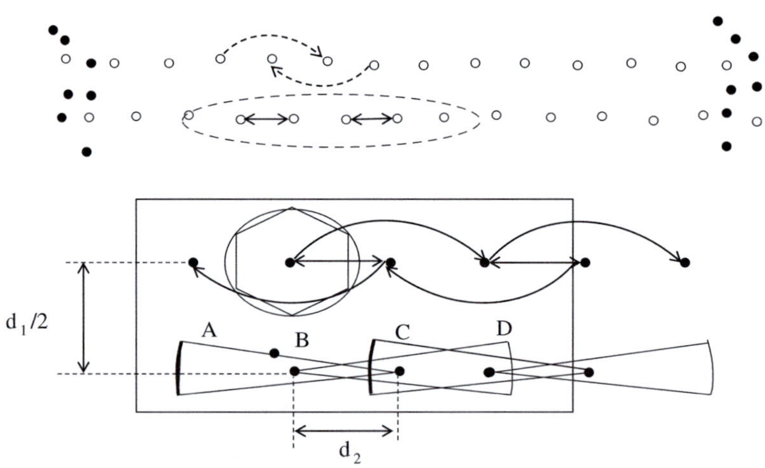

Fig. 14.30 Line patterns of directional antenna

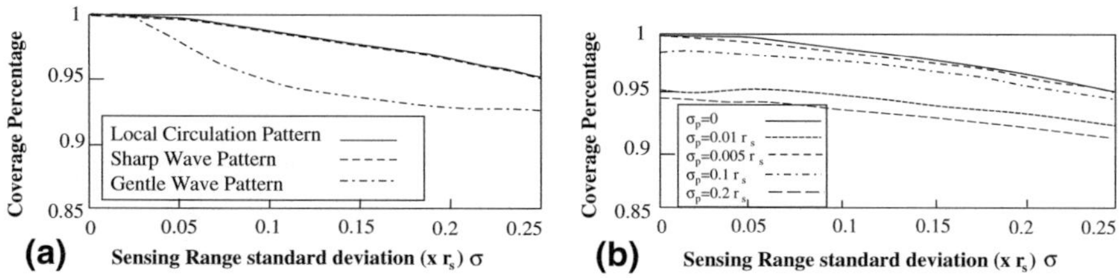

Fig. 14.31 Coverage of directional antenna under (**a**) anisotropic sensing model and (**b**) random node placement error

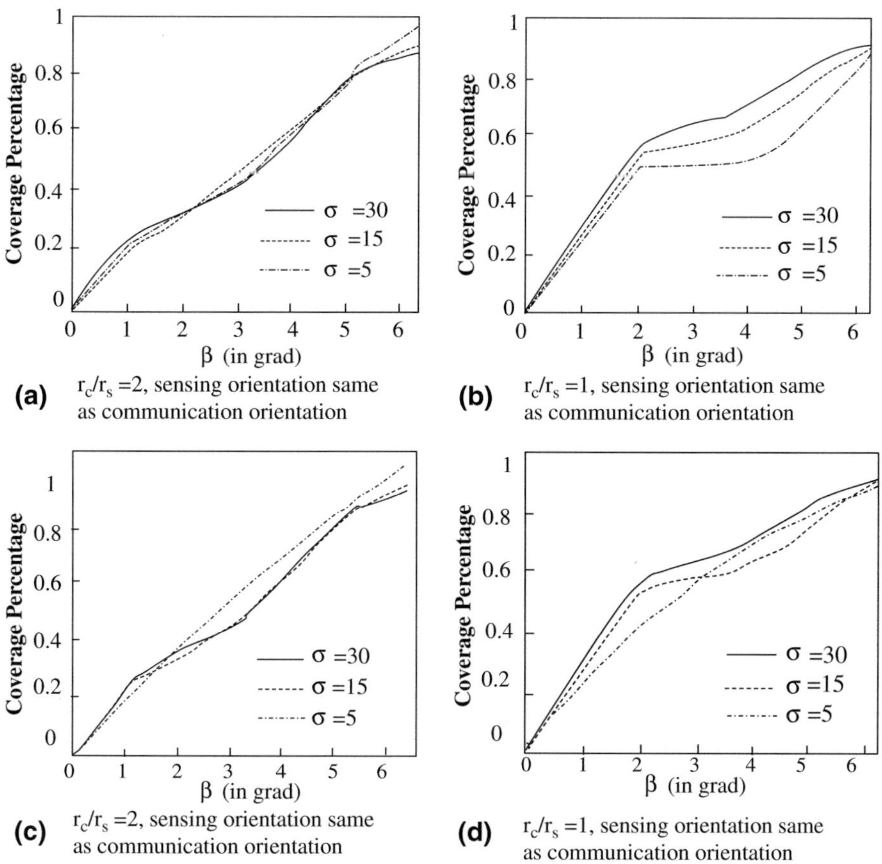

Fig. 14.32 Coverage of directional antenna with different orientations [12]

14.8 Questions

Q.14.1. What is the advantage of using a regularly placed WSN?

Q.14.2. What are the important features of a regular deployment of wireless sensor networks?

Q.14.3. What are the techniques you can use to reduce power consumption in a regular WSN?

Q.14.4. Assuming packet can be transmitted to cover multiple hops in one transmission? What is an optimal value of skipping hops in a regular WSN?

Q.14.5. A sensor network with four clusters is shown in the following diagram. Can you indicate appropriate locations of four CHs?

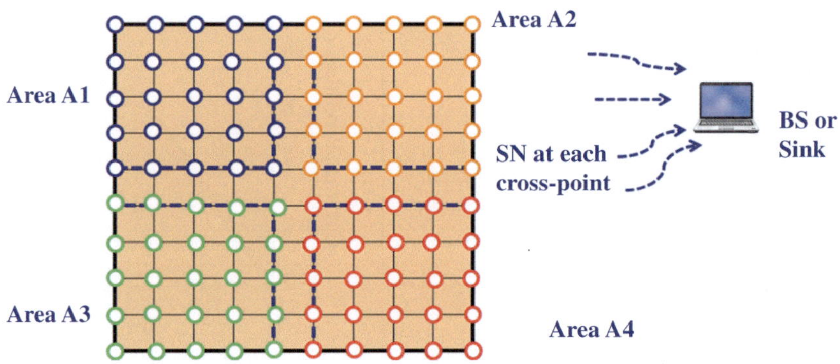

Q.14.6. How can you draw Voronoi diagram for a mutated topology of Fig. 14.19?

Q.14.7. What will be the use of Voronoi diagram of the previous question?

Q.14.8. How can you draw Voronoi diagram for a mutated topology of Fig. 14.21?

Q.14.9. What will be the use of Voronoi diagram of the previous question?

Q.14.10. How do you compare a rhombus topology with a rectangular topology?

Q.14.11. Why do you need more complex mutated topologies for WSNs?

Q.14.12. Why do you have to study WSNs with directional antennas?

Q.14.13. What are the practical implications of Q. 14.7?

Q.14.14. How does Voronoi diagram equivalent for directional antennas look like? Can you discuss the results?

Q.14.15. What will be your choice if 50% of SNs have omnidirectional antennas, while the remaining 50% have directional antennas?

Q.14.16. Can you draw the coverage for a selected topology?

References

1. Kenneth Stephenson, Introduction to Circle Packing: The Theory of Discrete Analytic Functions, Cambridge University Press, 1985.
2. Rabie A. Ramadan and Salah Abdel-Mageid, "Efficient Deployment of Connected Sensing Devices Using Circle Packing Algorithms," rabieramadan.org/papers/Circle%20**packing**-final.pdf.
3. Chih-Shun Hsu, Jang-Ping Sheu, and Yen-Jung Chang, "Efficient Broadcasting Protocols for Regular Wireless Sensor Networks," Proceedings of the 2003 International Conference on Parallel Processing (ICPP'03).
4. Yun Wang and Dharma P. Agrawal, "Optimizing Sensor Networks for Autonomous Unmanned Ground Vehicles," Proceedings of SPIE, Cardiff, Wales, United Kingdom | September 15, 2008.
5. Shuang Li, Alvin Lim, and Cong Liu, "Improving QoS-based Routing by Limiting Interference in Lossy Wireless Sensor Networks," International Journal of Wireless & Mobile Networks (IJWMN), vol. 2, no. 4, pp. 44–58, Nov. 2010.
6. Z. Junguo and Z. Feng, "Study on Optimal Regular Deployment Patterns of Wireless Sensor Network," Research Journal of Applied Sciences, Engineering and Technology, vol. 4, no. 15, Aug. 2012, pp. 2300–2303.
7. Xiaole Bai, Ziqiu Yun, Dong Xuan, Biao Chen, and Wei Zhao," Optimal Multiple-Coverage of Sensor Networks," INFOCOM 2011.
8. Philip L. Bowers, "Introduction to Circle Packing: A Review," www.math.fsu.edu/~aluffi/archive/paper356.pdf.
9. Xiaole Bai, Chuanlin Zhang, Dong Xuan, Jin Teng, and Weijia Jia, "Low-Connectivity and Full-Coverage Three Dimensional Wireless Sensor Networks," Proceedings of the tenth ACM international symposium on Mobile ad hoc networking and computing (MobiHoc), pp. 145–154, 2009.
10. Xiaole Bai, Ziqiu Yun, Dong Xuan, Weijia Jia, and Wei Zhao, "Pattern Mutation in Wireless Sensor Deployment," IEEE/ACM Transactions on Networking, vol. 20, no. 6, Dec. 2012.
11. S. M. Nazrul Alam and Zygmunt J. Haas, "Topology Control and Network Lifetime in Three-Dimensional Wireless Sensor Networks," Network Protocols and Algorithms, 2009, vol. 1, no. 2, pp. 85–98.
12. Z. Yu, J. Teng, X. Bai, D. Xuan, and W. Jia, "Connected Coverage in Wireless Networks with Directional Antennas," ACM Transactions on Sensor Networks, vol. 10, no. 3, April 2014.

Chapter 15
Routing and Performance of Regular WSNs

15.1 Introduction

We considered the usefulness of regular deployments of WSNs, and it is time now to quantify some of the specific parameters and compare how they perform as compared to a random scheme. An obvious reason to use regular scheme is that a desired area can be covered by fewer SNs and some underlying characteristics are shown in Table 15.1.

15.2 Routing in Regular Topologies

In a 2-D mesh network shown in Fig. 15.1a, routing is very simple as x and y coordinates of each SN can be used to determine a route from a source SN to a destination and can be easily extended to 3-D mesh networks. The location of BS is critical as all SNs need to transmit their sensed data to it. So, a natural choice is at the center of WSN as SN (0, 0). If you decided to have 4 BSs, then an intuitive solution of placing each of four BS_1 is the four corners. But a better solution is BS_2 as WSN can be divided in four parts as shown by a vertical and a horizontal dotted line such that the area of SNs reporting to different BSs can also be identified and each BS is in the center of their assigned area. That way, the distance from SNs to corresponding BS can be equalized among SNs. In a rhombus topology shown in Fig. 15.1b, routing can be done in a similar way to 2-D mesh.

SN addressing in 2-D hexagonal network is shown in Fig. 15.2a which is simple, but difficult to use for routing among SNs. An alternate addressing in 2-D hexagonal network is shown in Fig. 15.2b by dividing hexagon to equilateral triangles [1] and assigning address of each SN using 3 digits of (x, y, z) as illustrated in Fig. 15.2c [2]. In this scheme,

D.P. Agrawal, *Embedded Sensor Systems*,
DOI 10.1007/978-981-10-3038-3_15

Table 15.1 Comparing randomly deployed versus regular topology of WSNs

WSN topology	Randomly deployed	Regular topology
SNs distribution	Random	Uniform
Neighborhood discovery	Use beacon signals to find nearest neighbors within communication range, an involved process	As SNs are uniformly placed, the location of neighboring SNs is built in the system
Routing	Have to find the path with known local neighborhood information	Routing is relatively easy as each node with given address, is located at a location based on the basic topology
Voronoi diagram	Fairly involved process	Easy as SNs are regularly placed, complex if topology is complex

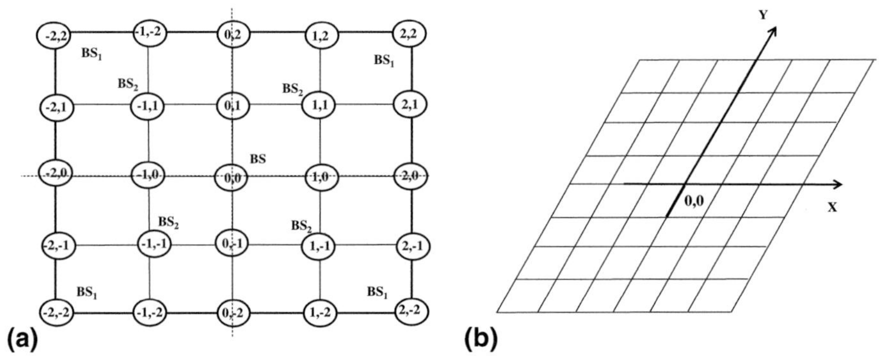

Fig. 15.1 **a** 2-D mesh network. **b** Rhombus topology

$$\mathbf{i}+\mathbf{j}+\mathbf{k}=0, \tag{15.1}$$

and

$$\mathrm{SN}(a,b,c)\ \text{is expressed as} = a\mathbf{i}+b\mathbf{j}+c\mathbf{k}. \tag{15.2}$$

As the addressing is not unique, it may be possible that

$$(a',b',c')=(a'',b'',c'') \Leftrightarrow a'\mathbf{i}+b'\mathbf{j}+c'\mathbf{k}=a''\mathbf{i}+b''\mathbf{j}+c''\mathbf{k}. \tag{15.3}$$

Optimal Routing in a hexagonal network is obtained by calculating difference between destination D and source S addresses as follows:

$$\left|\overrightarrow{SD}\right| = \min(|a-c|+|b-c|, |a-b|+|b-c|, |a-b|+|a-c|). \tag{15.4}$$

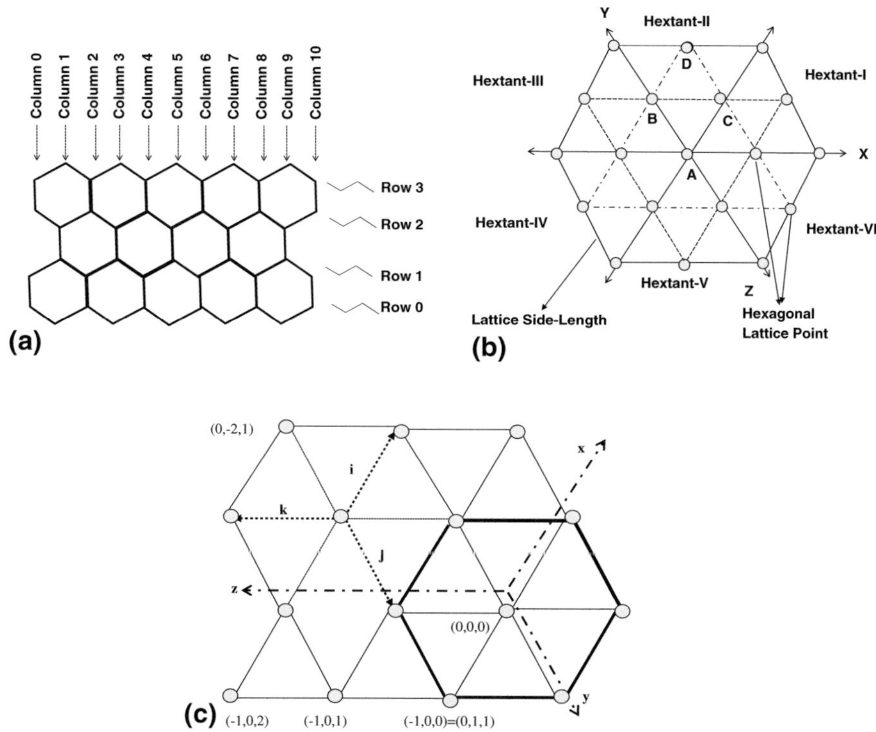

Fig. 15.2 **a** Addressing in hexagonal network. **b** Alternative way of SN addressing. **c** Use of coordinates shown for addressing SN

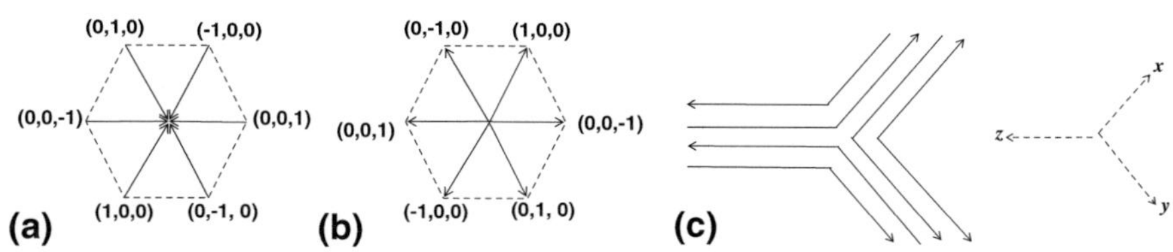

Fig. 15.3 **a** Addressing in hexagonal network. **b** Outgoing edges from (0, 0, 0). **c** Routing of the packets [3]

Optimal routing in a hexagonal network can be done by following Fig. 15.3c and calculating distance between S and D [3].

Addressing in 2-D triangular network can be done similar to a hexagonal network described earlier and is shown in Fig. 15.4a. So, routing can also follow hexagonal networks. Such a network has been simulated following a transmission schedule [4] of up to H transmissions are scheduled during every time slot using spatial reuse and Fig. 15.4b shows cumulative distribution of distance between receiver SNs and other sender SNs for networks of size $H = 5$ (90 SNs),

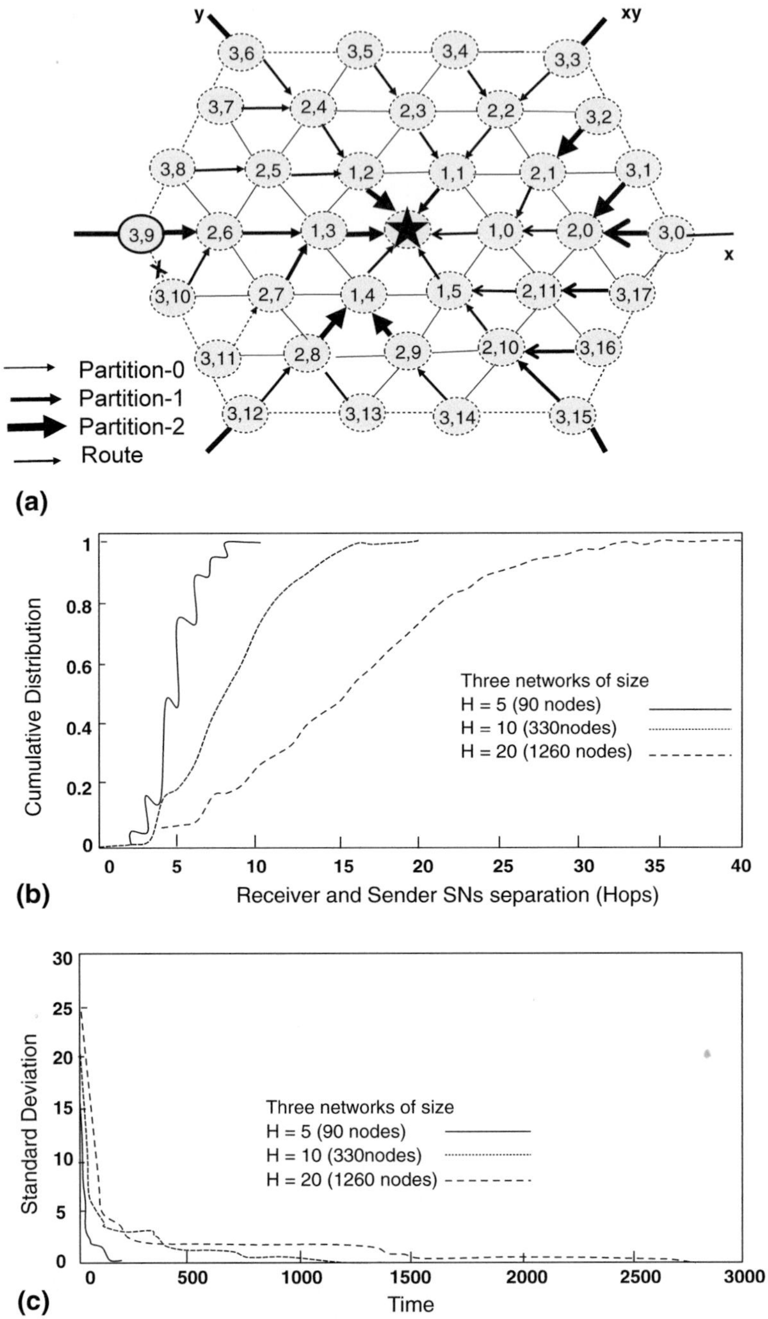

Fig. 15.4 **a** Addressing in triangular network. **b** Separation between receiver SNs and simultaneous sender SNs. **c** Convergence of clocks

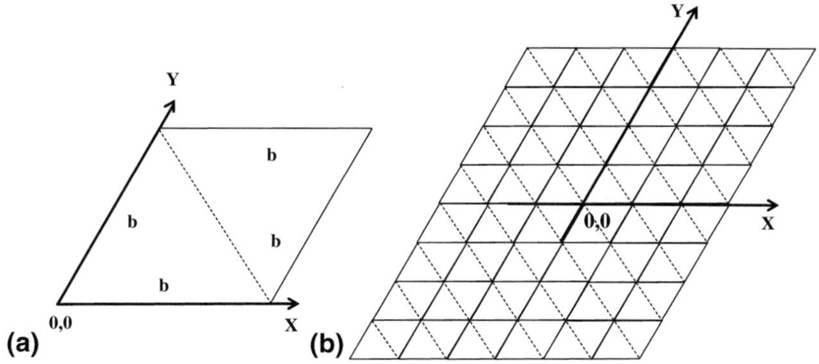

Fig. 15.5 **a** Rhombus divided into 2 triangles. **b** Routing in rhombus network

$H = 10$ (330 SNs) and $H = 20$ (1260 SNs), with more than 2 hop separation between SNs. A slow divergence of synchronization is observed in Fig. 15.4c. A rhombus topology can be obtained by combining two triangles (Fig. 15.5a), and thus, the same routing scheme can be utilized (Fig. 15.5b).

In WSNs, absolute address of each SN is not required and in fact relative address within the area is considered adequate. For relative location, some SNs are selected as reference points and are called anchor nodes. Selection of appropriate anchors is critical as they ought to be well-distributed inside the area where WSN has been deployed. Anchor placement in 2-D mesh network is done [1] by first determining virtual coordinates (VCs) of each SN that represents minimum hop distances with respect to a set of anchor nodes. Having two corner SNs at the bottom of a 2-D mesh makes all SNs to have unique VCs while their random placement leads to the same coordinates for several SNs. The effect of placing anchors at bottom two corners is shown in Fig. 15.6a, with all SNs have a unique VC. Random location of anchor modes (Fig. 15.6b) leads to several SNs with the same VC.

Figure 15.7 shows placement of anchor nodes in a triangular topology. Figure 15.7a has distance between two anchors of 6 hops following a zigzag path while anchors are placed on a straight line in Fig. 15.7b. Figure 15.8a indicates unique VCs in a triangular network if anchors are placed on straight edge and in Fig. 15.8b the dotted lines show region of identical VCs. Figure 15.9 identifies SNs region in triangular network using VCs. Dominating setin triangular topology is given Fig. 15.10a with density of 1/3 and Fig. 15.10b with density of 6/19 [5]. Figure 15.11 shows SNs deployed in hexagonal grid with dotted region shows SNs with identical VCs due to anchors placed on zigzag line in (a) and (b) illustrates different regions when anchors are placed on straight line. Identification of SNs region using VCs for hexagonal network [4] is shown in Fig. 15.12a while Fig. 15.12b shows spanning tree with a maximum distance of 3 to BS [6]. Figure 15.13a shows 1-to-all message broadcast in a hexagonal network, and all-to-all communication is given in Fig. 15.13b.

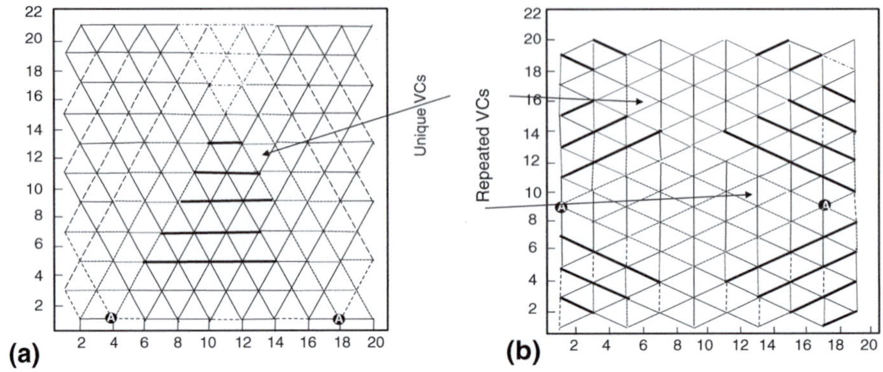

Fig. 15.6 Placement of anchor SNs in 2-D mesh at **a** Bottom corners give all SNs unique VCs. **b** Random location leads to several SNs with identical coordinates

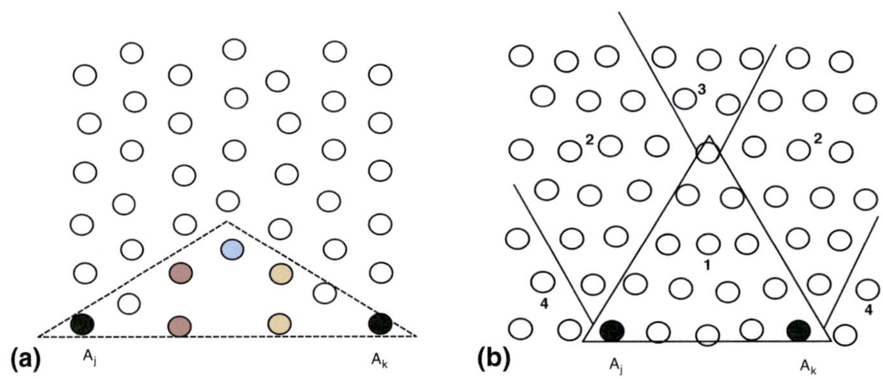

Fig. 15.7 Placement of anchor nodes in triangular topology. **a** 6 hops away following zigzag path. **b** Anchors placed on straight line

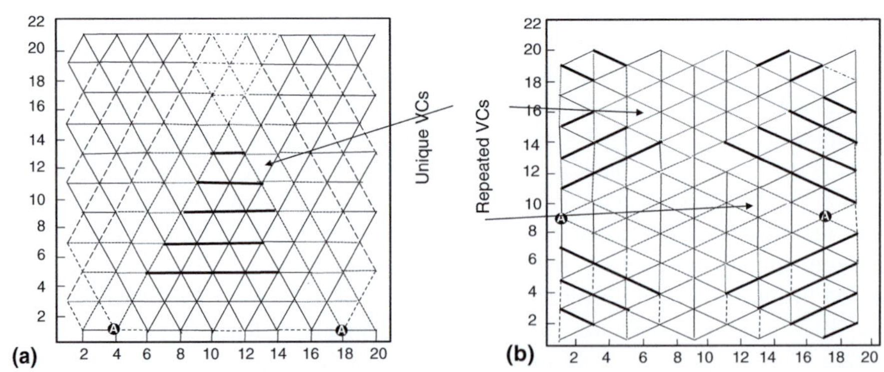

Fig. 15.8 **a** Unique VCs obtained if anchors in a triangular network are placed on straight edge. **b** *Dotted lines* showing region of identical VCs

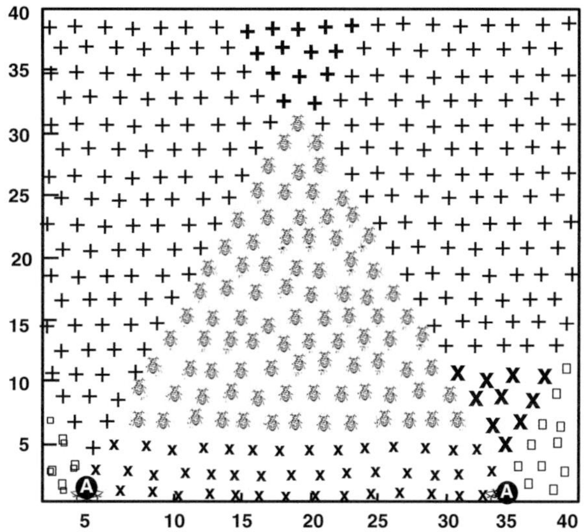

Fig. 15.9 Identification of SNs region in triangular network using VCs

15.3 Processing in Regular Topologies

Collaborative processing is another challenging area in WSNs and Power-Efficient Gathering in Sensor Information Systems (PEGASIS) [7] of Fig. 9.8 can be applied to regular WSN topologies without much difficulty. This can be done chaining all SNs together so that each SN receives only one message and transmits one till the message reaches the destination or the BS. Figure 15.14 shows examples of 2-D mesh and triangular topologies. Tree formation in a 2-D mesh is given in Fig. 15. 15a and broadcasting in 2-D mesh is shown in Fig. 15.15b, c.

Clustering of SNs can be done using SNs 1 hop or 2 hops away from CH and is shown for 2-D mesh, hexagonal, and triangular topologies in Fig. 15.16. Clustering of a large triangular WSN is shown in Fig. 15.17. In a WSN, if it is densely deployed, a set of sensors can be allowed to go to sleep mode while another set monitors the area. 4 sets of sleep–awake cycles in WSN are shown in Fig. 15.18 for three topologies. Figure 15.19 gives 4 sets of sleep–awake cycles in large WSNs with 4 BSs. The CH can define schedule for each SN so that collision from adjacent SNs can be eliminated. Figure 15.20 illustrates one such TDMA schedule in clustered WSNs.

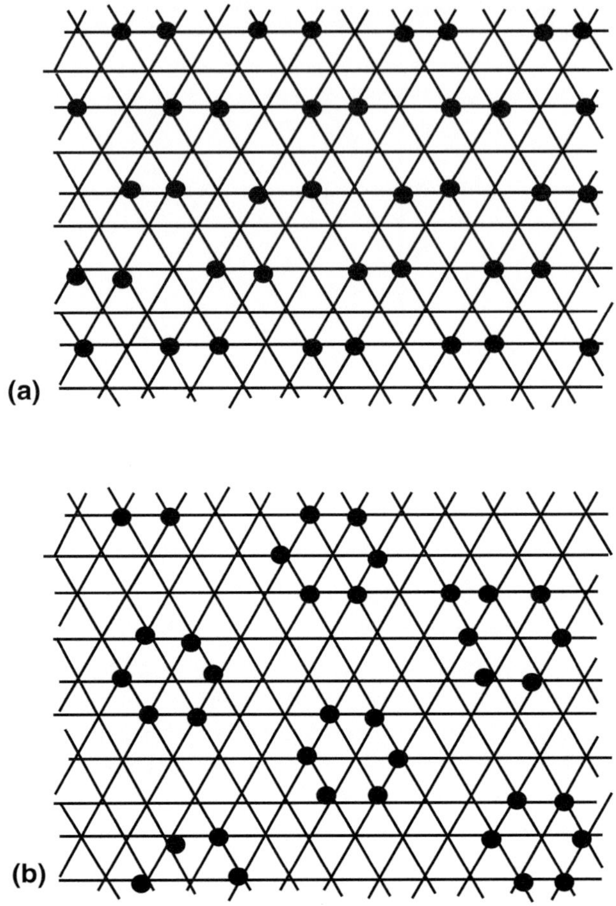

Fig. 15.10 Dominating set in triangular topology. **a** Density 1/3. **b** Density 6/19 [5]

15.4 Mobile Opportunistic Concept for Regular WSN Topologies

Most commonly, SNs forward their data to BS in multi-hop fashion so as to conserve energy. This increases delay in reaching BS while consumes energy in every transmission and reception. Another class of WSN concept is to use opportunistic network (MON) concept where randomly moving mobile relay nodes (RNs) pick up data from SN once within its communication range and RN eventually delivers data to BS. As mobile RNs are randomly moving and WSN has no control over the mobility of RNs, the connectivity between RN and SN as well as between RN and BS are intermittent (Fig. 15.21).

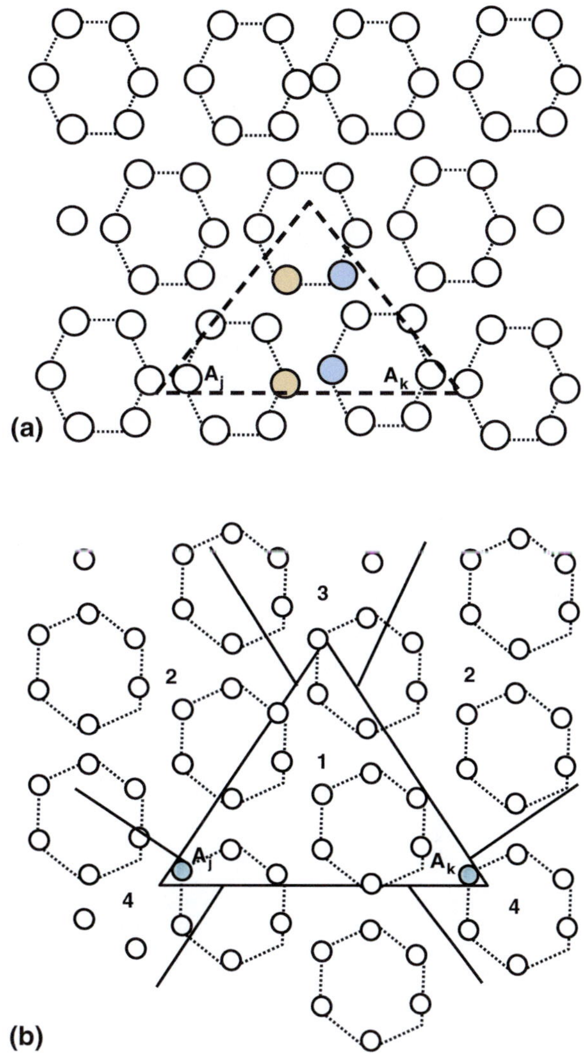

Fig. 15.11 SNs deployed in hexagonal grid. **a** *Dotted region* shows SNs with identical VCs due to anchors placed on *zigzag line*. **b** Different regions when anchors are placed on *straight line*

If T is the intercontact time, then the estimated time to deliver data to BS is $E(T) = N$. As the size on the WSN $N \rightarrow \infty$, then $P(T > n) \sim \text{constant}/n^{1/2}$, for large n where n is the number of SNs, a RN can contact in one transmission. Delivery times required in 2-D mesh of different sizes are given in Fig. 15.22. The RNs used for ferries have completely predictable routes through the geographic area. One example is an ideal spiral search process in a WSN [8] where, unlike *address–centric* routing in which the source node searches for a route to the destination node, the BS looks for a particular data object stored in an unknown subset of SNs. Such spiral is intended for *data-centric* routing as the routing problem is actually a query problem—the routing algorithm must search for the route to a SN with the

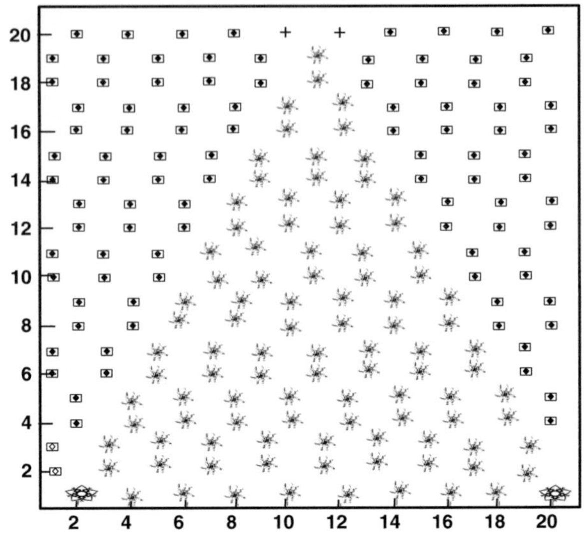

Fig. 15.12 Hexagonal network. **a** Identification of SNs region using VCs [4]. **b** Spanning tree for maximum distance of 3 to BS [6]

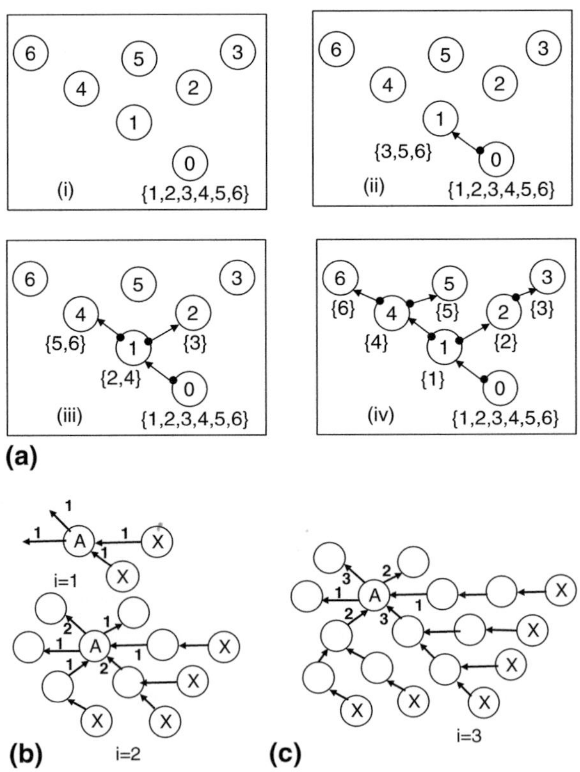

Fig. 15.13 Hexagonal network. **a** Three stages of the one-to-all communication. **b**, **c** 3 steps of all-to-all communication

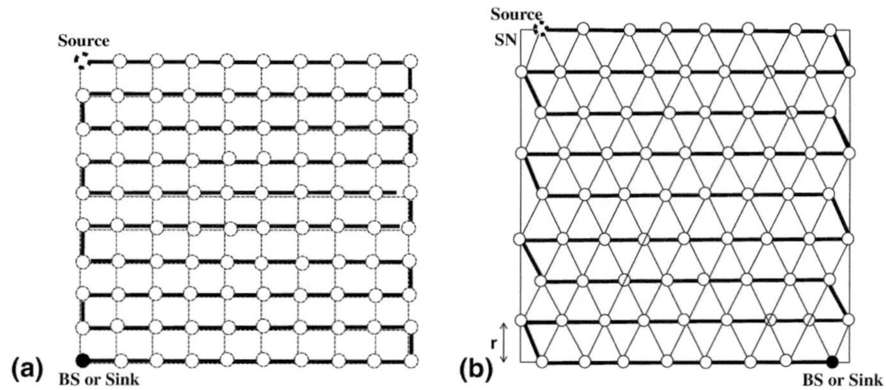

Fig. 15.14 PEGASIS algorithm applied to **a** 2-D mesh WSN. **b** triangular WSN

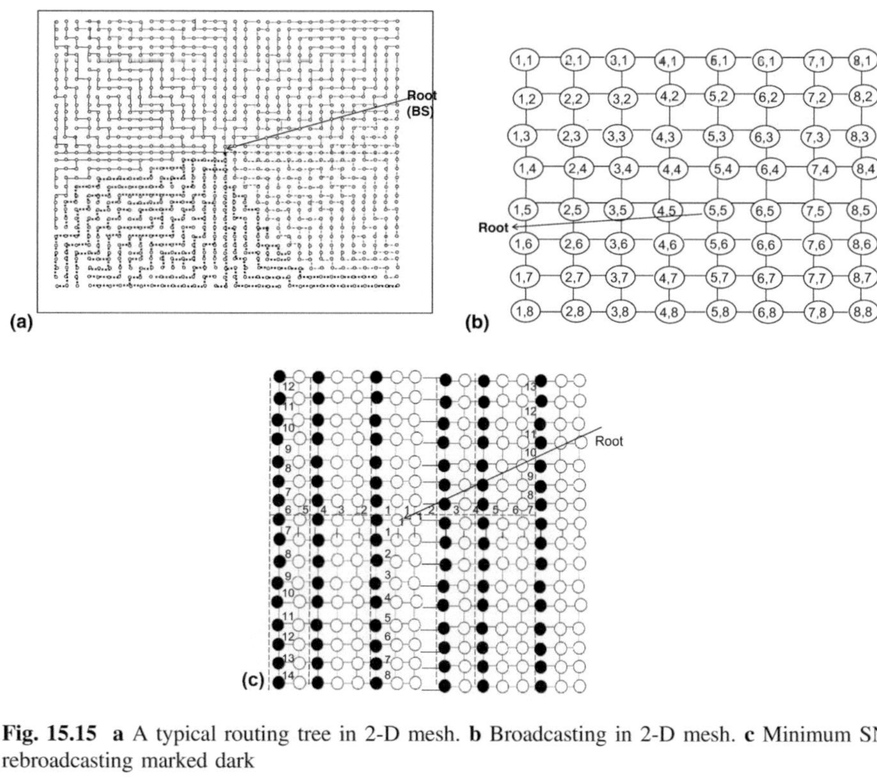

Fig. 15.15 **a** A typical routing tree in 2-D mesh. **b** Broadcasting in 2-D mesh. **c** Minimum SN rebroadcasting marked dark

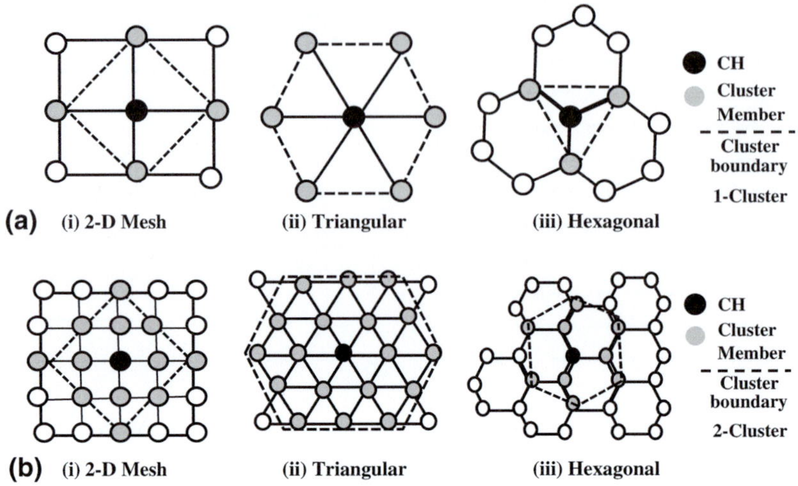

Fig. 15.16 Clustering of WSNs. **a** 1 hop. **b** 2 hops

Fig. 15.17 Clustering of a large triangular WSN

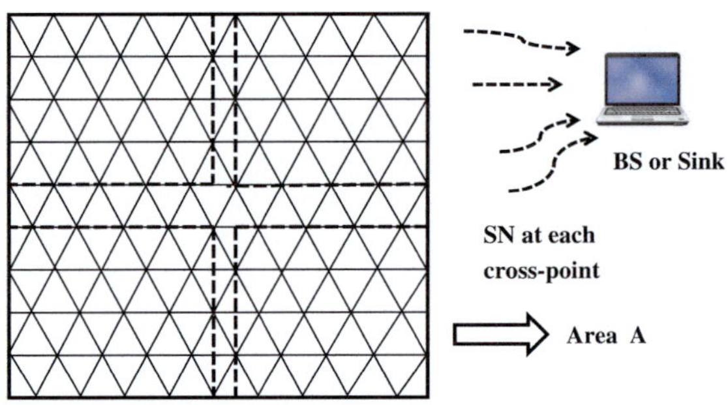

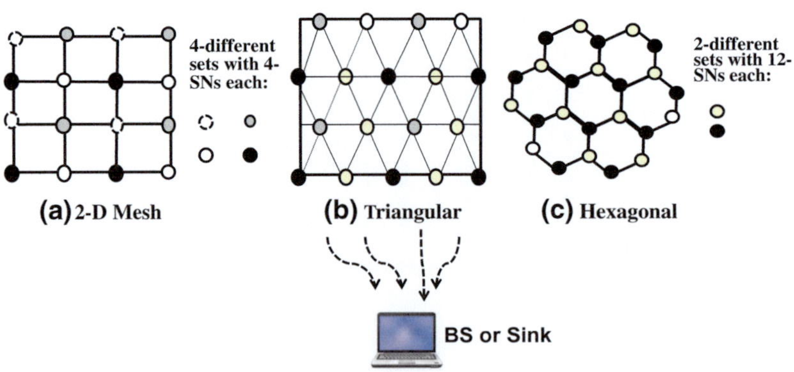

Fig. 15.18 4 sets of sleep–awake cycles in WSNs. **a** 2-D mesh. **b** Triangular. **c** Hexagonal

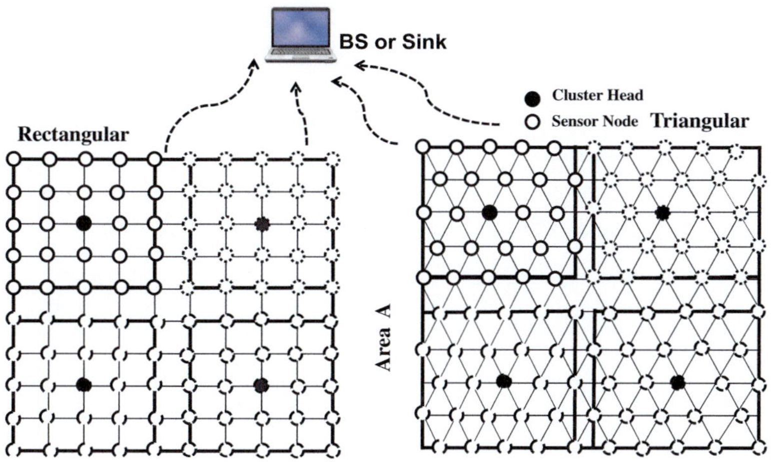

Fig. 15.19 4 sets of sleep–awake cycles in large WSNs with 4 BSs

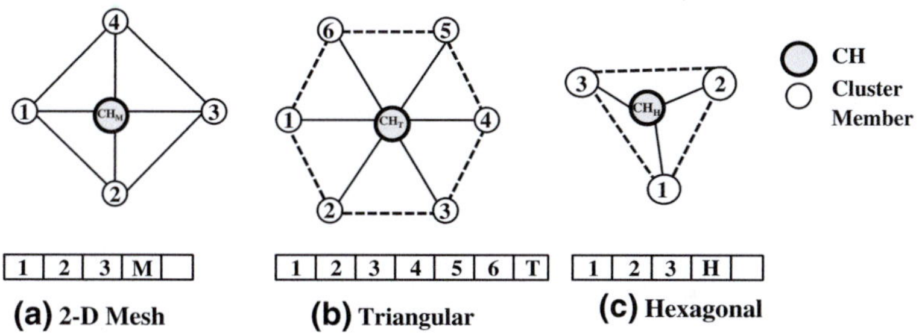

Fig. 15.20 TDMA schedule in clustered WSNs

Fig. 15.21 Random walk on Torus

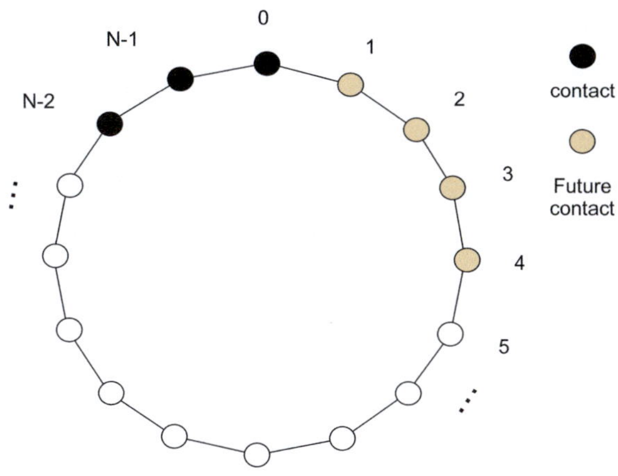

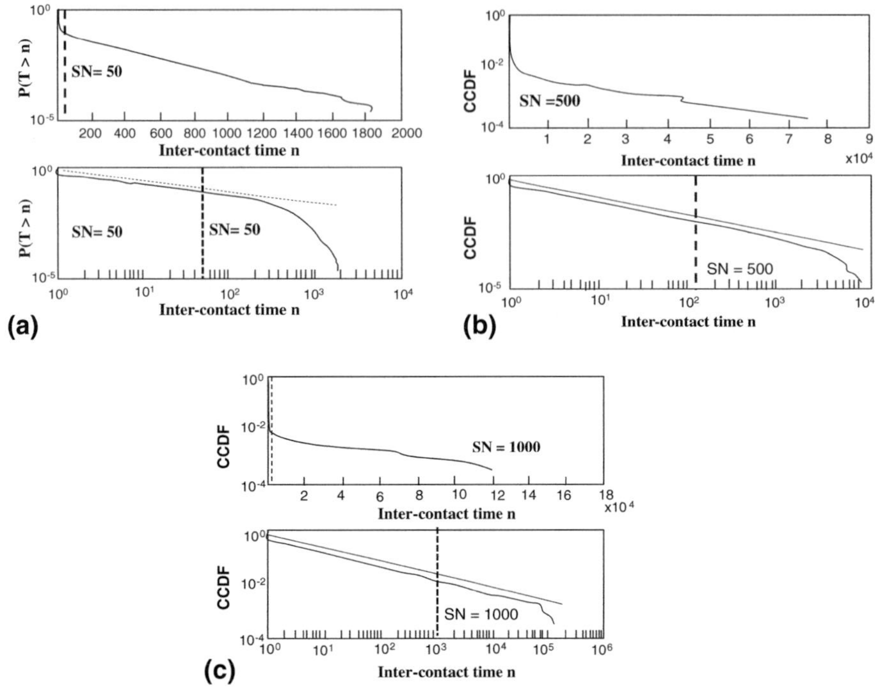

Fig. 15.22 Random walk on Torus with # SNs equal to **a** 50. **b** 500. **c** 1000

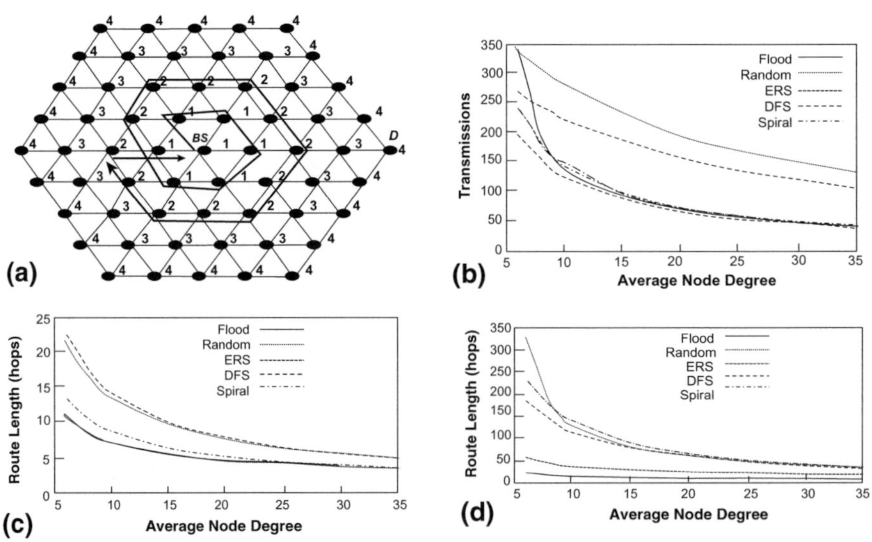

Fig. 15.23 **a** Spiral search process in a triangular WSN. **b** Route discovery cost versus network density. **c** Route length versus network density. **d** Search time versus network density

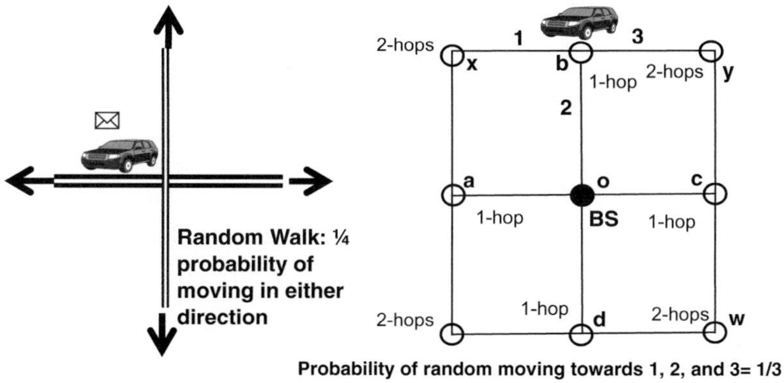

Fig. 15.24 Probability of random mobility in 2-D mesh

desired data and then data can be transmitted along the discovered route to the BS. Spiral reduces the communication overhead and the energy consumption for both route discovery and subsequent communication. One such spiral search in a triangular WSN is shown in Fig. 15.23a and the corresponding performance is shown in Fig. 15.23b, c. Spiral's total cost for 40 packets communication is only 72% to flooding, 81% of ERS (Expanding Ring Search), 74% of random walk, and 73% of DFS (Depth-First Search). In flood-based algorithms, the data query has to be flooded into entire WSN. The SNs with desired data transfer data using discovered the routes.

A MON basically elongates lifetime of WSNs as compared to traditional WSN. It provides constant throughput when number of SN approaches infinity and requires smaller number of SNs and practically no maintenance. The associated disadvantage is potentially long message delay due to intermittent connectivity as mobility of RNs is not controllable. If you have several RNs, you do not know which one SN should give a copy of data. This can be addressed satisfactorily if you know the mobility pattern of RNs. This may be a difficult proposition as SN may know in which direction RN is currently moving. The probability of randomly selecting a path depends on number of paths at a given SN (Fig. 15.24a, b). So, to select an appropriate RN, just use moving direction to determine the bias of each RN and select RN that is moving toward the BS. Thus, by slightly increasing the bias level of the message, forward data to RN with better bias toward the location of the BS. If there is a provision, SN could send many copies of the same data to multiple RNs (Fig. 15.25a). Select RNs such a way that will help to reduce the latency to BS. You could consider complex movement of the data (RN) relative to location of the BS. Latency of a single copy message forwarding algorithms as a biased random walking of message (RN) is shown in Fig. 11.25b, c.

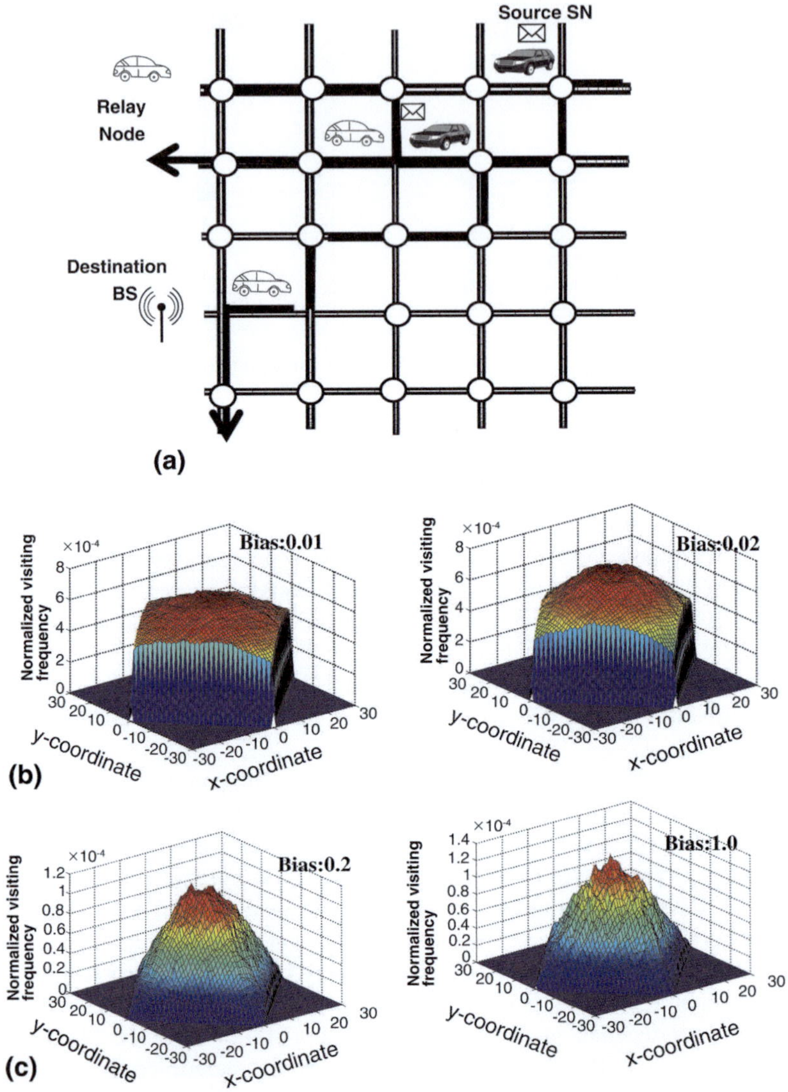

Fig. 15.25 **a** Random mobility in 2-D mesh. **b**, **c** Mobility with different bias to BS

So, there is a need to model a general random walking model that could possess both random walk and random waypoint walk which is difficult to do. So, an approximated formula of average message delivery time as a function of message bias level is feasible as there is a close relationship between mobility and bias. If RNs are independently moving randomly and delivering single copy to the BS via RNs, then that would lead to an optimal relay strategy. So, a basic scheme is to select a random destination point. Define a bias level between zero and one; once

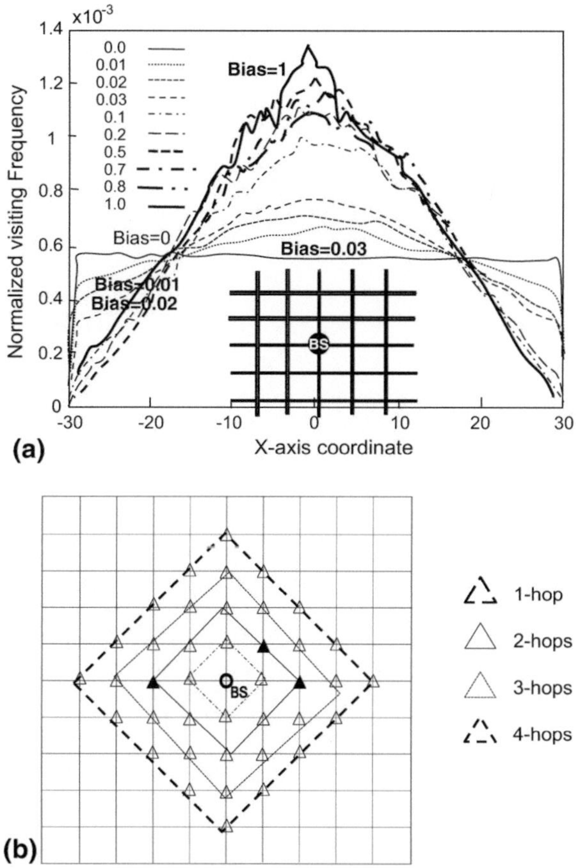

Fig. 15.26 2-D mesh. **a** Normalized visiting frequency on *x*-axis. **b** 1-, 2-, 3-, and 4-hop neighbors [9]

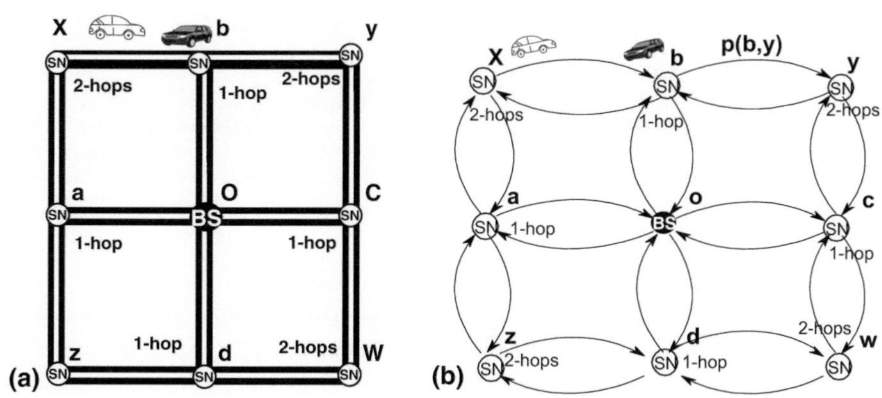

Fig. 15.27 **a** Analyzing relay time in 2-D WSN. **b** Converting to 2-D graph

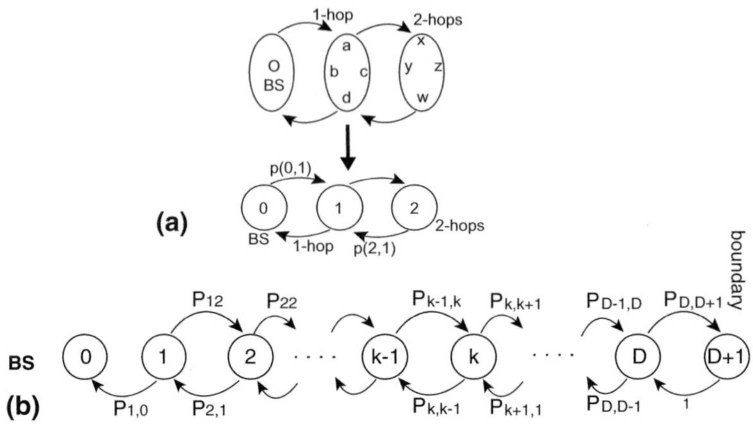

Fig. 15.28 **a** Converting 2-D model to 1-D. **b** Final 1-D Markov model

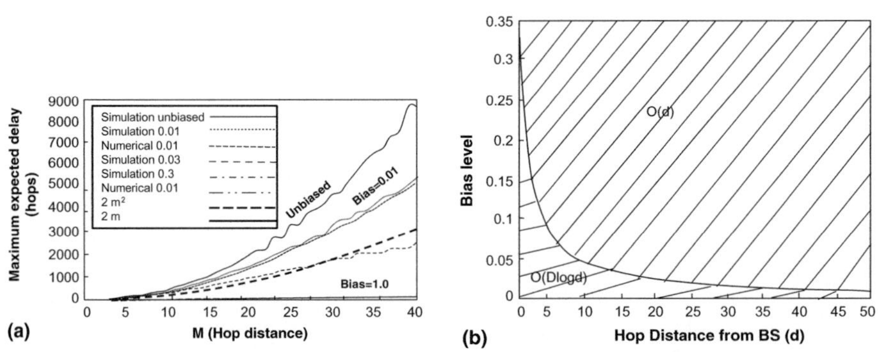

Fig. 15.29 **a** Analytical and simulation results for 2-D mesh. **b** Relationship between bias level and expected latency

the destination is reached, it selects another arbitrary destination (and bias level) and repeats these three steps. Normalized visiting frequency along *x*-axis is shown in Fig. 15.26a and various 1-, 2-, 3-, and 4-hop neighbors in a 2-D mesh from the BS are shown in Fig. 15.26b. If bias level—>0, the schemes become equivalent to a regular random walk; while bias level—>1, the scheme is equivalent to random

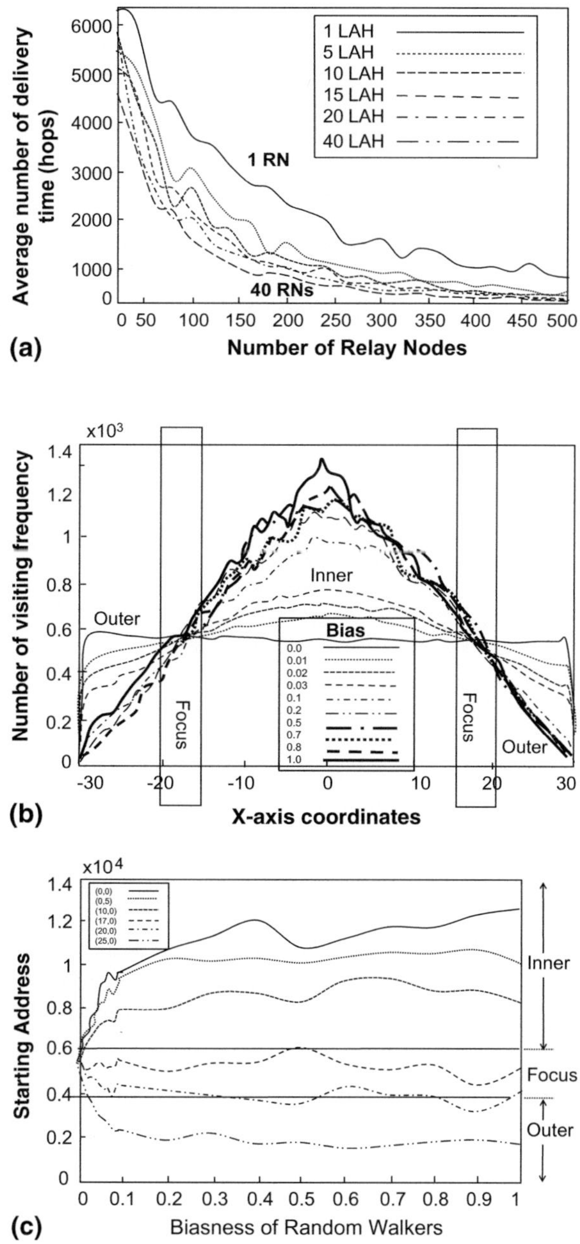

Fig. 15.30 **a** Simulation results with increasing number of RNs in 2-D WSNs. **b** Normalized visiting frequency on *x*-axis when biased changed. **c** Visiting frequency of different biased random walks

waypoint walk without pausing time. We can represent mixture of different random walkers by selecting different bias levels appropriately [9, 10] by modeling as a Markov chain in the case of unbiased walk ($a = 0$). The time taken by a message in reaching BS is given by

$$\begin{aligned} E[N_{d,0}] &= \sum_{i=1}^{d} \frac{(2D^2+4D+1)}{2i-2} - \sum_{i=1}^{d} \frac{2i^2}{2i-1} \\ &\approx (2D^2+4D+1)\log(2d-1) - \frac{1}{2}(d^2+d-2) \\ &= \Theta(D^2 \log d) \end{aligned} \tag{15.5}$$

where $(D + 1)$ is maximum distance of a SN from the BS along shortest path and d is the starting state of the random walk. The expected latency under this model is equivalent to generally known hitting time of a random walk in a grid and given by (n is the number of nodes in the grid):

$$\begin{aligned} n &= |V| = 2D^2 + 2D + 1 \\ \Theta(D^2 \log D) &= \Theta(n \log n). \end{aligned} \tag{15.6}$$

Suppose we have a 2-D mesh like topology, where destination BS is at the center of an area. In a biased random walk, with a bias level $0 \leq \alpha \leq 1$, the expected length of random walk path starting at d hops away from the BS follows the following bound:

$$E_{ax} = \begin{cases} O(D^2 \log d), & \text{for } ax = 0 \\ O(D \log d), & \text{for } 0 < ax < \frac{1}{2x+1} \\ O(\max\{d, D-d\}), & \text{for } ax = \frac{1}{2x+1} \\ O(d), & \text{for } \frac{1}{2x+1} < ax \leq 1, \end{cases} \tag{15.7}$$

where x is the current distance of the node from the BS during the random walk, α_x is the bias level at node, and d is the state from where random walk starts. This leads to relay time in 2-D mesh as shown in Fig. 15.27a and simplified to 1-D graph as Fig. 15.27b. The corresponding Markov model is obtained as shown in Fig. 15.28. This gives results of Fig. 15.29 for one RN. When the number of RNs is increased, the results obtained are summarized in Fig. 15.30. The expected delay of biased random walk reduces quickly as bias level is increased from 0 to 0.1 and the delay becomes almost linear when bias level is 0.3. So, a small bias for far away nodes works effectively in a 2-D mesh WSN.

15.5 Conclusions

Regular topologies offer many advantages such as no need for neighbor discovery, easy clustering and CH selection, and underlying TDMA schedule to eliminate any potential collisions. The use of regular topologies ought to be stressed further in civilian applications. Further reduction in energy consumption is possible if relay nodes (RNs) are deployed to collect data from SN in one hop and ultimately deliver to BS. The use of multiple RNs could prove to be very useful.

15.6 Questions

Q.15.1. How can you define a cluster in a regular WSN?

Q.15.2. What kind of clustering is possible in the following WSNs?

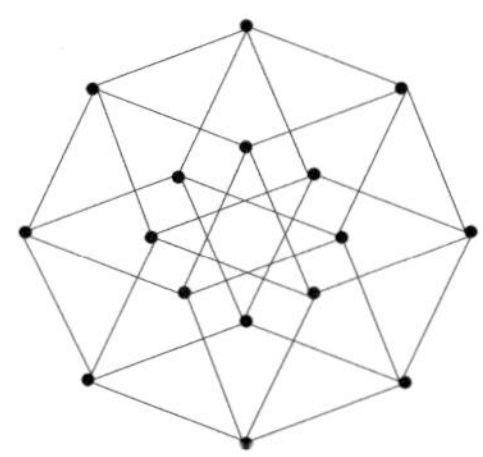

(i)

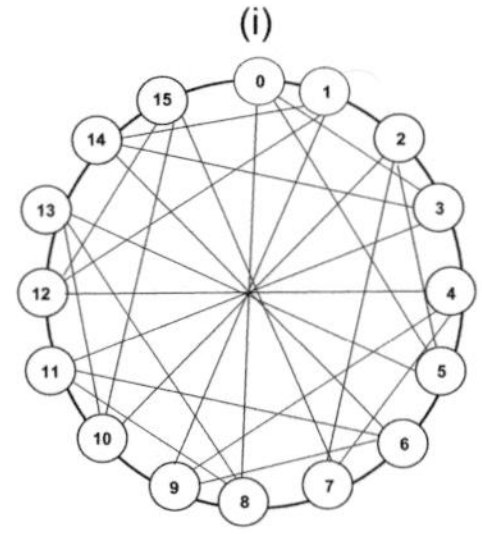

(ii)

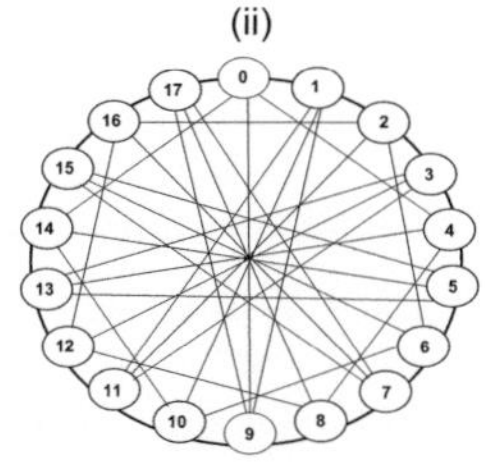

(iii)

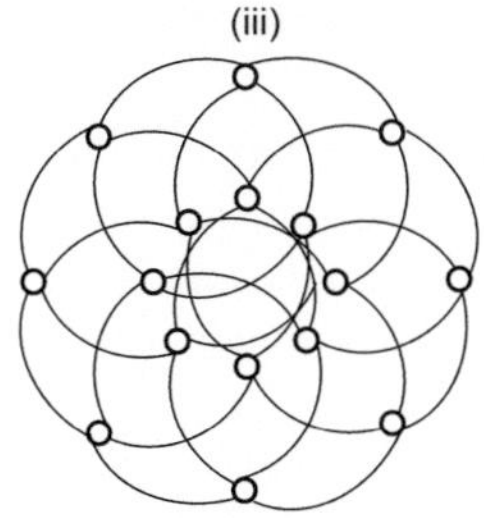

(iv)

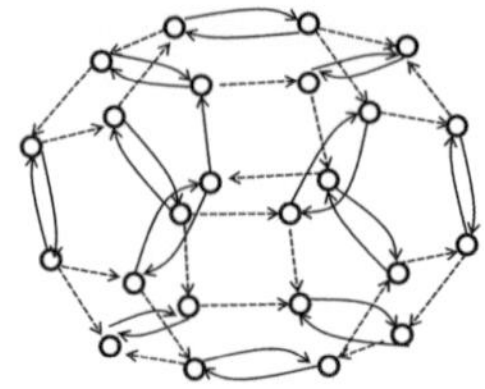

(v)

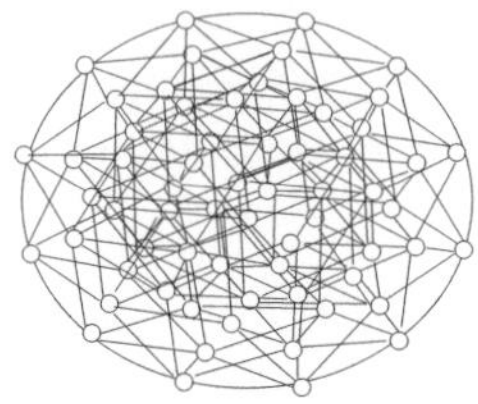

(vi)

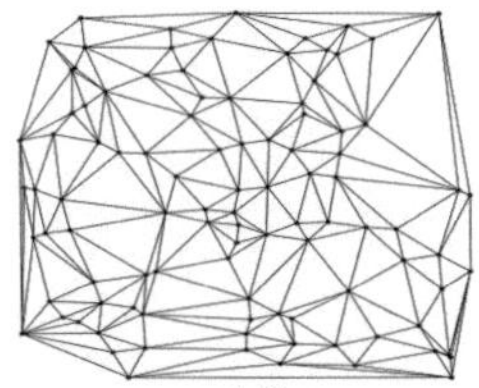

(vii)

Q.15.3. Can you use sleep–awake cycles for Problem 15.2?

Q.15.4. Can you determine a TDMA schedule for these examples?

Q.15.5. 4 RN will consume more power than 1 RN. What is the main advantage of using 4 RNs as compared to only 1 RN?

Q.15.6. Can you use RN just to collect data from CHs in place of all SNs? What are the advantages?

Q.15.7. In Q.15.6, can you aggregate data at CH before transferring to a RN? What is its impact?

Q.15.8. In a triangular topology, assume clusters have been defined such that SNs in a cluster are at the most 3 hops away. Can you determine how each cluster looks like?

Q.15.9. In Q.15.8, how do you compare this scheme if aggregation is done for the whole network at BS? What are the advantages and disadvantages of the two approaches?

Q.15.10. Does data gathering from CHs in opportunistic wireless sensor network having 4 RNs depend on the size of the cluster?

Q.15.11. Can you repeat Q.15.8 for hexagonal topology?

Q.15.12. Can you repeat Q.15.8 for rectangular topology for 6 hops away? A RN monitors these CH. Define one way of traversing the WSN so as to minimize visiting time?

Q.15.13. Repeat Q.15.10 when 2 and 4 RNs are used.

Q.15.14. How do you compare data gathering in opportunistic wireless sensor network having 1 RN with a CH-based scheme?

References

1. K. Shashi Prabh, Chinmay Deshmukh, and Shikhar Sachan, "A Distributed Algorithm for Hexagonal Topology Formation in Wireless Sensor Networks," Proceeding of ETFA'09 Proceedings of the 14th IEEE international conference on Emerging technologies & factory automation, pp. 675–681.
2. F. G. Nocetti, I. Stomenovic, and J. Zhang, "Addressing and routing in Hexagonal Networks with applications for tracking mobile users and connection, rerouting in Cellular Networks, IEEE Transactions on Parallel and Distributed Systems, vol. 13, no. 9, Sept. 2002, pp. 963–971.
3. Ignasi Sau and Janez Zerovnik, "*An Optimal Permutation Routing Algorithm on Full-Duplex Hexagonal Networks*," *Discrete Mathematics and Theoretical Computer Science* DMTCS vol. 10, no. 3, 2008, *April 1–3, 2014, pp. 49–62.*
4. K. Shashi Prabh and Tarek F. Abdelzaher, "On Scheduling and Real-Time Capacity of HexagonalWireless Sensor Networks," www.cs.virginia.edu/~ksp2q/…/PA07_hexnet.pdf.
5. Rex Kincaid, Allison Oldham, and Gexin Yu, "Optimal open-locating-dominating sets in infinite Triangular Grids," http://de.arxiv.org/pdf/1403.7061.
6. Bader Albader, Bella Bose, Mary Flahive, "Efficient communication algorithms in hexagonal mesh interconnection networks," Journal of Latex Class Files, vol. 6, no. 1, Jan. 2007, pp. 1–10.
7. Stephanie Lindsey and Cauligi S. Raghavendra, "PEGASIS: Power-Efficient Gathering in Sensor Information Systems," http://ceng.usc.edu/~raghu/pegasisrev.pdf.
8. Huilong Huang, John H. Hartman, and Terril N. Hurst, "Data-Centric Routing in Sensor Networks using Biased Walk," www.cs.arizona.edu/~jhh/papers/secon06.pdf.
9. Jung Hyun (Peter) Jun, W. Fu, and Dharma P. Agrawal, "Impact of Biased Random Walk on the Average Delay of Opportunistic Single Copy Delivery in Manhattan Area," *Ad Hoc & Sensor Wireless Networks,* pp. 195–222, 2011.
10. Pritam Shah, "Virtual Coordinate based technique for Wireless Sensor Networks: A simulation tool and localization & planarization algorithms," MS Thesis, Colorado State University, Fort Collins, Colorado, Summer 2013.

Chapter 16
Personal/Body Area Networks and Healthcare Applications

16.1 Introduction

A personal area network (PAN) is the interconnection of devices for information technology within the range of a single person, characteristically within a range of 10 m, and is typically coupled with wireless links and hence called wireless PAN (WPAN). These devices could be Bluetooth-based or ZigBee, or even new near-field communication components as pico-networks. The emphasis is to use groundbreaking data delivery schemes that could connect various transducers to form a comprehensive team and provide useful information for health care. Ever-increasing cost of health care has become a national concern as Medicare had 35 million members in 2003 and 35.4 million in 2004 (Fig. 16.1). Healthcare expenditures in the USA are projected to rise to 15.9% of the gross domestic product ($2.6 trillion) by 2010. The total cost for cancer treatment in 2020 is projected to be $173 billion, which represents a 39% increase from 2010. In 2013, 14.4 million Medicare beneficiaries are enrolled in Medicare Advantage Plans, an increase of more than 1 million (9.7%) from 2012. A vast majority of beneficiaries (98%) have access to Medicare Advantage Prescription Drug (MA-PD) Plans with no premium. Slightly more than half (55%) of beneficiaries are enrolled in a zero-premium plan in 2013. It is worth noticing that eligibility for Medicare = 65 = senior citizen/geriatric and "old" can be said between 65 and 85 years, while 85 + people can be classified as "very old." There are 700 million seniors worldwide (1.3 billion in 2040), and life expectancy in USA is around 78. So, people at 65 expected to live another 18.7 years. Women outnumber the man in the elderly population and consider a 70-year-old widow living at home all by herself. A lady may have mild cognitive impairment, but does most of the household work on her own and wants to remain as much independent as possible and even wants to help her friends with similar problems (Table 16.1). Multiple chronic illnesses require multiple medications, and if the lady is not taking medications, her condition may become acute. So, there could be one or more side

D.P. Agrawal, *Embedded Sensor Systems*,
DOI 10.1007/978-981-10-3038-3_16

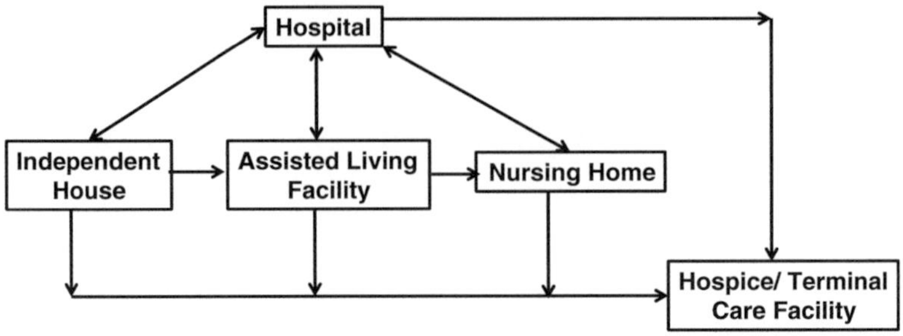

Fig. 16.1 Elderly people statistics

Table 16.1 Elderly people and daily activities

Age	Sensorimotor and cognitive abilities (average)	Deficit (average)	Technology support needed for daily activities
60–70	100–90	0–10	Minimal
70–80	70	30	Moderate
80–90	50	50	High
90+	0–10	90–100	Very high

effects. There could be a visit from a trained nurse once a week. So, the goal is to manage chronic conditions and delay a move to assisted living/nursing home for 10 years to save around $500,000 in nursing home expenses.

Table 16.2 [reproduced from 1] recapitulates features of different applications based on sampling rate desired, memory size required, communication bandwidth needed, migration of application components, network coverage, and preferred reliability. First of the five types is based on sampling periods varying from one second to few hours to cover a large area with no strict deadlines for the results. These include environmental and agricultural applications. Energy in SNs is expected to last from few days to months and possibly recharged with solar cells. Type 2 applications are defined for smaller space and intensive computing requirements with sampling period varying from 1 ms to 1 s, with possibility of processing being done after storing samples. Type 3 applications, processing of images are desired in Type 2 applications and SNs are synchronized and some mobile units. In type 4 applications, the space is restricted as compared to type 3 and SN energy is expected to last for a week and healthcare application belong to this type. Type 5 is primarily for industrial process control with restricted jitter, and sampling periods vary from 50 ms to few seconds.

Table 16.2 Application characterization and monitoring rates [1]

Application	Sampling rate	Possible applications	Computing capacity	Memory size	Communication bandwidth	Location	Mobility of components	Real-time	Network's coverage	Energy autonomy	Synchronization
Type 1	1 s to few hours	Agricultural and environmental applications	Low performance	Low	<256 kbps	Yes	No	Only measurement	Open space 10 km	Months	Yes
Type 2	1 ms to 1 s	Industrial and agricultural sectors	Low performance	Medium	<256 kbps	Yes	No	Only measurement	Confine space 100 m	There isn't restriction	Yes
Type 3	1 ms to 1 s	Agricultural sector for detection of pests, and environmental sector aimed at detecting fires	High performance	High	1 Mbps	Yes	Yes	Only measurement	Open space 10 km	Hours	Yes
Type 4		Diseases with body area networks	High performance	Medium	<256 kbps	Yes	Yes	Only measurement	Open space 1 km	Days	Yes
Type 5	50 ms to few seconds	Industrial control process	Low performance	Low	<256 kbps	No	No	End to end and minimum jitter variability	Confine space 100 m	There isn't restriction	Yes

16.2 Activities of Daily Living

Various activities of daily living (ADL) include food, hygiene, social needs, sleep, medications, managing chronic conditions, safety, and financial needs. Elderly people have increased susceptibility to falls, and if living alone, it could be hours or days before someone finds out. This could lead to more health complications, and any delay in treating such illness increases the severity. So, detection of falls is an important requirement, and few options for automatic detection of falls include estimation of posture and pressure on sensor-equipped floors. Visual fall detection is feasible along with context information and is suitable for sensor–motor and cognitive difficulties. One approach could be the use of wearable, portable, and implanted device to recognize the fall. Another simple scheme could be based on computers, Internet, Web sites, cell phones, and alarm system. An intermediate solution could be RFID-based emergency alarm system for medication and task reminder systems. An elaborate system could have a smart home with all clever devices that are reliable, smart and context-aware, personalized, robust, self-configuring, and causing no harm to the patients. The cognitive role could include executive function, decision making, and dual-task performance and could decline with age. An electronic patient record can be kept besides remote patient monitoring. So, for integration of wireless communication, networking and information technology is required and a large volume of medical information can be collected to define most effective strategies for treating chronic illness and reducing disability. Efforts should be made to improve health and reduce healthcare cost, and chronic disease must be managed by the effective use of clinical resources that necessitates a complete integration of IT.

Therefore, wireless communication, sensor platform, networking, and database need to be incorporated in a clinical practice with unequivocal security and privacy rules to protect end-to-end communication and limit access to sensitive medical information. A cellular 3G/4G technology can be used for such application as it provides real-time delivery with wide coverage adequate bandwidth and ability to work with other wireless technologies as they are widely used and are secure with possible location management. The only problems are the presence of dead spots, providing reliability is a real challenge, there is lack of broadcast/multicast, and pricing structure could impact of the commercial traffic. This forces us to consider WLAN for monitoring applications as it provides adequate bit rate, supports transmission from patients to access point (AP), and could prove to be handy for mobile patients as location management is feasible. The limitations are the coverage area, delays in monitoring, associated security, presence of colocated networks, no provision for multicasting, and reliability is questionable. Wireless LANs experience unpredictable coverage , and the data speed is variable as bandwidth needs to be shared and interference may be present with unlicensed ISM band. It may be possible that the device could not access the network as reaching to cellular phone sometimes is difficult and video quality may not be good due to variable delays. A generic telemedicine (Fig. 16.2a) ought to support utilization of different assets

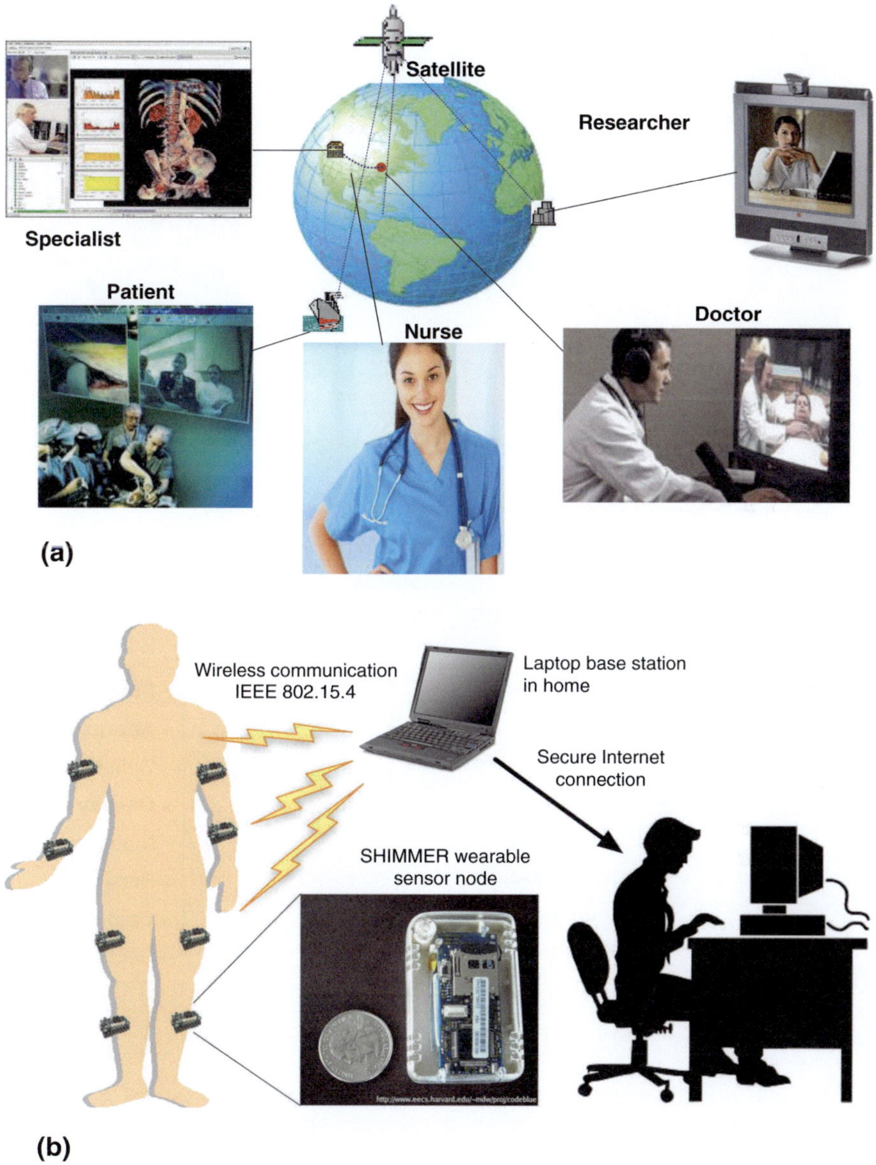

Fig. 16.2 **a** A generic telemedicine **b** Monitoring Mobility

independent of their geographical location, and there is a need for multidisciplinary collaboration. It is desirable to facilitate dissemination of medical knowledge to practicing doctors and medical students and enable doctors in remote and rural areas to refer with specialists in urban areas.

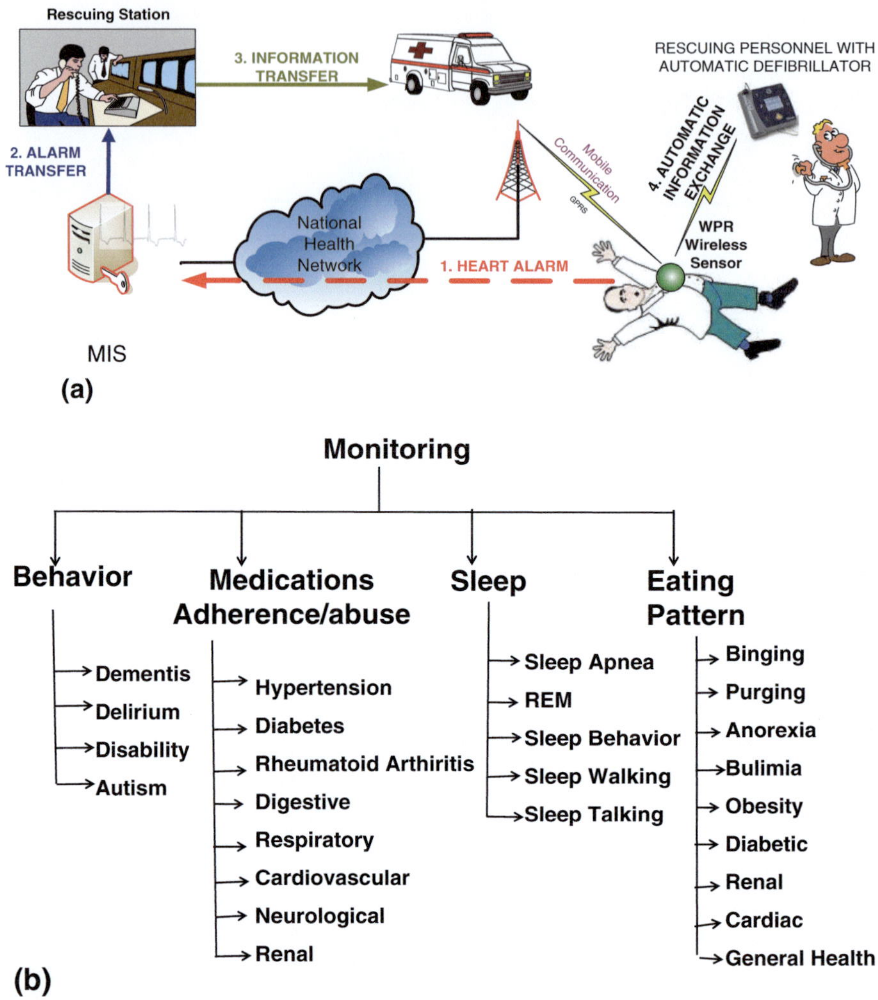

Fig. 16.3 **a** Emergency Scenario **b** Physiological parameters to be monitored

If the objective is to monitor mobility only, then sensor boards could be attached to legs and hands as shown in Fig. 16.2b. A more chaotic scene appears when an accident occurs (Fig. 16.3a), and every attempt is made to save human life. So, there is a need to monitor many different body parameters as shown in Fig. 16.3b. A personal health monitoring system can be envisioned as illustrated in Fig. 16.4a. Details of used different types of physiological parameters are shown in Fig. 16.4b, while a comprehensive health monitoring system is depicted in Fig. 16.5a, showing utilization of assets independent of their geographical location to constitute a multidisciplinary collaboration.

Functions to be performed and dissemination of knowledge at different facilities to practicing doctors and medical students are shown in Fig. 16.5b. Such an

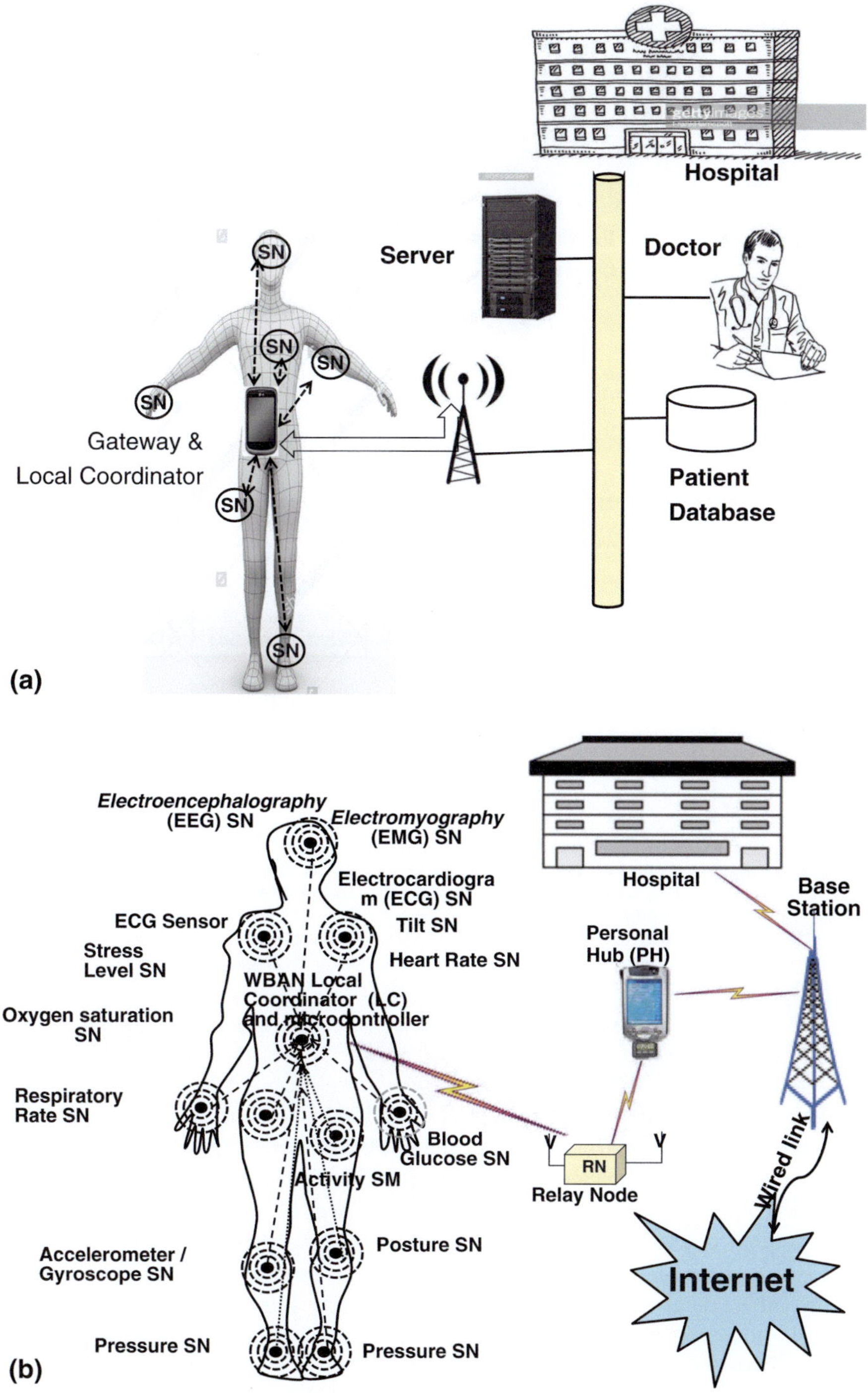

Fig. 16.4 **a** Personal health monitoring system **b** Detailed health monitoring system

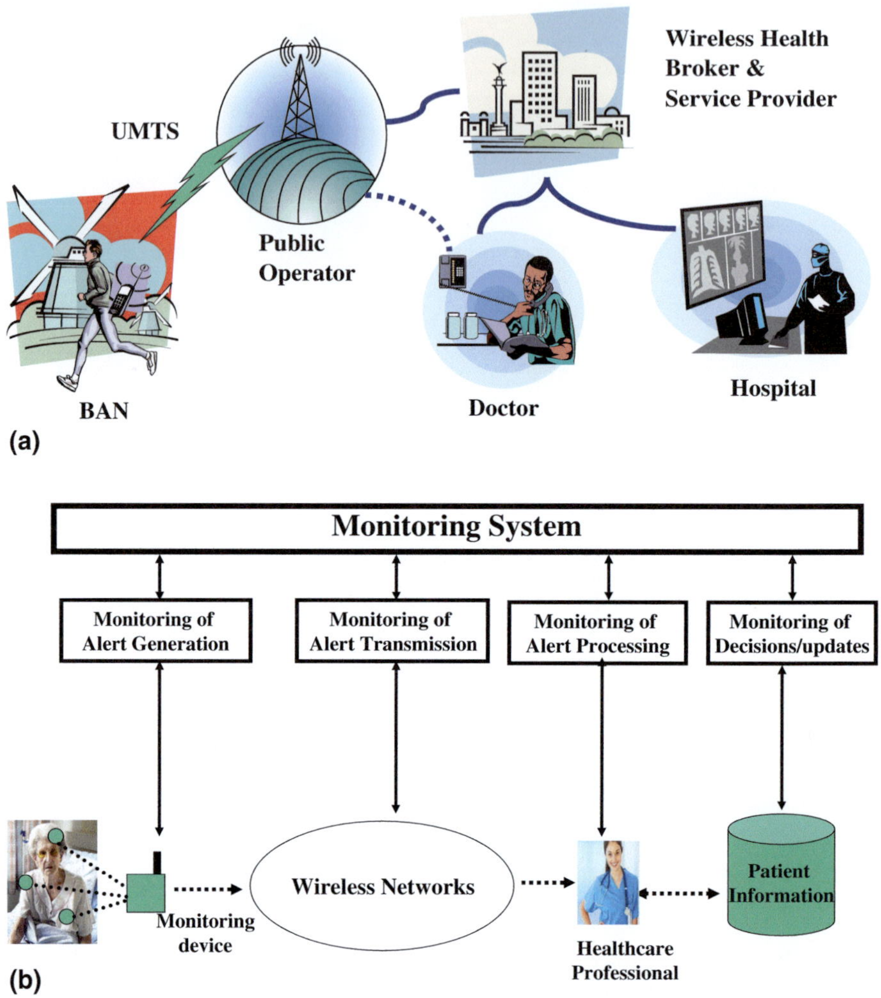

Fig. 16.5 **a** Scheme for personal health monitoring system **b** Functional details for personal health monitoring system

infrastructure allows doctors in remote and rural areas to consult with specialists in urban areas and take appropriate medical and clinical decisions. So, there is a clear need for intercommunication among medical devices and clinical information systems. This has been accomplished with a number of medical products such as infusion pumps and ventilators that commonly have RS-232 ports, and these devices can communicate with many physiological monitoring instruments. There are several medical equipments that could be easily linked for personal communication. However, virtually, all of these are specialized applications, and unique custom interfaces are needed. To address the medical device plug-and-play interoperability problem, a single communication standard is needed to provide

Table 16.3 Vital physiological parameters and associated characteristics [2]

Vital sign and parameters	Sampling rate	Quantization (bits/sample)	Total bit rate
Breathing rate	One sample/sec	4	4 bps
ECG	240 samples/sec	12–36	2.9–8.7 Kbps
Blood pressure	One sample/minute	64	1 bps
Oxygen saturation	One sample/sec	16	16 bps
Core body temperature	One sample/minute	16	0.3 bps

unobtrusive and persistent monitoring. So, different physiological parameters ought to be monitored, and associated characteristics are summarized in Table 16.3 [2]. Different factors need to be sensed and sent by SNs at dissimilar frequencies, depending on criticality of data.

16.3 Available Biomedical Transducers

In biomedical application, the most important issues to tackle are the quality of service (hand-over, interruption/delays in transmission, data loss bandwidth problems, etc.), social acceptance (health risks (cell phone usage), economic issues, ethical issues), and legal issues (accreditation of the devices and applications, protection of health-related data, privacy, security, and encryption of data, and medical responsibilities/liability). There are many commercial products, and they are helping human race. These include (Fig. 16.6) Nokia N810 Internet Tablet, Motion sensor (802.16.4), weight scale (Bluetooth) blood pressure monitor (Bluetooth) device.

Noninvasive technology is also being used to measure the heart rate (HR) and blood oxygen saturation (SpO_2) (Fig. 16.7a, b) as it projects infrared and near-infrared light through blood vessels near the skin and by determining the amount of light absorbed by hemoglobin in the blood at two different wavelengths, and the oxygen level can be determined. The heart rate can also be found as the blood vessels contract and enlarge with the patient's pulse. The *Pluto activity center* is useful for patients undergoing physical rehabilitation using long-lasting rechargeable battery.

Many other biomedical devices which use SN as an integral part are shown in Fig. 16.8. Areas that affect human health the most are the ECG (monitoring heart activity), EMG (electromyography), and for sensing motion (activity) and are discussed in some detail here (Fig. 16.9). Electrocardiogram (ECG) SNs require a large bandwidth as parallel transmission of many waveforms is needed. There could be environmental interferences, and patient's movement must be restricted during the test. The drug delivery mechanism is used to respond to any anomalies in the

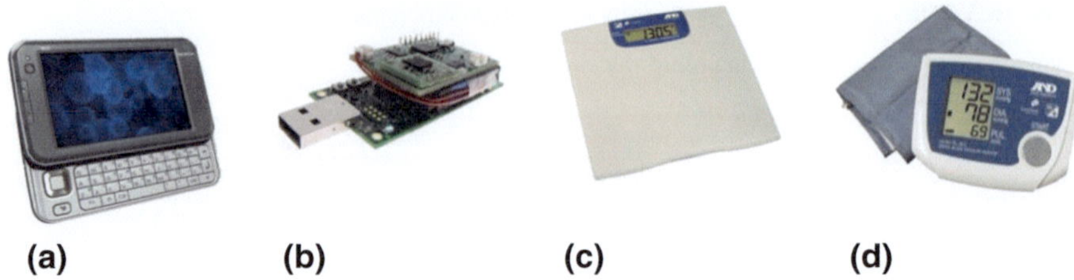

Fig. 16.6 **a** Nokia N810 tablet **b** Motion sensor **c** Weight scale **d** Blood pressure monitor

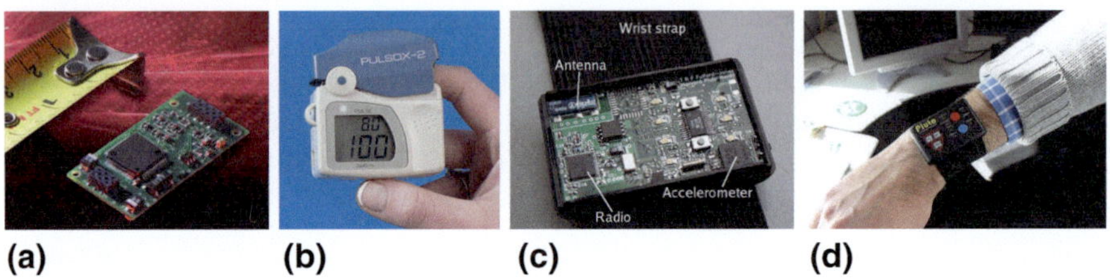

Fig. 16.7 **a b** Heart rate (HR) and blood oxygen saturation (SpO_2) **c d**: *Pluto* Activity Sensors

heart. The detection is based on SQRS Algorithm [3] on ECG waveforms, and misclassification due to noise can be avoided by having an appropriate threshold. The most common type of ECG involves the connection of several leads to a patient's chest, arms, and leg via adhesive foam pads (Fig. 16.10a). The device records a short sample, e.g., 30 s, of the heart's electric activity between different pairs of electrodes. When there is a need to detect irregular cardiac condition, an uninterrupted EKG measurement is adopted. This involves checking for extended period patient's cardiac activity utilizing 2 or 3 electrodes. The ECG signal is small (~ 1 mV peak-to-peak) and is amplified (gain >1000) using low-noise amplifiers and filtered to remove noise before being digitized. In Fig. 16.10b, P wave determines contractions of the atria and QRS is a series of waves linked with ventricular contractions. The T and U waves follow the ventricular contractions, and various sampling rates and quantization levels are used with sampling frequencies selected between 128 and 256 Hz. For rigorous details, higher sampling rates and bit rates, e.g., 16 bits, are adopted. IMEC (Fig. 16.10c) [4] has recently developed a wireless, flexible, stretchable EKG patch for continuous cardiac monitoring which can be placed on the arm or on the leg, and the same system can be used to monitor muscle activity (EMG). The patch of size 60×20 mm^2 includes a microprocessor, a 2.4 GHz radio link, and a miniaturized rechargeable lithium-ion battery. Data are sampled nonstop between 250 and 1000 Hz, and the battery has a capacity of 175 mAh adequate for several days.

Sensors for physiological conditions is designed for personal health and general environmental monitoring to measure temperature, pressure, humidity, and vibration/position and available as a wrist strap to make it wearable system

(Fig. 16.11a). Many different versions have been adopted in military, navy, and marine applications. Smart skin sensors (Fig. 16.11b) monitor health parameters such as heartbeat/pulse rate, body temperature, and acoustic waves and can be worn underneath soldiers' uniform. The RF-based wireless transmitters convey health parameters to health camp via a close-by vehicle. SHIMMER wearable mote has been developed by the Digital Health Group at Intel with TI MSP430 processor, CC2420 IEEE 802.16.4 radio, Triaxial accelerometer, rechargeable Li-polymer battery, and MicroSD slot supporting up to 2 GBytes of flash memory.

Sensor for optical motion capture (Fig. 16.12a) primarily employs reflective markers and multiple cameras that digitize different views of performance. Fiber-optic sensors (Fig. 16.12b) use rotation based on transmitted light. An embryonic method for transferring data from human body employs electronics textiles (e-Textiles) (Fig. 16.12c) as a communication medium. The medium consists of two electrically separate grids of conductive thread that could physically

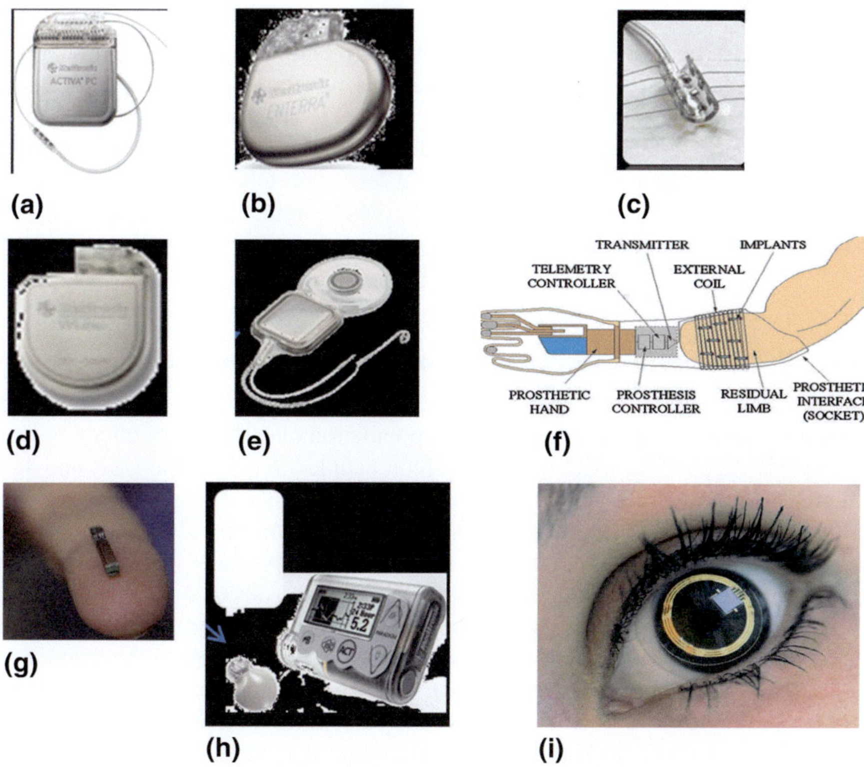

Fig. 16.8 **a** Deep brain Neuro-stimulator **b** Gastric stimulator **c** Foot drop implants **d** Cochlear implants **e** Cardiac defibrillator/pacemaker **f** Artificial hand **g** Implantable glucose sensor **h** Insulin Pump **i** Artificial Retina

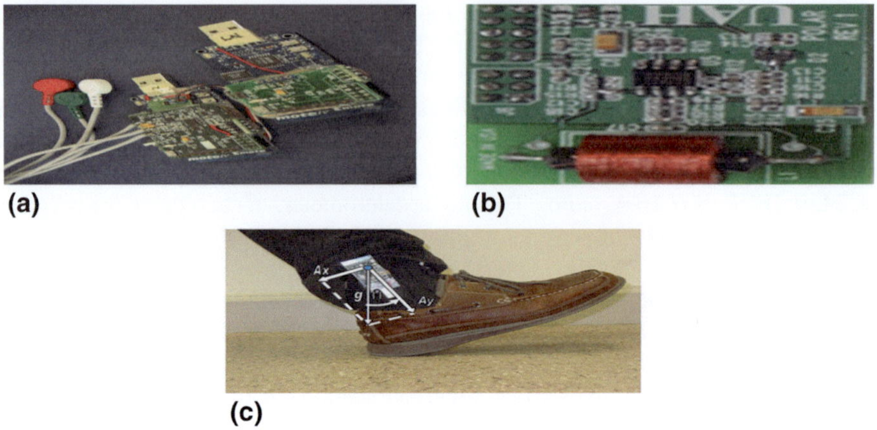

Fig. 16.9 **a** ECG **b** EMG **c** Motion sensing

connect SNs to the shared medium using metallic button-snaps and communicate via an e-Textiles transceiver chip. The use of a pair of physical low-impedance connections has the distinct advantage as it enables to connect signal differentially, permitting energy-efficient amplitude-modulation schemes that tolerate coupled interference and power SNs remotely from a local BS. Objects for posture determination shown in Fig. 16.13 can be for the whole human body, portions of the body, facial animation, and for animals or puppets.

16.4 Parkinson's Disease and Fatigue Level Detection

Parkinson's disease affects about 3% of the population over the age of 65 years, and NIH reports suggest that more than 0.5 million people affected in the USA and 4–6 millions in the world. There is no cure for Parkinson's disease. Even though Parkinson's disease is hereditary, early detection can improve the mortality rate. Early symptoms include mild tremors, problems with balanced walking, and no expression on face. Changes to patient's mobility can be detected with sensors and can be used to determine the onset of the disease. In a recent project [5], pressure sensors embedded in the patient's shoe soles can be used to measure the amount of pressure on the patient's feet (Fig. 16.14). The data obtained can be compared to that of a healthy person to determine the degree of unbalance during walking and freezing of gates (FoG) in Parkinson's patients can be determined with changes in pressure and data over time can be used to determine the progress of the disease. The sensor data can be recorded in real time, and the doctor can use the data for a better diagnosis. Pressure sensors embedded in the patient's shoe soles can be used to measure the freezing of gates as electronic circuit is hidden in the shoe sole (Fig. 16.15).

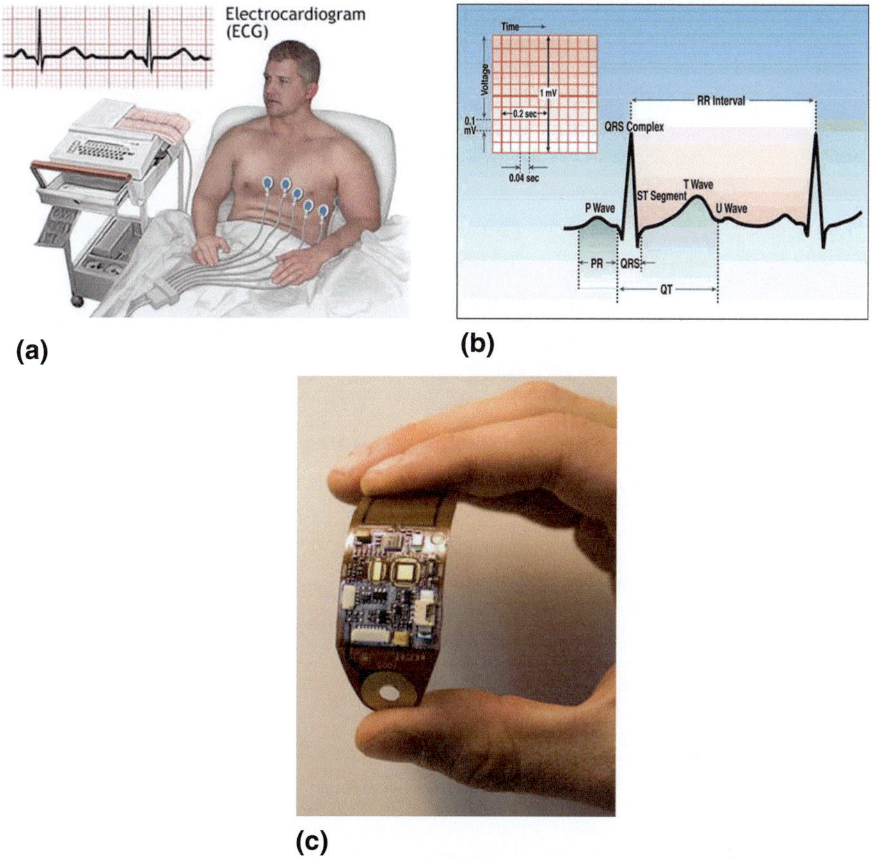

Fig. 16.10 **a** ECG set up **b** ECG waveform **c** IMEC patch

Monitoring Athlete's during their practice and game sessions is important for the captain and owners. It is critical to monitor how tired a player is and when to make a player rest by replacing by a new one and vice versa. Therefore, general health monitoring while in action is critical. A recent work [6] employs 9 pressure sensors embedded in the shoe soles that can be used to measure the amount of pressure on the player's feet and conveyed to the coach sitting on sidelines. Data obtained from two feet are compared to determine the degree of fatigue during playing, and changes in the pressure data over time can be used to determine time to change with a resting player. The sensor data can be obtained in real time by the coach and take a player out/in. The scheme can be useful for areas where uninterrupted long hours are needed such as nurses, doctors, army and defense personnel, and truck drivers.

In a similar way, helmet is being used by football players for avoiding concussion due to the impact on head by injuries (Fig. 16.16a) and integrated assembly equipped with thermal sensors, video cameras, and chemical and biological sensors is commercially available. So, efforts are being made in monitoring potential health

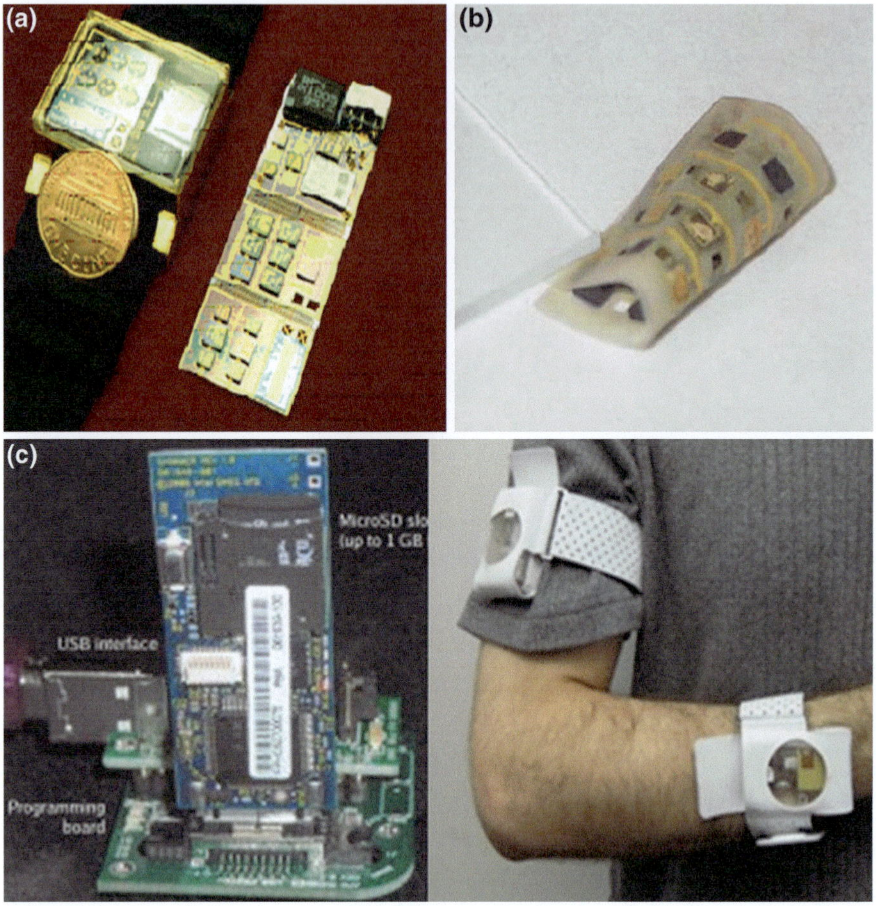

Fig. 16.11 **a** Wrist Strap for physiological conditions [15] **b** Smart skin sensors [16] **c** SHIMMER wearable mote [17]

problems due to different factors. But, once diagnosed, same amount of medicine is prescribed for duration of more than a week, commonly for a month, and no change in medication is done for extended period of time. Therefore, there is a need to monitor patient's condition 24/7 and accordingly adjust the medication doses as needed. Such a future system is illustrated in Fig. 16.16b.

16.5 Communication Through Skin

To monitor physiological parameters, SNs can be mounted on or implanted inside human body and transfer data to healthcare provider or doctor for analysis in multi-hop fashion. One such deployment is shown in Fig. 16.17, and irrespective of

Fig. 16.12 **a** Reflective Markers **b** Fiber optic sensors **c** e-Textile SNs

employed routing protocols (Table 16.4), SN in operation will cause heat rise in the surrounding vicinity as transmission, reception, and relaying of packets cause an increase in temperature. This is independent of underlying MAC protocol (Table 16.5). If heat rise is > thermoregulation, then tissues could be damaged. In order to obtain a mathematical mode for the heat rise, Pennes bio-heat equation [7] is given as follows:

$$C_P \frac{\partial T}{\partial t} = \nabla(K \nabla T) + A_0 + B_0(T - T_b) + \rho(SAR) + P_D, \tag{16.1}$$

where T is the temperature in °C, K is the thermal conductivity(J/(ms °C)), C is the specific heat(J/(kg °C)), A_0 is metabolic rate, B_0 is the blood perfusion constant(J/(m^3 s °C)), TB is the temperature of the blood in °C, and PD is the power dissipated

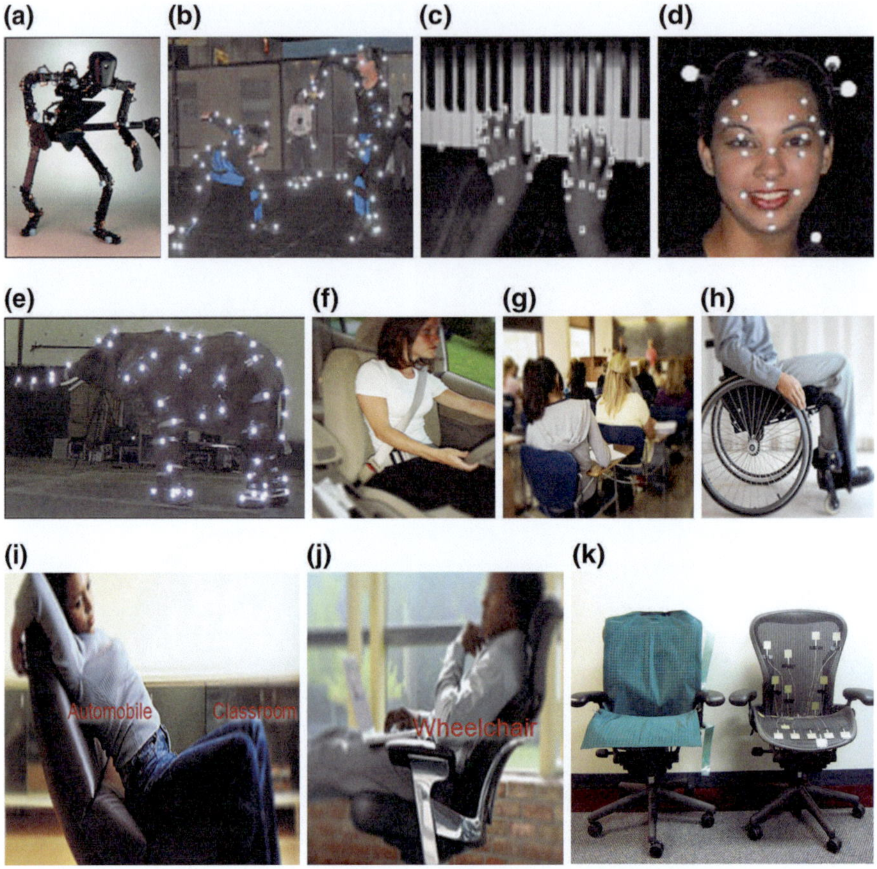

Fig. 16.13 **a**, **b** Human whole body **c** Portions of body **d** Facial animation **e**, **f**, **g**, **h**, **i** Sensors for seated postures **j** Original chair and chair with Sensors

over a volume. Temperature of different parts of the body does increase and is summarized in Table 16.6. Most existing work assumes each SN to have a fixed awake time (t_A). A sleep-awake cycle scheme has been proposed [8] which combines both thermal awareness and generating of efficient duty cycle by proposing a heat-based MAC protocol. Given a WBAN, set a fixed awake time for each of the SNs, termed as *sampling window* for each SN. Once a SN has its *sampling window period*, it goes to sleep. The duration of its sleeping period (t_s) is given by the following equation:

$$t_s = PDF(\Delta T), \tag{16.2}$$

where PDF is the probability distribution function. There are many possible sleep models that indicate the *amount of time SNs sleep* such as *Poisson*, *Binomial*,

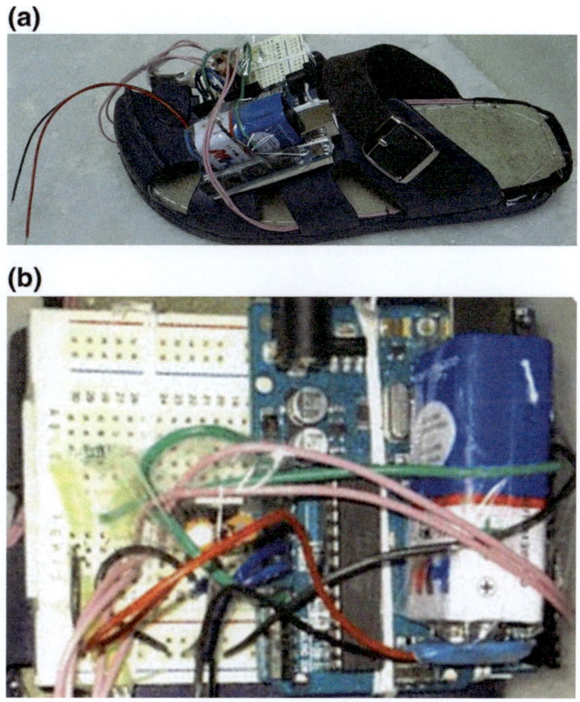

Fig. 16.14 **a** Shoe to determine FoG **b** Electronic circuitry inside shoe sole

lognormal, and Laplace. PDFs are able put the SNs to sleep for a longer period based on the temperature rise. These PDFs have been tried to check which provides a better throughput as a function of network size (Fig. 16.18a). Figure 16.18b shows the average temperature rise of the network when the sampling window time is increased. The network size is set to 3 × 3 2-D mesh and simulated for 1000 s by varying the value of sampling rate from 1 s to 5 s. Figure 16.18c shows the temperature rise of each node along the y-axis with a sampling rate of 5 s which clearly shows that Node 5 has a peak temperature increase as this is the busiest node in terms of this network. Figure 16.18d shows variation in sleep time over SNs. It is observed that Node 5 indeed sleeps for around 30% of the time in Poisson, Binomial, and Laplace distributions and 40% of time in case of the lognormal distribution. It may be noted that the grids are portions of human tissue that are able to transmit heat through convection and radiation constantly that makes SNs to be constantly heated up, even after packet loss (Fig. 16.19).

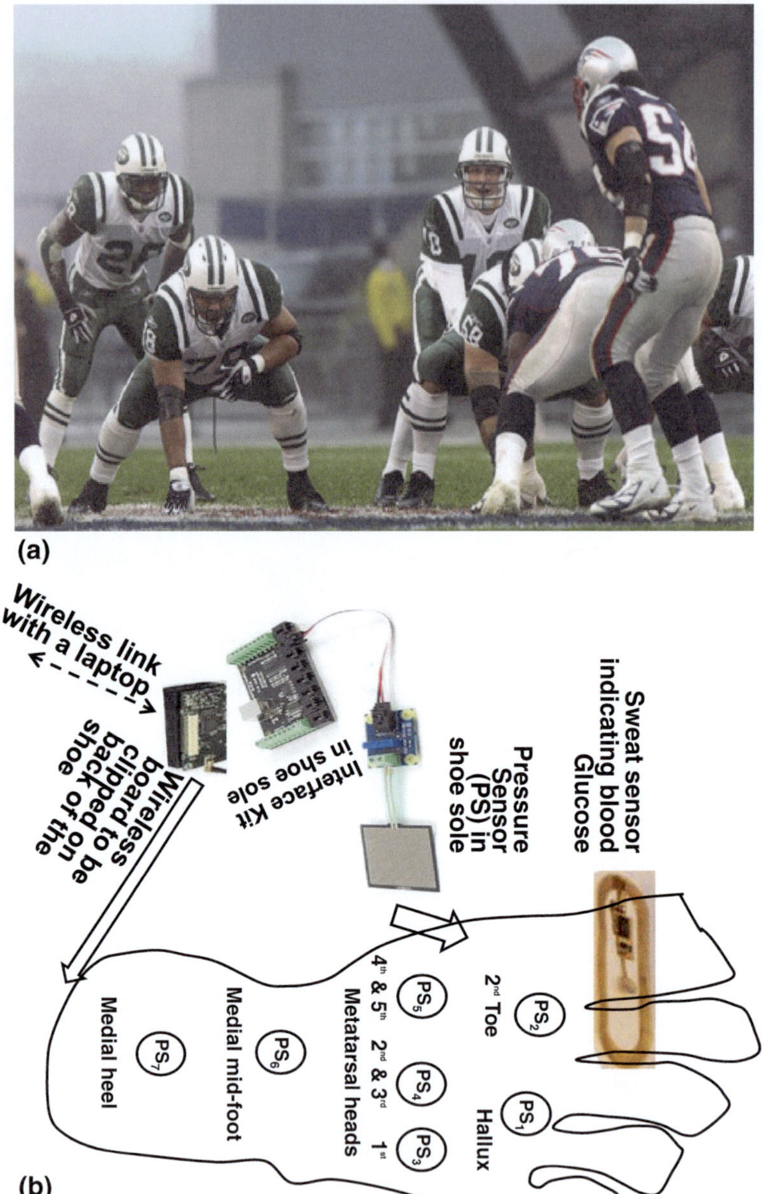

Fig. 16.15 **a** Players at football game **b** Shoe with 7 pressure sensors [6]

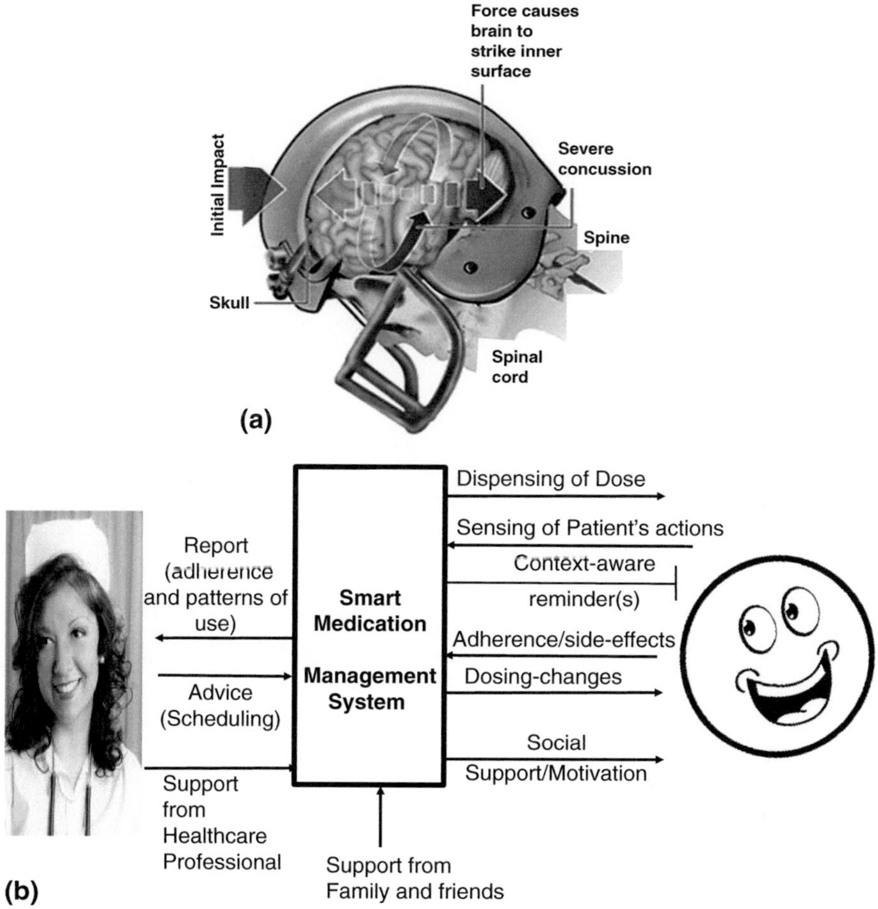

Fig. 16.16 **a** Helmet and concussion **b** Medication Monitoring

16.6 Interference in WBANs

When multiple WBANs are present close by, then there will be interference as other SNs could concurrently transmit within the transmission range of a sender SN. Two types of interference are possible: intra-WBAN interference and inter-WBAN interference (Fig. 16.20). Both intra-WBAN and inter-WBAN interferences are due to cochannel interference that could lead to critical data loss, which can prove to be life-threatening to patients using these devices for health monitoring. This is also a severe threat to reliability of the network functioning and is a security threat for the patient's personal data. Attempts have been made to increase throughput in the WBAN. These include opportunistic packet scheduling, variable TDMA scheduling, and random and incomplete coloring algorithm. The energy consumption of

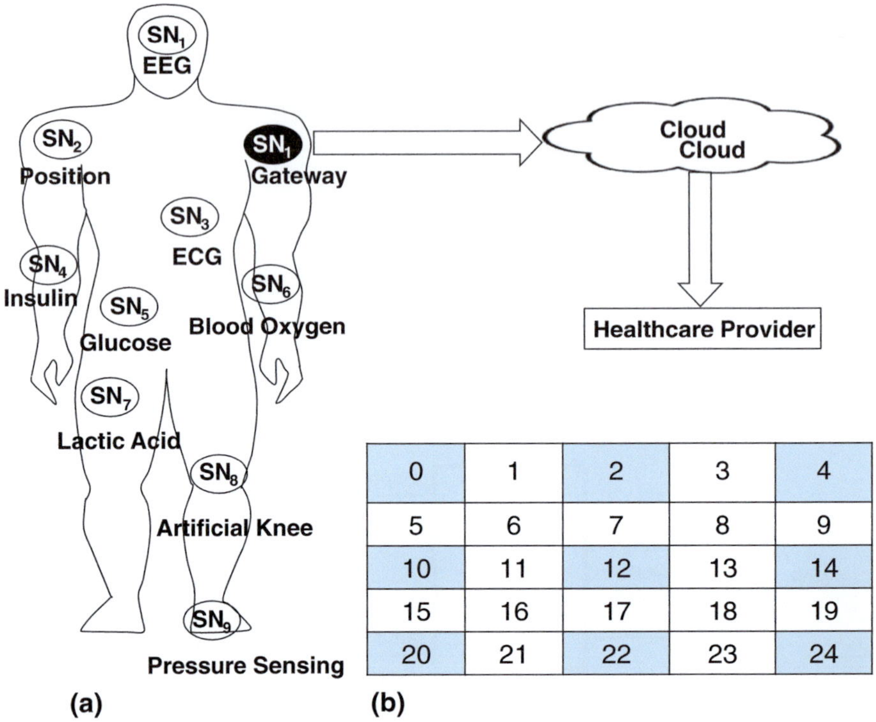

Fig. 16.17 **a** Transferring Physiological data using SNs **b** Coordinator Model on human body with 25 SNs

Table 16.4 Thermal aware routing in WBAN [18]

Protocol	Routing decision	Network temperature	Nodal temperature	Packet drops
TARA	Per-hop	High	Moderate	Yes
LTR	Per-hop	Moderate	Low, High[w]	Yes
ALTR	Per-hop	Moderate	Low, High[w]	No
LTRT	End-to-end	Moderate	Moderate	No
HPR	Per-hop	Moderate	Low	No
TSHR	Per-hop	Moderate	Low	No
SHR	Per-hop	High	High	No

[w]Worst case

SNs is too high in opportunistic scheduling, and most schemes do not address the issues of transmission losses due to unpredictable human mobility. None of the existing models use IEEE 802.16.6 standard which is designed especially for WBAN ensuring QoS. So, two schemes [9] have been suggested to mitigate interference in WBANs. Intra-WBAN interference mitigation is achieved by Fuzzy inference reasoning and decision making for allocating transmission slots

Table 16.5 MAC protocols and metrics [19]

Metric	CSMA/CA	TDMA
Power consumption	High	Low
Traffic level	Low	High
Bandwidth utilization	Low	Maximum
Scalability	Good	Poor
Synchronization	N/A	Required

Table 16.6 Temperature increase in human body with implanted coil dissipating 984 μW [20]

Tissue	Maximum temperature rise °C
Retina	0.025
Skin	0.089
Fat	0.152
Bone	0.018

(Fig. 16.20b, c), while inter-WBAN interference mitigation is achieved by a decentralized cooperative scheduling approach for mobile WBANS, among the interfering coordinators to ensure non-overlapping transmission slots.

Three input variables of sensor signal-to-noise ratio in dB, bit error rate ratio, and energy per bit to noise power spectral density ratio corresponding to all n SNs in the WBAN are used to have a unique output decision that maps to either of the three possibilities, namely "defer," "schedule," or "forward," data and defuzzifier is not required in this approach. The inputs are BER $\subset$ {too high, acceptable, good}; SNR $\subset$ {dangerous, just-okay, better}; and $E_b/N_0 \subset$ {critical, boundary, superior}, while the output decision is $\subset$ {defer, schedule, forward}. The Fuzzy interference table is given in Table 16.7. Decision considers two parameters BER and E_b/N_0.

BER and SNR are given in Fig. 13.21.

Inter-WBANs interference is present when two or more WBANs are within the interference range of one another Fig. (16.21). Each SN sends data to its coordinator in their own scheduled slot time. If two WBANs using the same frequency channel are overlapped, the received SINR of some SNs at the coordinator will be below threshold which is unacceptable. Thus, a cooperative scheduling among WBANs is required to mitigate interference (Fig. 16.22a). Shox network simulator used [9] to set up multiple WBAN scenarios with 25 SNs is strategically placed in 10 m × 10 m area. There are local coordinators (LCs), and other nodes are used for sensing, variable data rates up to 250 kbps. The properties of physical and MAC layers are set according to IEEE 802.15.4 standards. The SNs movement is decided by setting Random Waypoint Model (RWPM). A variable disc model is used in which the receiver receives packet with a signal strength of $s(rx) = s(tx)/d\hat{A}^2$, where d is the Euclidean distance between sender and receiver SNs pair. A number of messages are exchanged in the decision process as time elapses, which is given in Fig. 16.22b, c.

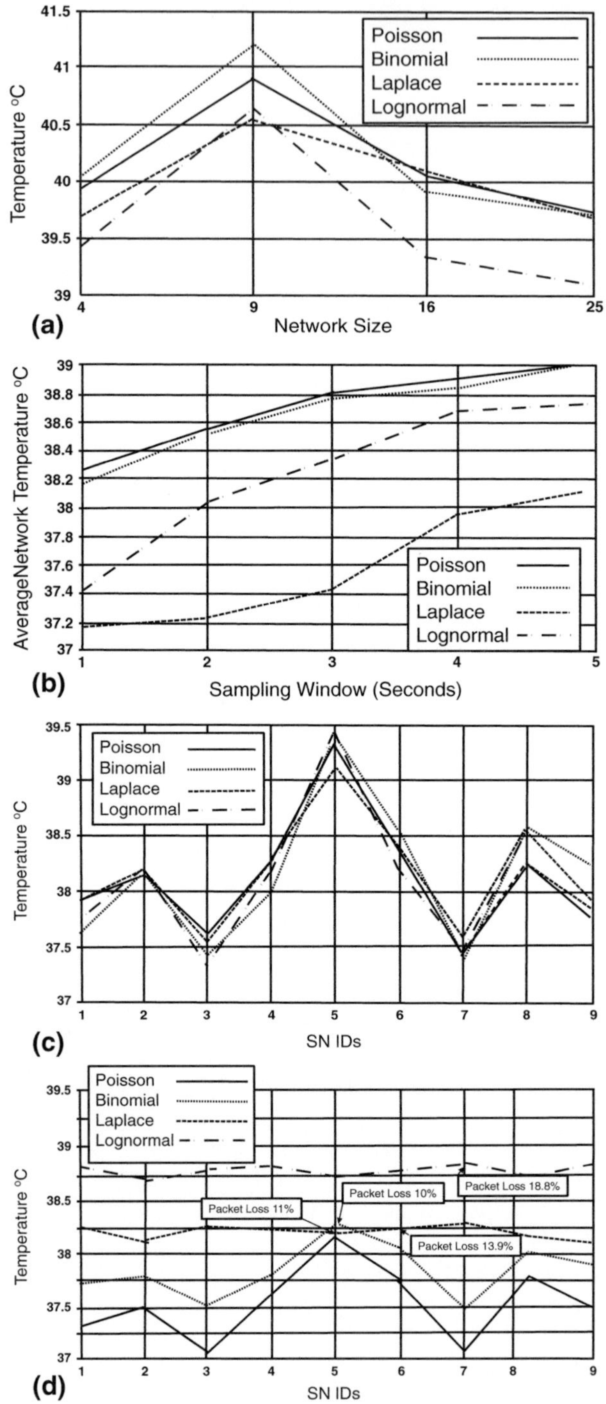

Fig. 16.18 **a** Variation of Temperature versus Network Size **b** Temperature versus sampling window **c** Temperature versus ID of SNs **d** Sleep Duration versus ID of SNs

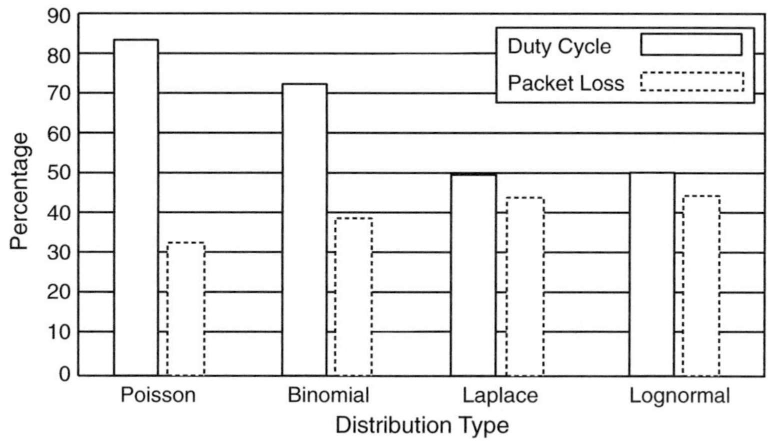

Fig. 16.19 Duty cycle and packet loss

16.7 Data Reduction Schemes

Table 16.8 shows encoding of two physiological signals: pulmonary artery pressure (PAP) and ECG. These could be sent through SMS messages. As there is a limit of 160 textual characters in SMS, 3600 samples for 10 secs would need 24 messages. One possibility is to utilize short-duration PAP and ECG signals. Another alternative is to skip some of the frames without affecting the final outcome, and a sample reduction by a factor of 5 would lead to only 5 SMS messages. To enable skipping of frames, architecture has been proposed [10] as shown in Fig. 16.23.

A WBAN with a coprocessor is used as an additional microcontroller for data logging, processing and temporary storage of data samples. Smartphone is used as the Coordinating Sink Station (CSS). Wireless extension/add-on for SN communicates with the CSS over GSM. SNs sense and process the physiological data, encode, pack as a text message or as a voice-coded data message, and pass on to GSM extension. The extension transmits the data to the smartphone CSS. CSS can make decisions regarding a need-based use of voice/data network instead of WBAN links. Another important functionality is the use of speech signal encoding of the physiological sensor data using PSK and transmission as a voice call. The coprocessor sends encoding request to the signal processor which acknowledges the request and generates a digitally modulated output of the compressed sensor data using BPSK. This encoding uses human speech frequencies (100 Hz–3.3 kHz) in digital modulation. The generated output is a human voice signal, and Arduino microcontroller board with a GSM shield extension provides necessary interface. Four different subsets are derived from original PAP and ECG signals, with the first subset retains alternate samples, the second contains every third sample, the third set has every fourth sample, and the fourth has every fifth sample (Fig. 16.24). The reduced samples by skipping frames is compressed, encoded, and transmitted as

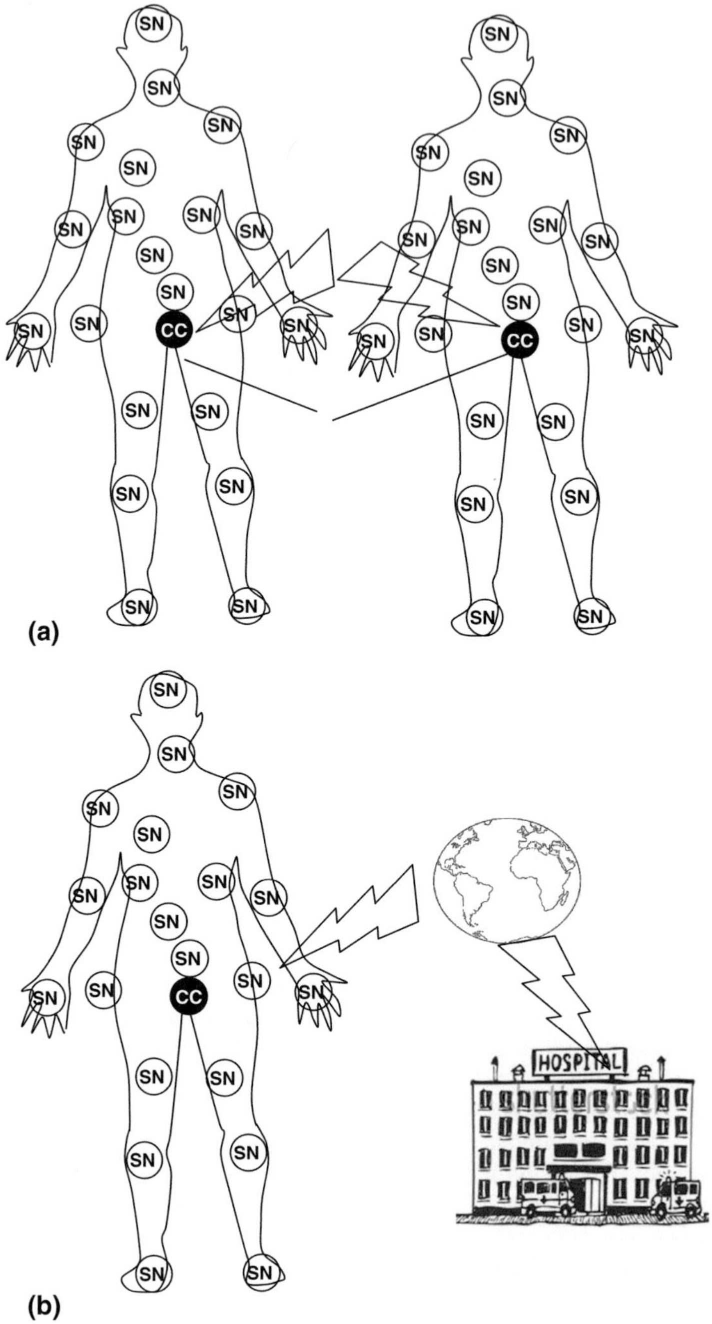

Fig. 16.20 **a** Interference in WBANs **b** Intra-WBAN Interference **c** A General Fuzzy Logic System

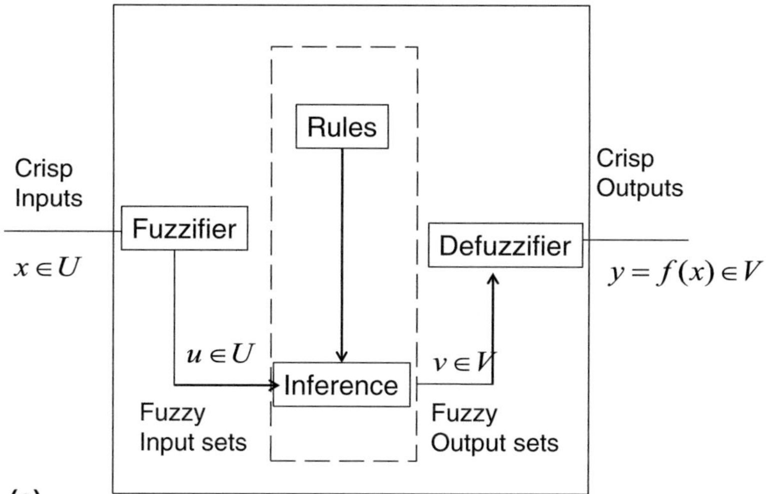

Fig. 16.20 (continued)

Table 16.7 Fuzzy inference table

BER	SNR			E_b/N_0
	Dangerous	Just-okay	Better	
Too high	Defer	Defer	Defer	Critical
Too high	Defer	Defer	Defer	Boundary
Too high	Defer	Defer	Schedule	Superior
Acceptable	Defer	Schedule	Schedule	Critical
Acceptable	Defer	Schedule	Schedule	Boundary
Acceptable	Forward	Schedule	Forward	Superior
Good	Forward	Forward	Forward	Critical
Good	Forward	Forward	Forward	Boundary
Good	Forward	Forward	Forward	Superior

text messages over GSM network. At the receiving end, the encoded and compressed BAN data is processed to rebuild the original samples. Missing samples are recreated using five numerical interpolation techniques. The nearest-neighbor interpolation algorithm has higher errors, while linear spline interpolation performed better on data sets (Fig. 16.24).

Another approach for minimizing physiological data is to do aggregation in time domain [11] using regression polynomial discussed in Chap. 10. A fourth-order polynomial can be written as:

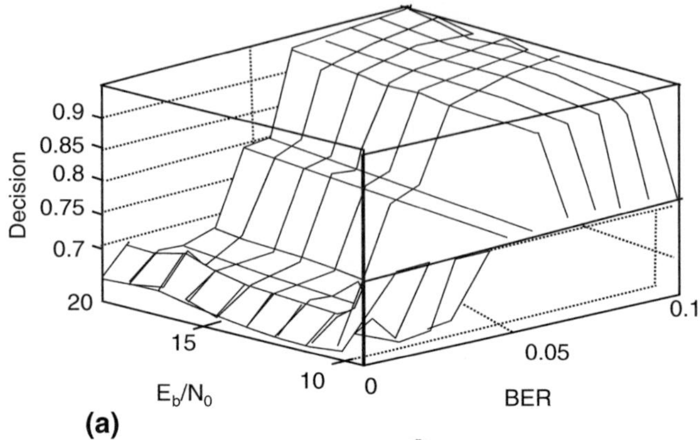

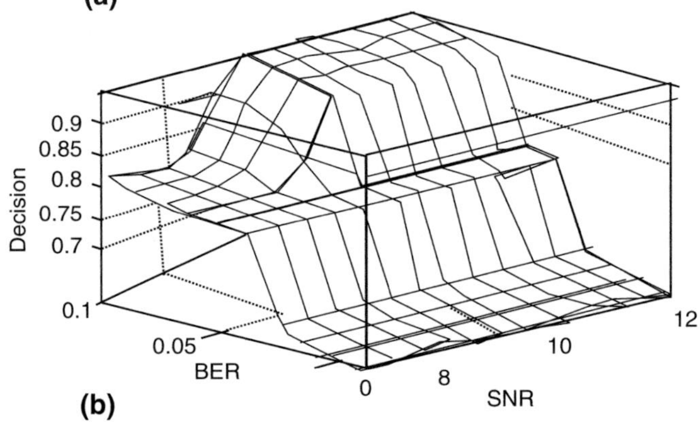

Fig. 16.21 Intra-cluster Decision based on various Input parameters **a** BER and E_b/N_0 **b** BER and SNR

$$f(t) = \beta_0 + \beta_1 t + \beta_2 t^2 + \beta_3 t^3 + \beta_4 t^4. \tag{16.3}$$

and eighth-order polynomial as:

$$f(t) = \beta_0 + \beta_1 t + \beta_2 t^2 + \beta_3 t^3 + \beta_4 t^4 + \beta_5 t^5 + \beta_6 t^6 + \beta_7 t^7 + \beta_8 t^8, \tag{16.4}$$

where β' s are constant coefficients and f(t) is the value at time t.

A fourth-order polynomial can be easily created by the following matrix:

$$\begin{bmatrix} y_1 \\ y_2 \\ \vdots \\ y_n \end{bmatrix} = \begin{bmatrix} 1 & t_1 & t_1^2 & t_1^3 & t_1^4 \\ 1 & t_2 & t_2^2 & t_2^3 & t_2^4 \\ \vdots & \vdots & \vdots & \vdots & \vdots \\ 1 & t_n & t_n^2 & t_n^3 & t_n^4 \end{bmatrix} \begin{bmatrix} \beta_0 \\ \beta_1 \\ \beta_2 \\ \beta_3 \\ \beta_4 \end{bmatrix} + \begin{bmatrix} \varepsilon_1 \\ \varepsilon_2 \\ \vdots \\ \varepsilon_n \end{bmatrix} \Leftrightarrow \vec{y} = T\vec{\beta} + \vec{\varepsilon}. \tag{16.5}$$

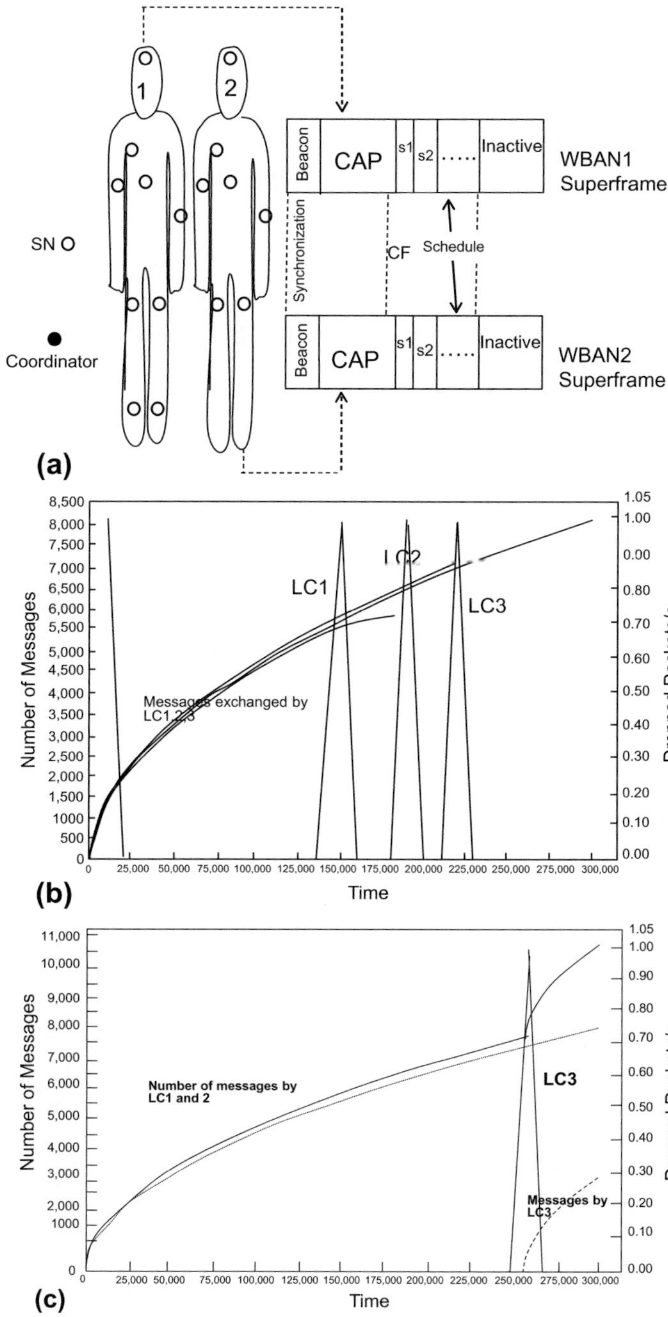

Fig. 16.22 **a** Inter-WBAN Interference **b** Interference effects with no MAC scheduling **c** Interference effects with TDMA -MAC scheduling

Table 16.8 Encoding for the two physiological signals

Physiological parameter	Range (mV)	Span (mV)	Step size (8bit encoding) (MuV)	Max encoding error (MuV)
PAP	20–45	25	97.6	48.8
ECG-II	−0.75–0.0	1.75	6.83	3.41

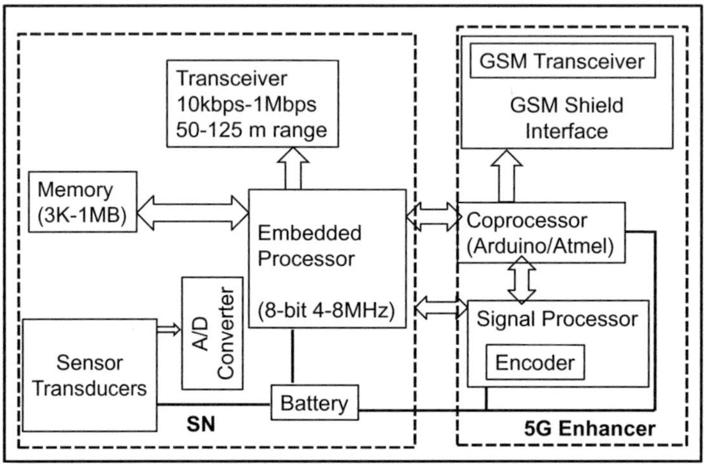

Fig. 16.23 Proposed architecture for skipping frames

Then,

$$\widehat{\vec{\beta}} = \left(T^{\mathbb{T}}T\right)^{-1}T^{\mathbb{T}}\vec{y}, \tag{16.6}$$

where $\widehat{\vec{\beta}}$ is an estimate of the original coefficients. This has been used for 4 types of biological data, blood pressure, EEG scalp readings, motor movement signals, and motor movement signals in patients with neurodegenerative disorders. The accuracy is measured for both fourth-order and eighth-order polynomials (Fig. 16.25). Eighty blood pressure samples are taken at 0.01 s intervals and 80 samples in μV are taken for motor movement at 0.0039062 s intervals, while the sample size of Degenerative Gait EEG Readings is 300, and measures signals in mV are obtained from the left leg at 0.0033333 s intervals.

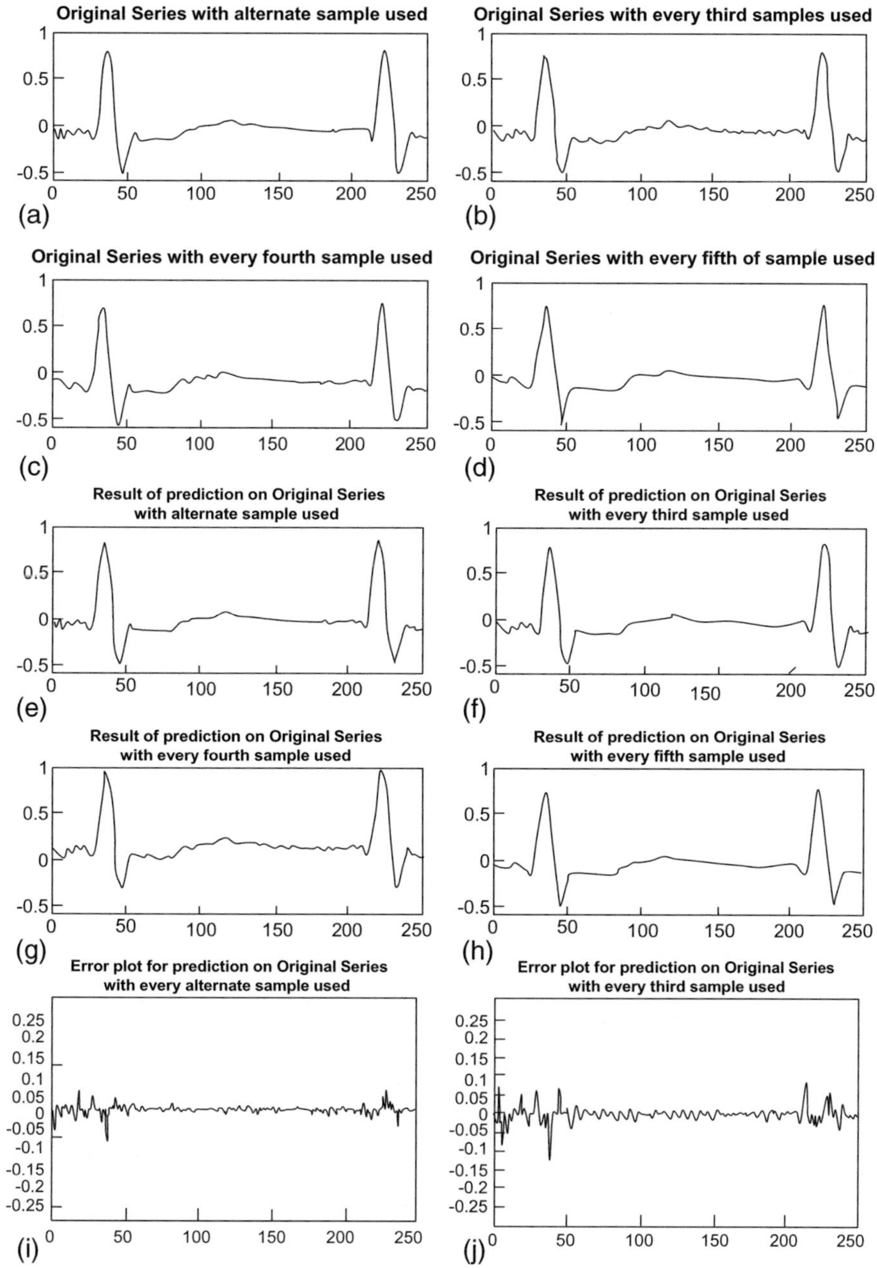

Fig. 16.24 **a**, **b**, **c**, **d** Original ECG samples **e**, **f**, **g**, **h** Rebuilt ECG Signals **i**, **j**, **k**, **l** Error for Recreated ECG data **m**, **n**, **o**, **p** Original PAP signals **q**, **r**, **s**, **t** Error for Recreated PAP data

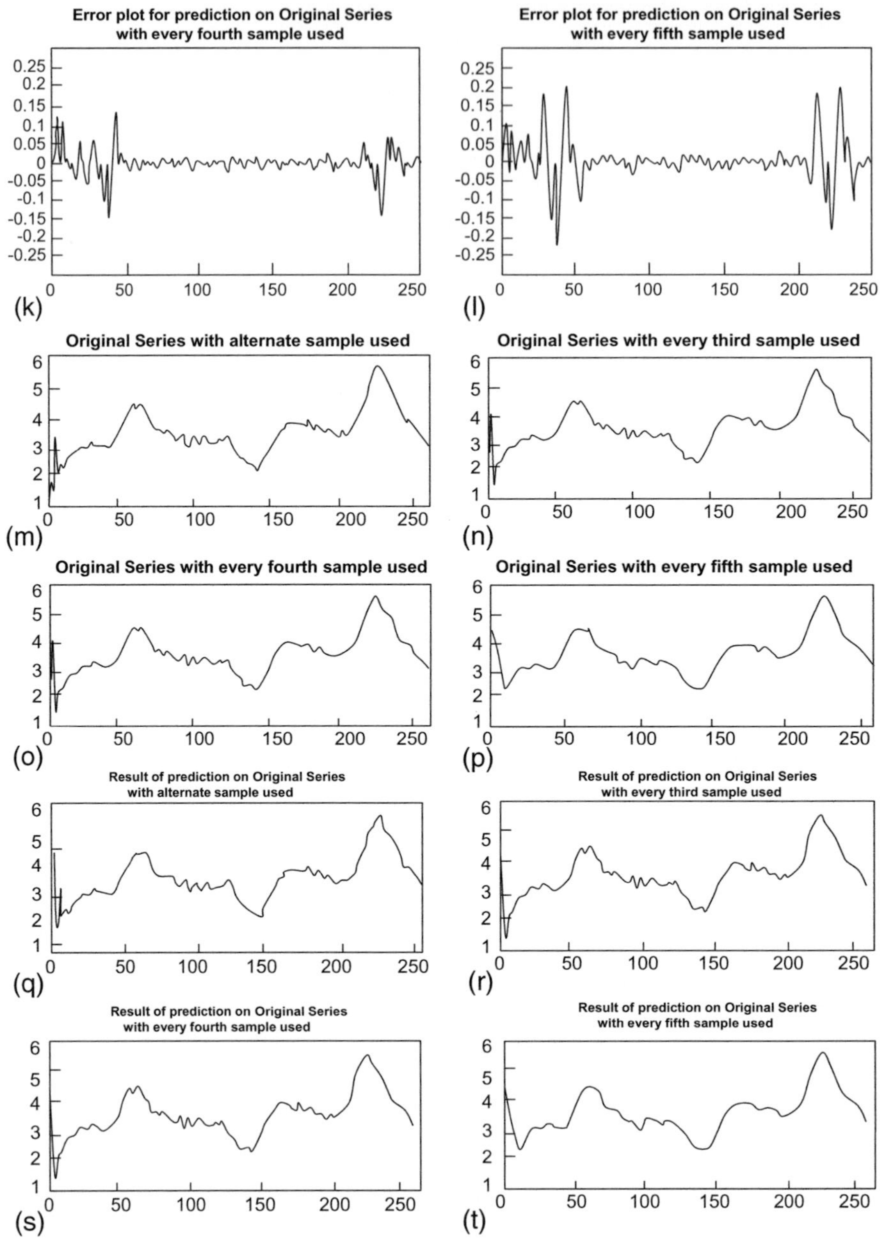

Fig. 16.24 (continued)

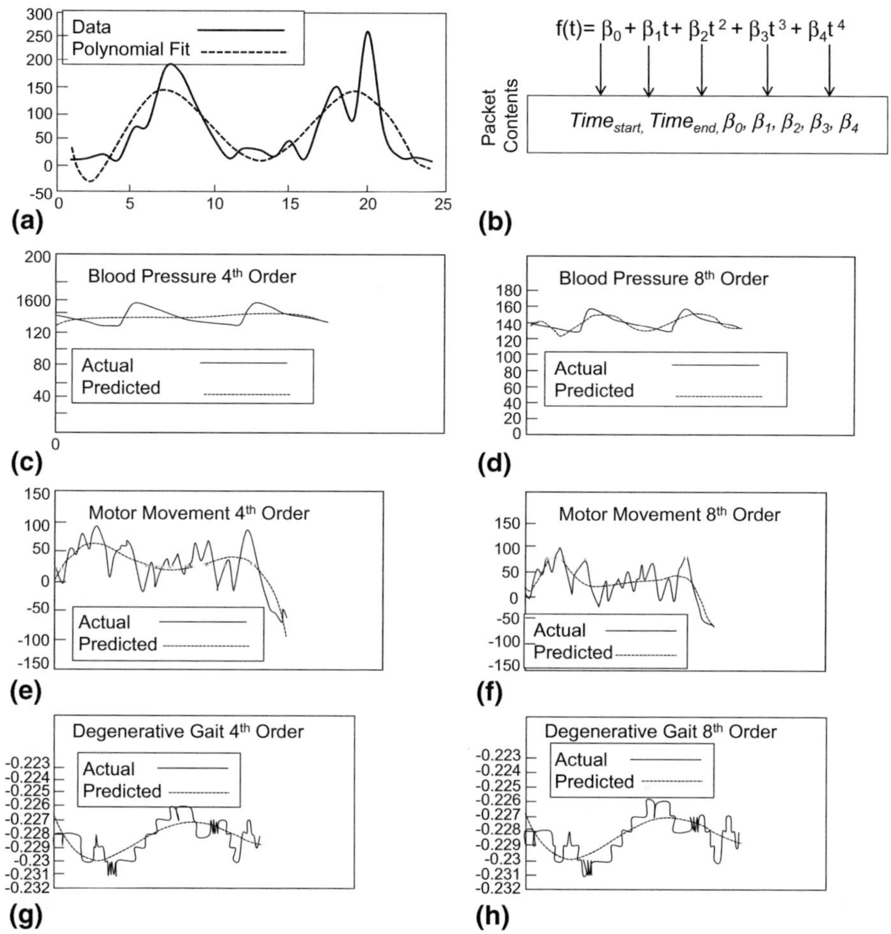

Fig. 16.25 **a** A generic Aggregation with Polynomial Regression **b** 4th order example **c** Blood Pressure **d** EEG Scalp **e**, **f** Motor movement **g**, **h** Neurodegenerative Motor Movement

16.8 Physiological Parameters for Identification Secured Communication

The traditional biometric approaches make use of distinct physiological characteristics of a person and use the same to determine their identity. This entire process is termed as biometric authentication [12]. The most conventional parameters involve physiological characteristics ranging from noninvasive features such as facial and hand geometry to invasive techniques such as impression from a finger, the distinction of an iris, or the structure of the DNA. Some behavioral patterns also find application in identity association such as voice modulation and acoustics, the mechanics of locomotion, keystroke dynamics, and one's penmanship. In general, biometric parameters are qualified by: *invariance*, *measurability*, *singularity*, *acceptance*, *reducibility*, *reliability*, *and privacy.* Given such considerations, the numbers of such parameters which find applicability are few. A point to be taken into consideration is that biometric systems are not perfect nor are they designed to be so; additionally, there are two error metrics commonly associated with biometrics: FAR (false accept rate) and FRR (False Reject Rate) [13]. FAR refers to the probability of generating a false positive, which is wrongly identifying and accepting an impostor for a genuine user. FRR refers to the probability of mistakenly rejecting a valid user. These two parameters jointly serve as tools which can gauge the overall performance of a biometric feature in action. A third parameter at which the false rejection rate and the false acceptance rate are equal also acts as a metric for determining the accuracy of a biometric system known as equal error rate.

Over the years, several researchers have concluded that the study of a person's gait is adequate to determine their gender and identity. Most of the research in the analysis of gait has been limited to the usage of photography and video capturing devices, limiting the study to spatiotemporal components. Gait recognition techniques can be broadly classified into three categories: (a) machine vision (MV) based, (b) floor sensor (FS) based, and (c) wearable sensor (WS) based. MV-based gait analysis techniques usually incorporate studying the silhouette of a person as capture on reel. Usually, the parameters of interest are stride, cadence, height, proportions of bodily features and, the overall silhouette of the person. The cameras used for studies could be video or infrared or a combination of both, depending upon use. The main idea behind such monitoring model is to breakdown movement into a collection of joints and their functioning, thus computing the angular motion of each component during motion. These involve the placement of floor-mounted load transducers, commonly referred to as force sensors. Such a platform is responsible for measuring the ground reaction forces along with the direction, magnitude, and location of the applied pressure. While this technique

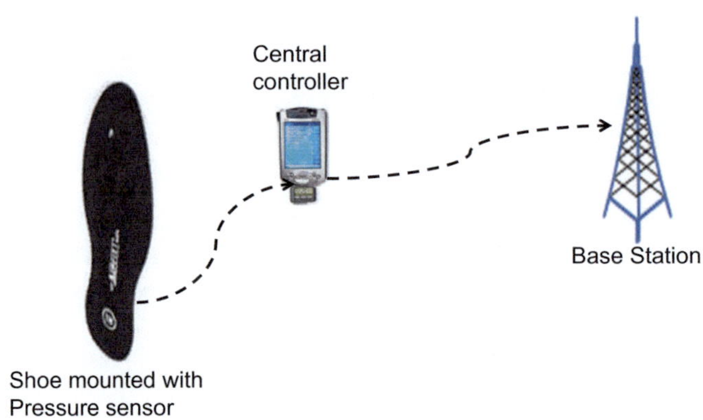

Fig. 16.26 Architecture for Gait Monitoring

does satisfy the properties that make a good biometric, the infrastructure costs involved render this technique economically unfeasible, thus preventing it from finding application in real life. WSN-based techniques usually involve the usage of inertial sensors such as accelerometers and gyroscopes to study the human form while in motion. The purpose of this technique is to apply the motion sequence generated by the lights to identify the wearer, thus providing a large scope for biometric applications.

Person's stride interval is used as a biometric and can be achieved by placing pressure sensors in the sole of the person's shoe (Fig. 16.26). By virtue of its placement, every footstep taken is recorded; as such, the time lapse between steps can be monitored and processed. A standard pressure sensitive floor is not the preferred tool of choice given its high susceptibility to external noise and subjectivity to exterior motion. As such, on-body sensors allow the sensed data to be localized to only the user concerned, allowing for improved identification and authorization. A critical feature of this model is to insure that the sensor cannot be felt by the user, which may lead to consciousness and discomfort. Given its inconspicuous positioning, this system can be easily deployed on humans, ready to adapt to everyday living.

While the limitations of gait are known, we propose using gait as a passive biometric, paired with another feature rather than being used in isolation. However, before we investigate the merits of the latter, in this research we will be discussing the benefits of using gait alone.

The data set is analyzed by employing statistical measures for the purposes of establishing consistency and uniqueness and to derive its characteristics precisely. Based on the available data, distribution of the stride interval values for (a) a user and (b) across multiple users has to be determined. In the former case, the stride interval generated by each of 10 users follows a normal distribution (Fig. 16.27). This implies that a particular user tends to follow a rhythmic motion, and each step is almost equally spaced apart, following a normal distribution. Theoretically, the probability density function is given by:

$$f(x) = \frac{1}{\sigma\sqrt{2\pi}} \exp^{-\frac{(x-\mu)^2}{2\sigma^2}} \tag{16.7}$$

The parameter μ ψ is the expectation value against the sensed data; σ ψ refers to the standard deviation about the mean. It can be concluded that humans in motion tend to adopt a two-dimensional Gaussian distribution. As shown in Fig. 16.27, a Gaussian density function is made available at each data point, and over the range of the data, the sum of density functions is computed. Considering the randomness of the sensed data, data from different users exhibits a bell-curve distribution, with the trend for each user being localized around a mean. With every user in consideration, the distribution plot is plotted and a normal distribution can be obtained. Although the distribution function for each group is not uniform, the three means for each user are largely centric about a similar mean. The *stride interval* exhibits long-range power-law correlations which indicate a fractal process model

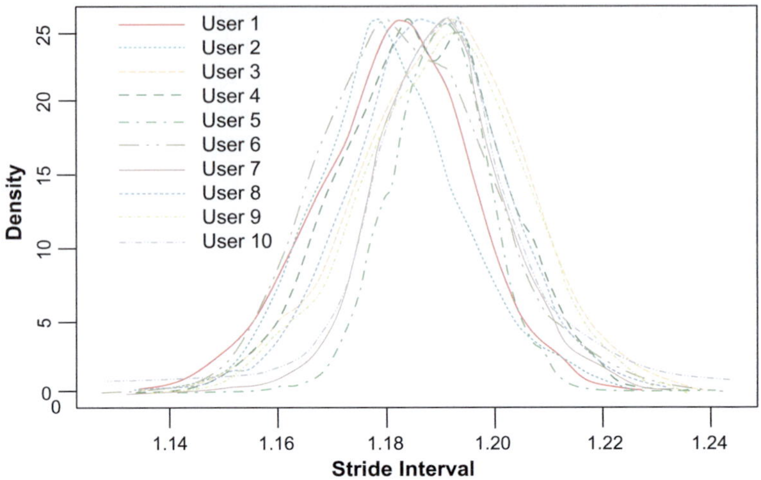

Fig. 16.27 Demonstration of Density Estimation

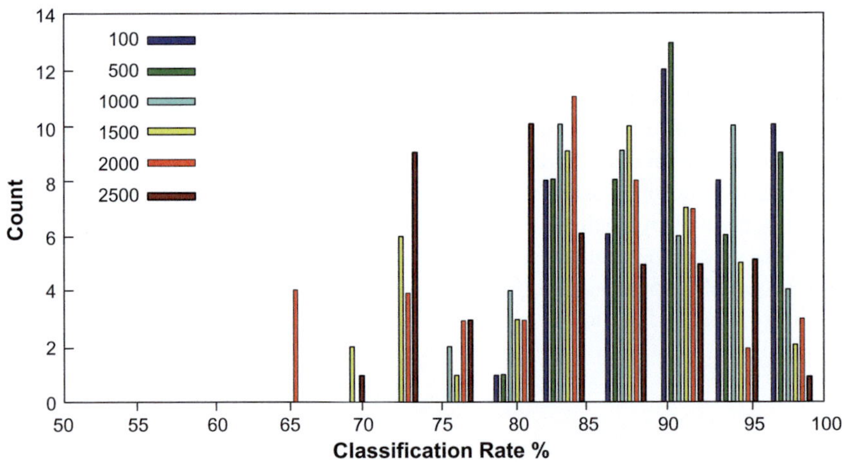

Fig. 16.28 Variance of positive classfication rate for various sample sizes

assumptions that we make when we apply gait as a biometric is that each recorded value can be modelled as a normal random variable that each occurrence of a statistic is statistically independent and, finally, that each result obtained traces back to a population having the same variance. To perform the summary analysis, a one-way ANOVA (analysis of variance) was carried out with alpha $>\psi$ 0.05, exhibiting statistical significance. The results from the ANOVA runs are very promising and confirm the fractal nature of human strides.

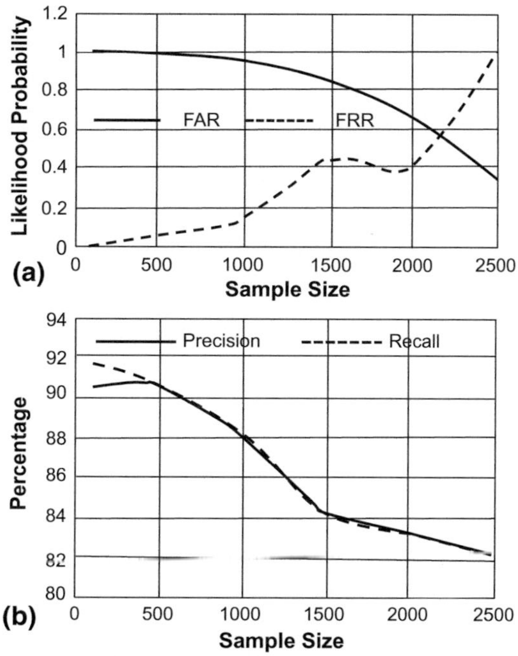

Fig. 16.29 **a** Results of GMM classification **b** Precision vs Recall from GMM classification

Given that authentication is the main goal behind applying Gaussian Mixture Models (GMMs) [14], the gait model should be such that it can produce a low FRR and FAR. Additionally, confusion matrix and classification performance rate are generated which are used to compute the acceptance/rejection parameters. The results for FRRs and FARs for 220 samples (obtained from Monte Carlo methods) are shown in Fig. 16.28. A key feature to be noted is that the algorithm should fare well for computing FRR and FAR over time. The results depicted show the variance of FRR and FAR over a varied sample space (from 100 samples to 2500 samples) (Fig. 16.29). These values span across a range from 0.1 to 8.0% using the aforementioned features. Our conclusion is that smaller sample spaces exhibit better modelling prediction(s). The classification algorithm averaged 87% positive classification rate, which is reasonably high (Fig. 15.29). The trend that we observe is that FRR and FAR showed a consistent variance, thus illustrating the advantage of the adaptive modeling.

16.9 Conclusions

Numerous biomedical applications of transducers and SNs are feasible, and the demand and need for new areas are continuously growing. This is relatively new application area even though this has a long-lasting impact on human health. The key is to develop transducers that could function with 100% reliability and would not provide any false alarms. There is also a need for 24 × 7 unattended monitoring that poses a lot more constraints and needs to be addressed carefully. The number of SNs on human body could be limited to around 25 and could be placed not totally randomly, but not in a 2-D mesh structure either. This calls for further investigation of WPASNs and is an open area of research.

16.10 Questions

Q.16.1. How can you ascertain reliability in the monitoring of patient vital signals from sensors?

Q.16.2. What is the impact of stability on the results of biomedical monitoring?

Q.16.3. What impact do you expect on the WBAN performance if a patient is also moving around?

Q.16.4. What specific measures are needed to ensure security of patients' physiological data?

Q.16.5. What are the parameters that you utilize in WBAN for a biomedical application?

Q.16.6. In a WBAN, transducers are connected using wireless links to the local coordinator. Can they be hardwired to the coordinator?

Q.16.7. What are the limitations and advantages of approach suggested in Q. 16.5?

Q.16.8. What will be the impact on accuracy, reliability, delay, and interference if the number of SNs is doubled in a WBAN?

Q.16.9. In order to minimize interference, distance covered by individual SN in a WBAN ought to be reduced. On the other hand, new Bluetooth scheme is increasing the communication distance. What may be an appropriate approach for future biomedical applications?

Q.16.10. What will be the impact on performance and reliability if some transducers are embedded inside body skin while others are outside on the body?

Q.16.11. What other physiological parameters that could be used for identification of a person and associated security issues?

References

1. Diego Martinez, Francisco Blanes, Jose Simo, and Alfons Crespo, "Wireless Sensor and Actuator Networks: Characterization and Case Study for Confined Spaces Healthcare Applications," Proceedings of the IEEE International Multiconference, Computer Science and Information Technology, 2008, pp. 687–693.
2. S. Arnon, D. Bhastekar, D. Kedar, and A. Tauber, "A comparative study of wireless communication network configurations for medical applications," IEEE Wireless Communications, vol. 10, no. 1, pp. 56–61, Feb. 2003.

3. Esteban Pino, Lucila Ohno–Machado, Eduardo Wiechmann, and Dorothy Curtis, "Real–Time ECG Algorithms for Ambulatory Patient Monitoring," Proceedings of AMIA Annual Symposium, pp. 604–608, 2005.
4. http://www2.imec.be/be_en/press/imec-news/ecgpatch.html.
5. Anagha Jamthe, Suryadip Chakraborty, Saibal K Ghosh, and Dharma P. Agrawal, "Mobility monitoring of Parkinson's Disease patients using Received Signal Strength Indicator (RSSI) in Wireless Sensor Networks," IEEE-CPS The 13th International Conference on Computational Science and Its Applications (ICCSA), June 24–27, 2013, Ho Chi Minh City, Vietnam.
6. Dharma P. Agrawal, Abhinav Prakash, Suryadip Chakraborty, Anagha Jamthe, and Saibal Kumar Ghosh, "Real-time Personnel Fatigue level Monitoring," Invention Disclosure, provisional patent confirmation no. 14/846,851 dated Sept. 7, 2015.
7. https://en.wikipedia.org/wiki/Bioheat_transfer.
8. Ashwin Ganesh Krishnamurthy, Junghyun Jun and Dharma Agrawal, "Temperature gradient search for temperature-aware routing in Bio-medical Sensor networks," 9th International Conference on Body Area Networks, September 29–October 1, 2014 London, Great Britain.
9. A. Jamthe, A. Mishra, and D. P. Agrawal, "Scheduling schemes for Interference Suppression in Healthcare Sensor Networks," *in Proceedings of IEEE International Conference on Communication*, June 2014.
10. Amitabh Mishra and Dharma P. Agrawal, "Continuous Health Condition Monitoring by 24×7 Sensing and Transmission of Physiological data over 5-G Cellular Channels," Invited Paper, International Conference on Computing, Networking and Communications, (ICNC 2015) Anaheim, California, USA, Feb. 16–19, 2015, pp. 584–590.
11. Andrew Knox, Suryadip Chakraborty, and Dharma P. Agrawal, "Evaluating Polynomial Regression Based Data Aggregation in Body Area Networks," IEEE ANTS 2014, New Delhi, Dec. 14–17, 2014.
12. R. Bolle, J. Connell, S. Pankanti, N. Ratha, A. Senior, Guide to biometrics book, WorldCat.org, 2004.
13. D. Gafurov, "Behavioral Biometrics for Universal Access and Authentication," *Annual Norwegian Computer Science Conference*, pp. 19–21, 2007.
14. Pallavi Meharia and Dharma P. Agrawal, "The Human Key: Identification and Authentication in Wearable Devices using Gait," report CDMC, University of Cincinnati, 2016.
15. A. Mason, N. Yazdi, A. Chavan, K. Najafi, and K. Wise, "A Generic Multielement Microsystem for Portable Wireless Applications," *Proceedings of the IEEE,* vol. 86, no. 8, pp. 1733–1746, 1998.
16. W. Khalid, I. Hossain, S. M. Mahmud and Y. Xu, "Intelligent Vehicle Based Architecture for Real-Time Monitoring of Soldiers' Health using MEMS Flexible Smart Skin Sensors," Proc. 5th Annual Intelligent Vehicle Systems Symposium of National Defense Industries Association (NDIA), National Automotive Center and Vectronics Technology, pp. 59–64, June 13 –16, 2005, Traverse City, Michigan.
17. https://en.wikipedia.org/wiki/Shimmer_Sensing.
18. M Tabandeh, M Jahed, F Ahourai, and Saber Moradi, "A thermal-aware shortest hop routing algorithm for in vivo biomedical sensor networks," In Information Technology: New Generations, 2009. ITNG'09. Sixth International Conference on, pages 1612–1613. IEEE, 2009.
19. Sana Ullah, Henry Higgins, Bart Braem, Benoit Latre, Chris Blondia, Ingrid Moerman, Shahnaz Saleem, Ziaur Rahman, and Kyung Sup Kwak, "A comprehensive survey of wireless body area networks," Journal of medical systems, 36(3):1065–1094, 2012.

20. Gianluca Lazzi, "Thermal effects of bioimplants," Engineering in Medicine and Biology Magazine, IEEE, vol. 24, no. 5, pp. 75–81, 2005.
21. R. Weist, E. Eils, and D. Rosenbaum, "The Influence of Muscle Fatigue on Electromyogram and Plantar Pressure Patterns as an Explanation for the Incidence of Metatarsal Stress Fractures," American Journal of Sports Medicine, December 2004, vol. 32, no. 8, pp. 1893–1898, available online after November 23, 2004.

Part IV
Security and Actuator Issues

Chapter 17
Authentication, Encryption, and Secured Communication

17.1 Introduction

WSNs are impacting our daily life and are deployed 24×7 unattended monitoring a given environment. Therefore, many possible attacks are possible. Moreover, authentication of a user is an important factor that needs special consideration. This ensures that the user that is claiming is the correct one and is applied to different SNs. Then, when data are transferred from SNs to BS, the data must be transferred in a secured way. This also applies to transfer query from BS to SNs, and encrypting transmitted data is critical. These are the topics considered in this chapter.

17.2 Possible Attacks

Multi-hop paradigm is prevalent in WSNs, assuming all SNs faithfully forward any received messages. If a SN is compromised, it might refuse to forward packets, unless neighboring SNs might start using another route. It is more dangerous to het a compromised SN forward selected packets. Various attack models include spoofed, altered, or replayed routing information; selective forwarding; Sinkhole attacks; Sybil attacks; Wormholes attacks; HELLO flood attacks; and Acknowledgment spoofing. In spoofed, altered, or replayed routing information, a loop is created either to attract or repel network traffic that extend or shorten the source routes. Moreover, false error messages could be generated, and there could be partitioning of the network. Selective forwarding includes black-hole attack that refuses to forward certain messengers and simply drop them. This is done by either "in-path" or "beneath path" by deliberately jamming and can be prevented by using either frequency hopping sequence or spread spectrum. A simple approach adopted by such attack is achieved by advertising a path of least hop count to BS and include itself in it.

D.P. Agrawal, *Embedded Sensor Systems*,
DOI 10.1007/978-981-10-3038-3_17

Sinkhole attacks tries to lure nearly all traffic from a particular area through a compromised SN by making selective forwarding trivial and specialized communication pattern causes this problem. In Sybil attack assumes mapping between an identities to multiple entities, and hence, multiple identities are forged with a set of faulty entities representing a larger set of identities. Sybil Attack undermines assumed mapping between identity and entity and hence a number of faulty entities. Wormholes tunnel the messages over alternative low-latency links, confuse the routing protocol by creating sinkholes, and exploit the routing race condition. In Hello flood attack, an attacker sends or replays a routing protocol's hello packets with more energy. Acknowledgment spoofing employs link layer acknowledgement to trick other nodes to believe that a link or SN is either dead or alive.

Wormhole tunnels packets received on one part of the network to another, and a well-placed wormhole can completely change the routing. Wormholes may convince distant SNs that they are close to sink which may lead to sinkhole if SN on the other end advertises high-quality route to the sink. Wormholes can exploit routing race condition which happens when SN takes routing decisions based on the first advertised route. An attacker may influence network topology by delivering routing information to the nodes before it would really reach them by multi-hop routing. Even encryption cannot prevent this attack as wormholes may convince two SNs that they are neighbors, when on fact, they are far away from each other. Wormholes may be used in conjunction with Sybil attack.

All proposed sensor network routing protocols are susceptible to attacks, and the routing protocols are analyzed: TinyOS beaconing, directed diffusion, geographic routing, minimum cost forwarding, LEACH: low-energy adaptive clustering hierarchy, rumor routing, and energy-conserving topology maintenance (GAF, SPAN). TinyOS beaconing operation is based on spanning tree construction by BS broadcasting route update periodically. Packets are forwarded to parent SN until they reach BS. Attacks are possible against TinyOS beaconing as routing updates are not authenticated and any device can claim to be the BS. In Wormhole/sinkhole attacks, HELLO packets are flooded causing routing loops. WSN Protocols can be attacked such as directed diffusion operation (Sinks flood query, and a gradient is set up for SNs to propagate data back to the sink). It is rather hard to attack during flooding phase, but control messages can be spoofed by suppression and eavesdropping enables cloning while control messages can be spoofed influencing the path taken. Selective forwarding and tampering are possible even though multi-path version is more robust against attacks, while Sybil attacks are possible.

In LEACH, there are two phases: setup (CHs are randomly picked for energy savings) and steady state (CH send/receive data from SNs using TDMA schedule and send aggregated data to BS using CDMA. Possible attacks against LEACH include HELLO flood attacks as an adversary SN acts as CH by sending powerful signal. Small-size WSNs are liable to selective forwarding, and Sybil attacks are possible and attacks are aimed higher layers.

In rumor routing operation, a probabilistic selection of next hop allows agents to carry the events, TTL (time to live), etc., and energy is saved as compared to uncontrolled flooding. Rumor routing can be attacked by selective forwarding and is easier by creating wormhole attacks. Energy-conserving topology maintenance essentially has more deployed SNs than required. GAF (Geographical Adaptive Fidelity), SPAN, CEC (Cluster-based Energy Conservation), and AFECA (Adaptive Fidelity Energy Conservation Algorithm) are some of the examples. Attacks against GAF and SPAN are possible by broadcasting high-ranking control messages, and allowing selective forwarding. HELLO flood attack is possible such as Sybil attack. Possible countermeasures that can be taken are called as outsider attacks and link layer (LL) security. Major outsider attacks can be prevented via link layer encryption and authentication using global shared key. The Sybil attack is irrelevant as identity can be checked using public key cryptography which generates digital signatures. Sinkhole attack and SF (selective forwarding) is not possible. LL mechanisms are not sufficient for wormhole or HELLO flood attacks. Link layer security cannot encounter against insider attacks.

Other relevant attacks include Sybil attack, bogus routing information, selective forwarding , and no wormholes and sinkholes attack. In Fig. 17.1, when data being transferred from SN B to SN D, A forges a wrong information to claim B is in (2,1), so C will send packets back to B which causes a loop. Is a back off-based cost field algorithm (Fig. 17.1b) for efficiently forwarding packets from SNs to BS, the message, carrying dynamic cost information moves along the minimum cost path as each intermediate SN forwards the message only if it is on the optimal cost path. Relevant attacks are Sinkhole attack, Hello flood attack, Bogus routing information, Selective forwarding, and Wormholes. General typical WSN routing protocols primarily follow flooding, gradient evaluation, clustering and cellular system, geographic routing, and energy-aware routing.

Fake routing control packets are injected into the network, such as attract/repeal traffic, or generate false error messages (Fig. 17.2a). This leads to routing loops, increased latency, decreased lifetime of the network, and low reliability. A captured SN attracts traffic by advertising shortest path to BS, high battery power, etc. Many WSN routing protocols require SNs to broadcast HELLO packets after deployment (Fig. 17.2b), which is a sort of neighbor discovery based on radio range of the SN. Laptop-class attacker can also broadcast HELLO message to SNs

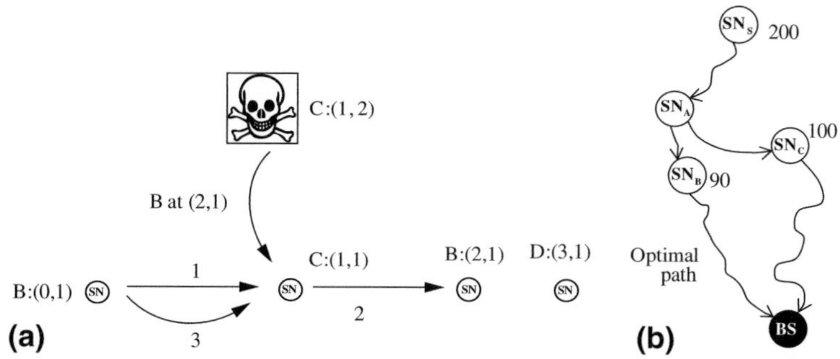

Fig. 17.1 Attack in **a** geographic routing and **b** minimum cost forwarding

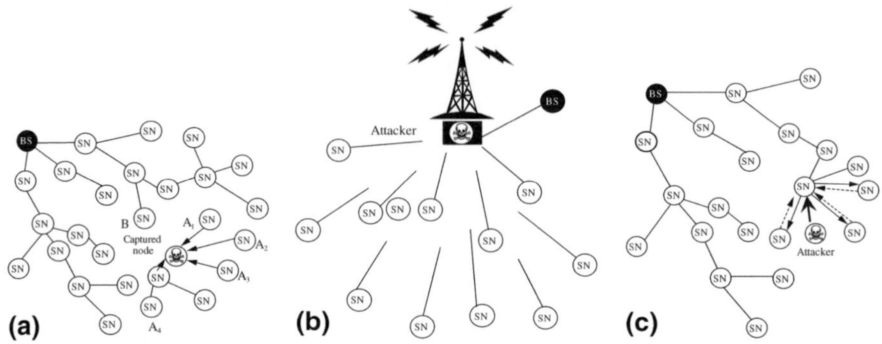

Fig. 17.2 **a** False routing information, **b** hello flood attack, and **c** acknowledgment spoofing

by advertising high-quality route to the BS. Some routing protocols use link layer acknowledgments (Fig. 17.2c). An attacker can spoof ACKs to persuade that a weak link is strong or a dead SN is alive. This encourages weak link to be selected for routing, and packets send through that link may be lost or corrupted.

To summarize, link layer encryption could prevent majority of attacks such as Bogus Routing Information, Sybil Attacks, and Acknowledgment Spoofing. This means having appropriate key management architecture is of a great importance. Wormhole attack, HELLO flood attacks, and some others are still possible as an attacker can tunnel legitimate packets to the other part of the network or broadcast large number of HELLO packets. Multi-path routing and bidirectional link verification can also be used to prevent particular types of attacks such as selective forwarding and HELLO flood. Geographic routing contains Sybil attack (Fig. 17.3a), Bogus routing information, Selective forwarding, and Wormholes and sinkholes attack (Fig. 17.3b) where an adversary may present multiple identities to other nodes. The Sybil attack can disrupt geographic and multi-path routing protocols by "being in more than one place at once" and reducing diversity. In a HELLO flood attack, SNs broadcast HELLO messages to neighbors as a powerful laptop-class device can be used to convince network that the adversary is SN's neighbor. SNs hearing this message will use this route. Acknowledgement spoofing is possible as several routing protocols rely on link layer acknowledgement, and adversary spoofs these messages to notify neighboring SNs a weak link is strong, or a dead node is alive and spoofed messages can be used to launch selective forwarding attack.

17.3 Attacks in Routing Schemes

In TinyOS beaconing, the BS broadcast routes periodly and SNs receive update and mark the BS as their parent. Relevent possible attacks include Bogus routing information, Selective forwarding, Sinkholes, Sybil, Wormholes, and Hello floods. These shafts received packets in one place of the WSN and replay them in another

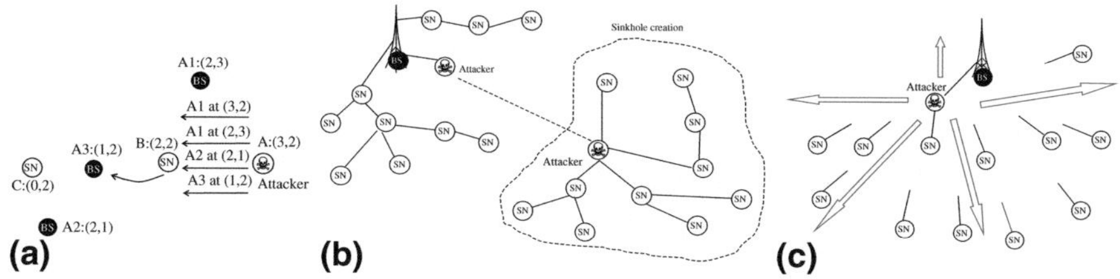

Fig. 17.3 Geographic routing **a** sybil attack, **b** wormhole attack, and **c** HELLO flood attack

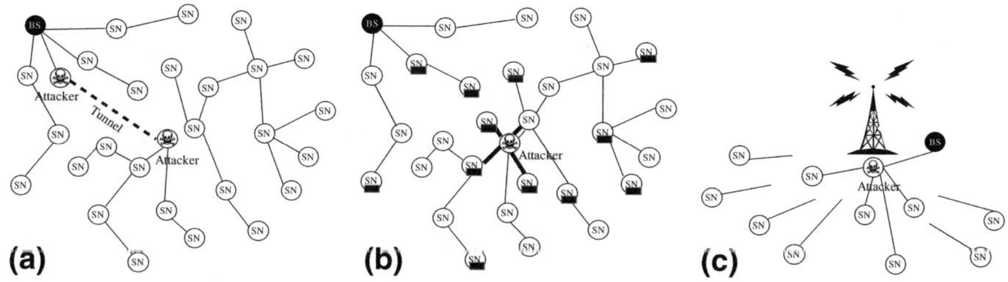

Fig. 17.4 **a**, **b** Wormhole and sinkhole combination [1] **c** HELLO flood attack

place. The attacker does not have any key material. All it needs is two transceivers and one high-quality out-of-band channel. Wormhole and sinkhole (Fig. 17.4a, b) can be combined together as most packets will be routed to the wormhole, and the wormhole can drop packets directly (sinkhole) or more delicately selectively forward packets to avoid detection. In HELLO flood attack (Fig. 17.4c), a laptop-class adversary can retransmit a routing update with high power to be received by the entire WSN.

In LEACH (Low-Energy Adaptive Clustering Hierarchy), a randomized, self-configuration is done and low-energy media access control is used. CH collects data and performs processing before transmitting to BS. Relevant attack modes are Hello floods, Selective forwarding, and Sybil attack. In Hello floods, CH selection is based on signal strength and a powerful advertisement can make the malicious attacker be its CH. Sybil attack combines HELLO floods if SNs try to randomly select CH instead of strongest signal strength.

Rumor routing (Fig. 17.5) is designed for reducing query/event ratios between query and event flooding and lowers the energy cost of flooding. It is observed that two lines in a bounded rectangle have a 69% chance of intersecting, while 5 lines have more than 99%. Relevant attack modes are Bogus routing information, Selective forwarding, Sinkholes, Sybil, and wormholes. In traditional computing, physical security is often taken for established, while SNs in a WSN are likely to be placed in open areas and an attacker might capture SNs, extract cryptographic secrets, modify their programming, and possibly replace them with malicious SN. A tamper-resistant packaging may be one defence while it is expensive. In order to

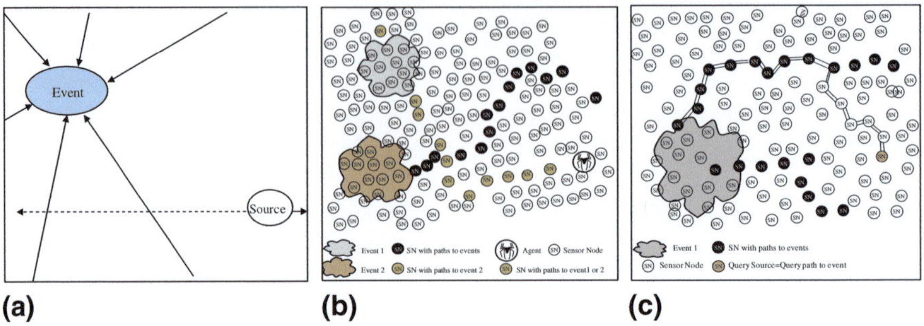

Fig. 17.5 **a**, **b**, **c** Rumor routing

have resilience against SN capture, attempt is made to build WSNs that operate correctly, even in the presence of SNs that might behave in an arbitrarily malicious way. One approach could be to replicate state across the network and use majority voting to detect inconsistencies, and this way, you gather redundant views of the environment and crosscheck them for consistency. SN capture is one of the most challenging problems in WSN security, and there is no complete solution.

Denial of service in WSNs can be initiated by creating attacks at the Network Layer by using Neglect and greed (Malicious SNs randomly drops packets as a neglectful SN, malicious SN as a greedy element, gives priority to its own messages, and Defense (multiple routing path and redundant messages are present), and Homing (Passive attack to identify critical components—e.g., BS, CHs and then launch an active attack and can be addressed if shared cryptographic keys are used). DoS attacks at the network layer could cause misdirection (in an active attack, messages are forwarded along wrong paths and can be avoided by utilizing egress filtering approach) and black holes (malicious SNs advertise zero-cost routes, making them attractive for traffic path and SNs around the malicious SN have exhausted their energy, causing a hole or partition; this is easy to detect, but is very disruptive, and the defense is to detect inconsistent advertisements). DoS defenses at the network layer could be either authorization (defense against misdirection and black holes), monitoring (monitoring proper routing and is simple and less expensive), probing (probes should be indistinguishable from normal traffic and is defense against neglect and greedy attacks), or redundancy (diversity coding and is less expensive).

DoS attacks at the transport layer could cause flooding (memory is exhausted in useful connections), and two possible defenses are either limiting the number of connections or using computationally expensive client puzzles), desynchronization (forged messages, such as changed sequence numbers and control flags for end systems for retransmission, can be stopped only through appropriate authentication. Protocol vulnerabilities include adaptive rate control (high BW traffic generated by malicious SNs is given priority) or real-time location-based protocols (RAP) (flooding the network with high-velocity packets).

17.4 Encoding Schemes

Security levels in a WSN is important as security mechanism takes a majority of systems time such as data transfer time for communication-intensive applications (encoding/decoding time for secured communication) and generic security approaches (asymmetric public/private key, symmetric, and authenticated key establishment Approach). A simple encryption scheme could be achieved by performing Ex-OR at the sending end and doing similar EX-OR at receiving end, or adding a positive number before sending data and performing subtraction at the receiving end, and many such process can be done using other operations such as permuting the bits in a prespecified manner before transmitting them and reconstructing by using reverse steps ("Data Encryption Standard (DES) and AES" on input bits) (Fig. 17.6).

Key establishment is a process through which a shared secret becomes available to two or more parties, for subsequent cryptographic use, and could be based on master key approach, public-key an private-key pairs (RSA algorithm), naïve and random distribution-based symmetric schemes, random distribution from a pool of keys, matrix-based 2-D and 3-D approaches, or polynomial-based approaches, and generalization scheme summarized in Table 17.1. In public key cryptography (Fig. 17.7a), two interrelated keys are used, a public key that is known to everyone and another private key that stays with BS. The two keys can be interchanged in values while maintaining secrecy of private key. The SN can encrypt message using its public key, and when received by BS, it can be easily decrypted using private keys and neighboring SNs receiving encrypted message cannot easily decrypt the message. Even the decryption process in BS is quite involved. So, to reduce the computation complexity, elliptical curve scheme has been proposed [2] as shown in Fig. 17.7b which is plotted as a function $y^2 = x^3 - x + 1$. This makes adding two points an easy step. Modulo p needed for addition and multiplication can be executed much faster if the prime p is a pseudo-Mersenne prime and NIST recommends the following five primes [3]:

$$\begin{aligned}
p_{192} &= 2^{192} - 2^{64} - 1 \\
p_{224} &= 2^{224} - 2^{96} + 1 \\
p_{256} &= 2^{256} - 2^{224} + 2^{192} + 2^{96} - 1 \\
p_{384} &= 2^{384} - 2^{128} - 2^{96} + 2^{32} - 1 \\
p_{521} &= 2^{521} - 1.
\end{aligned} \tag{17.1}$$

Asymmetric-key cryptography consumes a lot more energy in a WSN, and possibly BS can satisfy needed resources, but not the CHs and SNs. Various parameters are compared in Table 17.2, and power consumed by various processors are given in Table 17.3. So, for RSA-type algorithm, high-performance CPU with a large memory size is needed. Moreover, it is lot more time-consuming and is 1000 times slower than symmetric encryption. The computational cost is 2–3 orders of

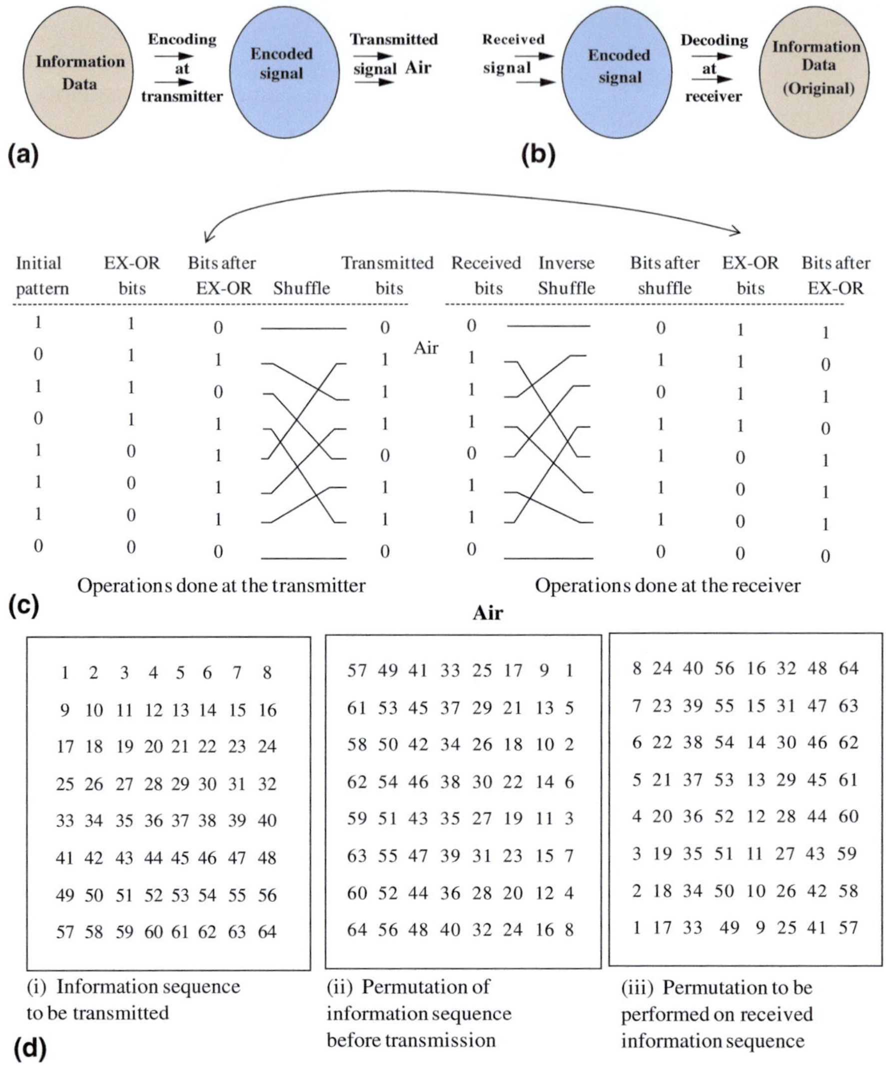

Fig. 17.6 **a** Generic encoding at transmitter, **b** decoding at receiver, **c** example operations at transmitter and corresponding steps at receiver, and **d** an example 64-DES sequence used

magnitude larger than symmetric encryption. In symmetric-key cryptography, shorter key is needed for the same level of security in asymmetric approach. For example, 64-bit AES provides the same level of security as 512-bit RSA and 128-bit AES are equivalent to 1024-bit RSA, while the calculation isare faster with lower computational overhead. The challenge in a WSN is how to establish a secret key between two entities in a WSN, especially when SNs are deployed randomly using low flying airplane. As SNs have limited resources, one logical approach is to

Table 17.1 Summary of key establishment process

Scheme	Description
0. No encoding	Plain text, not secured
1. Common master key approach	A single network-wide master encryption key used by all members
2. Public-key cryptography approach	ECC-based TinyECC is designed and implemented to provide a ready-to-use, publicly available software package in wireless sensor networks
3. Random key predistribution mechanism	Randomly pre-selecting and storing a subset of keys from a very large size key pool. Each pair of neighboring nodes needs to find their shared keys via a common-key discovery process after deployment
4. "q-composite" random key predistribution scheme	"q-composite" scheme aims to improve the network resilience against node capture attack by requiring any two communicating nodes to share at least q common keys to establish a secure link between them
5. Symmetric matrix-based key predistribution with "λ-secure"	Key pre-distribution method on symmetric matrix which can ensure any two members in a group to compute a common key between them
6. Bivariate polynomial key predistribution scheme	Any two members in a group can generate a unique pair-wise key based on their preloaded bivariate polynomial shares. The established key is secure when the number of compromised nodes is less than the degree of the polynomial
7. Matrix-based key predistribution scheme	Each node is randomly preloaded a row and a column of keys from a large size two-dimensional asymmetric key matrix. Any two nodes share at least two common keys after deployment
8. Hierarchical polynomial key predistribution mechanism	Based on polynomial key generation method, hierarchical network structure is applied to efficiently establish secure links between cluster members and cluster heads. Most energy consumptions and key storages are kept on cluster heads, and sensor nodes can achieve better resource efficiency
9. Diffie–Hellman-based asymmetric key approach	A key distribution scheme based on the Elliptic Curve Diffie–Hellman key exchange mechanism. Each node has a unique key shared with the sink node
10. Polynomial-based scheme (PBS) for authentic associations	Establishes authentic associations among mobile entities of IEEE 802.16 m wireless mesh network on the fly by using bivariate polynomials. Also provides security against traffic analysis and node capture attacks

generate a large key pool P with $2^{17} - 2^{20}$ keys [5] and corresponding key identifiers and create n key rings by randomly selecting k keys from P. Then, load key rings into memory of SNs and have key identifiers of a key ring and associated identifier. For each SN, load a key which it shares with other SNs (Fig. 17.8a).

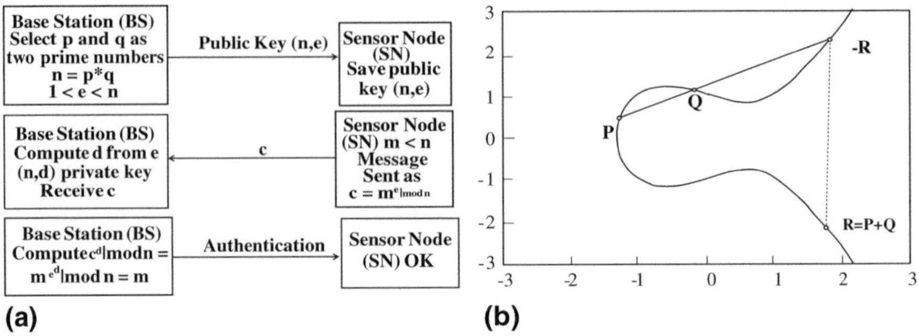

Fig. 17.7 **a** Authentication using public/private keys and **b** *elliptic curve* cryptography (ECC)

Table 17.2 Comparing asymmetric key versus symmetric key approach

Field	Value
Outdoor transmission range	500–1000 ft
Center frequency	868/916 MHz
Battery	3 V coin cell
Program flash memory	128 k bytes
Measurement (serial) flash	512 k bytes
EEPROM	4 k bytes
Effective data rate	12.4 kbps
Energy to transmit	59.2 μJ/byte
Energy to receive	28.6 μJ/byte

During the key establishment phase, path keys are assigned to selected pairs of SNs that are within communication range of each other, but do not share a key. SN may broadcast the message with its id, id of intended SN, and some key that it possesses but not currently uses, to all SNs with which it currently has an established link. Those SNs rebroadcast the message to their neighbors. Once this message reaches the intended SN (possible through a long path), this SN contacts the initiator of path key establishment. Analysis shows that after the shared-key discovery phase, a number of keys on a key ring are left unused. This is more robust than approaches that use a single key. In case a SN is captured, $k << n$ keys are obtained. This means that the attacker SN has a probability of k/P to attack successfully any other WSN link, where p is the probability that a link between any two SNs exists. Since keys are drawn out of a pool P without replacement, the number of key rings with k members each can be expressed as follows (Fig. 17.8b):

$$c = \binom{P}{k} = \frac{P!}{k!(P-k)!} \tag{17.2}$$

Table 17.3 Power consumption in processors for different algorithms [4]

Processor	Clock speed (MHz)	Max power load (mW)	Computational energy consumption (mJ)								
			RSA sign	RSA verify	DSA sign	DSA verify	Diffie–Hellman	El gamal sign	El gamal verify	XTR sign	XTR verify
MIPS R4000	80	230	16.7	0.81	9.9	20.	15.9	9.94	134	1.91	4.5
SA-1110 "StrongARM"	133	240	15.0	0.74	9.1	18.2	14.6	9.1	123	1.71	4.1
Z-180	10	300	3700	184	2300	4500	3640	2300	31,000	420	102
MC68328 "DragonBall"	16	52	840	42	520	1040	829	520	7000	96	230

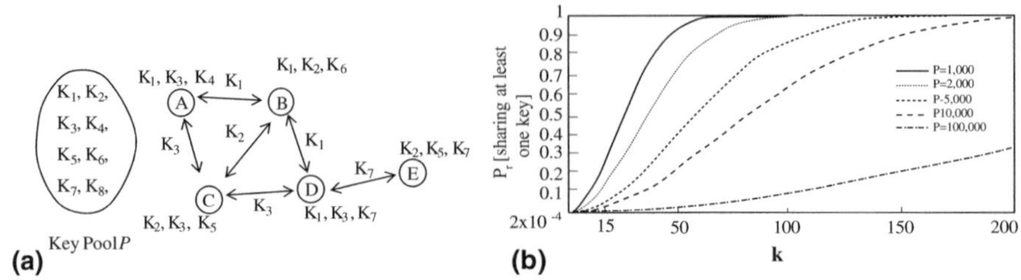

Fig. 17.8 **a** Key pool and random key predistribution and **b** probability of sharing at least one key when two SNs choose k keys from a pool of size P

Let us pick the first key ring, the total number of possible key rings that do not share a key with this key ring is the number of key rings that can be drawn out of remaining $P - k$ unused keys in pool, which is:

$$\binom{P-k}{k} = \frac{(P-k)!}{k!(P-2k)!} \tag{17.3}$$

Consequently, the probability p' that no key is shared between the two rings is the ratio of the number of rings without a match by the total number of rings:

$$p' = \frac{\binom{P-k}{k}}{\binom{P}{k}} = \frac{k!(P-k)!(P-k)!}{p!k!(P-2k)!} \tag{17.4}$$

Since P is very large, Stirling's approximation can be used to derive the final expression for p':

$$p' = 1 - \frac{\left(1 - \frac{k}{P}\right)^{2(P-k+1/2)}}{\left(1 - \frac{2k}{P}\right)^{2(P-2k+1/2)}} \tag{17.5}$$

WSN contains $n = 10{,}000$ SNs, desired probability of network connectivity is $P_c = 0.99999$, and communication range supports 40 SNs neighborhoods. According to formula (17.5), $c = 11.5$; therefore, $p = 2 \times 10^{-3}$ and $d = p \times P_c = 2 \times 10^{-3} \times 9999 = 20$. This means that if each node can communicate with on average 20 other nodes, the network will be connected if

$$p' = d/(2d - 1) = 20/(40 - 1) = 0.5. \tag{17.6}$$

According to formula (17.5), k can be set to 250 and P can be set to 100,000. Then, the prior scheme needs to exchange 3–4 hops neighboring key information. This extension approach can achieve "full connectivity" with only one hop key information exchange. An enhancement of the random key predistribution scheme proposed by having two nodes should share at least $q(q \geq 1)$ keys to have a secured link. "*q-composite*" approach [6] has better resiliency against node capture attack as for a network with 10,000 SNs inside, when $q = 2$, 50 SN captures only compromise 5.74% communications of the WSN. This is an enhancement of the basic probabilistic approach as SNs should share q keys instead of only one. The approach is to have a key pool P as an ordered set, and during initialization phase, SNs broadcast ids of keys that they have. After discovery, each SN identifies the neighbor with which it shares at least q keys. The communication key is computed as a hash of all shared keys. The keys appear in hash in the same order as in key pool. q-composite approach has greater resiliency to node capture than the basic approach if small number of SNs were captured. Simulations show that for $q = 2$, the amount of additional communications compromised when 50 SNs (out of 10,000) have been compromised is 4.74%, as opposed to 9.52% in the basic scheme. However, if a large number of SNs have been compromised, q-composite scheme exposes larger portion of network than the basic approach. A larger q is harder if it is to obtain initial information. The parameter q can be customized to achieve required balance for a particular WSN. An enhancement is possible id pseudorandom number generator is used to improve security of key discovery algorithm. Also, it is used for secret sharing jointly with logical paths that allow SNs to establish a pairwise key that is *exclusively* known to two SNs. You basically utilize a pseudorandom number generator that produces a sequence of values based on a seed. Given the same seed, the algorithm will always output the same sequence of values. A key pool P of size l is generated, and for each SN u, pseudorandom number generator is used to generate the set of m distinct integers between 1 and l (key ids). SNs unique id u is used as a seed for the generator. Each SN is loaded with key ring of size m. Keys for the key rings are selected from key pool P in correspondence with integers (key ids) generated for a particular SN by pseudorandom number generator. This allows any SN u that knows another nodes v id to determine the set of ids of keys that v poses. Established on the previous step, keys are not exclusive and consequently not secure enough; however, they can be used to establish exclusive key. During the network initialization phase, SNs discover the so-called logical paths. SNs can establish a direct path in case they share a common key on their key rings. This can easily be accomplished as was described earlier by discovering common key id. In case SNs do not share a key, authors propose a path-key establishment algorithm similar to one in basic probabilistic approach, and the difference is that SNs try to establish several logical paths, which later should help in establishing a pairwise key. The next step of network initialization is pairwise key establishment. A sender SN randomly generates a secret key ks. Then, $n - 1$ random strings $\mathrm{sk}_1, \mathrm{sk}_2, \ldots, \mathrm{sk}_{n-1}$ are derived. sk_n is computed as follows:

$$\mathrm{sk}_n = \mathrm{ks\ XOR\ sk_1 XOR\ sk_{2\,XOR}}, \ldots, \mathrm{XOR\ sk}_{n-1} \tag{17.7}$$

This way, a recipient has to receive all n shares in order to derive a secret key ks. After secret shares are computed, each of them is send to the recipient using different logical paths. Once all shares are received, the recipient can confirm the establishment of pairwise key by sending a HELLO message encoded with a new key. A framework [7] is provided according to which number of shares and the way they are send is decided. So far, all the discussed approaches have used either key id notification, challenge response, or pseudorandom key id generation. These algorithms work well against the so-called oblivious attacker, the one that randomly selects next SN to compromise. What if attacker selects SNs that will allow to compromise the WSN faster, based on already obtained information? This is the case of the so-called smart attacker which can be defined as follows: At each step of the attack sequence, the next sensor to tamper is SNs, where s maximizes $E[G(s)|I(s)]$, the expectation of the key information gain $G(s)$ gives the information $I(s)$ the attacker knows on SNs key ring.

17.5 Symmetric Matrix-Based Scheme

l-degree bivariate polynomial key predistribution schemes allow a symmetric matrix-based shared secret key generation. In Fig. 17.9a, SN A stores $f(a, y)$ and node B stores $f(b, y)$ and they exchange their node id a–b first. SN A evaluates $f(a, b)$, and SN B evaluates $f(b, a)$, *establishing a shared secret key.* Bivariate polynomial $F^t_{i,j}(x, y)$ of degree j satisfies the following property:

$$F^t_{i,j}(x, y) = F^t_{i,j}(y, x). \tag{17.8}$$

For example, $F^2_{x,y}(x, y) = 5x + 5y + x^2y^2$ and many such values are possible. Polynomial-based key predistribution scheme can be generalized to any λ by changing polynomials in the following way:

$$f(x, y) = \sum_{i=0}^{i=\lambda} \sum_{j=0}^{j=\lambda} a_{i,j} x^i y^j \mod p;\ f(x, y) = f(y, x).$$
$$g_u(x) = f(x, r_u) \mod p = \sum_{i=0}^{\lambda} a_{u,i} x^i \tag{17.9}$$

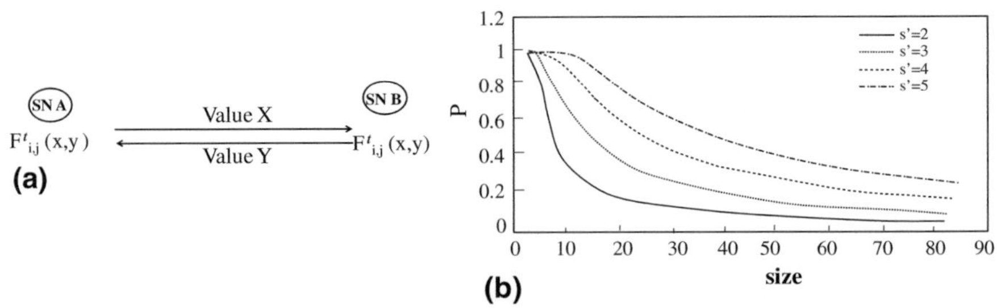

Fig. 17.9 **a** Bivariate polynomial and **b** probability p that 2 sensors share a polynomial versus number of polynomials

where $f(x, y)$ is a randomly generated, bivariate λ-degree, symmetric polynomial over finite field Z_p, $p \geq n$ being prime.

Liu-Ning approach [8] combined polynomial-based key predistribution with the key pool that increases network resilience to SN capture as it can tolerate no more than λ compromised SNs, where λ is constrained by the size of memory of a SN. The idea is to use a pool of randomly generated polynomials. When pool contains only one polynomial, the approach degenerates to basic polynomial-based key predistribution scheme. When all polynomials are of degree 0, the approach degenerates to key pool approach. Three phases of the scheme are setup, direct key establishment, path key establishment. Set F of bivariate λ-degree polynomials over finite field F_q is generated, and each polynomial is assigned a unique id. For each SN, a subset of s' polynomial is randomly chosen from F and is loaded into SNs' memory. During path establishment, all possible direct links are established. A SN can establish a direct link with another SN if they both share a polynomial. If direct connection establishment fails, SNs have to start path key establishment phase by finding a path such that each intermediate SNs share a common key (Fig. 17.10b). SN may broadcast the message with polynomials ids that it possesses to all SNs with which it currently has an established link. Oncc this message reaches the intended SN, a key is computed and contacts the initiator of path key establishment. It may be note that the scheme may have a considerable communication overhead.

17.6 Matrix-Based Scheme (EPKEM) [9]

In this approach, m^2 keys are placed in 2-D matrix of size m × m and each of m^2 SNs are assigned a row and column of (2 m^{-1}) keys. This is done before deployment during initialization step, and let each SN remember their row and

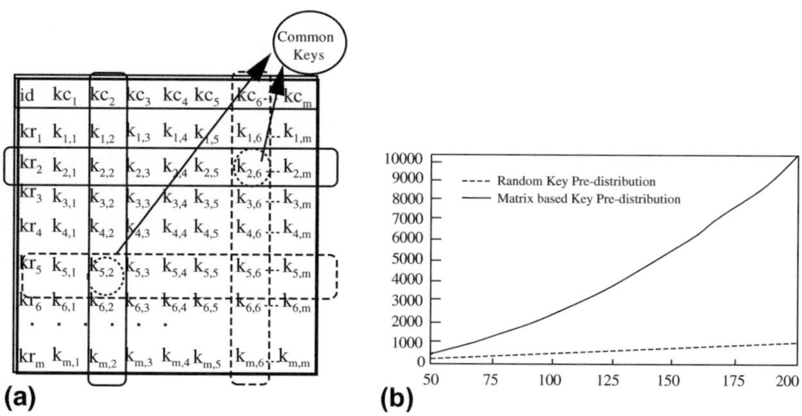

Fig. 17.10 **a** Matrix-based approach for establishing shared secret key and **b** maximum network size versus number of keys in each SN

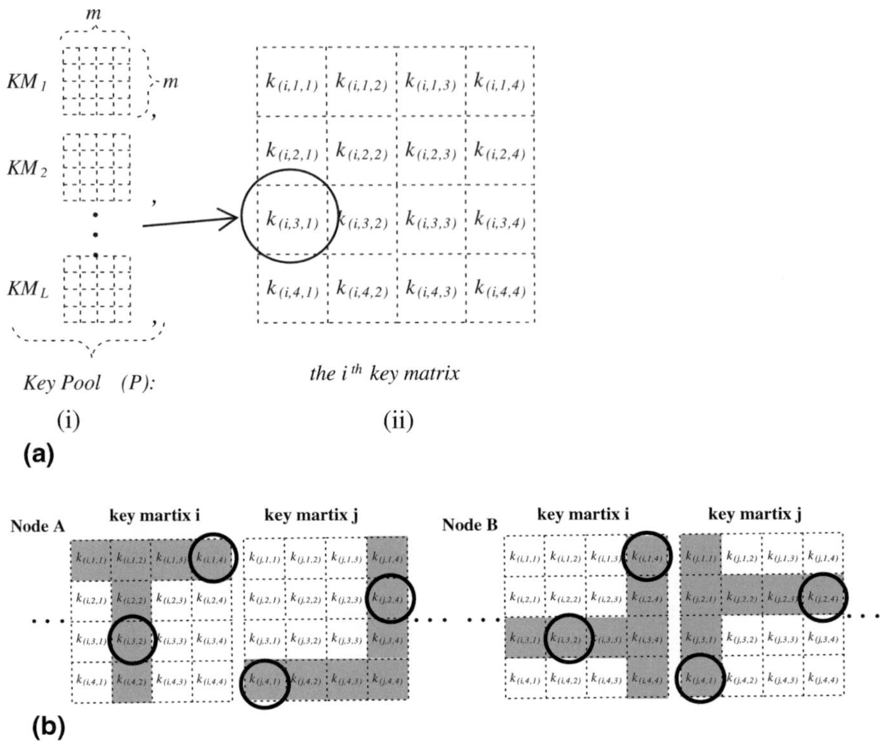

Fig. 17.11 **a** Key pool and keys in 3-D matrix approach and **b** common key discovery

column number. Then, the SNs are deployed. Once SNs discover their neighboring SN, they exchange their row and column number to have two common keys between them. Either of these two keys can be used as a shared secret key or can be combined together by EX-OR operation. To further reduce the key storage overhead and key pool size and increase the security of the established pairwise key, 3-D extension of EPKEM has been proposed [10] that allows each SN broadcasts message its node ID and the identifiers of its preloaded keys to its neighbors. After exchanging handshake messages, *A* and *B* will know they have 4 common keys (Figs. 17.11 and 17.12).

Further generalization is done by placing m^2 bivariate polynomials in the forms of 2-D matrix [10] and uses that in establishing a shared secret key by exchanging seed values between SNs *A* and *B*. This also provides security against traffic analysis and SN capture attacks.

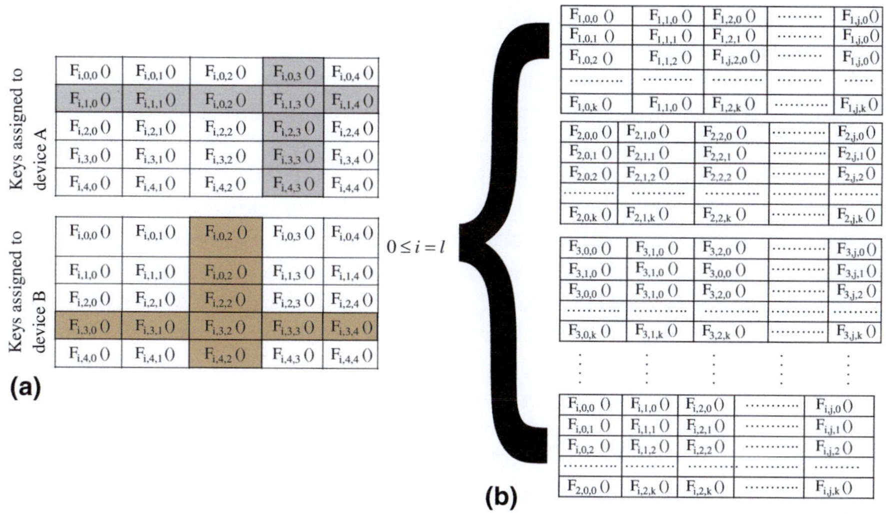

Fig. 17.12 a Polynomial-based matrix scheme and b 3-D bivariate polynomial

17.7 Authenticated Key Agreement Based on Identity-Based Cryptography (IBC)

Public key is derived from some unique identity of the user IBC, which is applied to establish a key in a WSN. It is possible to have a distributed authenticated key with multiple domains. Steps to be followed are as follows: setup security parameters, register SNs and CHs, establish pairwise master key, and then generate session key. So, the question is what kind of delay is involved in authentication, how many levels in the hierarchy from authentication point of view, how to change the key obtained from bivariate polynomial between BS and CHs and SNs in a WSN, and how can you establish a secured path between two arbitrary entities. These are important as security mechanism takes a majority of systems time such as data transfer time for communication intensive applications, and encoding/decoding time for secured communication. Security of WSN could be easily comprised as it utilizes a shared wireless medium and vulnerability of channel and network entities are possible.

Authentication-related security attacks include unauthorized access, replay attack, spoof attack, and denial of service. Therefore, authentication is important to provide access control to WSN. Network side authentication includes a guarantee that only legitimate users have access and network infrastructure and services are protected. On the device area, it is critical to avoid subscription fraud. In wired networks, traffic and computation are typically monitored and analyzed for anomalies at various concentration points. WSNs require a solution that is fully distributed and inexpensive in terms of communication, energy, and memory requirements. In order to look for anomalies, applications and typical threat models must be understood. It is particularly important for researchers and practitioners to understand how cooperating adversaries might attack the system. The use of secure groups may be a promising approach for decentralized intrusion detection. Now, we consider high-level security mechanisms where a trusted third party is employed to establish key protocols that involve a centralized or trusted party, for either or both initial system setup and online actions. In a WSN Domain, BS is owned and operated by a user, including CHs and SNs. AAA (*authentication*, *authorization, and accounting*) server is the central administrative center that assigns authorization policies for different entities, defines their role, permissions, and may be used for accounting, and there could be home AAA (BS) and visiting AAA (another BS). BS is assumed trustable, to some extent CHs as well. Limitations of a WSN are fake registration, traffic analysis attack, and SN capture attack. The use of more CHs can reduce the delay from SNs as the network traffic has to be propagated in a multi-hop fashion from the SNs to the AAA server *through other SNs, CHs, and BS for authentication.*

Individual domain function generation can be defined as follows:

$$f_i(v,w,x,y) = \sum_{i,j,m,n=0}^{t} a_{i,j,m,n} v^i w^j x^m y^n \tag{17.10}$$

$$f_i(v,w,x,y) = f_i(v,w,y,x) \tag{17.11}$$

Individual function initialization and distribution can be defined as follows:

$$F_i(v,w,x) = \sum_{k=1}^{N} f_k(v,w,x,i) \tag{17.12}$$

$$F_{\mathrm{CH}(l,i)}(v,x) = F_i(v,w,x)|_{w=1} = F_i(v,l,x) \tag{17.13}$$

$$F_{\mathrm{SN}(b,j)}(w,x) = F_j(v,w,x)|_{v=b} = F_j(b,w,x) \tag{17.14}$$

Authenticated pair-wise master generation can be given by:

$$F_{\mathrm{SN}(a,i)}(w,x) = F_i(a,w,x) \tag{17.15}$$

$$F_{\mathrm{CH}(b,j)}(v,x) = F_j(v,b,x) \tag{17.16}$$

Symmetric polynomial between AAA and SN can be given by:

$$F(x,y) = F(y,x) = \sum_{i,j=0}^{k} a_{i,j} x^i y^j = \sum_{i,j=0}^{k} a_{i,j} y^i x^j \tag{17.17}$$

In this way, SN i having $f(i, y)$ calculates $f(i, j)$, *while* SN j having $f(j, y)$ calculates $f(j, i)$ to have a shared secret keys. Security parameters are set up by

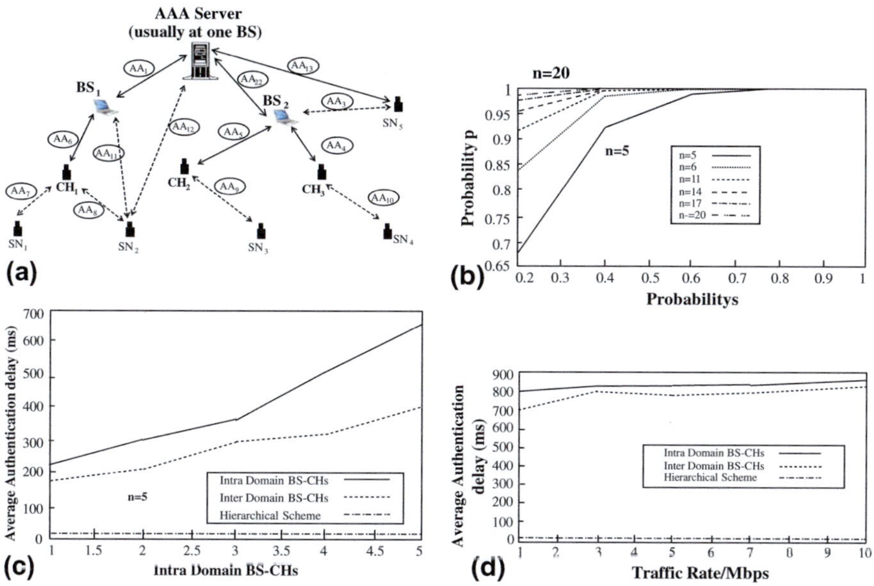

Fig. 17.13 **a** Authentic associations in a WSN, **b** n AAAs exchange $(n - 1)$ polynomials, **c** hierarchical scheme against intra/inter-domains with varying number of hops, and **d** hierarchical scheme against intra/inter-domains with varying load

having ISPs/AAAs agree with globally security parameters, including P and G_1. Then, AAA_1 computes $(P, s_1P \in G_1, s_1 \in Z * q)$ and s_1 is selected by AAA_1 kept secret for AAA_1. In a similar way, AAA_2 calculates $(P, s_2P \in G_1, s_2 \in Z * q)$ and s_2 is selected by AAA_2 and kept secret. SNs and CHs can check their registration accordingly. One-way hashing function $H_f()$ is applied to all seeds to generate a unique secure symmetric key *(USSK) for communication.*

$$\text{USSK} = H_f(\text{SNID}, 1, \text{SNID}, 2, \ldots) \tag{17.18}$$

n AAAs will exchange $(n - 1)$ polynomials, leading to $n(n - 1)$ exchanges. Transmit index value of nonzero coefficients is the probability of coefficient being nonzero. Authentication latency with varying SN-BS hops is given in Fig. 17.13c, while delay with varying load is given in Fig. 17.13d.

17.8 Conclusions

Secured communication in WSNs is a very important aspect of any given application, and adequate investment is essential for a successful operation of a system. Therefore, a careful attention is not only desirable but essential. Complex authentication measures are required based on criticality of information exchanged between any two entities. The matrix-based approach offers many flexibility as robustness of security scheme depends on critically of data exchanged. To sum up, security in WSN not an option and and it must be incorporated in a WSN. This calls for further investigation in this area.

17.9 Questions

Q.17.1. Is it appropriate to address each SN by its unit address or location from security point of view?
Q.17.2. How can you qualitatively and quantitatively measure efficient and secure communication in a WSN?
Q.17.3. What are the similarities and differences between authentication, access control, data confidentiality, data integrity, and non-repudiation?
Q.17.4. How do you define resiliency due to SN capture? Explain carefully.
Q.17.5. What should you do to ensure repudiation?
Q.17.6. What are the advantages and limitations of q-composite approach?
Q.17.7. Can you adopt matrix-based approach among CHs of a WSN? If so, what would you suggest between CHs and associated SNs?
Q.17.8. Can you utilize matrix-based scheme for regular topologies of 2-D mesh, triangular, and hexagonal networks?
Q.17.9. What are the limitations and advantages of question 17.8?
Q.17.10. What is meant by a trivariate polynomial? Explain with an example.
Q.17.11. What is meant by node replication attack and how can you address this?
Q.17.12. What is meant by denial of sleep attack and what can you do to avoid this?
Q.17.13. What is meant by secured data aggregation?

References

1. www.cs.virginia.edu/crab/wormhole.ppt.
2. http://www.linuxjournal.com/content/elliptic-curve-cryptography?page=0,1.
3. Craig Costello, Patrick Longa, and Michael Naehrig, “A brief discussion on selecting new elliptic curves,” http://csrc.nist.gov/groups/ST/ecc-workshop-2015/papers/session4-costello-craig.pdf.
4. Goran Panić, Thomas Basmer, Oliver Schrape, Steffen Peter, Frank Vater, and Klaus Tittelbach-Helmrich, “Sensor Node Processor for Security Applications,” http://www.ics.uci.edu/~steffenp/files/GPanic-ICECS2011_final.pdf.
5. Falko Dressler, “Key Management in Wireless Sensor Networks,” book chapter, Security in Wireless Mesh Networks, December 15, 2006.
6. Jun Zhao, Osman Yagan, and Virgil Gligor, “On Topological Properties of Wireless Sensor Networks under the q-Composite Key Predistribution Scheme with On/Off Channels,” IEEE International Symposium on Information Theory (ISIT), 2014.

7. Baojiang Cui, Ziyue Wang, Bing Zhao, Xiaobing Liang, and Yuemin Ding, "Enhanced Key Management Protocols for Wireless Sensor Networks," Mobile Information Systems, Volume 2015.
8. D. Liu and P. Ning, "Establishing Pairwise Keys in Distributed Sensor Networks,: 10th ACM CCS '03, Washington D.C., October, 2003.
9. Yi Cheng and Dharma P. Agrawal, "An improved key distribution mechanism for large-scale hierarchical wireless sensor networks," Ad Hoc Networks, vol. 5, pp. 35–48, 2007.
10. Y. Cheng, M. Malik, B. Xie, and Dharma P. Agrawal, "Enhanced Approach for Random Key Pre-Distribution in Wireless Sensor Networks," *International Conference on Communication, Networking and Information Technology*, Amman, Jordan, 6–8 December 2007.

Chapter 18
Interaction with Actuators and WSN Test Beds

18.1 Introduction

SN in a WSN senses the environment and sends data to BS which makes decision about corrective action that needs separate unit(s) called actuators. In this way, SNs sense the environment and measure multiple physical properties, thereby acting as an interface between physical world and devices. SNs' counterpart is the actuator(s) that convert electrical signal into a physical phenomenon. The BS can serve the purpose of a controller, and *system actuators* potentially control the environment. In brief, SNs do monitoring while action is facilitated by actuator(s) and a WSN is incomplete without actuator(s).

18.2 A Generic WSN–Actuator Organization

A generic actuator can be either an open-loop system or a closed-loop system as illustrated in Fig. 18.1. In open loop, the actuator is activated by the sink such as periodic watering of plants or coffee machine making same tasting coffee when activated. In a closed-loop system, you can compare measured value with a reference and decide what to do. For example, in a plant watering system, you can compare existing humidity near the plant, measure current temperature, and accordingly adjust the amount of water to be delivered. In coffee machine, you can adjust the type and amount of coffee, sugar, and milk and provide those in appropriate ratio before making coffee. The corresponding models are shown in Fig. 18.2.

It is important to remember that there is no feedback in open-loop system while the parameters sensed by SNs are compared with desirable threshold values before taking any appropriate action(s). In an open-loop system shown in Fig. 18.3, initially lights in all rooms are off, and if someone enters the room # 2, the light turns on and

D.P. Agrawal, *Embedded Sensor Systems*,
DOI 10.1007/978-981-10-3038-3_18

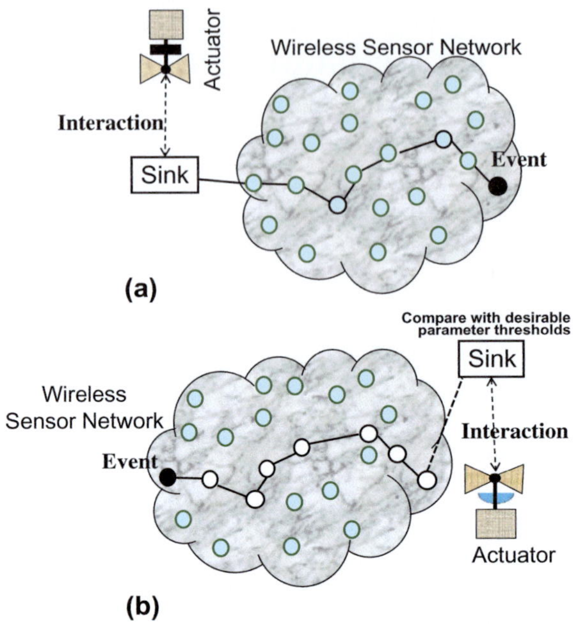

Fig. 18.1 **a** Open-loop system **b** closed-loop system

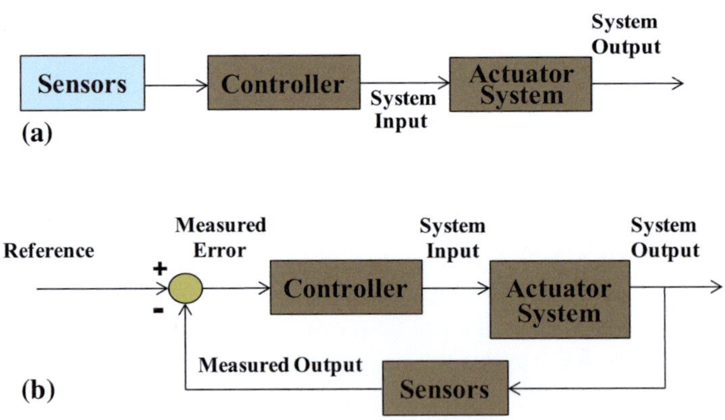

Fig. 18.2 Model of an **a** open-loop system and **b** closed-loop system

remains on indefinitely, illustrating open-loop actuator. If the system employs a closed loop, then if the person leaves the room # 2, it is detected by SNs and the light is turned off. Thus, the role of a WSN and a system actuator is clear as decision making is based on the determination of the object of interest from each SN that directly represents phenomenon of interest to BS, SNs sense unattended environment autonomously, deployed sensor network approximates a physical space for an

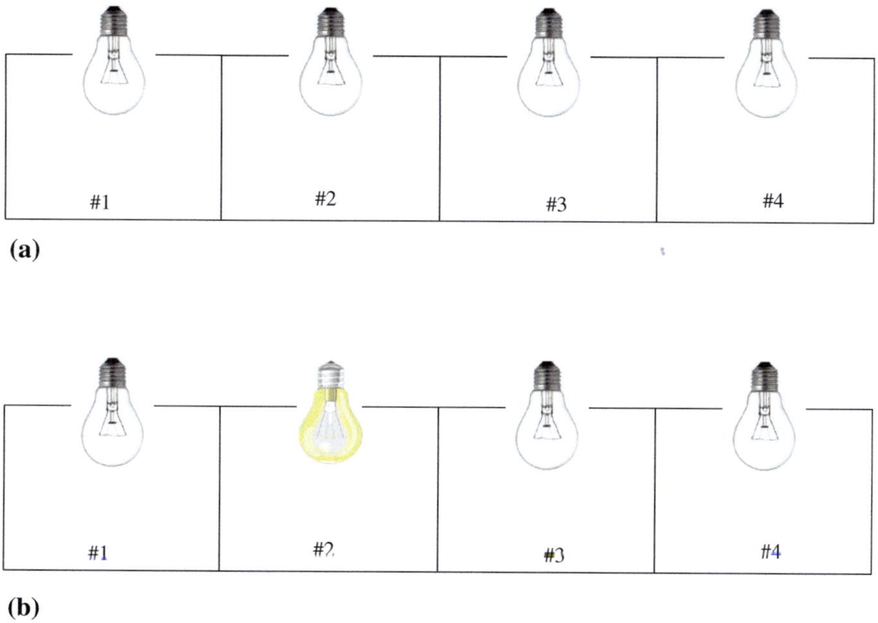

Fig. 18.3 Events in an open-loop and closed-loop systems **a** no one and **b** someone enters room 2

application, and the space ought to be programmed, rather than the network. Adequate number of SNs must be functioning correctly in order to obtain relevant data by equally weighing and combining each SN data through data aggregation process and store the result at the BS.

An actuator performs action(s) based on data collected from SNs, and a logical choice is the data at the BS. *Actuators* hypothetically control the environment, while SNs do monitoring while action is facilitated by the actuator(s). SNs and actuators, respectively, take care of real-time requirements and coordination of activities. Thus, by combining SNs and actuators, it is possible to realize a variety of applications. We can have a powerful and yet cost-effective hybrid wireless sensor–actuator Network (WSAN). The question is how to associate SNs to the actuators so as to correctly utilize their measured values and how can we determine individual effect of each actuator on SNs? Actual control mechanism is at a system actuator; e.g., adjusting a controller based on SN data is handled by a central controller such as base station(s) and is independent of underlying protocols. When sensing and actuation are combined, the next logical step in the evolution of WSNs is achieved. So far, there are a very few examples of sensor networks that actually integrate actuators. Actuator network performance critically affects the behavior, stability, and safety of the controlled physics, while SNs' data set is stored in a BS. So, as for an actuator is concerned, a logical organization could be one actuator associated with one WSN as indicated earlier. In general, one BS could control many actuators or there could be many BSs controlling many actuators.

For example, in Fig. 18.3, initially lights in all rooms 1, 2, 3, and 4 are turned off and two people enter rooms 2 and 4, respectively, turning light on in those rooms by different actuators. If the person leaves room # 2, the light turns off while room 4 remains lighted. If another person leaves room # 4, the light is turned off. Such control could be extended to air conditioners in each room being controlled by separate actuators. Such a generic system is shown in Fig. 18.4. (Fig. 18.5).

Sensed data are collected at the BS power actuators that affect the sensed environment illustrating interaction between the BS and system actuators. An actuator operates in an environment with readily available energy while communication between BSs and actuators can use wired or wireless media. In general, extensive computation may have to be done at the BS as physical control

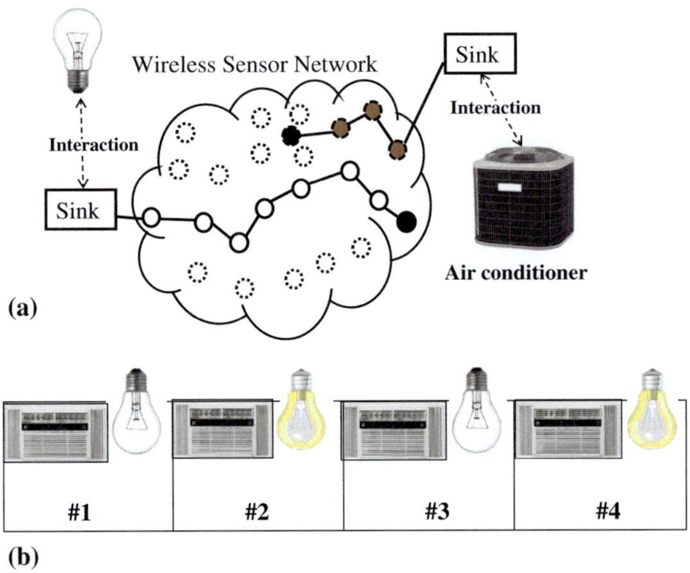

Fig. 18.4 **a** Generic multiple sink multiple actuator and **b** rooms with both lights and air conditioners

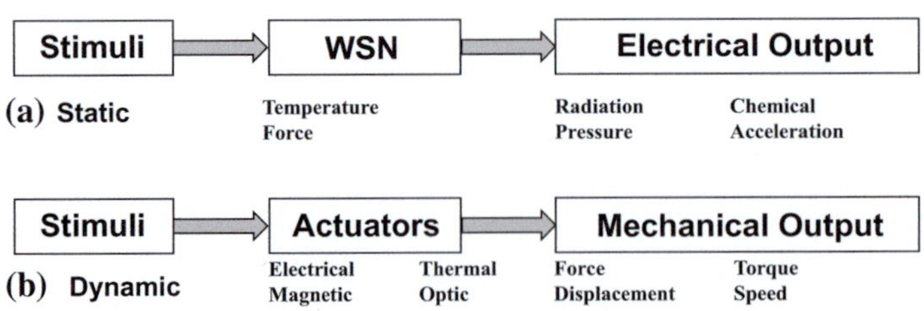

Fig. 18.5 Stimuli for **a** sensors and **b** actuators [1]

of actuators could impose real-time constraints. So, there is a close relationship between sensing and actuation process. Methods for sensing include optical, electrostatic, magnetic, piezoelectric, and thermal schemes and have been covered in detail in Chap. 2. Common actuation methods include shape-memory alloys, piezoelectricity, magnetostrictive, electrostatic, thermal, and few others that are compared in Table 18.1.

There are many applications of actuators for flow control, including noise control, jet vectoring, separation control, and heat transfer [2], and MEMS being small in size and low cost offers an answer to these difficulties as they can be manufactured into the large numbers as required. It is possible to have either direct or indirect actuation methods. For example, acoustic vibration can shake a microsensor to have indirect actuation. For direct methods, the most common method is to use a *p*iezoelectric *z*-axis actuating transducer, which accurately vibrates at different frequencies based on the driving voltage. Micro-actuators have been made up of numerous different processes. For example, pressure sensor is based on a simple idea which requires careful design to possess sensitivity and desired frequency response. A square membrane has a piezoresistor on each side and a pair of unproven membranc intersecting opposite edges with largest stress and other pair on the firm substrate used as a reference pair. The actuators are active depressions, and electroactive polymers are used as the actuating medium and lie flush with the surface. When actuated, these deflect downward to form a dimple. Associated characteristics are summarized in Table 18.2 (Fig. 18.6).

Table 18.1 Summary of an actuator mechanism [1]

Actuation mechanism	Advantages	Disadvantages
Electromagnetic	• Low actuation voltage • Relative large displacement	• Difficult in fabrication of magnetic material with CMOS technology • Challenge in minimizing size of devices
Piezoelectric	• High switching speed • Low power consumption	• Small displacement range • High actuation voltage
Electro thermal	• Easy fabrication • Low actuation voltage	• High power consumption • Slow response time • Thermal fatigue due to thermal cycle
Electrostatic	• Low power consumption • Fast response time • Easy to integrate and implement with CMOS technology • Compatible with most fabrication methods	• High actuation voltage • Limited operation range due to pull-in

Table 18.2 Type of actuators and associated materials [3]

Type of actuator	Stress (MPa)	Strain (%)	Strain rate (Hz)	Power density (W/kg)	Efficiency (%)
Electrostatic (microscopic composite)	0.04	>10	>1	>10	>20
Cardiac muscle (human)	0.1	>40	4	>100	>35
Polymer (polyacrylic acid/polyvinyl alcohol)	0.3	>40	0.1	>5	30
Skeleton muscle (human)	0.35	>40	5	>100	>35
Polymer (polyaniline)	180	>2	>1	>1000	>30
Piezoelectric polymer (PVDF)	3	0.1	>1	>100	<1
Piezoelectric ceramic	35	0.09	>10	>1000	>30
Magnetorestrictive (Terfenol-D)	70	0.2	1	>1000	<30
Shape-memory alloy (NiTi bulk fiber)	>200	>5	3	>1000	>3

18.3 Actuators for a Vineyard

In Napa Valley vineyard [4], sensing of humidity and temperature at vine plants is done by SNs and a dog goes around in receiving data from SNs. The objective for actuation is to remove fog, and localized heaters and monitoring of micro-climates at the vines are done using appropriate query from the BS. Information about temperature, lighting levels, humidity, the movement and presence of people, and many other aspects of the environment can be obtained from SNs, and BS takes the decision when to actuate the actuator to provide adequate amount of water to the plants. Sensor readings could be taken less frequently during night hours. In a similar way, changes in the conditions throughout the monitored area might be important than RF transmission range in determining the locations of SNs. The risk of mildew risk can be determined from the temperature data over time.

Cell arrangement is a special case of process-oriented layout [5], and it utilizes the fact that *different* machines and steps are followed in making a finished product. This reduces inventory, floor space, and direct labor costs while increasing equipment utilization, employee participation, and quality of work. If the task can be divided into 6 independent subtasks, then cost in moving parts in adjacent areas could be one unit while between non-adjacent units can be double. There are 6! or 720 possibilities, and it is not possible to consider all of them. Two different options are shown in Fig. 18.7. Such a procedure allows specialized focus on only one skill, and a worker can watch several machines at once, thereby allowing high level of product flexibility. This encourages large lot sizes while makes cross-training challenging.

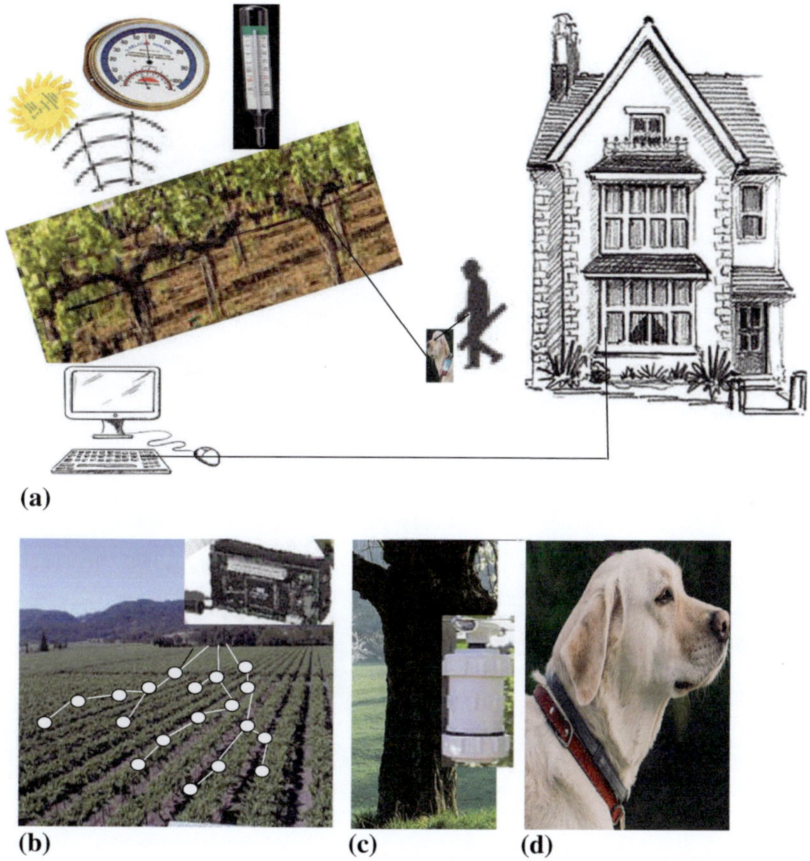

Fig. 18.6 **a** Vineyard monitored by SNs and data collected at central database, **b** Napa Valley vineyard, **c** SN sensing data, and **d** dog collecting SN data [4]

18.4 Mobile BS and Anchor Node Collecting Data from SNs

As discussed earlier, a mobile BS or sink node could be used to collect data from SNs. When BS is within the communication range of a SN, it can collect data from the SN and move on. This type of arrangement is feasible in some applications, while in others, the actuator can do the job of collecting data from SNs by having a mobile actuator [6] moving around the monitored area. Such schemes are illustrated in Fig. 18.8a, b and are redrawn from [6]. The idea behind placing SNs randomly in places such that placement of new/additional SNs increases the sensing area coverage by SNs. Placement of SNs can be decided by the BS or can be done by an actuator and is summarized below.

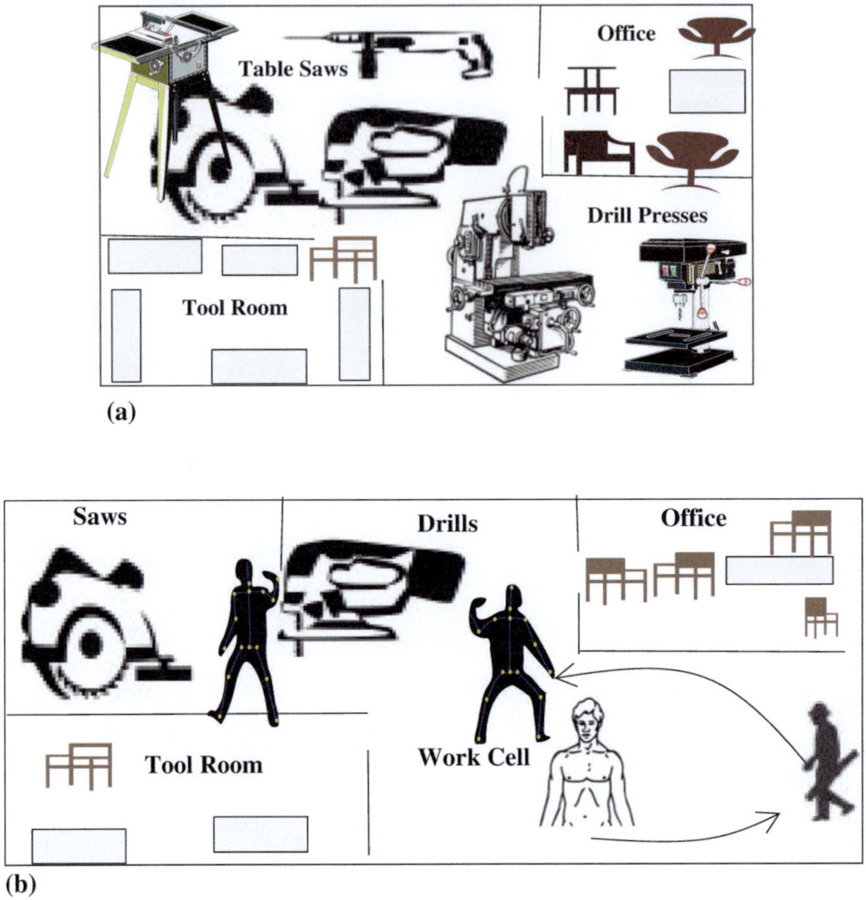

Fig. 18.7 **a** One arrangement of areas and **b** an alternative arrangement

If SNs are to be placed based on decision at the actuators, then least recently visited approach is used. Each SN determines least recently visited direction to the actuator when in its communication range. The actuator travels a predefined distance in recommended direction, and if the chosen direction is blocked, it asks for a new suggested direction. Supply chain maintenance by system actuators has not yet been well studied how to repair/maintain sensor coverage using actuators. Most existing solutions are direct application of SN clustering and flooding the WSN with messages. It is assumed that sufficient spare SNs are present. In cluster-based approach, replacing failed SNs in a WSAN by following either a centralized, or a distributed or a dynamic protocol. In a centralized protocol, an actuator is selected as a central manager which is responsible for SN failure reports as the BS broadcasts its location to all SNs and other actuators. BS maintains the latest position of each actuator, and SNs monitor each other and report noticed node failures to the

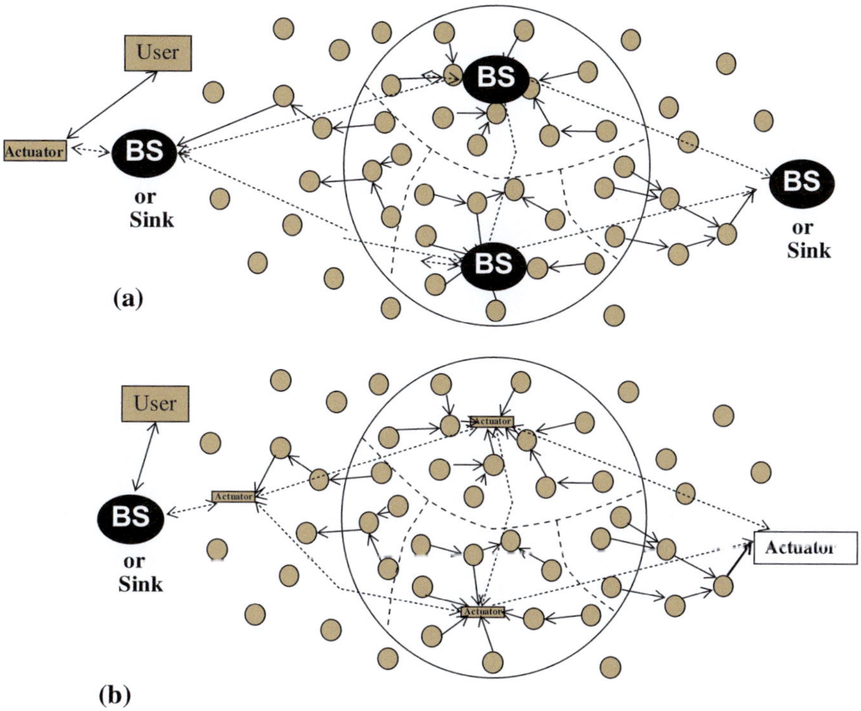

Fig. 18.8 **a** Mobile BS/sink collecting data from SNs and **b** mobile actuator collecting data from SNs

BS, which then ships closest actuators to replace failed SNs with their spare ones. Multiple requests are served on first come, first served basis. Once an actuator moves to its assigned new location, it updates the central manager with its latest position.

In a distributed protocol, the monitored field is partitioned into equal-sized subregions, with each region assigned to one actuator that acts as a manager to collect information about failed SN. It is also responsible for failed SN replacement in its own subregion, and the centralized algorithm is run within each subregion. In a dynamic protocol, the monitored field is dynamically partitioned according to the current position of each robot that broadcasts location information. SNs receiving messages from multiple robots rebroadcast only the one from closest robot. A Voronoi diagram is constructed based on hop count, and detected SNs' failures are reported for creating actuators of their home Voronoi cells, which is used to replace the failed SN with their spare ones. If the actuators are moving, they broadcast their latest location to update the Voronoi diagram .

In a perimeter-based approach, actuators are required to form a connected network and need densely deployed SNs in a small area and then spread out using self-deployment approach. When a hole is detected by a SN, it informs the actuator to take corrective action. This approach lacks destination information and could cause failure in routing process. The actuator has to take responsibility of filling the reported hole.

18.5 Role of a System Actuator

An actuator network needs to operate in an environment with whatever energy is available. If the communication is based on wired media, extensive computation can be done. The physical control dynamics primarily requires real-time constraints. The sampling period can be periodic as a control system behaves totally different in different regions to be monitored. WSN/actuator could work as event-triggered and self-triggered, and SNs could be static or dynamic. It is important to have data paths between BS and actuators and could be achieved using wired line or wireless link. SNs and WSNs are used in smart building applications to support heating, ventilation, air-conditioning systems, lightning, shading, air quality and window control, system switching-off devices, metering, standard household applications, security, and safety [7].

18.6 Conclusions

Interaction between SNs and an actuator is important, and closed-loop control ought to be preferred. BS is expected to contain and represent data measured by SNs in the monitored area. Several actuators can be controlled by one or more BSs using data from SNs. Many actuators can use the same physical parameter from different areas or different parameters in a given area. Cascade scheme with interstage actuators is very useful in numerous practical applications. Coordination among interstage delay enhances the system performance.

18.7 Questions

- Q.18.1. What is the difference between a sensor, an actuator, and a transducer?
- Q.18.2. Can a SN work as an actuator and vice versa?
- Q.18.3. Is there any interaction between a BS and an actuator?
- Q.18.4. What are some real-world examples of what an actuator can do?
- Q.18.5. How can multiple BSs control a single actuator? Explain clearly with suitable examples.
- Q.18.6. What is the impact of the goal-driven, self-propelling process learning in a WSN with an actuator system?

Q.18.7. How does resolution, accuracy, and precision of SNs affect the actuator actions?

Q.18.8. Can you connect several SNs to one communication module?

Q.18.9. Can you develop complex robotic applications using manipulators and mobile robots with the help of high-end robotic SNs

Q.18.10. How can autonomous navigation of mobile robot be planned for robot manipulators?

Q.18.11. What is the effect of SNs and actuator quality on a robot performance?

Q.18.12. Two different materials, with different coefficients of thermal expansion, are joined together in such a way that a temperature change will cause the entire structure to deform in a desired way. Can you estimate the deflection of such a device?

Q.18.13. Where in everyday life do you find piezodevices?

Q.18.14. Why is the event rate/frequency of the reading from SNs limited to some fixed quantity and what is the impact on the accuracy of an actuator?

References

1. J. Sam Jebar Kumar, Enoch Amoatey Tetteh, E. Paul Braineard, "A study of why Electrostatic actuation is preferred and a simulation of an electrostatically actuated cantilever beam for MEMS applications," International Journal of Engineering Sciences & Emerging Technologies, vol. 6, no. 5, pp. 441–446, April 2014.
2. G.G. Arthur, B.J. McKeon, S.S. Dearing, J.F. Morrison, and Z. Cui, "Manufacture of micro-sensors and actuators for flow control," Microelectronic Engineering 83, 2006, pp. 1205–1208.
3. Qing-Ming Wang, "Physical Mechanisms of MEMS devices," University of Pittsburgh, http://www.pitt.edu/~qiw4/Academic/ME2080/Lecture18.pdf.
4. Jenna Burrell, Tim Brooke, and Richard Beckwith, "Vineyard Computing: Sensor Networks in Agricultural Production," IEEE PERVASIVE computing, pp. 38–45, January–March 2004.
5. Ron Lembke, "Processes Selection and layout operations management," http://slideplayer.com/slide/4855301/.
6. Ivan Stojmenovic, "**Tutorial:** Data communication and coordination in wireless sensor and sensor-actuator networks," SENSORCOMM, Valencia, Spain, 15:45–18:30 pm, October 15, 2007.
7. Smart Sensor Networks: Technologies and Applications for Green Growth, Dec. 2009, https://www.oecd.org/sti/ieconomy/44379113.pdf.

Part V
Research Directions

Chapter 19
Deployed Large-Scale WSNs and Associated Design Steps

19.1 Introduction

SNs and WSNs are being used for numerous civilian and defense areas and numerous applications have been covered in Chap. 2. In most of these presentations, few SNs have been deployed to show usefulness in different areas and very few projects deal with a large number of SNs for continuing monitoring over few months. Although advantages of WSNs for monitoring the environment have been established, deployment of WSNs in real environment remains to be challenging. Very few groups have deployed WSNs with the goal of using them for long-term monitoring.

19.2 Deployed WSNs

With optimization for precise needs, there have been many WSN deployments in the past few years. However, these cannot reliably observe general behavior. A number of practical network deployments have been reported during the last decade [1–4]. Even though these are important, measurements done usually at tens of SNs can hardly reveal generic network conducts, such as routing dynamics and topology evolution, which exist only in large-scale WSNs. Researchers have considered indoor medium scale test bed such as MoteLab [5] and Kansei [6] which cannot fully capture the characteristics of realistic environments. Deploying WSNs at large scale is important because increase in the network size introduces new set tasks.

Bennett et al. [7] consider the issues in using WSNs in an underground environment based on two field case studies of the Prague Metro and the London Underground. MICAz boards made by Crossbow, operating at 2.4 GHz were used in both trials were and using the XMesh routing protocol. The network design and the radio propagation are described in deploying a WSN in these two tunnels. The

D.P. Agrawal, *Embedded Sensor Systems*,
DOI 10.1007/978-981-10-3038-3_19

time stamped messages are received at a Gateway, the Crossbow MIB600 linked via Ethernet cable to a single-board computer with a GPRS modem. For the Prague Metro, a total of 28 motes were installed; 10 inclinometer motes, 2 crack meter motes and 16 relay motes. In the London Underground, a total of 25 motes were installed; 16 inclinometer motes, 5 crack meter motes and 4 relay motes. In both the cases, a data packet was sent every 3 min from each SN.

SensorScope with about 100 SNs and communication range of 200 m outdoors is used during the summer 2007, is capable of weather monitoring in the wild [3]. 6 different outdoor deployments, with 6–97 SNs, each capable of measuring 9 distinct environmental quantities: air temperature and humidity, surface temperature, incoming solar radiation, wind speed and direction, precipitation, soil water content, and soil water suction have been run during 15 months. VigilNet with 200 SNs in an area of 200 m × 300 m [8] is designed to support long-term defense observation by having redundant distribution of SNs, partitioning them into non-overlapping sets and keeping only one set active at a time.

A large-scale solar powered WSN "Trio" with 557 solar-powered mote SNs over an area of approximately 50,000 square meters are deployed for multi-target tracking [9] during last four months of 2005. ExScal with 1000 *regular, hierarchical structured* SNs and 200 backbone SNs [10] covers 1300 * 300 m^2 *in a remote area of Florida, during December 2004*. More SNs are deployed at the boundary to classify intruder as a person, a Sport Utility Vehicle or an All-Terrain-Vehicle by intruder's magnetic effect and acoustic power. Despite substantial efforts made, real-world applications of WSNs are still limited, and often burdensome. Prominent examples of WSNs and associated limitations are described in [11] and methods to be used to pinpoint failures are described.

A measurement study led by Prof. Liu [11] employs 330 SNs WSN named GreenOrbs and has been regularly operating since July 2008 to monitor all-year-round ecological surveillance in the forest, collecting various sensory data, such as temperature, humidity, illumination, and carbon dioxide. Figure 19.1 shows the real topology of with the sink deployed at the upper left corner. Each SN of

Fig. 19.1 Placement of SNs in GreenOrbs project [4]

GreenOrbs in 2-D geographical location continuously monitors the environment, supporting fine-grained real-time fire risk prediction. System performance is observed when 100 SNs to 200, and then to 330 nodes when each SN generates three packets per hour. The traffic load is increased in a stepwise manner by shortening the cycle lengths to 3600, 400, and 200 s and increasing the traffic load severely degrades the system performance from over 60% to less than 10%. It is observed that among all packet losses are evenly distributed for different intensities, Transmit_Timeout accounts for 61.08% and the rest 38.92% for Receive_Pool_Overflow. No Send_Queue_Overflow is detected. Many associated issues such as secured communication [4] and other areas are also being investigated.

19.3 Forest Fire with Regularly Deployed SNs and Following Gaussian Distribution

In a WSN, each application requires handling of various types of data, following different steps and diverse interpretations. It is rather difficult to uniquely generalize steps needed or desirable for implementation. In general, cause and effect paradigm can be used in defining steps to be followed in having a useful system that can do the job and provide required functionalities. In this chapter, we define steps to be followed to determine exact boundary of wild forest fire while minimizing required number of SNs when deployed randomly from low flying airplane or unmanned aerial or ground vehicle.

Example of forest fire is illustrated in Fig. 19.2. Wild fires often happen in wild, unpopulated areas, but they can occur anywhere. Wildfires could be caused by natural reasons such as lava or lightning constitute only 10% of the events. But, it is interesting to note that most forest fire are caused by humans and as many as 90% of wild fires in United States are result of campfires left unattended, burning of debris, carelessly throwing away cigarettes and some intentional acts of burning. Many factors, affect the fire such as geography, climate, weather, and topography [12]. Time of year impacts the effects of fire. For example, in the western U.S, wildland fire season is from June through October, while March through May is the fire time in the southeastern U.S. Most fires occur in the New England states in late fall. Composition of many factors such as moisture level, chemical makeup and density, governs the degree of flammability. A normal tree contains good amount of moisture, while died tree has negligible moisture. Soil moisture also influences the fire intensity. The nature of fire is greatly affected by wind (oxygen supply), temperature and humidity.

Whenever fire starts, it is important to determine boundary of the fire and origination point of fire. An approximate of wild forest fire can be determined by satellite images which gives an approximate region of fire (Fig. 19.2a). Surrounding smoke may cause major errors in determining fire area and the origination point

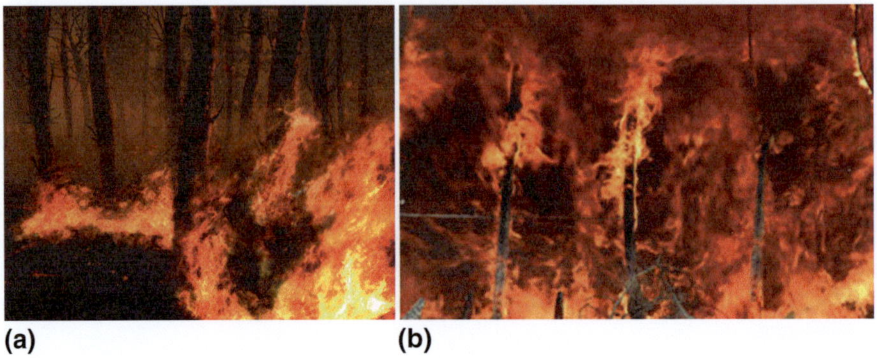

Fig. 19.2 Wild forest fire a forest [10]. b Outskirts [18]

need to be guessed based on the flames and it is hard to estimate severity in different parts of the fire area. Moreover, the wind causes major change in estimating fire area as wind causes the smoke to change the area drastically. Exact estimate of fire area and temperature distribution can be determined if SNs are randomly deployed over the fire area using low flying airplanes or drones (Fig. 19.2b). Such an approach requires a large number of SNs, assuming that they are fireproof and are not destroyed due to fire heat. It is rather difficult to model an arbitrary irregular boundary either analytically or using simulation tools. So, for simplicity, the fire boundary is assumed to be circular in nature and is shown in Fig. 19.3a.

Even circular area to monitor and detect using SNs is not that easy. So, the boundary detection is transformed to bin filling problem as shown in Fig. 19.3. A circular area that could be covered by SNs is within a distance of $2r_s$, and this circular area is divided into sectors of equal size. For example, Fig. 19.3a shows 8-sectors for simplicity and could be increased to large value m depending on r_s and fire radius. SN can be placed in any of the sectors and the issue is transformed to linear bin-filling problem as illustrated in Fig. 19.3b. It is important to remember that it is not necessary to have one SN in each bin as interpolation between close by SNs can be used with minimum error and as shown in Fig. 19.3b, the question becomes filling n bins out of a total of m bins in the circular sectored area. As SNs are dropped by low flying UAVs or fighter planes, the problem of covering X% of a shape with minimum SNs. As the shape is considered as circular, SNs are dropped in a uniform way in the area. So, we look at the probability that a SN falls within the boundary of the circular area. By dividing the circular area into m bins, the probability of filling a bin can be estimated if SNs deployed in a uniform rate. This makes SNs as balls and bins as the boundary area and is comparable to filling m bins with n balls. So, essentially the coverage problem is translated to a bin filling problem: Given we have m bins representing the boundary, we estimate no. of SNs needed to fill at least n bins.

This leads to n-state Markov model as shown in Fig. 19.4a where each state represents a bin being occupied by a SN and probability for changing state from i to

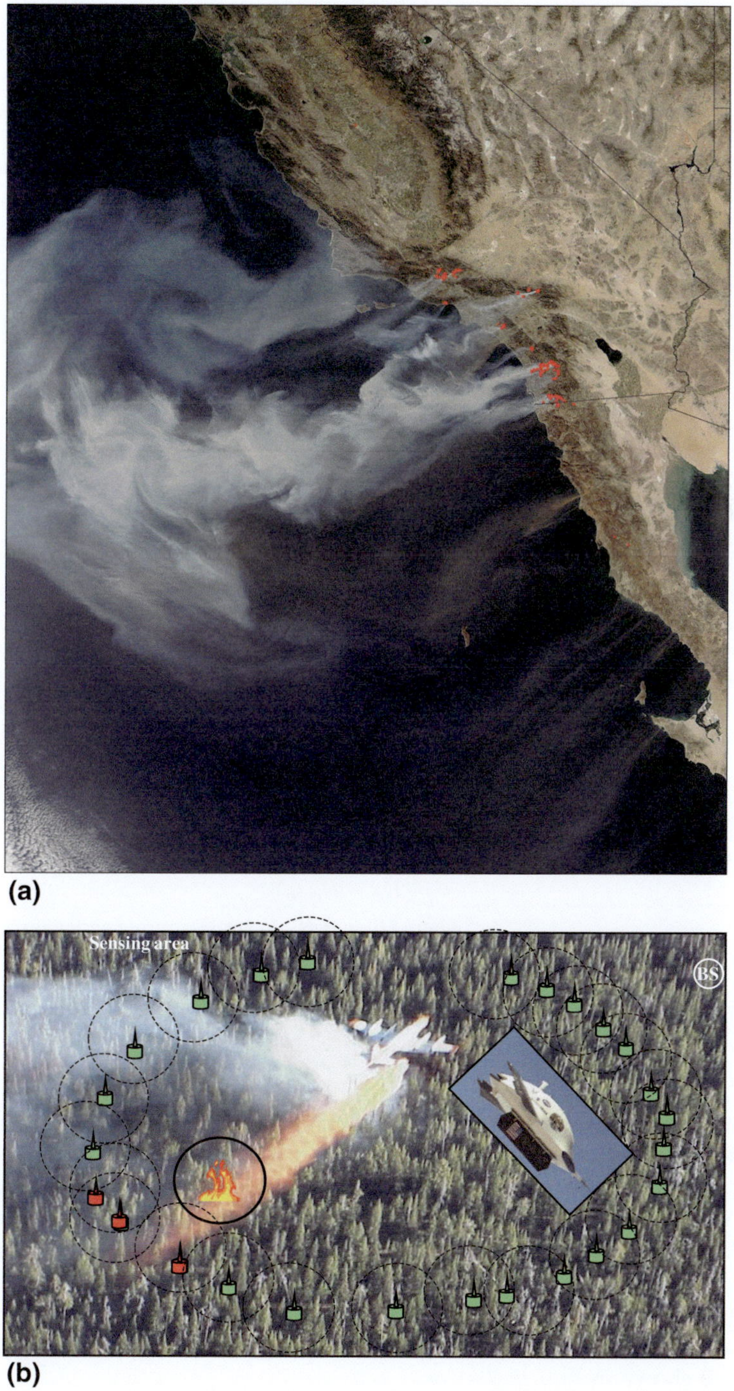

Fig. 19.3 **a** Approximate area of wild forest fire [19]. **b** SNs can be deployed with low flying Airplane or Drone [20]

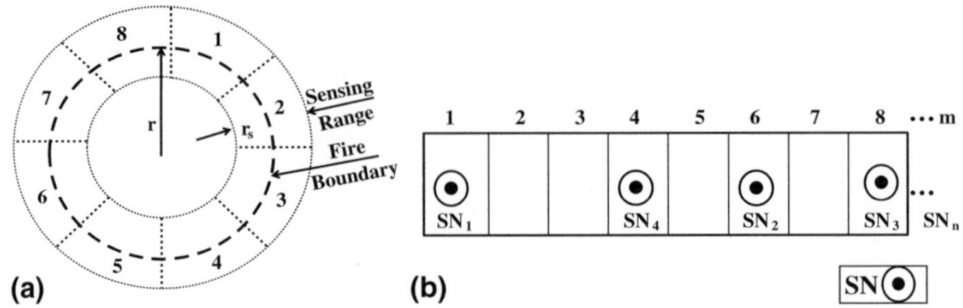

Fig. 19.4 **a** Fire boundary, sensing range and boundary detection. **b** Transforming boundary to linear bins

$i + 1$ depends on m. The corresponding transition matrix from one state to another is shown in Fig. 19.4b when SNs are being dropped. Thus, m-bins represent the boundary and n-bins need to be filled out of a total of m-bins can be obtained as follows:

$$\begin{aligned}
\text{Estimated number of balls, say } K &= \sum_{u=0}^{n-1} \frac{m}{m-u} \\
&= m \sum_{u=0}^{n-1} \frac{1}{m-u} \\
&\text{Since } m \geq u \\
&= \left(\sum_{i=1}^{n-1} \frac{1}{i} - \sum_{j=1}^{m-n} \frac{1}{j} \right) \\
&= m(H_m - H_{m-n}) \\
&\text{where } H_m \text{ and } H_{m-n} \text{ are Harmonic series respectively} \\
&\approx m(\ln(m) - \ln(m-n)) \\
&= m \ln\left(\frac{m}{m-n}\right)
\end{aligned} \tag{19.1}$$

A uniform distribution of SNs is illustrated in Fig. 19.5a and required number of randomly deployed SNs to cover n bins out of a total of 31 bins is shown in Fig. 19.5b. There is slight error in analytical estimation as the SN is assumed to fall in the middle of the bin while it could be anywhere inside the bin as illustrated in Fig. 19.5c. A correction factor has been suggested to incorporate this discrepancy [13] and is shown in Fig. 19.6 when r_s is changed for two different fire areas and different coverage percentage. Rather than deploying SNs uniformly, SNs can be scattered following a Gaussian distribution in the fire area and the corresponding results are shown in Fig. 19.7.

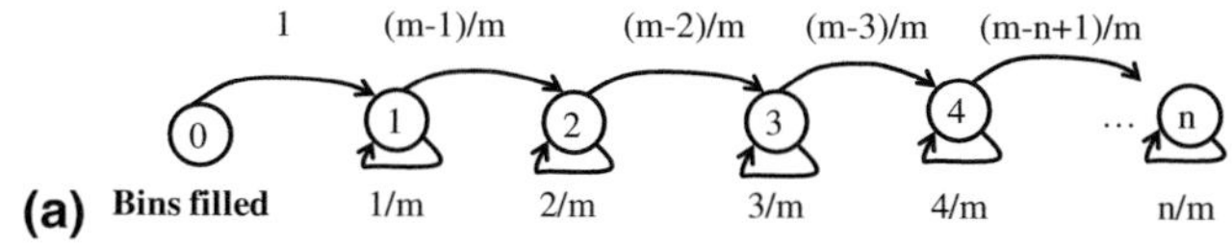

	0	1	2	3	4	5		......			m bins
0	0	1	0	0	0	0	0				0
1	0	1/m	(m-1)/m	0	0	0	0				0
2 3	0	0	2/m	(m-2)/m	0	0	0				0
4	0	0	0	3/m	(m-3)/m	0	0				0
.	0	0	0	0	4/m	.	0				0
.	0	0	0	0	0	.	.				0
.	0	0	0	0	0	0	.				0
.	0	0	0	0	0	0	0				0
.	0	0	0	0	0	0	0				0
.	0	0	0	0	0	0	0				m-(n-1)/m
N balls	N balls	0	0	0	0	0	0.				n/m

(b)

Fig. 19.5 **a** Markov chain model of filling n-bins by SNs. **b** Transition matrix for a SN to fall in each Bin

19.4 Use of Controlled Deployment and Needle-Comb Approach to SNs

Up till now, all SNs are deployed at one time with the hope that an exact 100 or 80% of the fire boundary could be determined (Fig. 19.8). This will be based on both coverage of the boundary area and connectivity to the BS. In a SNs deployment, coverage of fire boundary is defined as percentage of fire boundary angle θ covered and is given by $\theta/360 * 100$. The SNs connectivity does not totally depend on deployed SNs, but is defined as connected (SNs/required SNs) * 100. Detectability is defined as % of boundary also connected to BS and = % delectability = $\varphi/360 * 100$. Here, $\theta = \Sigma$ degrees of the SNs sensing radius r_s on the boundary calculated by the intersection method and φ = degrees covered by SNs connected to the BS on the perimeter or fire border. Connectivity is important as fire event may be covered by SNs; if it does not have connectivity or delectability to the BS it is of no use hence we need connectivity at the earliest with an optimal distribution strategy of SNs. Several distribution patterns are analyzed [14, 15] and dynamic distribution strategies is proposed [16, 17] to optimize the coverage and connectivity thereby minimizing the deployment time and resources (SNs). The entire fire area of 100 by 100 units is divided into grids such that the width of each grid is equal to the sensing radius r_s (5 units) of the SN such that SNs in adjacent grids can communicate with each other and to the BS located at one corner. The BS

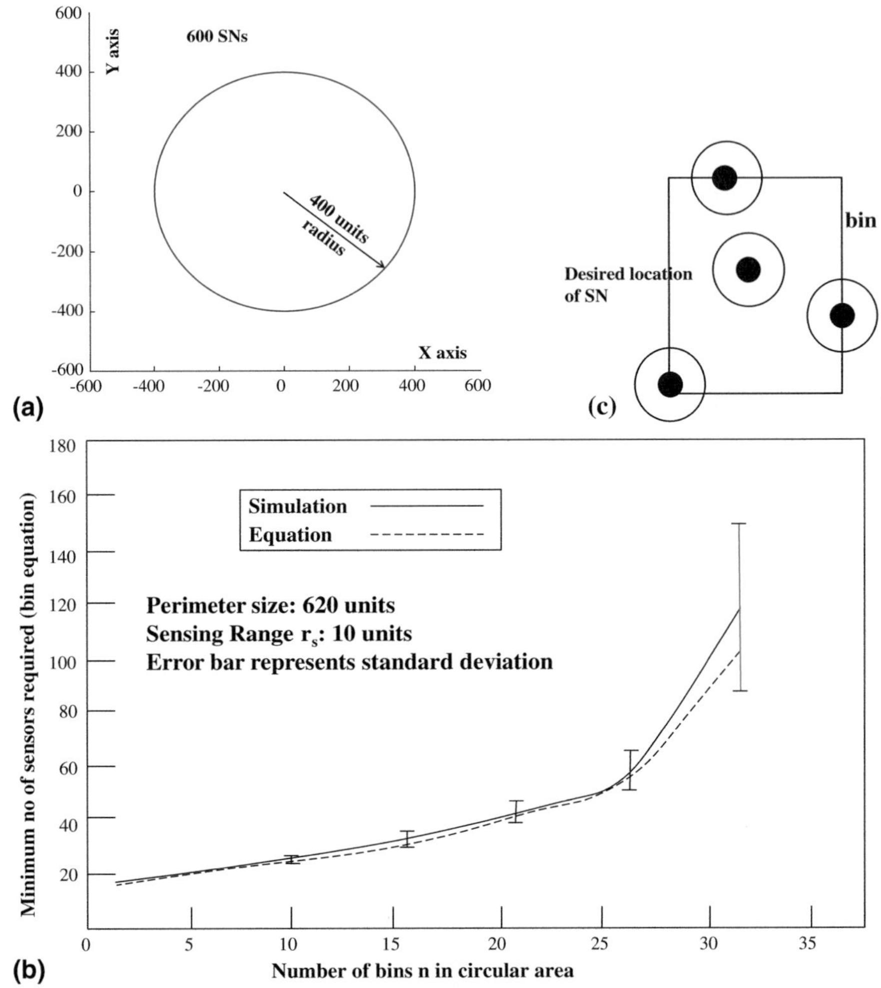

Fig. 19.6 **a** Uniform distribution with 600 SNs and boundary radius 400 units. **b** SNs needed for required coverage with m = 31 bins [13]. **c** SN could fall anywhere within the bin

gets information on event through connected SNs. The dropping agent drops SNs and is capable of making decisions on where to drop next SN based on information from the BS as the boundary of fire radius $r = 50$. The number of sensors needed for 80% connectivity and 100% connectivity on the boundary of a forest fire is determined and heuristically compare their performance using simulation tools. SNs are dropped starting from the BS such that they are connected and can communicate to the BS. Possible moves of next SN placement can be based on chosen grids as shown in the Fig. 19.9. If the current grid happens to have a border sensor, then more SNs are dropped around the boundary.

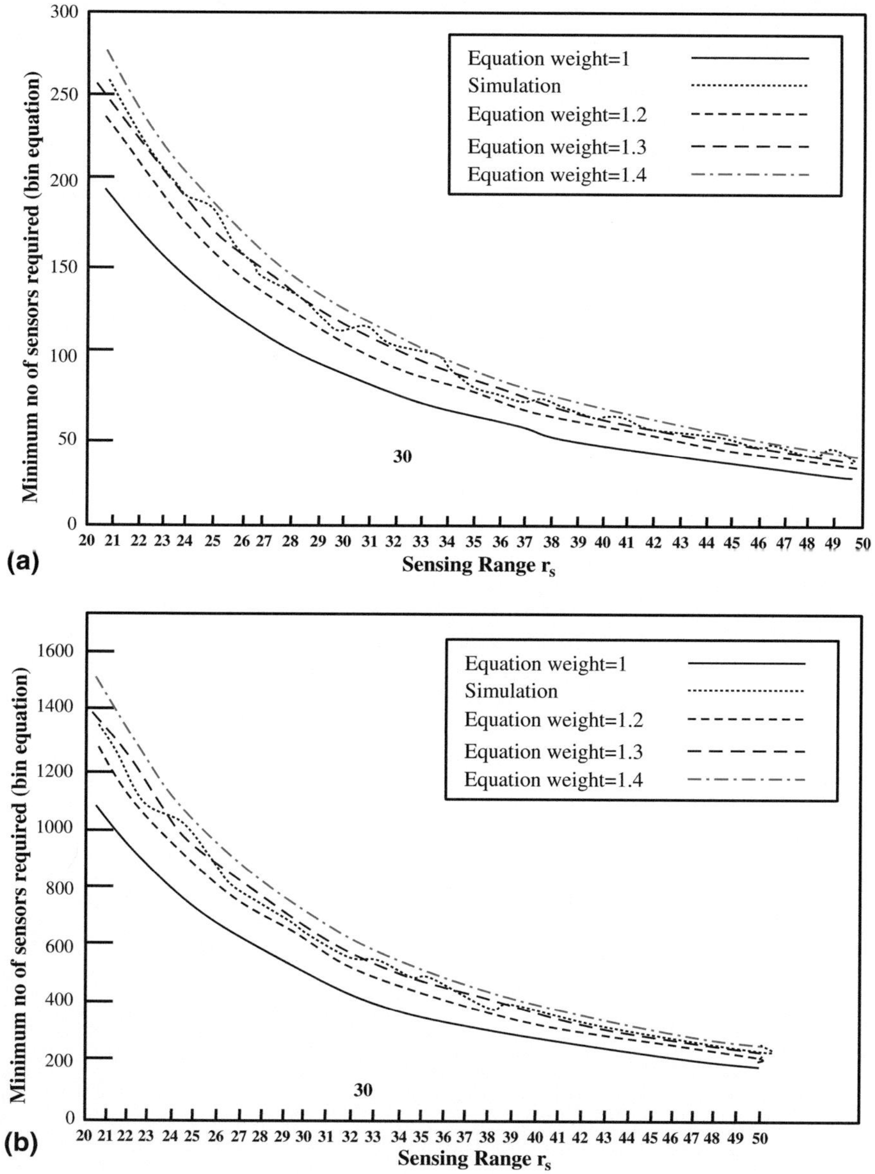

Fig. 19.7 Edge detection with different weighing factors and different sensing range. **a** 40% coverage with fire area of radius 500. **b** 70% coverage with fire area of radius 200

The primary objectives for a controlled deployment is to minimize wastage of SNs and if partial fire boundary can be detected, then remaining actual boundary can be estimated using interpolation and estimate the actual fire area using controlled deployment of SNs. Therefore, determining X% of the event boundary in

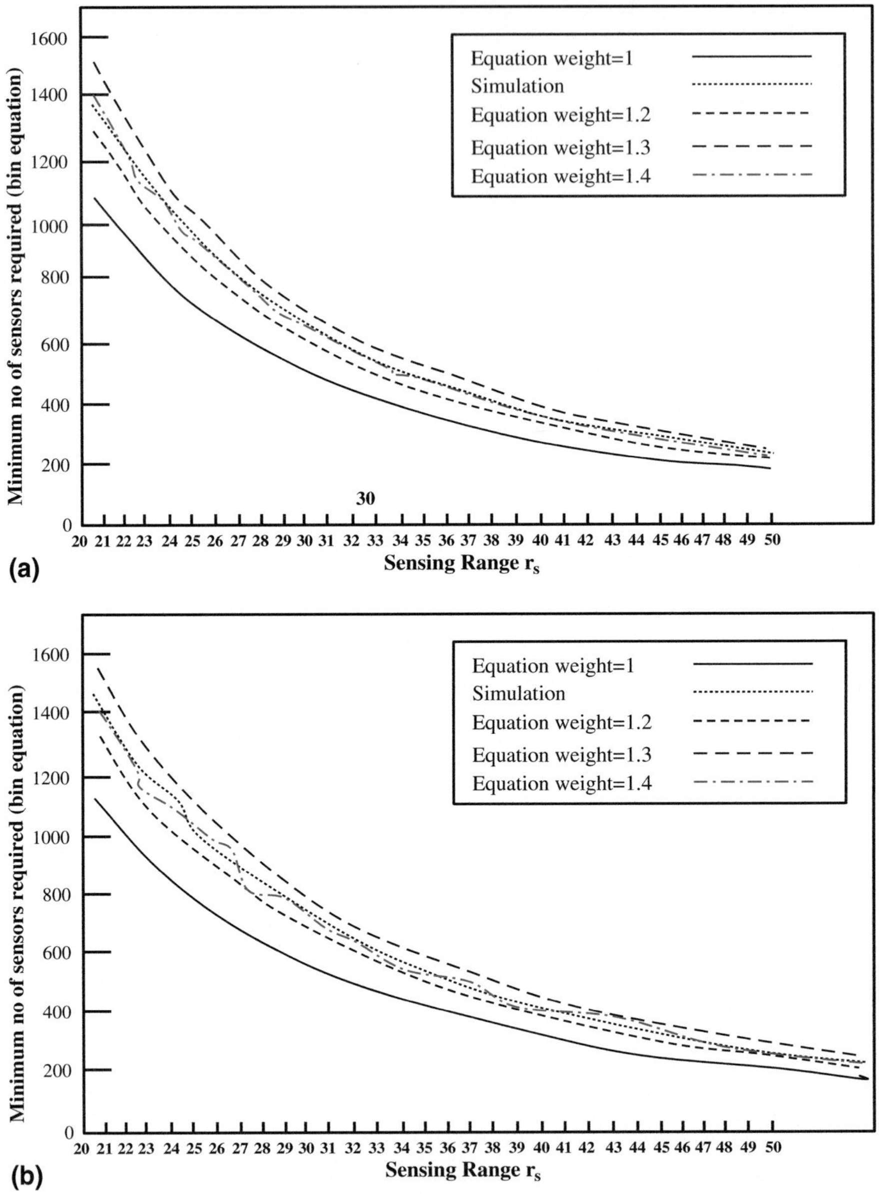

Fig. 19.8 Edge detection with SNs following Gaussian distribution and different sensing range. **a** 40% coverage with fire area of radius 500. **b** 70% coverage with fire area of radius 200

reasonable time is acceptable if the number of SNs could be minimized while reducing the detection time to decrease potential losses. The SNs detecting X% of the boundary must be connected to the BS as all decisions are made there. Therefore, SNs need to be dropped starting BS so that SNs remain connected to the

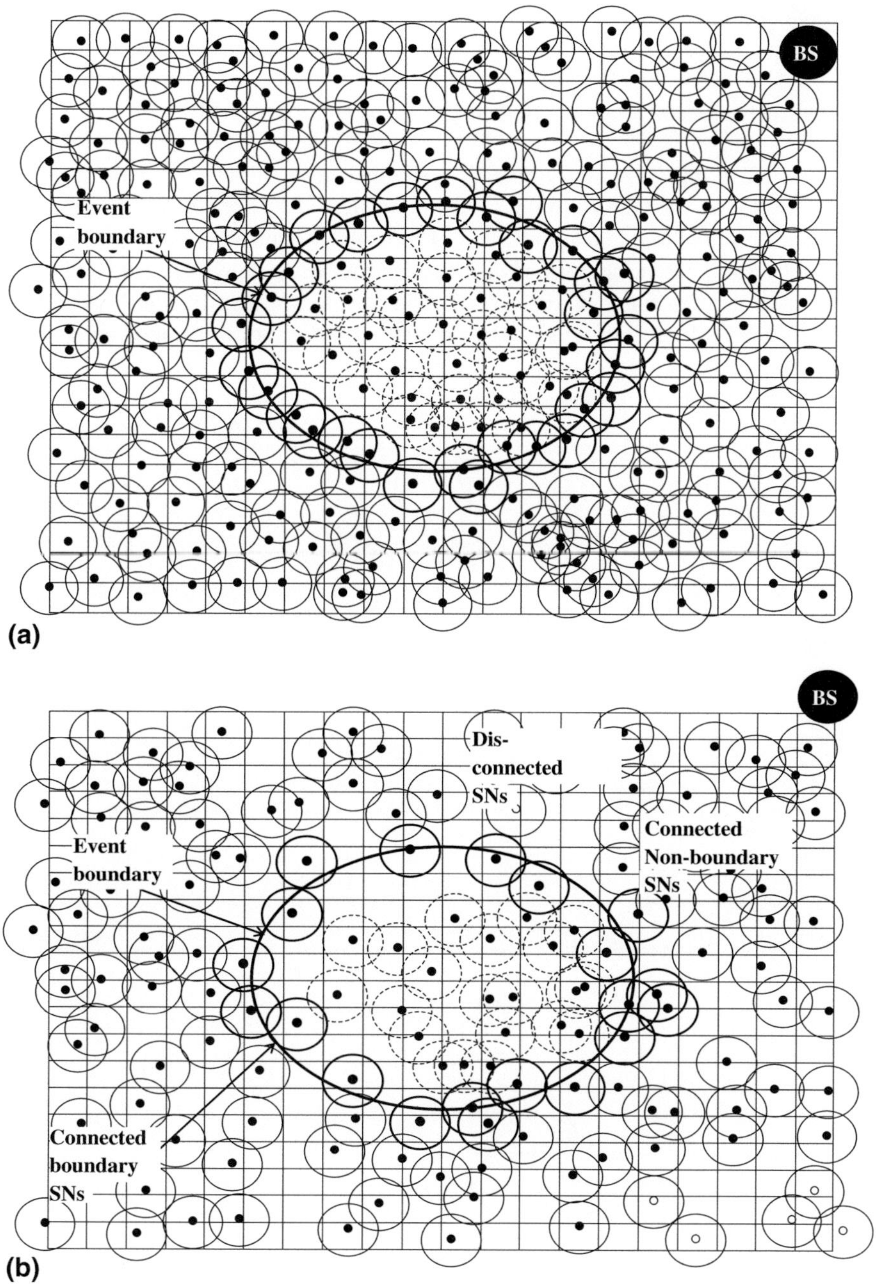

Fig. 19.9 **a** One shot random deployment with 100% connectivity. **b** One shot random deployment with 80% connectivity

BS. If the current SN happens to have on the border, location of the next SN can be selected around the boundary detecting SN and the possible moves can be to chosen grids as shown in Fig. 19.9b. This minimizes wastage of SNs and increases potential of the boarder being detected by the next deployed SN.

So, the bottom line is to place SNs such that at least one SN detects the fire boundary. This splits boundary detection problem into Initial boundary detection (when first SN on boundary is able to communicate to BS) and drop SNs at random until one SN is able to connect to BS and then drop by controlled random approach. This can be done in random way as discussed earlier or strategy of needle and comb can be used as illustrated in Fig. 19.10. The idea is to drop N Random vertical needles of length l (strip of SNs) and then a horizontal rib of a comb of appropriate

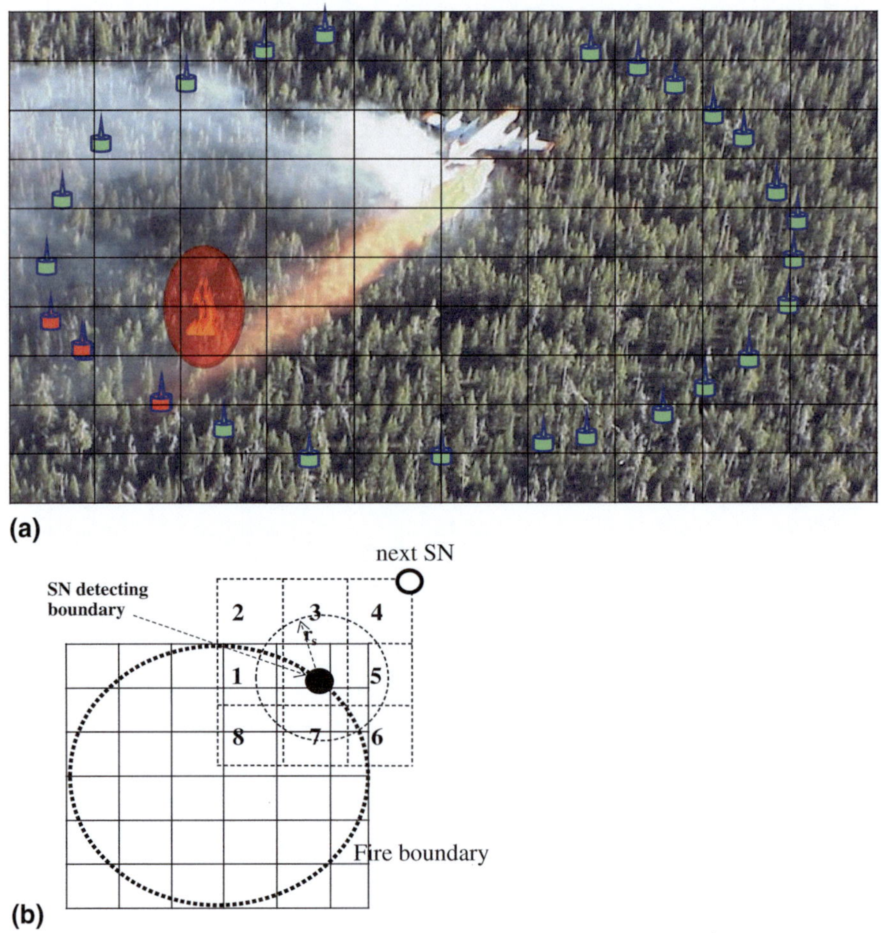

Fig. 19.10 **a** Diving fire area by 2-D mesh. **b** Potential next SN placement

equal to locate the boundary as quickly as possible. Do this randomly until first SN on the fire boundary is able to connect to BS and then drop by controlled random approach. Continue this process for N Random sequential SNs as needle and then N random single SNs as comb and alternate between these till the boundary of fire is detected by a SN and reported to the BS. Once a boundary point is found, then either continue with the same approach or after initial boundary detection, a weighted grid approach can be used to give priority to the grids around detected boundary area. The drawback with sequential approach is that only one point of connectivity may not enough as it will be a partial picture that cannot be extrapolated. SNs are dropped at random, considering a line of sight towards the estimated boundary such that we are able to pick 2 farthest points on the estimated boundary and work around from there so that we just place SNs around the boundary and not inside the fire area. So, weighted approach can be used to target 2 farthest points on the boundary. A Gaussian diminishing or increasing factor can be adopted to reduce or increase the number of SNs inside the fire area. This approach gives equal probability to all the points on the boundary, instead of giving importance to one point as observed in the sequential approach (Figs. 19.11 and 19.12).

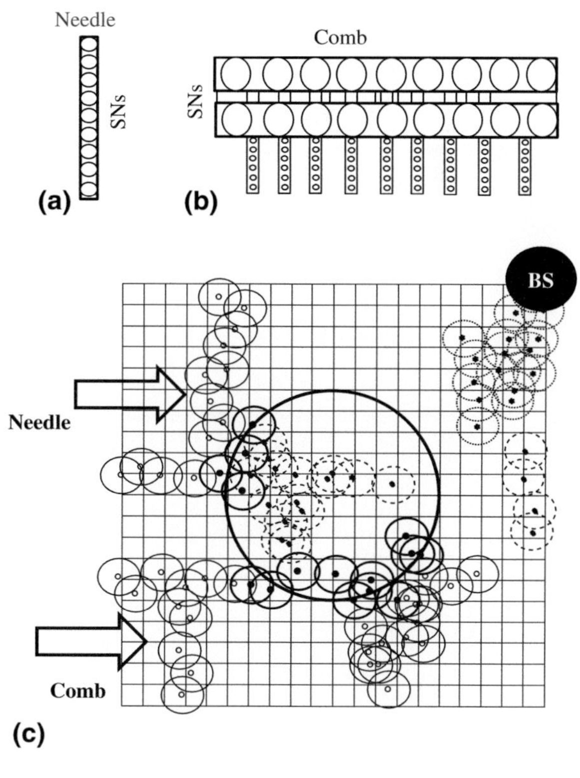

Fig. 19.11 **a** SNs arranged as a needle. **b** SNs arranged as a comb. **c** SN deployed as needle and comb in fire boundary area [15]

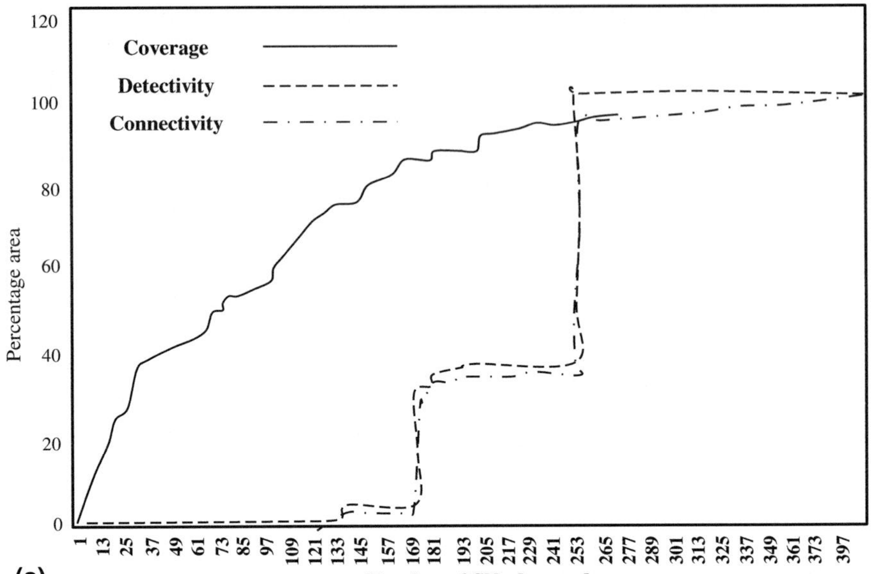

(a)

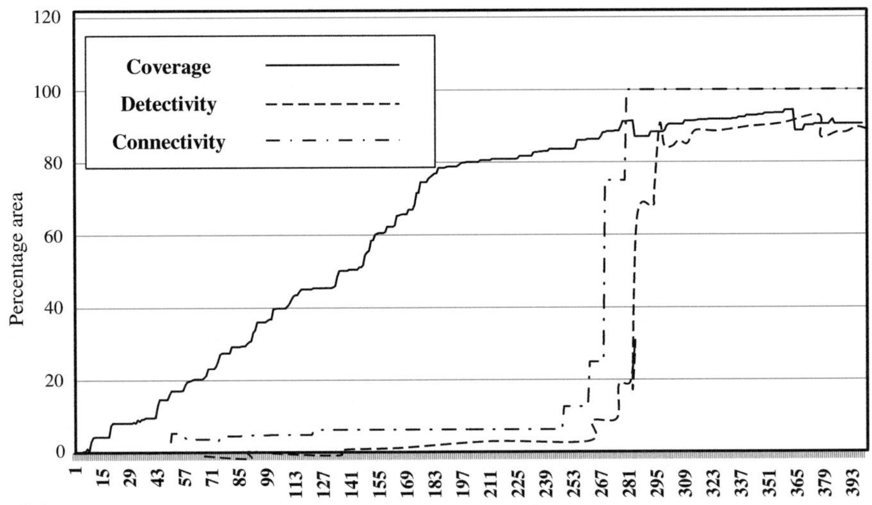

(b)

Fig. 19.12 **a** Variation of connectivity and detectivity with random SN 80% coverage and distribution with $s_r = 5$, fire area radius = 50 in 100 × 100 area. **b** Results with random needle and comb [15]

19.5 Conclusions

SNs are useful in identifying boundary area of an event such a wild forest fire. WSNs are useful in identifying boundary area of an event, not just wild forest fire and could include detection of oil-spill boundary, detection of an iceberg boundary, detection of Ozone layer boundary, detection of Radon: radioactive decay of radium on earth, detection of epidemic area boundary and detection of a Battle-field activities boundary.

In such applications, placement of BS plays an important role as there may not be too many possible options. For detecting the boundary, both coverage and connectivity are important for effectiveness of SNs as data need to reach BS where all decisions are made. It is observed that controlled deployment of SNs saves the required numbers. There are many open issues such as how to avoid deploying SNs inside an event area as SNs may get destroyed by the event. So, if SNs placed inside an event is destroyed, what are other options that ought to be explored is not clear.

19.6 Questions

Q.19.1. What are the differences between a WSN and largely-deployed WSN?

Q.19.2. What are the performance parameters that differentiate WSN from a largely-deployed WSN?

Q.19.3. Can boundary determination be useful for monitoring growth of a given crop? Explain clearly.

Q.19.4. How do you justify the use of “needle and comb” approach in determining boundary of an event like wild forest fire?

Q.19.5. How can you set sampling rate in a boundary determination application?

Q.19.6. Can you utilize directed diffusion scheme in a boundary determination application? If no, why not; and if yes, what changes ought to be made in the deployment scheme?

Q.19.7. Can you utilize data aggregation for boundary determination? Explain clearly.

Q.19.8. Can boundary determination be useful for other civilian applications?

Q.19.9. Radio transmission consumes more energy than processing in a WSN. But, time delay is also important in a boundary determination problem. What will you do under these circumstances?

Q.19.10. Can boundary determination be useful for other civilian applications?

Q.19.11. Can you incorporate security in data communication for a boundary determination problem? Explain clearly with suitable examples.

Q.19.12. Can boundary determination be useful for other civilian applications?

Q.19.13. What kind of routing protocol could be appropriate for boundary determination problem?

Q.19.14. What kind of query you expect in a boundary determination application?

Q.19.15. Can you use multiple BSs in a boundary determination problem?
Q.19.16. How can you address energy-hole problem in boundary determination applications?

References

1. G. Tolle, J. Polastre, R. Szewczyk, D. Culler, N. Turner, K. Tu, S. Burgess, T. Dawson, P. Buonadonna, D. Gay, and W. Hong, "A Macroscope in the Redwoods," in *Proc. of ACM SenSys*, 2005.
2. G. Werner-Allen, K. Lorincz, J. Johnson, J. Lees, and M. Welsh, "Fidelity and Yield in a Volcano Monitoring Sensor Network," in *Proc. of OSDI*, 2006.
3. Peter J. Bennett, Kenichi Soga, Ian Wassell, Paul Fidler, Keita Abe, Yusuke Kobayashi, and Martin Vanicek, "Wireless sensor networks for underground railway applications: case studies in Prague and London," Smart Structures and Systems, vol. 6, no. 5–6, pp. 619-639, 2010.
4. Yunhao Liu, Yuan He, Mo Li, Jiliang Wang, Kebin Liu, Lufeng Mo, Wei Dong, Zheng Yang, Min Xi, Jizhong Zhao, and Xiang-Yang Li, "Does Wireless Sensor Network Scale? A Measurement Study on GreenOrbs," IEEE Transactions on Parallel and Distributed Systems (TPDS), vol. 24, no. 10, Pages 1983–1993, October 2013.
5. E. Ertin, A. Arora, R. Ramnath, M. Nesterenko, V. Naik, S. Bapat, V. Kulathumani, M. Sridharan, H. Zhang, and H. Cao, "Kansei: A Testbed for Sensing at Scale," in *Proc. of ACM/IEEE IPSN*, 2006.
6. Lan Zhang, Xiang-Yang Li, and Yunhao Liu, "Message in a Sealed Bottle: Privacy Preserving Friending in Social Networks," 2013 IEEE 33rd International Conference on Distributed Computing Systems (ICDCS), pp. 327–336, 8-11 July 2013.
7. Anish Arora, Rajiv Ramnath, Emre Ertin, Prasun Sinha, Sandip Bapat, Vinayak Naik, Vinod Kulathumani, Hongwei Zhang, Hui Cao, Mukundan Sridharan, Santosh Kumar, Nick Seddon, Chris Anderson, Ted Herman, Nishank Trivedi, Chen Zhang, Mikhail Nesterenko, Romil Shah, Sandeep Kulkarni, Mahesh Aramugam, Limin Wang, Mohamed Gouda, Young-ri Choi, David Culler, Prabal Dutta, Cory Sharp, Gilman Tolle, Mike Grimmer, Bill Ferriera, and Ken Parker, "*ExScal*: Elements of an Extreme Scale Wireless Sensor Network," 11th IEEE International Conference on Embedded and Real-Time Computing Systems and Applications (RTCSA'05), pp. 102 – 108, 17–19 Aug. 2005.
8. T. He, P. Vicaire, T. Yan, Q. Cao, G. Zhou, L. Gu, L. Luo, R. Stoleru, J. A. Stankovic, and T. F. Abdelzaher, "Achieving Long-Term Surveillance in VigilNet," in *Proc. of IEEE INFOCOM*, 2006.
9. T. He, P. Vicaire, T. Yan, Q. Cao, G. Zhou, L. Gu, L. Luo, R. Stoleru, J. A. Stankovic, and T. F. Abdelzaher, "Achieving Long-Term Surveillance in VigilNet," ACM Transactions on Sensor Networks, vol. 5, no. 1, pp. 1–39, Feb. 2009.
10. http://cms.firehouse.com/web/online/Photo-Stories/Wildfire-Scenes-from-Acro.
11. P. Dutta, J. Hui, J. Jeong, S. Kim, C. Sharp, J. Taneja, G. Tolle, K. Whitehouse, and D. Culler, "Trio: Enabling Sustainable and Scalable Outdoor Wireless Sensor Network Deployments," in *Proc. Of ACM/IEEE IPSN*, 2006.
12. https://www.nps.gov/fire/wildland-fire/learning-center/fire-in-depth/fire-behavior.cfm.
13. Harshvardhan Kelkar, "Boundary Marking of Phenomenon using Wireless Sensor Networks," MS Thesis, University of Cincinnati, November 16, 2009.
14. Aparna Venkataraman, "Dynamic Deployment Strategies in Ad Hoc Sensor Networks to optimize Coverage and Connectivity in unknown Event Boundary Detection," MS Thesis, University of Cincinnati, July 08, 2011.

15. G. Barrenetxea, F. Ingelrest, G. Schaefer, M. Vetterli, O. Couach, and M. Parlange, "SensorScope: Out-of-the-Box Environmental Monitoring," in *Proc. of ACM/IEEE* Information Processing in Sensor Networks, pp. 332–343, 2008.
16. Aparna Venkataraman, Jung Jun, and Dharma P. Agrawal, "Sensor Deployment in Detecting Event's Boundary," The Ninth IEEE International Conference on Mobile Ad-hoc and Sensor Systems (IEEE MASS 2012), October 8–11, Las Vegas, Nevada.
17. G. Werner-Allen, P. Swieskowski, and M. Welsh, "MoteLab: A Wireless Sensor Network Testbed," in *Proc. of ACM/IEEE IPSN*, 2005.
18. www.spotfireimages.com.
19. http://earthobservatory.nasa.gov/IOTD/view.php?id=40011.
20. http://www.smokejumpers.com/gallery/v/Bases/album22/album25/RathboneFire.jpg.html.

Chapter 20
Recent Advances

20.1 Introduction

The use of WSNs is being explored in new application areas, and they offer endless possibilities. Researchers across the globe are busy finding new areas of research, and development is progressing at an unprecedented rate. New research areas are being presented, and altogether, different approaches are being introduced. Quest for higher data rate has been growing, and new schemes are being discovered to effectively process volume of data. Some technology considered harmful could be useful for others. It is also desirable to be prepared for future innovative technologies, and quantum computing is knocking our door and seems to be feasible in the near future.

20.2 Visual Sensor Networks

A wireless Visual Sensor Node (VN) typically consists of a smart camera, processing chips, and wireless transceiver (Fig. 20.1). The smart camera is capable of extracting information from images, while processing chips could do further image processing operations. Wireless transceiver takes care of forwarding images. Uniform distribution of VNs introduces "energy holes" in WSN (Fig. 20.2a). So, non-uniform deployment such as Gaussian has been suggested, and this offers a new option. To handle volume of data, addition of relay nodes has been suggested (Fig. 20.2b) and two-tier wireless visual sensor network has been suggested (Fig. 20.2c) [2]. The question that comes up is how many visual sensor nodes (VNs) should be deployed for monitored region, how many relay nodes (RNs) should be deployed within tier-1 sensing network, and how to configure the parameters of Gaussian distribution so as to optimize the network performance?

D.P. Agrawal, *Embedded Sensor Systems*,
DOI 10.1007/978-981-10-3038-3_20

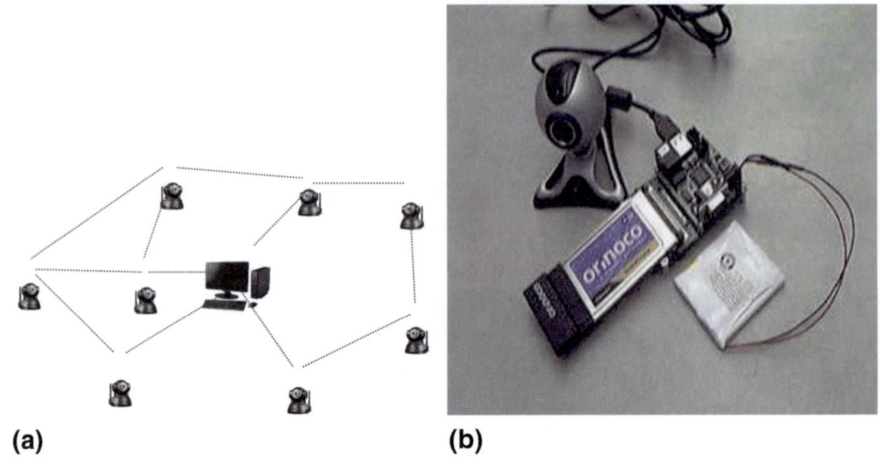

Fig. 20.1 **a** Wireless visual sensor network and **b** prototype visual sensor node

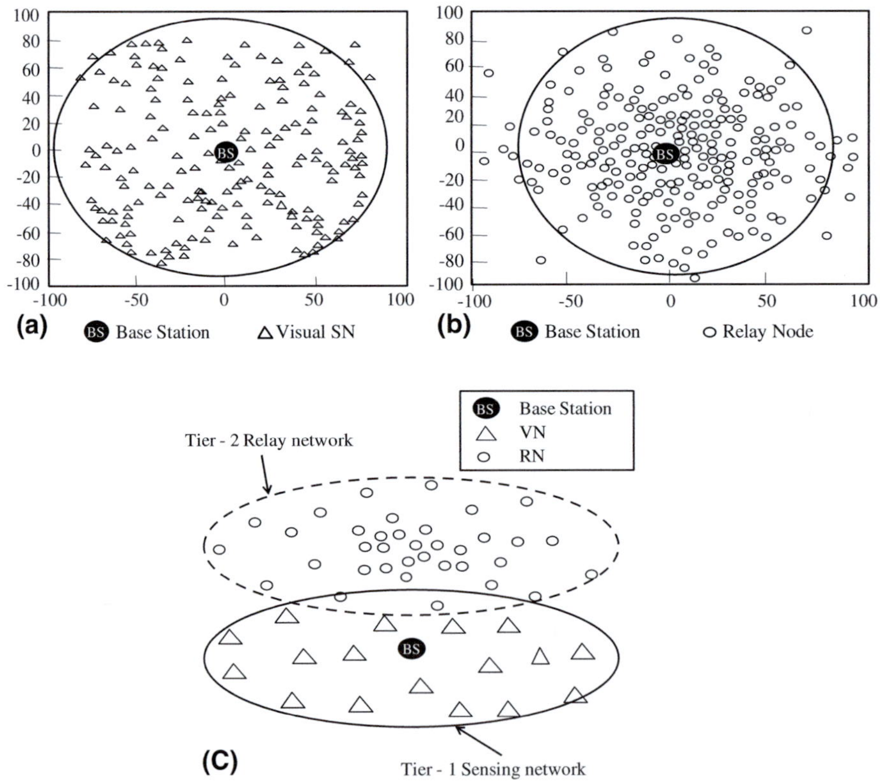

Fig. 20.2 **a** Visual sensor network in 100 × 100 area, **b** with relay nodes, and **c** two-tier network shown

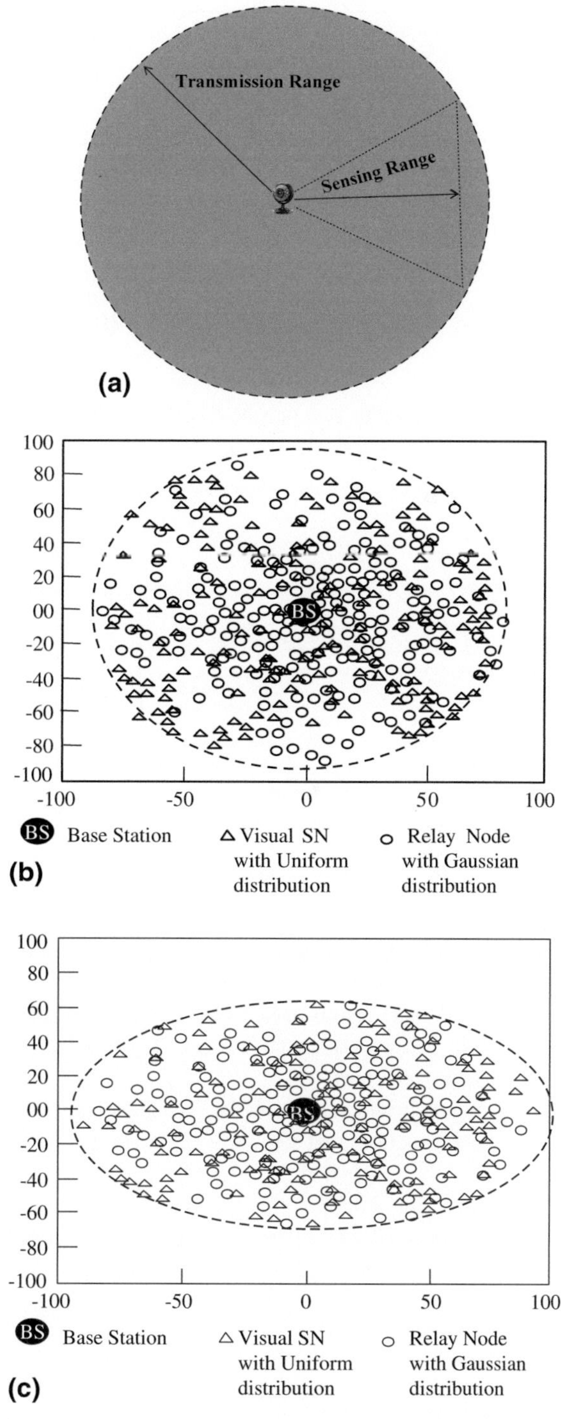

Fig. 20.3 **a** Coverage and connectivity of a visual node (VN), **b** circular VSN, and **c** elliptical VSN

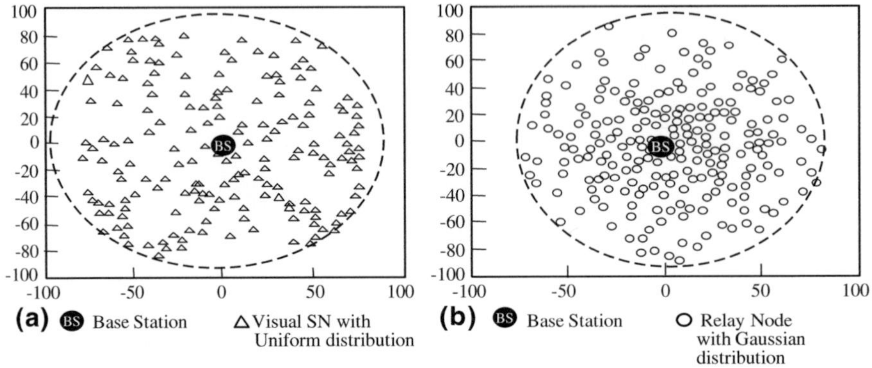

Fig. 20.4 **a** Fig. 20.3b with only VNs and **b** only with RNs

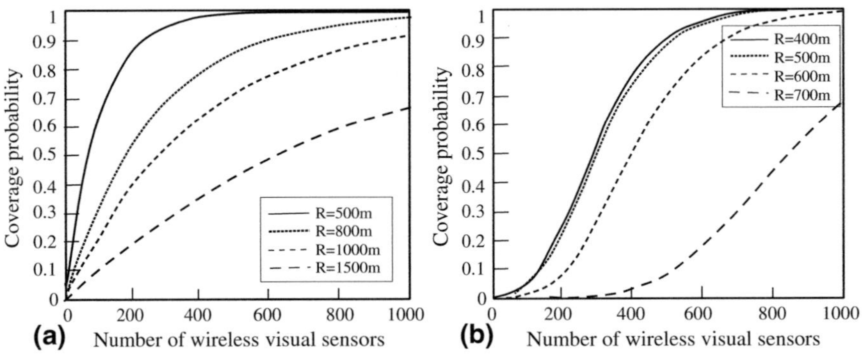

Fig. 20.5 2-tier network **a** circular visual sensor coverage and **b** relay node connectivity

The two networks can be deployed following a regular distribution or following a Gaussian distribution.

In two levels, visual nodes (VNs) are distributed following a uniform distribution, while relay nodes (RNs) are deployed with Gaussian rule so as to avoid energy-hole problem.

These are shown separately in Fig. 20.4 and have been analyzed in [2] following the sectoring scheme of [3]. σ is the constant for Gaussian distribution. The coverage by VN is shown in Fig. 20.5a, while the connectivity between RNs is observed in Fig. 20.5b.

Performance of 2-tier network is compared with a single uniformly and Gaussian distributed SNs and is given in Fig. 20.6.

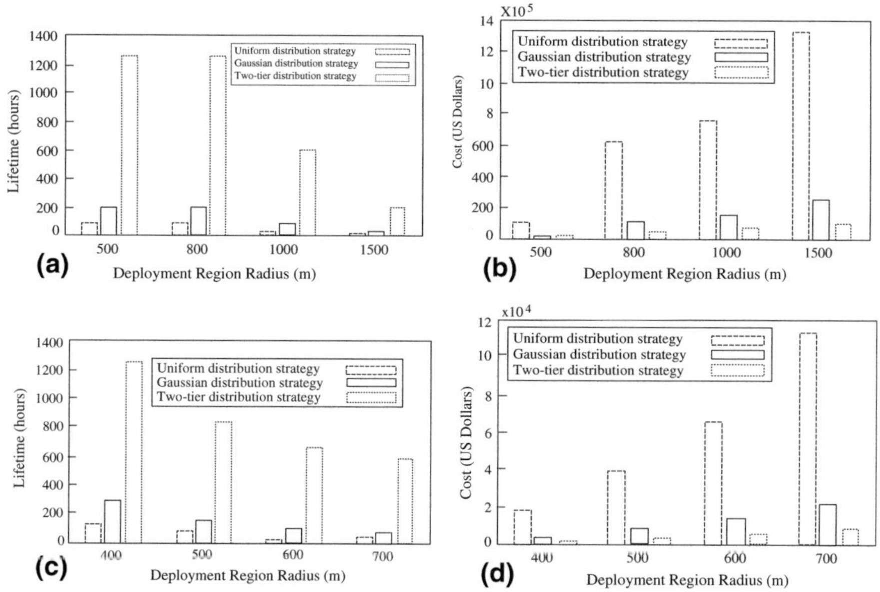

Fig. 20.6 **a** Performance comparison of circular deployed two-tier network, **b** cost comparison of circular deployed two-tier network, **c** performance comparison of elliptical relay nodes for two-tier network, and **d** cost comparison of elliptical relay nodes for two-tier network

20.3 WSNs in the Context of IoT, WoT, and SWoT

WSNs are being used and deployed in many locations for different applications. In a WSN, SNs connect with each other using wireless proprietary protocols, such as Bluetooth or ZigBee. However, the main challenge is establishing a connection between the BS and the Internet. This is now feasible with IPv6 over low-power wireless personal area networks (6LoWPANs). SNs, actuators, and RFID are key constituents of Internet of Things (IoT) (Fig. 20.7). A generic vision of IoT is to interconnect smart devices and objects anytime and anywhere throughout the world, leading to Web of Things (WoT) (Fig. 20.8a). While the IoT addresses the mechanisms to connect smart objects, the WoT (Fig. 20.8b) addresses the incorporation of embedded systems into the Web. Therefore, Social Web of Things Social Web of Things (SWoT) can be considered as a merging paradigm of social network site (SNS) and WoT [4]. The basic idea is to bring SNS services and features, such as social charts into an integrated system.

Integration of IP-enabled devices is a real challenge in IoT (Fig. 20.9). As IP is too far for low-power and resource-constrained objects, IPv6 over low-power wireless personal area networks (6LoWPANs) make it possible to realize IoT and machine-to-machine communication (Fig. 20.10). Bluetooth IEEE 802.15.1 is a short-range, low-power communication protocol used as wireless personal area networks (WPAN). IEEE 802.15.3a—ultra-wide band (UWB)—is a short-range

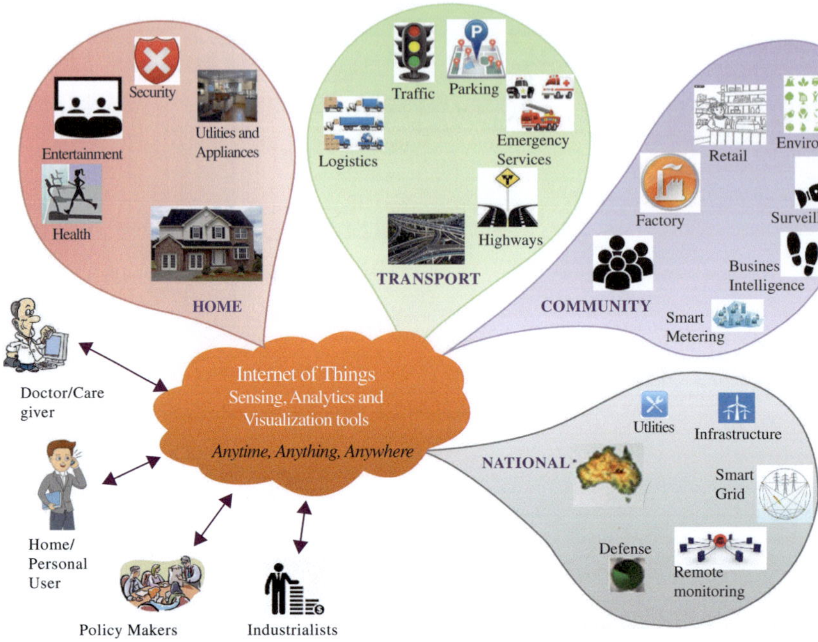

Fig. 20.7 Vision for Internet of Things (IoT)

high-speed wireless communication scheme for an indoor setting [5]. The bandwidth ranges from 110 to 480 Mbps covering a distance of 10 m and provides extremely high data rate areas for multimedia applications. IEEE 802.15.4 is a standard for low-rate wireless personal area network (LRWPAN) [6], with a device to utilize either a 64-bit address or a 16-bit short address and permitting either star or peer-to-peer topologies.

IEEE 802.15.4 ZigBee standard only describes physical and MAC layers using unique 64-bit physical address or 16-bit short address (Fig. 20.10). The network layer is responsible for addressing and routing (Fig. 20.11). Thousands of ZigBee devices can form a mesh network. Neither IEEE 802.15.4 nor ZigBee can support direct interworking with an IP network. A simple solution is to use IPv6 over low-power wireless personal area network (6LoWPAN) [7] which supports IPv6 communication over the IEEE 802.15.4 network (Fig. 20.13). Z-Wave is a wireless communication protocol designed especially for home automation [8], using low-power radio waves around 900 MHz ISM bands, and operates at data rate of 9.6 kbps. Z-Wave is now supported in 2.4 GHz with 200 kbps bit rates with a single network containing up to 232 devices (Fig. 20.12).

The 6LoWPAN supports two kinds of routing, namely mesh-under and route-over, allowing SNs to be addressable using IPv6 addresses. The communication between 6LoWPAN SNs and IP-enabled device is quite simple. SN sends its

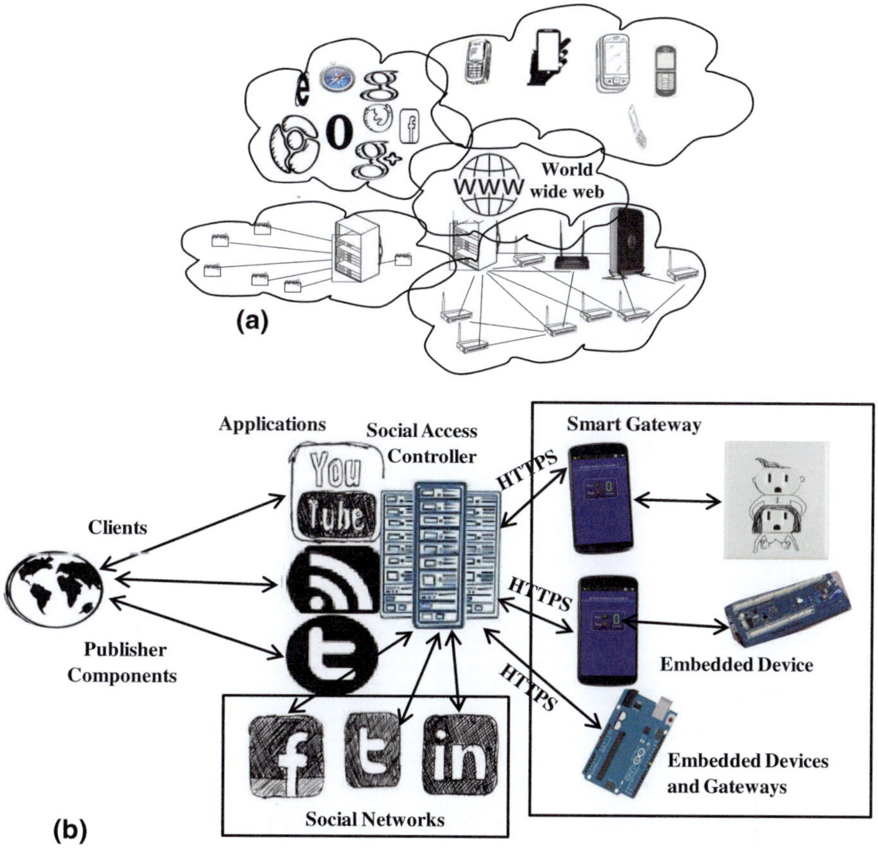

Fig. 20.8 **a** An overview of Web of Things (WoT) and **b** an overview of Social Access Controller in Social WoT (SWoT)

data to a router , which in turn forward it to 6LoWPAN gateway. When routing data is inside WSN, there is no need to send full IP addresses, thereby reducing the overhead. The 6LoWPAN gateway uses the destination IP address to forward packets to the destination, enabling possible integration of WSNs with heterogeneous IPv6 networks. However, implementing network layer protocol is complicate and difficult. Moreover, 6LoWPAN cannot be utilized for WSN using other non-compatible communication protocol, such as ZigBee protocol. It may be noted that 6LoWPAN is still not standardized yet.

SNSs are online platforms that facilitate individuals to publish, collaborate, and share information and experiences with others and build social associations. SNSs are emerging area of Web 2.0, where content and information are provided by the users. Today, numerous SNSs are available such as Facebook, Flickr, Google+,

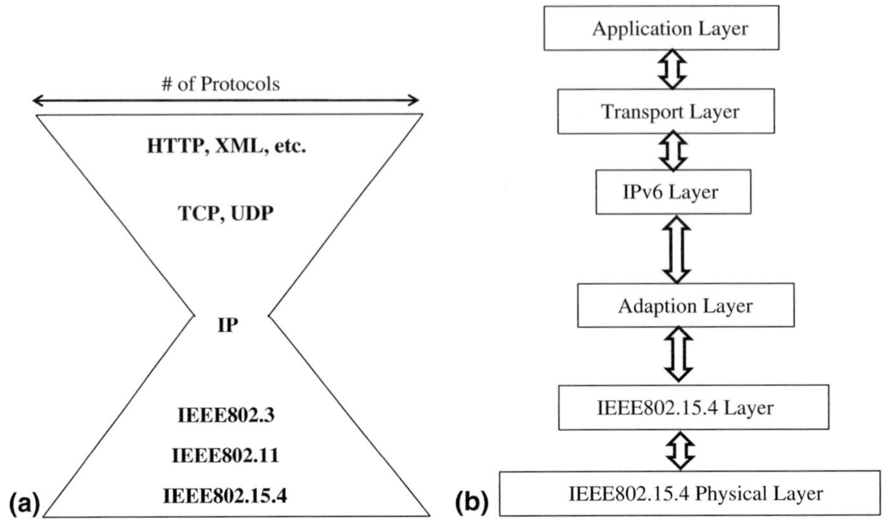

Fig. 20.9 **a** Convergence of IP and **b** TCP/IP overlay for low-power personnel area network (WSN)

MySpace, Twitter, Friendster, Orkut, YouTube, and LinkedIn. Users create unique pages called profiles by answering questions. Millions of users worldwide utilize Internet to browse the Web, play games, and use SNS and applications, enabling a global platform to have interconnection between machines and objects to have smart environment what is called an IoT paradigm. Thus, IoT first used in 1999 is enabling things to be connected anytime, anyplace, with anything, and anyone using any network and any service. However, there is still no standard definition of IoT [9, 10]. The Social Web of Things (SWoT) enables users to manage, share, and access web-enabled devices [11]. SNS increases the popularity of Web 2.0 technology, and interactive facility has permitted users to communicate and exchange contents with each other. These Web services can extend the social relations [12]. Figure 20.14 shows the architecture of Social Access Controller (SAC) [13]. In order to have combination of physical objects and social networks, new elements and concepts ought to be incorporated using some of the SNS features. SNS can be used as authentication server, monitoring tools, or a place to share smart object data.

20.4 Conclusions

WSN is useful in monitoring a given area and usually has volume to data to handle. So, network that senses data ought to be separated from the network that forwards data to the BS. So, from that point of view, two-tier WSN is desirable to handle visual data. The data so obtained need to be interfaced with Internet so that it is

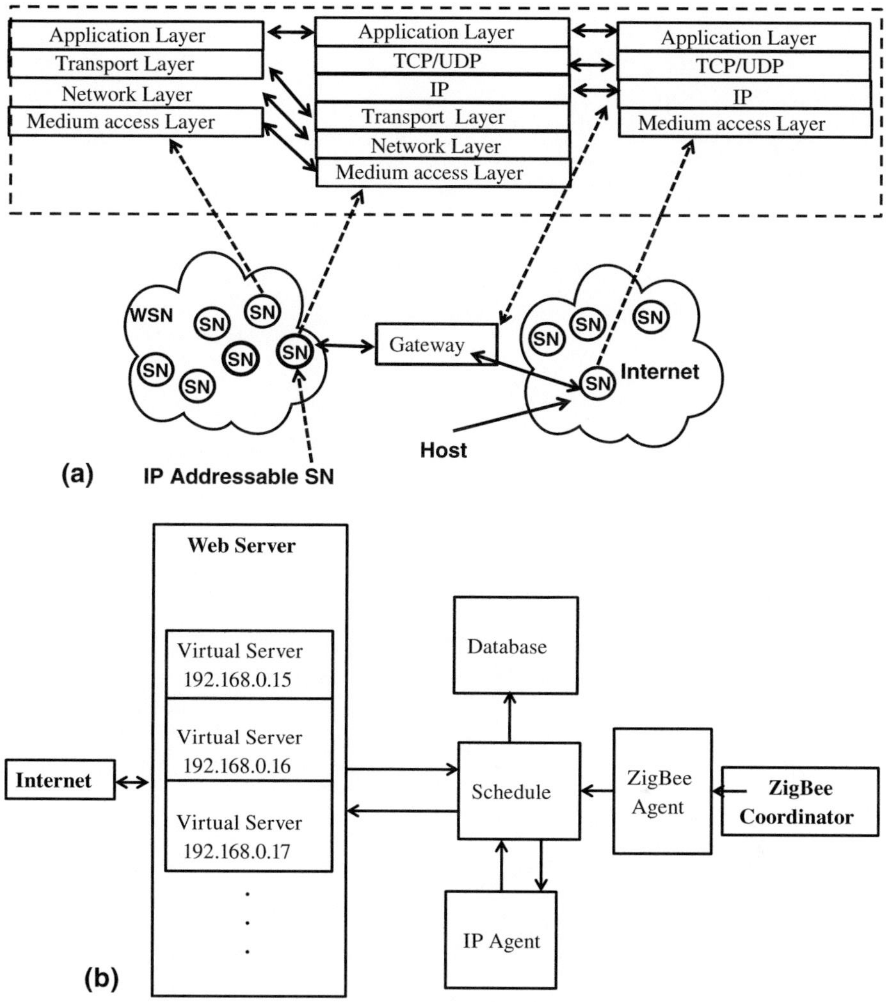

Fig. 20.10 **a** TCP /IP overlay for low-power WSNs and **b** web server to ZigBee coordinator

available to anyone, anywhere, and at any time. From that point of view, IPv6 needs to be interfaced with WSN and IoT, WoT, and SWoT paradigms ought to be explored so that there could be good interactions between different objects.

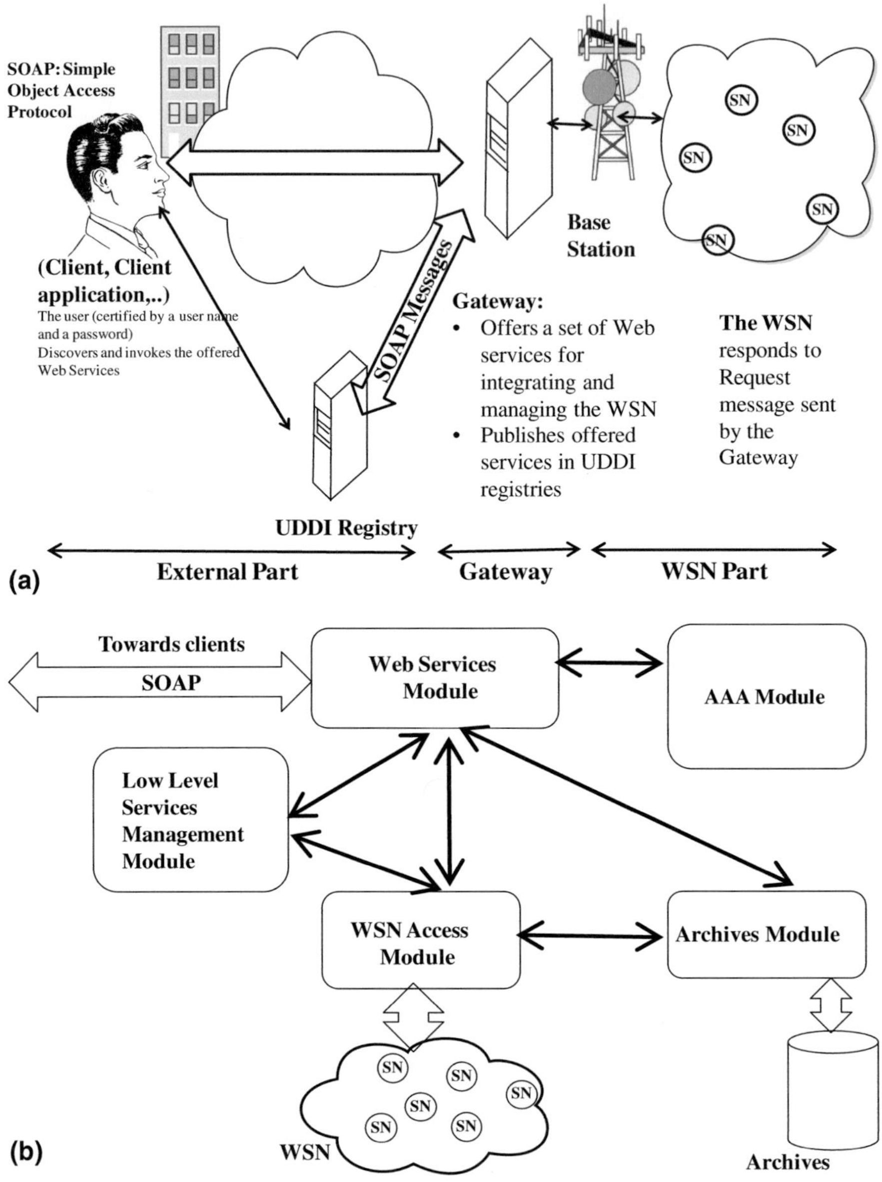

Fig. 20.11 **a** Signal exchange between IP and a WSN and **b** details of signal exchange between IP and a WSN

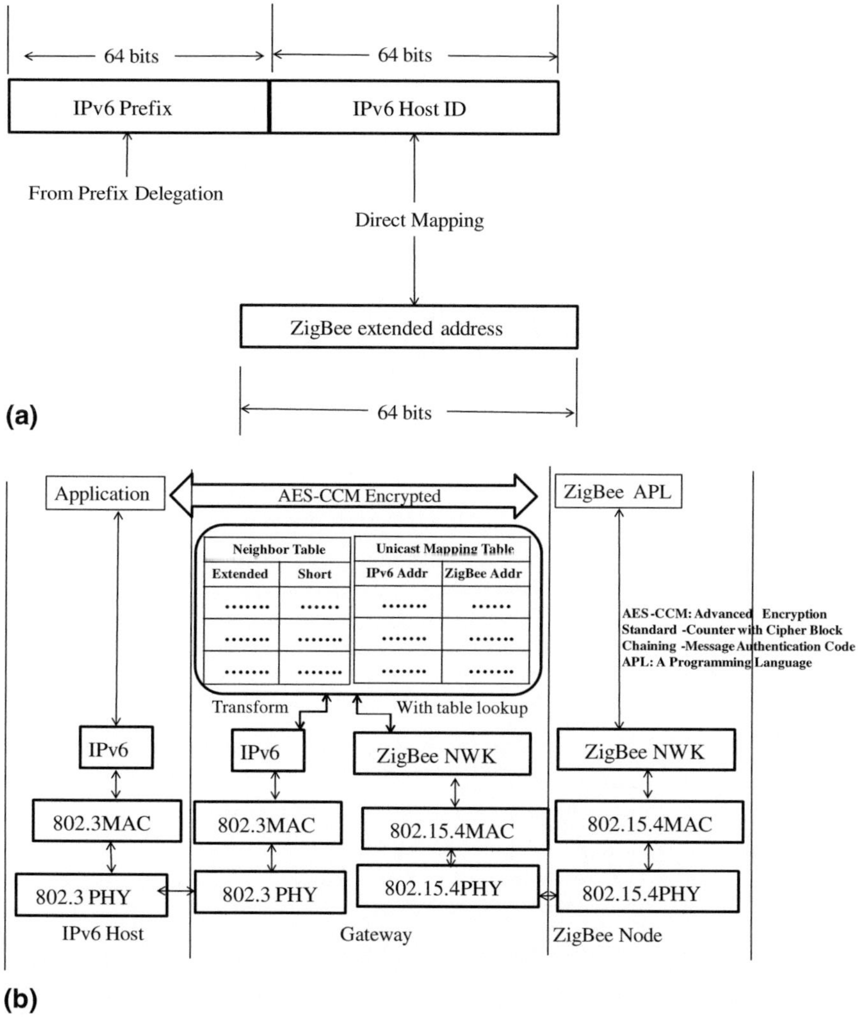

Fig. 20.12 **a** Signal exchange between IP and a WSN and **b** details of signal exchange between IP and a WSN

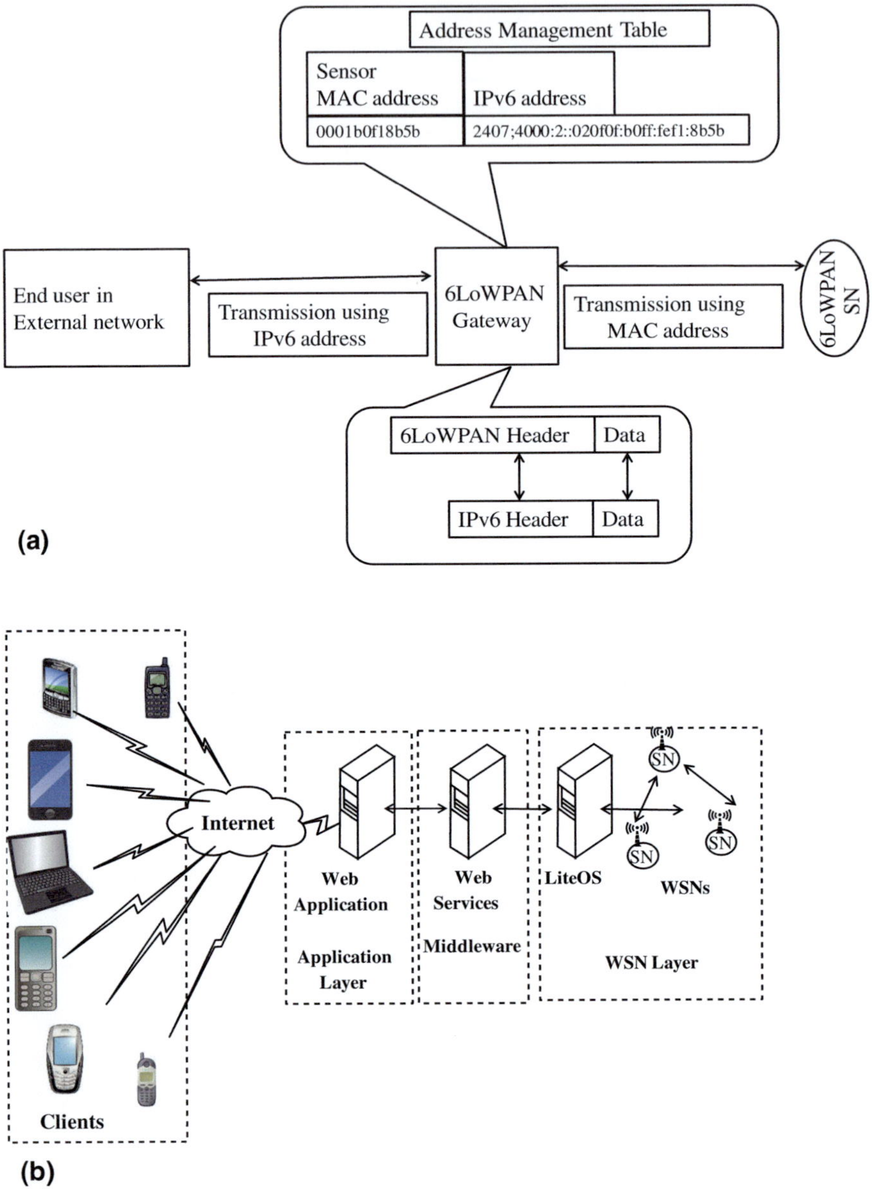

Fig. 20.13 **a** Interface between users and a 6LoWPAN WSN and **b** interface between users and a 6LoWPAN WSN

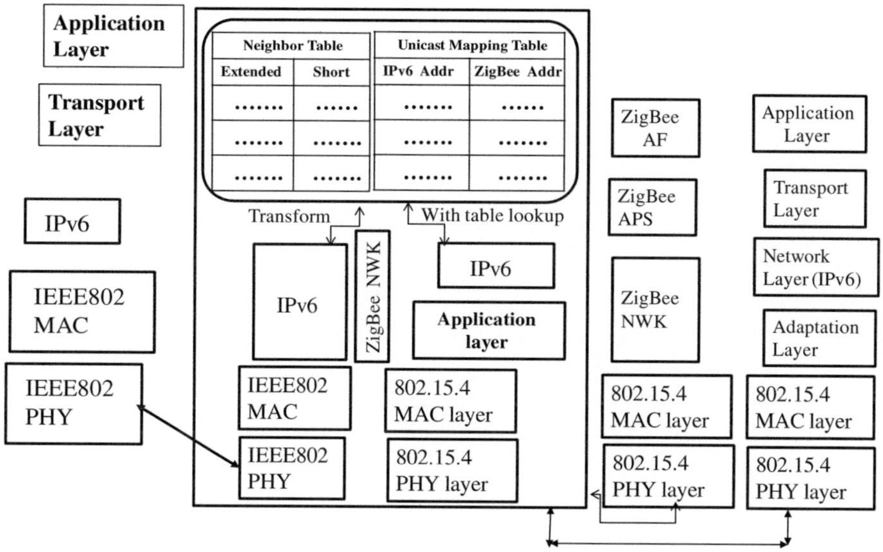

Fig. 20.14 Architecture of Social Access Controller (SAC)

20.5 Questions

Q.20.1. Is it possible to simulate a wireless visual sensor network? What are the challenges you face?

Q.20.2. What are the advantages of a "Gaussian distribution" of relay nodes in a visual sensor network?

Q.20.3. Visual data are much bigger and more complicated. So intelligent schemes are required to capture/process/transmit visual data in limited resources. What are the hurdles you face in doing that? Explain clearly.

Q.20.4. What type of collaborative solutions should be devised to tackle challenging issues in wireless visual sensor networks?

Q.20.5. Visual sensor networks are becoming increasingly popular in a number of applications. Can you say that there is a need for work similar to coverage efforts for directional sensor networks?

Q.20.6. In a visual sensor networks (VSNs), a network of smart cameras will process video data in real time, extracting features and 3-D geometry from the raw images of cooperating cameras. What kind of query mechanism can mediate access to such distributed data storage?

Q.20.7. Energy consumption in a VSN is affected by the number of camera sensors deployed, as well as the number of captured video frames. What strategy should you adopt to minimize energy consumption?

Q.20.8. Object recognition, face recognition, and image retrieval are important in a VSN. What feature you will emphasize in a VSN, given associated limitations?

Q.20.9. What is the major concern in adopting IoT or WoT?
Q.20.10. What types of SNs are used in IoT applications?
Q.20.11. What is meant by IoT and WoT and what are underlying differences?
Q.20.12. What do you consider as the most important utilization of the IoT?
Q.20.13. What are the major limitations in IoT?
Q.20.14. Will things connected by WoT be interoperable?
Q.20.15. What is an important impact will IoT have on our daily lives?
Q.20.16. What shouldn't be connected in IoT and WoT and why?
Q.20.17. Will IoT actually employ Internet or do you recommend using a dedicated network?
Q.20.18. What is the impact of IoT and WoT on big data and privacy?
Q.20.19. Will devices be secured in IoT and WoT?
Q.20.20. What can be done with data obtained in WoT?
Q.20.21. Is SWoT desirable? If yes, what are its implications? If no, then why not and is feasible to stop this?
Q.20.22. How much privacy and personal autonomy will be lost due to WoT and SWoT?

References

1. Cintia B. Margi, Vladislav Petkov, Katia Obraczka and Roberto Manduchi, "Characterizing energy consumption in a visual sensor network testbed," *2nd International Conference on Testbeds and Research Infrastructures for the Development of Networks and Communities, TRIDENTCOM, 2006.*
2. Hailong Li, Vaibhav Pandit, Dharma P. Agrawal, "Deployment Optimization Strategy for a Two-Tier Wireless Visual Sensor Network," Wireless Sensor Network, vol. 4, pp. 91–106, 2012.
3. C. Bettstetter, "On the minimum node degree and connectivity of a wireless multihop network," in *Proceedings of the 3rd ACM international symposium on Mobile ad hoc networking & computing*, ser. MobiHoc '02. New York, NY, USA: ACM, 2002.
4. Ibrahim Mashal, Osama Alsaryrah, Tein-Yaw Chung, Cheng-Zen Yang, Wen-Hsing Kuo, and Dharma P. Agrawal, "Choices for Interaction with Things on Internet and Underlying Issues," Ad Hoc Networks Journal, *6 January 2015*, available online http://dx.doi.org/10.1016/j.adhoc.2014.12.006.
5. Ian F. Akyildiz, Tommaso Melodia, and Kaushik R. Chowdhury. "A survey on wireless multimedia sensor networks," Computer networks, 51.4 (2007): 921–960.
6. J. T. Adams, "An introduction to IEEE STD 802.15.4," 2006 IEEE Aerospace Conference, pp. 8, doi:10.1109/AERO.2006.1655947
7. G. Mulligan, "The 6LoWPAN architecture," in Proceedings of the 4th Workshop on Embedded Networked Sensors, (EmNets '07), pp. 78–82, June 2007.
8. Gomez, Carles, and Josep Paradells, "Wireless home automation networks: A survey of architectures and technologies," Communications Magazine, IEEE 48.6 (2010): 92–101.
9. The Internet of Things 2012 - New Horizons, (European Research Cluster on the Internet of Things) IERC Books.
10. Lu Tan, and Neng Wang, "Future internet: The Internet of Things," International Conference on Advanced Computer Theory and Engineering (ICACTE), 2010, 3rd, vol. 5, pp. V5376-V5380, 20–22 Aug. 2010

11. Andrei Ciortea, Olivier Boissier, Antoine Zimmermann, and Adina Magda Florea, "Reconsidering the social web of things," position paper. In Proceedings of the 2013 ACM conference on Pervasive and ubiquitous computing adjunct publication (UbiComp '13 Adjunct), 2013.
12. A. Kamilaris, D. Papadiomidous, and A. Pitsillides, "Lessons Learned from Online Social Networking of Physical Things," 2011 International Conference on Broadband and Wireless Computing, Communication and Applications (BWCCA), pp. 128,135, 26–28 Oct. 2011
13. D. Guinard, M. Fischer, and V. Trifa, "Sharing using social networks in a composable Web of Things," 2010 8th IEEE International Conference on Pervasive Computing and Communications Workshops (PERCOM Workshops), pp. 702–707, March 29, 2010-April 2, 2010.

Questions and Ideas for Design Projects

Project Questions: The following questions have been formulated as a guideline for project. You need to either simulate or do experiments with sensor boards to obtain results as shown in different chapters. Once you have the results, you need to prepare 5–10 pages of double space report describing project title, team members, abstract, introduction with objective, project details, design decisions, results, summary, and references. Indicate reasons for differences between your results and those shown in the slides, and prepare 10–15 PowerPoint slides providing an overview, circuit diagram, design decisions, results, difficulties, and future work.

Project 1 Design an Arduino system to control $3 \times 3 \times 3$ LED cube. Assuming each LED to be a SN and by providing one coverage, connect to your laptop and measure different physical quantities such as temperature, humidity, light intensity, and sound level. How many additional LEDs you need to deploy to have 2 and 3 coverage of the area?

Project 2 Design an Arduino system to control $3 \times 3 \times 3$ LED cube. Assuming each LED to be a SN and connect to your laptop and measure different physical quantities such as temperature. BS is assumed to be located at the center. How many alternate paths are available from each SN to BS, assuming each LED to be a SN. The path length should not exceed shortest path between SN and BS.

Project 3 Design an Arduino system to control $3 \times 3 \times 3$ LED cube. Assuming each LED to be a SN and by providing one coverage, connect to your laptop and measure different physical quantities such as temperature. How much overlap you will have if 10% wiggle effect is present?

Project 4 Can you design a "pong" game with two Ardunio controllers and two 10 KΩ potentiometers?

Project 5 How can you use accelerometer to determine 2-D acceleration of a car? How can you use this information for possible overturning of a car?

Project 6 Photosensor can detect the distance of a car you are following. Can you design a system to avoid any potential collision?

Project 7 How can you determine RSSI and LQI in a system, with a SN used as a BS and a second SN is moved around at different distance?

D.P. Agrawal, *Embedded Sensor Systems*,
DOI 10.1007/978-981-10-3038-3

Project 8 Can you design Arduino-powered Trinket system? What are other potential applications of such a setup?

Project 9 How can you measure temperature and humidity of different rooms in a house? What do you need to add if value exceeds lower and upper limits?

Project 10 Can you design an efficient sprinkler controller which takes temperature and humidity into account?

Project 11 Design an Arduino system to control $3 \times 3 \times 3$ LED cube. Assume each LED to be a SN and BS is located at the center. How much time is needed to have asymmetric and symmetric encryption between a SN and BS? What will be the impact if a triangular topology is used?

Project 12 A camera is used as a SN in triangular and hexagonal topologies. How can you determine the number of cameras needed for full coverage and 3 coverage? Ref slide 68 and slide 19 of Chapter 13.

Project 13 Design an Arduino system to control $3 \times 3 \times 3$ LED cube. Assume each LED to be a SN, BS is located at the center, and these are randomly deployed WSN. Show that a matrix-based key distribution approach works in such a system.

Project 14 How can you put 8 pressure sensors in a rectangular, triangular, and hexagonal grid configurations? Get the pressure reading on your laptop.

Project 15 Design an Arduino system to control $3 \times 3 \times 3$ LED cube. Assume each LED to be a SN and BS is located at the center. For such a rectangular WSN, there are two options to establish a shared symmetric key: either use Diffie–Hellman algorithms or use Bivariate polynomial. Show that both of them can work.

Index

D.P. Agrawal, *Embedded Sensor Systems*,
DOI 10.1007/978-981-10-3038-3

MIX
Papier aus verantwortungsvollen Quellen
Paper from responsible sources
FSC® C105338

If you have any concerns about our products,
you can contact us on
ProductSafety@springernature.com

In case Publisher is established outside the EU,
the EU authorized representative is:
Springer Nature Customer Service Center GmbH
Europaplatz 3, 69115 Heidelberg, Germany

Printed by Libri Plureos GmbH
in Hamburg, Germany